Low-Energy Lunar Trajectory Design

DEEP-SPACE COMMUNICATIONS AND NAVIGATION SERIES

The Deep-Space Communications and Navigation Systems
Center of Excellence
Jet Propulsion Laboratory
California Institute of Technology

Joseph H. Yuen, Editor-in-Chief

Published Titles in this Series

Radiometric Tracking Techniques for Deep-Space Navigation
C. L. Thornton and J. S. Border

Formulation for Observed and Computed Values of Deep Space Network Data Types for Navigation
Theodore D. Moyer

Bandwidth-Efficient Digital Modulation with Application to Deep-Space Communications
Marvin K. Simon

Large Antennas of the Deep Space Network
William A. Imbriale

Antenna Arraying Techniques in the Deep Space Network
David H. Rogstad, Alexander Mileant, and Timothy T. Pham

Radio Occultations Using Earth Satellites: A Wave Theory Treatment
William G. Melbourne

Deep Space Optical Communications
Hamid Hemmati

Spaceborne Antennas for Planetary Exploration
William A. Imbriale, Editor

Autonomous Software-Defined Radio Receivers for Deep Space Applications
Jon Hamkins and Marvin K. Simon, Editors

Low-Noise Systems in the Deep Space Network
Macgregor S. Reid, Editor

Coupled-Oscillator Based Active-Array Antennas
Ronald J. Pogorzelski and Apostolos Georgiadis

Low-Energy Lunar Trajectory Design
Jeffrey S. Parker and Rodney L. Anderson

Low-Energy Lunar Trajectory Design

Jeffrey S. Parker
Rodney L. Anderson
Jet Propulsion Laboratory
California Institute of Technology

WILEY

Published by John Wiley & Sons, Inc., Hoboken, New Jersey
Published simultaneously in Canada

Library of Congress Cataloging-in-Publication Data:

Parker, Jeffrey S.
Low-energy lunar trajectory design / Jeffrey S. Parker and Rodney L. Anderson.
pages cm
Includes index.
ISBN 978-1-118-85387-0 (cloth)
1. Lunar probes—Trajectories. 2. Space flight to the moon—Cost control. I. Anderson, Rodney L. II. Title.
TL1075.P37 2014
629.4'11—dc23 2014001158

Printed in the United States of America.

10 9 8 7 6 5 4 3 2 1

Jeffrey Parker:
I dedicate the majority
of this book to my wife
Jen, my best friend and
greatest support
throughout the
development of this book
and always. I dedicate
the appendix to my son
Cameron, who showed up
right at the end.

Rodney Anderson:
I dedicate this book to
my wife Brooke for her
endless support and
encouragement.

We both thank our
families and friends for
their support throughout
the process.

CONTENTS

FOREWORD

The Deep Space Communications and Navigation Systems Center of Excellence (DESCANSO) was established in 1998 by the National Aeronautics and Space Administration (NASA) at the California Institute of Technology's Jet Propulsion Laboratory (JPL). DESCANSO is chartered to harness and promote excellence and innovation to meet the communications and navigation needs of future deep-space exploration.

DESCANSO's vision is to achieve continuous communications and precise navigation—any time, anywhere. In support of that vision, DESCANSO aims to seek out and advocate new concepts, systems, and technologies; foster key technical talents; and sponsor seminars, workshops, and symposia to facilitate interaction and idea exchange.

The Deep Space Communications and Navigation Series, authored by scientists and engineers with many years of experience in their respective fields, lays a foundation for innovation by communicating state-of-the-art knowledge in key technologies. The series also captures fundamental principles and practices developed during decades of deep-space exploration at JPL. In addition, it celebrates successes and imparts lessons learned. Finally, the series will serve to guide a new generation of scientists and engineers.

Joseph H. Yuen, DESCANSO Leader

PREFACE

The purpose of this book is to provide high-level information to mission managers and detailed information to mission designers about low-energy transfers between the Earth and the Moon. This book surveys thousands of trajectories that one can use to transfer spacecraft between the Earth and various locations near the Moon, including lunar libration orbits, low lunar orbits, and the lunar surface. These surveys include conventional, direct transfers that require 3–6 days as well as more efficient, low-energy transfers that require more transfer time but which require less fuel. Low-energy transfers have been shown to be very useful in many circumstances and have recently been used to send satellites to the Moon, including the two *ARTEMIS* spacecraft and the two *GRAIL* spacecraft. This book illuminates the trade space of low-energy transfers and illustrates the techniques that may be used to build them.

ACKNOWLEDGMENTS

We would like to thank many people for their support writing this book, including people who have written or reviewed portions of the text, as well as people who have provided insight from years of experience flying spacecraft missions to the Moon and elsewhere. It is with sincere gratitude that we thank Ted Sweetser for his selfless efforts throughout this process, providing the opportunity for us to perform this work, and reviewing each section of this manuscript as it has come together. We would like to thank Al Cangahuala, Joe Guinn, Roby Wilson, and Amy Attiyah for their valuable feedback and thorough review of this work in each of its stages. We would also like to thank Tim McElrath for his feedback, insight, and excitement as we considered different aspects of this research.

We would like to give special thanks to several people who provided particular contributions to sections of the book. We thank Ralph Roncoli for his assistance with Sections 2.3 and 2.4, as well as his feedback throughout the book. Kate Davis assisted with Sections 2.6.3 and 2.6.11.3, most notably with the discussions of Poincaré sections. Roby Wilson provided particular assistance with Section 2.6.5 on the subject of the multiple shooting differential corrector. We would like to sincerely thank Andrew Peterson for his contribution to the development of Chapter 4. Finally, George Born and Martin Lo provided guidance for this research as it developed in its early stages, leading to the authors' dissertations at the University of Colorado at Boulder.

Jeffrey Parker's Ph.D. dissertation (J. S. Parker, *Low-Energy Ballistic Lunar Transfers*, Ph.D. Thesis, University of Colorado, Boulder, 2007) provides the backbone to this manuscript and much of the dissertation has been repeated and amplified in this book. Much of the additional material that appears in this manuscript has been presented by the authors at conferences and published in journals. Such material has been reprinted here, with some significant alterations and additions. Finally, a number of additional journal articles and conference proceedings directly contributed to each chapter in the following list. In addition to the listing below, they are cited in the text where the related material appears.

Chapter 2:

- J. S. Parker, K. E. Davis, and G. H. Born, "Chaining Periodic Three-Body Orbits in the Earth–Moon System," *ACTA Astronautica*, vol. 67, pp. 623–638, 2010.
- M. W. Lo, and J. S. Parker, "Chaining Simple Periodic Three-Body Orbits," *AAS/AIAA Astrodynamics Specialist Conference* (Lake Tahoe, California), Paper No. AAS 2005-380, August 7–11, 2005, vol. 123, *Advances in Astronautical Sciences* (B. G. Williams, L. A. D'Amario, K. C. Howell, and F. R. Hoots, editors), AAS/AIAA, Univelt Inc., San Diego, CA, 2006.
- R. B. Roncoli, *Lunar Constants and Models Document,* JPL D-32296 (internal document), Jet Propulsion Laboratory, California Institute of Technology, Pasadena, California, September 23, 2005.
- R. L. Anderson and J. S. Parker, "Survey of Ballistic Transfers to the Lunar Surface," *Journal of Guidance, Control, and Dynamics,* vol. 35, no. 4, pp. 1256–1267, July–August 2012.

Chapter 3:

- J. S. Parker, "Monthly Variations of Low-Energy Ballistic Transfers to Lunar Halo Orbits," *AIAA/AAS Astrodynamics Specialist Conference,* (Toronto, Ontario, Canada), Paper No. AIAA 2010-7963, August 2–5, 2010.
- J. S. Parker, "Targeting Low-Energy Ballistic Lunar Transfers," *AAS George H. Born Special Symposium,* (Boulder, Colorado), May 13–14, 2010, *American Astronautical Society,* 2010.
- J. S. Parker, "Targeting Low-Energy Ballistic Lunar Transfers," *Journal of Astronautical Sciences,* vol. 58, no. 3, pp. 311–334, July–September, 2011.

- J. S. Parker, "Low-Energy Ballistic Transfers to Lunar Halo Orbits," *AAS/AIAA Astrodynamics Specialist Conference,* (Pittsburgh, Pennsylvania, Paper No. AAS 09-443, August 9–13, 2009, *Advances in Astronautical Sciences, Astrodynamics 2009* (A. V. Rao, A. Lovell, F. K. Chan, and L. A. Cangahuala, editors), vol. 135, pp. 2339–2358, 2010.

- J. S. Parker, and G. H. Born, "Modeling a Low-Energy Ballistic Lunar Transfer Using Dynamical Systems Theory," *AIAA Journal of Spacecraft and Rockets,* vol. 45, no. 6, pp.1269–1281, November–December 2008.

- J. S. Parker and G. H. Born, "Direct Lunar Halo Orbit Transfers," *Journal of the Astronautical Sciences,* vol. 56, issue 4, pp. 441–476, October–December 2008.

- J. S. Parker and G. H. Born, "Direct Lunar Halo Orbit Transfers," *AAS/AIAA Spaceflight Mechanics Conference* (Sedona, Arizona, January 28–February 1, 2007), Paper No. AAS 07-229, *Advances in Astronautical Science,* vol. 127, pp. 1923–1945, 2007.

- J. S. Parker, "Families of Low-Energy Lunar Halo Transfers," *AAS/AIAA Spaceflight Dynamics Conference,* (Tampa, Florida, January 22–26, 2006) Paper No. AAS 06-132, (S. R. Vadali, L. A. Cangahuala, J. P. W. Schumacher, and J. J. Guzman, editors), vol. 124 of *Advances in Astronautical Sciences,* San Diego, CA, AAS/AIAA, Univelt Inc., 2006.

- J. S. Parker and M. W. Lo, "Shoot the Moon 3D," Paper AAS 05-383, *AAS/AIAA Astrodynamics Conference* held August 7–10, 2005, South Lake Tahoe, California, (originally published in) AAS publication, *Astrodynamics 2005* (edited by B. G. Williams, L. A. D'Amario, K. C. Howell, and F. R. Hoots) American Astronautical Society (AAS) *Advances in the Astronautical Sciences,* vol. 123, pp. 2067–2086, 2006, American Astronautical Society Publications Office, San Diego, California (Web Site: http://www.univelt.com), pp. 2067–2086.

Chapter 4:

- J. S. Parker and R. L. Anderson, "Targeting Low-Energy Transfers to Low Lunar Orbit," *Astrodynamics: Proceedings of the 2011 AAS/AIAA Astrodynamics Specialist Conference,* (Girdwood, Alaska, July 31–August 4), Paper AAS 11-459, edited by H. Schaub, B. C. Gunter, R. P. Russell, and W. T. Cerven, Vol. 142, *Advances in the Astronautical Sciences,* American Astronautical Society, Univelt Inc., San Diego, California, pp. 847–866, 2012.

- J. S. Parker, R. L. Anderson, and A. Peterson, "A Survey of Ballistic Transfers to Low Lunar Orbit," *21st AAS/AIAA Space Flight Mechanics Meeting,* (February 13–17, 2011, New Orleans, Louisiana), Paper AAS 11-277, Vol. 140, *Advances in the Astronautical Sciences* (edited by M. K. Jah, Y. Guo, A. L. Bowes, and

P. C. Lai), American Astronautical Society, Univelt Inc., San Diego, California, pp. 2461–2480, 2011.

Chapter 5:

- R. L. Anderson, and J. S. Parker, "Survey of Ballistic Transfers to the Lunar Surface," *Journal of Guidance, Control, and Dynamics,* vol. 35, no. 4, pp. 1256–1267, July–August 2012.
- R. L. Anderson and J. S. Parker, "Comparison of Low-Energy Lunar Transfer Trajectories to Invariant Manifolds," *Celestial Mechanics and Dynamical Astronomy,* vol. 115, DOI 10.10075 10569-012-9466-3, pp. 311–331, published online February 16, 2013.
- R. L. Anderson, and J. S. Parker, "Comparison of Low-Energy Lunar Transfer Trajectories to Invariant Manifolds," *AAS/AIAA Astrodynamics Specialist Conference* (Girdwood, Alaska, July 31–August 4, 2011), Paper AAS 11-423, edited by H. Schaub, B. C. Gunter, R. P. Russell, and W. T. Cerven, Vol. 142, *Advances in the Astronautical Sciences,* American Astronautical Society, Univelt Inc., San Diego, California, pp. 333–352, 2012.
- R. L. Anderson, and J. S. Parker, "A Survey of Ballistic Transfers to the Lunar Surface," *Proceedings of the 21st AAS/AIAA Space Flight Mechanics Meeting* (New Orleans, Louisiana, February 13–17, 2011), Paper AAS 11-278, edited by M. K. Jah, Y. Guo, A. L. Bowes, and P. C. Lai, Vol. 140, *Advances in the Astronautical Sciences,* vol. 140, American Astronautical Society, Univelt Inc., San Diego, California, pp. 2481–2500, 2011.

Chapter 6:

- J. S. Parker, "Targeting Low-Energy Ballistic Lunar Transfers," *Journal of Astronautical Sciences,* vol. 58, no. 3, pp. 311–334, July–September, 2011.
- J. S. Parker and R. L. Anderson, "Targeting Low-Energy Transfers to Low Lunar Orbit," *Astrodynamics 2011: Proceedings of the AAS/AIAA Astrodynamics Specialist Conference* (Girdwood, Alaska, July 31–August 4, 2011), Paper AAS 11-459, edited by H. Schaub, B. C. Gunter, R. P. Russell, and W. T. Cerven, Vol. 142, *Advances in the Astronautical Sciences,* American Astronautical Society, Univelt Inc., San Diego, California, pp. 847–866, 2012.
- J. S. Parker, "Targeting Low-Energy Ballistic Lunar Transfers," AAS 09-443, *AAS George H. Born Special Symposium* (Boulder, Colorado, May 13–14), American Astronautical Society, 2010.

A large portion of the research in this book, and the compilation of related research documentation from other sources, were carried out at the Jet Propulsion Laboratory,

California Institute of Technology, under a contract with the National Aeronautics and Space Administration. This work has been supported through funding by the Multimission Ground System and Services Office (MGSS) in support of the development of the Advanced Multi-Mission Operations System (AMMOS).

Reference herein to any specific commercial product, process, or service by trade name, trademark, manufacturer, or otherwise, does not constitute or imply its endorsement by the United States Government or the Jet Propulsion Laboratory, California Institute of Technology.

Jeffrey S. Parker & Rodney L. Anderson

AUTHORS

Jeffrey S. Parker received his B.A. in 2001 in physics and astronomy from Whitman College (Walla Walla, Washington) and his M.S. and Ph.D. in aerospace engineering sciences from the University of Colorado at Boulder in 2003 and 2007, respectively. Dr. Parker was a member of the technical staff at the Jet Propulsion Laboratory (JPL) from January 2008 to June 2012. While at JPL he supported spacecraft exploration as a mission design and navigation specialist. He worked both as a spacecraft mission designer and as a navigator on the GRAIL mission, which sent two spacecraft to the Moon via low-energy ballistic lunar transfers. He supported India's *Chandrayaan-1* mission to the Moon, also as a mission designer and spacecraft navigator. Dr. Parker led the mission design development for numerous design studies and mission proposals, including missions to the Moon, near-Earth objects, the nearby Lagrange points, and most of the planets in the Solar System. At present, Dr. Parker is an assistant professor of astrodynamics at the University of Colorado at Boulder, teaching graduate and undergraduate courses in many subjects related to space exploration. His research interests are focused on astrodynamics and the exploration of space, including the design of low-energy trajectories in the Solar System, the optimization of low-thrust trajectories in the Solar System, autonomous spacecraft operations, and use of these engineering tools to provide new ways to achieve scientific objectives.

Rodney L. Anderson received his B.S. in 1997 in aerospace engineering from North Carolina State University at Raleigh and his M.S. and Ph.D. in aerospace engineering sciences from the University of Colorado at Boulder in 2001 and 2005, respectively. Upon the completion of his Ph.D., he worked as a research associate at the University of Colorado at Boulder, conducting a study for the U.S. Air Force that focused on understanding the effects of atmospheric density variations on orbit predictions. Dr. Anderson has been a member of the JPL technical staff since 2010, where he has participated in mission design and navigation for multiple missions, and continues to work on the development of new methods for trajectory design. His research interests are concentrated on the application of dynamical systems theory to astrodynamics and mission design. Some specific applications that he has focused on are the design of lunar trajectories, tour and endgame design in the Jovian system using heteroclinic connections, missions to near-Earth asteroids, and low-energy trajectories in multi-body systems. He has worked closely with multiple universities and has taught at both the University of Colorado at Boulder and the University of Southern California, with an emphasis on the intersection of dynamical systems theory with astrodynamics.

CHAPTER 1

INTRODUCTION AND EXECUTIVE SUMMARY

1.1 PURPOSE

This book provides sufficient information to answer high-level questions about the availability and performance of low-energy transfers between the Earth and Moon in any given month and year. Details are provided to assist in the construction of desirable low-energy transfers to various destinations on the Moon, including low lunar orbits, halo and other three-body orbits, and the lunar surface. Much of the book is devoted to surveys that characterize many examples of transfers to each of these destinations.

1.2 ORGANIZATION

This document is organized in the following manner. The remainder of this chapter first provides an executive summary of this book, presenting an overview of low-energy lunar transfers and comparing them with various other modes of transportation from near the Earth to lunar orbit or the lunar surface. It then provides background information, placing low-energy lunar transfers within the context of historical lunar missions. The chapter describes very high-level costs and benefits of low-energy transfers compared with conventional transfers.

Chapter 2 provides information about the methods, coordinate frames, models, and tools used to design low-energy lunar transfers. This information should be sufficient for designers to reconstruct any transfer presented in this book, as well as similar transfers with particular design parameters.

Chapter 3 presents information about transfers from the Earth to high-altitude three-body orbits, focusing on halo orbits about the first and second Earth–Moon Lagrange points. The chapter includes surveys of the transfer types that exist and discussions about how to construct a particular, desirable transfer.

Chapter 4 presents information about transfers from the Earth to low-altitude lunar orbits, focusing on polar mapping orbits. The techniques presented may be used to survey and construct conventional direct lunar transfers as well as low-energy transfers.

Chapter 5 presents information about transfers from the Earth to the lunar surface, including discussions and surveys of transfers that intersect the lunar surface at a steep 90 degree (deg) angle, as well as transfers that target a shallow flight path angle. The techniques illustrated in Chapter 5 may be used to generate conventional direct transfers as well as low-energy transfers.

Chapters 3–5 also include discussions about the variations of these transfers from one month to the next. The discussions are useful for mission designers and managers to predict what sorts of transfers exist in nearly any month and what sorts of transfers are particular to specific months.

Chapter 6 discusses several important operational aspects of implementing a low-energy lunar transfer. The section begins with a discussion of the capabilities of current launch vehicles to inject spacecraft onto low-energy trajectories. The section then describes how to design a robust launch period for a low-energy lunar transfer. Additional discussions are provided to address navigation, station-keeping, and spacecraft systems issues.

1.3 EXECUTIVE SUMMARY

This book characterizes low-energy transfers between the Earth and the Moon as a resource to mission managers and trajectory designers. This book surveys and illustrates transfers between the Earth and lunar libration orbits, low lunar mapping orbits, and the lunar surface, including transfers to the Moon and from the Moon to the Earth.

There are many ways of transporting a spacecraft between the Earth and the Moon, including fast conventional transfers, spiraling low-thrust transfers, and low-energy transfers. Table 1-1 summarizes several of these methods and a sample of the missions that have flown these transfers.

The vast majority of lunar missions to date have taken quick, 3–6 day direct transfers from the Earth to the Moon. The Apollo missions took advantage of 3–3.5 day transfers: transfers that were as quick as possible without dramatically

Table 1-1 A summary of several different methods used to transfer between the Earth and the Moon.

Transfer Type	Typical Duration	Benefits	Example Missions[a]
Direct, conventional	3–6 days	Well known, quick	Apollo, *LRO*, others
Direct, staging	2–10 weeks	Quick, many launch days	*Clementine*, *CH-1*
Direct to lunar L_1	1–5 weeks	Staging at L_1	None to date
Low-thrust	Many months	Low fuel, many launch days	*SMART-1*
Low-energy	2.5–4 months	Low fuel, many launch days	*Hiten*, *GRAIL*, *ARTEMIS*

[a]Missions referred to include *Lunar Reconnaissance Orbiter (LRO)*, *Chandrayaan-1 (CH-1)*, *Small Missions for Research in Technology 1 (SMART-1)*, and *Mu Space Engineering Spacecraft (MUSES 1, Hiten)*

increasing the transfers' fuel requirements. The *Lunar Reconnaissance Orbiter (LRO)* followed a slightly more efficient 4.5-day transfer. The additional transfer duration saved fuel and relaxed the operational timeline of the mission. The Apollo missions and *LRO* had very limited launch opportunities: they had to launch within a short window each month. *Clementine* and *Chandrayaan-1* implemented phasing orbits about the Earth to alleviate this design constraint and expand their launch periods. *SMART-1* was also able to establish a wider launch period using low-thrust propulsion. The low-thrust system requires less fuel mass than conventional propulsion systems, but the transfer required significantly more transfer time than any typical ballistic transfer.

The *Gravity Recovery and Interior Laboratory (GRAIL)* mission was the first mission launched to the Moon directly on a low-energy transfer. *GRAIL*'s low-energy transfer required much less fuel than a conventional transfer, though it required a longer cruise that traveled farther from the Earth. The longer cruise (~90–114 days) made it possible to establish a wide, 3-plus week long launch period and significantly relaxed the operational timeline. Furthermore, *GRAIL* launched two satellites on board a single launch vehicle and leveraged the longer cruise to separate their orbit insertion dates by more than a day. Finally, *GRAIL*'s low-energy transfer reduced the orbit insertion change in velocity (ΔV) for each vehicle, permitting each spacecraft to perform its lunar orbit insertion with a smaller engine and less fuel.

In general, a low-energy transfer is a nearly ballistic transfer between the Earth and the Moon that takes advantage of the Sun's gravity to reduce the spacecraft's fuel requirements. The only maneuvers required are typical statistical maneuvers needed to clean up launch vehicle injection errors and small deterministic maneuvers to target specific mission features. A spacecraft launched on a low-energy lunar transfer travels beyond the orbit of the Moon, far enough from the Earth and Moon to permit the gravity of the Sun to significantly raise the spacecraft's energy. The spacecraft remains beyond the Moon's orbit for 2–4 months while its perigee radius rises. The spacecraft's perigee radius typically rises as high as the Moon's orbit, permitting the spacecraft to encounter the Moon on a nearly tangential trajectory. This trajectory has a very low velocity relative to the Moon: in some cases the spacecraft's two-body energy will even be negative as it approaches the Moon, without having performed any maneuver whatsoever. As the spacecraft approaches the Moon, it may target a trajectory to land on the Moon, to enter a low lunar orbit, or to enter any number of three-body orbit types, such as halo or Lissajous orbits. No matter what its destination, the spacecraft requires less fuel to reach it than it would following a conventional transfer.

Low-energy transfers provide many benefits to missions when compared with conventional transfers. Six example benefits include the following:

1. They require less fuel. A low-energy transfer to a lunar-libration orbit saves 400 meters per second (m/s) of ΔV and often more. This is a significant savings, which is fully demonstrated in Chapter 3. A low-energy transfer to a 100-kilometer (km) lunar orbit saves more than 120 m/s of ΔV for cases when a mission *can* use an optimized conventional transfer. The savings are far more dramatic for missions that cannot use an optimized conventional transfer.

2. Low-energy transfers are more flexible than conventional transfers and may be used to transfer spacecraft to many more orbits on a given date. It is shown in Chapter 4 that low-energy transfers may be used to reach polar orbits with any node at any arrival date—conventional transfers may only target specific nodes at any given date.

3. Low-energy transfers have extended launch periods. It requires very little fuel to establish a launch period of 21 days or more for a mission to the Moon that implements a low-energy transfer. Conventional transfers may be able to accomplish similar launch periods, but they require multiple passes through the Van Allen Belts, necessitating improved radiation protection. The low-ΔV costs of establishing a launch period for a low-energy transfer are discussed in Chapter 6.

4. Low-energy transfers have a relaxed operational timeline. Modern launch vehicles, such as the Atlas V family with their Centaur upper stages, place spacecraft on their trajectories with small errors. Missions such as *GRAIL*, which launched aboard a Delta II launch vehicle, may be able to wait 6 days or more before performing a maneuver. In fact, *GRAIL* was able to cancel the first trajectory correction maneuver (TCM) for both spacecraft; the first TCM performed was executed 20 days after launch. In this way, a spacecraft operations team has a great deal more time to prepare the spacecraft before requiring a maneuver, when compared to conventional transfers that typically require a maneuver within a day or less.

5. Low-energy transfers may place several vehicles into very different orbits at the Moon using a single launch vehicle. The *GRAIL* mission separated two lunar-orbit insertions by over a day using very little fuel. Chapter 3 illustrates how to place multiple spacecraft in many different orbit types using a single launch vehicle. This typically requires a large amount of fuel when using conventional transfers.

6. Low-energy transfers may be used to transfer a spacecraft from the Moon directly to any location on the surface of the Earth. Typical conventional transfers, for example, those used by the Apollo missions, return spacecraft to a near-equatorial landing site. Low-energy transfers may be used to target any location (such as the different hemispheres of the Utah Test and Training Range in North America and the Woomera Weapons Testing Range in South Australia) using relatively small quantities of fuel.

The typical drawbacks of low-energy transfers between the Earth and the Moon are the longer transfer durations for missions that are very time-critical and the longer link-distances, as the spacecraft travels as far as 1.5–2 million kilometers away from the Earth.

The next sections define *direct* and *low-energy* transfers to provide a clear understanding of what trajectories are presented in this book.

1.3.1 Direct, Conventional Transfers

A direct lunar transfer is a trajectory between the Earth and the Moon that requires only the gravitational attraction of the Earth and Moon. A spacecraft typically begins from a low altitude above the surface of the Earth as a result of an injection by a launch vehicle, as a result of a maneuver performed by the spacecraft, or as a result of some intermediate orbit. The spacecraft then cruises to the Moon on a trajectory that typically remains within the orbit of the Moon about the Earth. It is a trajectory whose dynamics are dominated by the gravitational attraction of the Earth and Moon, and all other forces (such as the Sun or any spacecraft events) may be considered to be perturbations. The spacecraft then enters some orbit about the Moon via a maneuver. Direct transfers may be constructed from the Moon to the Earth in much the same way as they are constructed to the Moon.

Figure 1-1 illustrates a 3-day transfer nearly identical to the one the *Apollo 11* astronauts used to go from the Earth to the Moon in 1969 [1]. The mission implemented a low-Earth parking orbit with an inclination of approximately 31.38 deg. From there, the launch vehicle was required to attain a trans-lunar injection energy (C_3) of approximately -1.38 km^2/s^2 to reach the Moon in approximately 3.05 days. Upon arrival at the Moon, the vehicle injected into an elliptical orbit with a periapse altitude of approximately 110 km and an apoapse altitude of approximately 310 km, followed soon after by a circularization maneuver [1]. In order to compare the *Apollo 11* transfer with the transfers in the surveys presented here, the *Apollo 11* transfer would have a velocity of approximately 2.57 kilometers per second (km/s) at an altitude of 100 km above the mean lunar surface, requiring a hypothetical, impulsive ΔV of approximately 0.94 km/s to insert into a circular 100-km orbit.

Direct transfers may be constructed between the Earth and the Moon with durations as short as hours or as long as a few weeks. In general, the most fuel-efficient direct transfers require about 4.5 days of transfer duration. Any longer duration typically sends the spacecraft beyond the orbit of the Moon before it falls back and encounters the Moon.

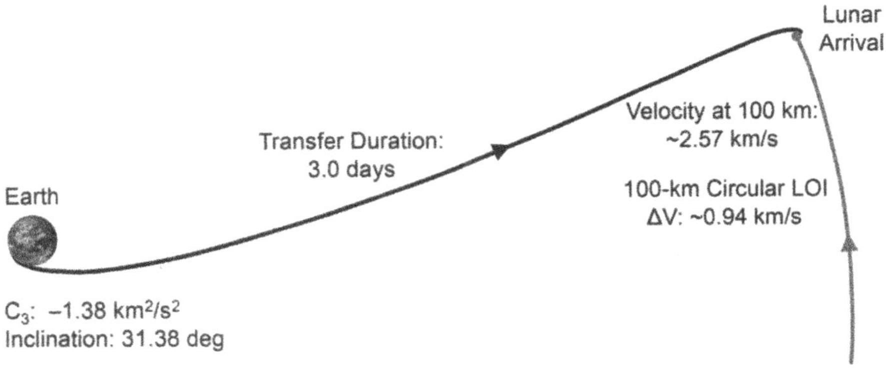

Figure 1-1 A modified version of the Apollo 11 Earth–Moon transfer, as if it had performed an impulsive lunar-orbit insertion (LOI) maneuver directly into a circular 100-km lunar orbit [2]. (Copyright © 2011 by American Astronautical Society Publications Office, all rights reserved, reprinted with permission of the AAS.)

Direct transfers may also be constructed between the Earth and lunar libration orbits for similar amounts of fuel as required to transfer directly to low lunar orbits. The launch energy requirement is very similar for missions to the Moon, to Lagrange 1 (L_1), and to Lagrange 2 (L_2), and to a first order may be treated as equal. A direct transfer requires 400–600 m/s of ΔV to insert into a lunar libration orbit about either L_1 or L_2, though a powered lunar flyby en route to a libration orbit about L_2 may be used to reduce the total transfer cost by 100–200 m/s. These transfers are examined in Chapter 3.

Several missions have added Earth phasing orbits to their mission itineraries, such that they launch into a high-altitude, temporary Earth orbit and remain in that orbit for several orbits before arriving at the Moon. A mission designer may add these orbits to a flight plan for several reasons. First, they may be used to establish an extended launch period, since the mission planners can adjust the size of the phasing orbits to compensate for varying launch dates. Second, they may be used to reduce the operational risk of the mission by increasing the amount of time between each maneuver en route to the Moon. They may also be used if the launch vehicle is not powerful enough or accurate enough to send the spacecraft directly to the Moon, such as *Chandrayaan-1* [3]. Drawbacks of Earth phasing orbits include additional passes through the Van Allen Belts and an extended transfer duration.

1.3.2 Low-Energy Transfers

Low-energy transfers take advantage of the Sun's gravity to reduce the transfer fuel costs. They involve trajectories that take the spacecraft beyond the orbit of the Moon, where the Sun's gravity becomes more influential. The Sun's gravity works slowly and steadily, gradually raising the spacecraft's periapse altitude until it has risen to the altitude of the Moon's orbit about the Earth. When the spacecraft falls back toward the Earth, it arrives at the Moon with a velocity that closely matches the Moon's orbital velocity. The result is that the spacecraft's lunar orbit insertion requires much less fuel than required by a conventional, direct lunar transfer. Figure 1-2 illustrates an example 84-day low-energy transfer that arrives at the Moon when the Moon is at its first quarter. More explanation of these transfers is provided in Section 1.7 and in later chapters.

Low-energy transfers typically travel far beyond the orbit of the Moon; hence, they may be designed to take advantage of one or more lunar flybys on their outbound segment. The lunar flybys may be used to reduce the injection energy requirements, or to change the spacecraft's orbital plane, similar to the flight of each of the two *Acceleration, Reconnection, Turbulence and Electrodynamics of the Moon's Interaction with the Sun* (*ARTEMIS*) spacecraft [4]. If a mission takes advantage of a lunar flyby immediately after launch, it may be useful to add one or more Earth phasing orbits into the design, as described above.

1.3.3 Summary: Low-Energy Transfers to Lunar Libration Orbits

Low-energy transfers may be used to save a great deal of fuel when a mission's destination is a lunar libration orbit, such as a halo orbit, a Lissajous orbit, or

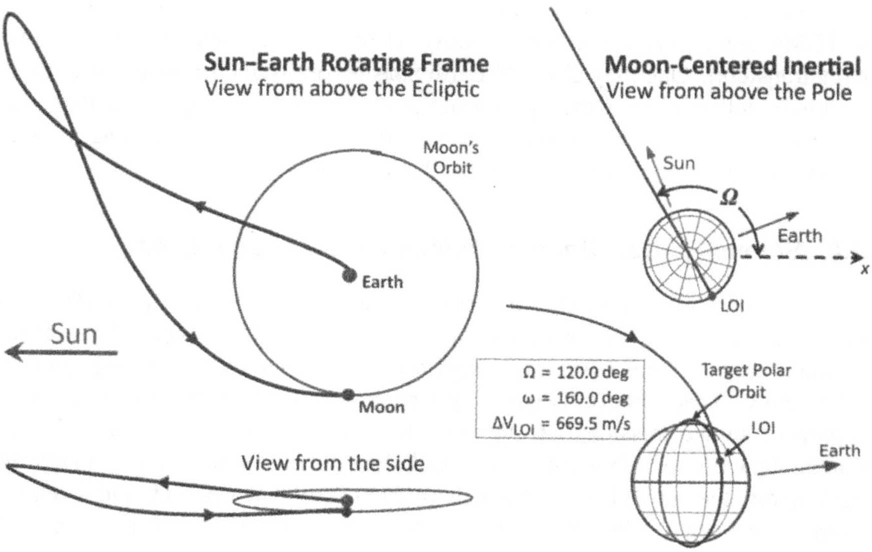

Figure 1-2 An example 84-day low-energy lunar transfer to a low, polar lunar orbit [2]. (Copyright © 2011 by American Astronautical Society Publications Office, all rights reserved, reprinted with permission of the AAS.)

some other three-body orbit. Many studies have demonstrated practical applications of lunar libration orbits, including locations for communication satellites [5–7], navigation satellites [8–13], staging orbits [14–18], and science orbits [4, 19]. The *ARTEMIS* mission took advantage of the geometries of several orbits about both the lunar L_1 and L_2 points, and it used two different low-energy transfers to arrive at those orbits.

Chapter 3 presents a full study of the characteristics and performance of low-energy transfers to lunar libration orbits. The results demonstrate that a typical transfer requires 70–120 days to travel from Earth departure to an arrival state that is within 100 km of the target libration orbit. The transfers arrive asymptotically, such that they do not require any insertion maneuver. This is an extraordinary benefit: it saves a mission upwards of 500 m/s of ΔV when compared to conventional, direct transfers to lunar libration orbits. The typical transfers studied in Chapter 3 depart the Earth with a C_3 of -0.7 to -0.3 km^2/s^2, which is higher than the conventional transfer that has a C_3 of approximately -2.0 km^2/s^2, but the low-energy transfer requires only small TCMs after the Earth-departure maneuver. Studies show (Section 6.5) that two or three deterministic maneuvers with a total of only ~70 m/s of ΔV may be used to depart the Earth from a specific inclination (such as 28.5 deg), and from any day within a 21-day launch period, and arrive at a particular location in a specified libration orbit.

Figures 1-3 and 1-4 illustrate two example direct transfers and two example low-energy transfers to lunar libration orbits, respectively. One can see that these transfers are ballistic in nature: they require a standard trans-lunar injection maneuver, a few TCMs, and an orbit insertion maneuver (which is essentially zero for the low-energy transfers). One may also add Earth phasing orbits and/or lunar flybys to the trajectories, which change their performance characteristics. Figure 1-5 illustrates two transfers that a spacecraft may take to depart the libration orbit using minimal fuel and transfer to a low lunar orbit or to the lunar surface.

1.3.4 Summary: Low-Energy Transfers to Low Lunar Orbits

Robotic spacecraft may take advantage of the benefits of a low-energy transfer when transferring to a low lunar orbit, such as *GRAIL*'s target lunar orbit. The transfer duration is about the same as a low-energy transfer to a lunar libration orbit, namely, 70–120 days. This duration is typically far too long for human occupants, unless the purpose of the mission is to demonstrate a long deep-space transfer. There are many benefits for robotic missions, including smaller orbit insertion maneuver requirements, the capability to establish an extended launch period, and a relaxed operational schedule. The *GRAIL* mission took advantage of these benefits, as well as the characteristic that it requires very little ΔV to separate the two spacecraft from their joint launch. *GRAIL*'s two spacecraft flew independently to the Moon and arrived 25 hours apart: a feat that requires a great deal more ΔV and/or operational complexity when implementing direct lunar transfers. Low-energy transfers may also access a much broader range of lunar orbits for a particular arrival date than direct transfers.

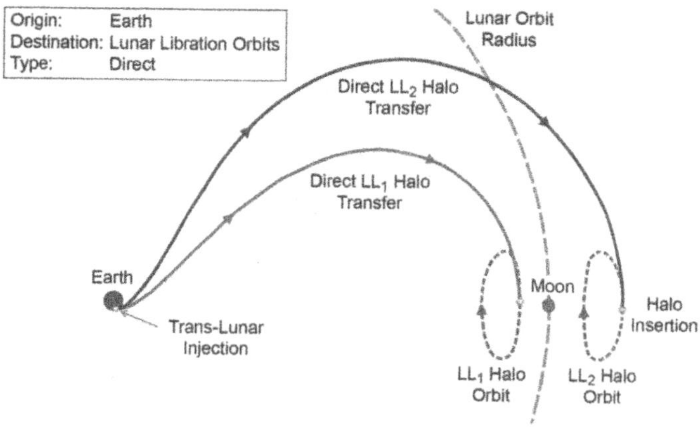

Figure 1-3 The profile for a simple, direct transfer from the Earth to a lunar libration orbit about either the Earth–Moon L_1 or L_2 point, viewed from above in the Earth–Moon rotating coordinate frame.

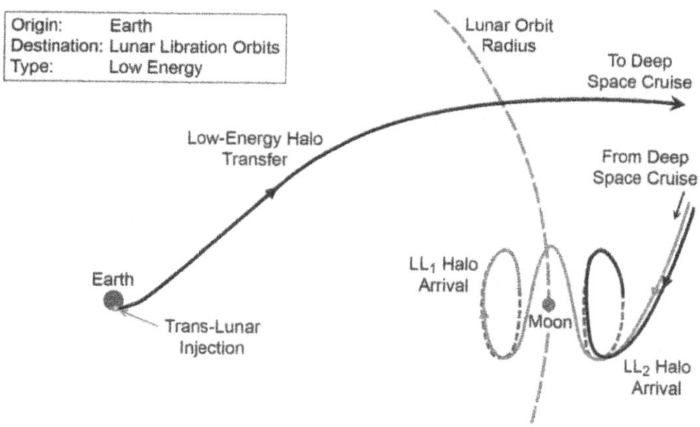

Figure 1-4 The profile for a simple, low-energy transfer from the Earth to a lunar libration orbit about either the Earth–Moon L_1 or L_2 point, viewed from above in the Earth–Moon rotating coordinate frame.

Chapter 4 presents a full study on the characteristics and performance of low-energy transfers to low lunar, polar orbits. The examination uses 100-km circular, polar orbits as the target orbits to simplify the trade space. It remains relevant to practical mission design since many spacecraft missions have inserted into very similar orbits, including *Lunar Prospector, Kaguya/ Selenological and Engineering Explorer (SELENE), Chang'e 1, LRO,* and *GRAIL*, among others. The results of the study indicate that low-energy transfers typically depart the Earth with an injection C_3 of –0.7 to –0.3 km^2/s^2, much like low-energy transfers to lunar libration orbits,

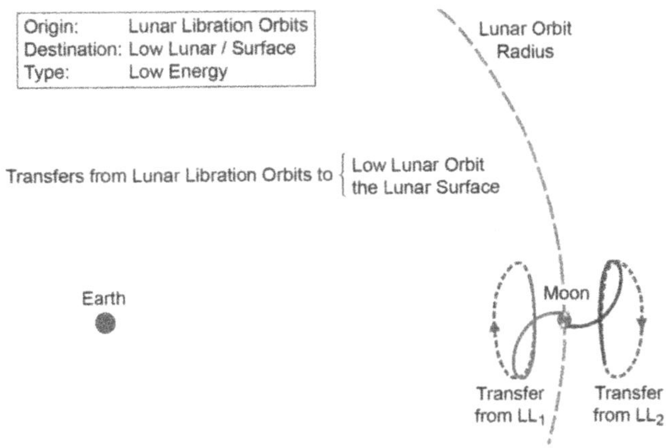

Figure 1-5 The profile for a simple, low-energy transfer from a libration orbit to either a low lunar orbit or the surface of the Moon, viewed from above in the Earth–Moon rotating coordinate frame.

and require 70–120 days to reach the Moon. A spacecraft may implement a lunar flyby on the outbound segment to reduce the launch energy requirement, but such an event would increase the complexity and operational risk of the mission. When the spacecraft arrives at the Moon, it arrives traveling at a slower relative speed than if it had used a direct lunar transfer. The examination shows that the lunar-orbit insertion maneuver is at least 120 m/s smaller for any low-energy mission; the ΔV savings are often much greater.

Low-energy transfers may also be used in such a way that a spacecraft transfers to a lunar libration orbit, or some other three-body orbit, before transferring to the target orbit. This strategy was used in the *ARTEMIS* mission and has been used in a number of spacecraft proposals.

Figure 1-6 illustrates an example direct transfer and an example low-energy transfer to two low lunar orbits. The transfers are very similar to those presented in the previous section, except of course that these target low lunar orbits instead of lunar libration orbits.

1.3.5 Summary: Low-Energy Transfers to the Lunar Surface

Low-energy transfers from the Earth to the lunar surface may be constructed in much the same way as transfers to low lunar orbit. They have the same sorts of benefits and drawbacks as other low-energy transfers.

Chapter 5 presents a full study on the characteristics and performance of low-energy transfers to the lunar surface. There are two main classes of missions studied: those that arrive at the surface with a high impact angle and those that arrive at the surface with a shallow flight path angle. The shallow angles are useful for missions that aim to land on the surface, and then it is useful that the low-energy transfers

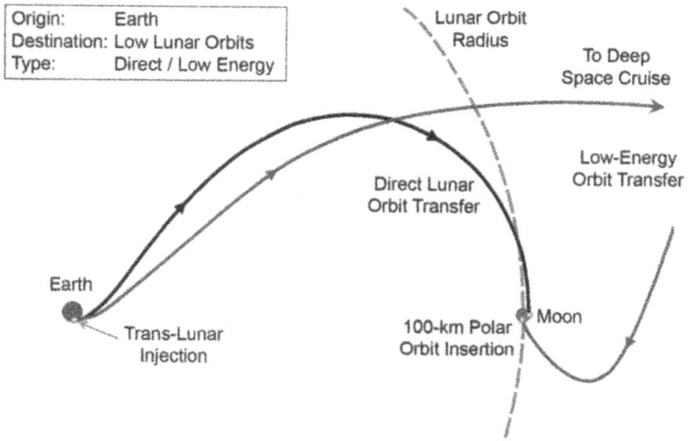

Figure 1-6 The profiles for both a direct and a low-energy transfer from the Earth to a low lunar orbit.

yield trajectories that arrive at the surface with lower velocities. The steeper arrival conditions are useful for lunar impactors, such as the *Lunar Crater Observatory and Sensing Satellite (LCROSS)*. In this case, higher velocities are typically preferred. Low-energy transfers may not result in the highest impact velocities achievable, but they do offer the capability of targeting any location on the surface of the Moon with ease.

As with the low-energy transfers studied in Chapters 3 and 4, the typical transfers to the lunar surface require 70–120 days. They typically depart the Earth with C_3 values between −0.7 and −0.3 kilometers squared per square second (km^2/s^2) and only require small trajectory correction maneuvers after launch. The same sort of two- or three-burn strategies may be used to target a particular low-energy transfer from a specified low Earth parking orbit, and from any day within a 21-day launch period.

The lunar surface may also be accessed from a lunar libration orbit or from a low lunar orbit. Hence, a mission may implement a low-energy transfer to either type of orbit studied in Chapters 3 or 4 and then follow a transfer to the lunar surface. This sort of trajectory design is also studied in Chapter 5.

Figure 1-7 illustrates an example direct transfer and an example low-energy transfer to the lunar surface. Again, the transfers are very similar to those presented in the previous two sections, except (of course) that these target the lunar surface.

1.4 BACKGROUND

This section reviews historical lunar missions as a reference for the discussions about designing future lunar missions, including future missions that use direct transfers as well as low-energy transfers. Nearly one hundred spacecraft have flown conventional,

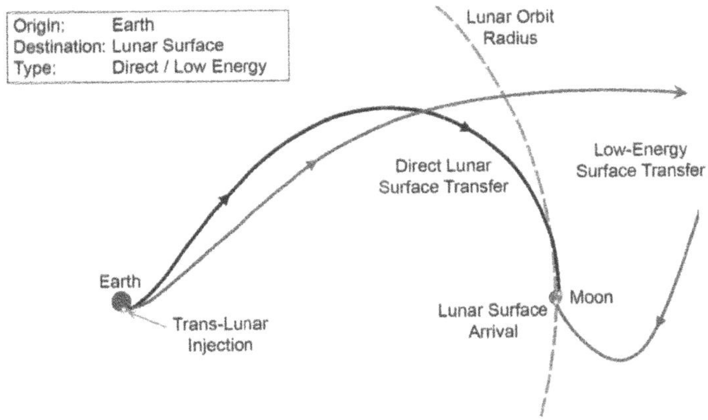

Figure 1-7 The profiles for both a direct and a low-energy transfer from the Earth to the lunar surface. Transfers may be constructed to arrive with a shallow or steep flight path angle.

direct transfers between the Earth and the Moon, including the Union of Soviet Socialist Republics' (USSR's) Luna spacecraft, the USA's Apollo spacecraft, and the most recent international missions. Only five spacecraft have flown low-energy lunar transfers, though several others have followed low-energy transfers to other destinations near the Earth. The complexity of lunar missions has gradually grown, as has the need to save money and collect a greater scientific return using less fuel. Modern flight operations, spacecraft hardware, and infrastructure have opened the door to low-energy techniques as a method to reduce costs.

The first two missions to implement low-energy transfers—*Hiten* and *ARTEMIS*—demonstrated the technique as a method to extend their missions to the Moon, despite not having the fuel to reach lunar orbit using conventional techniques. The *GRAIL* mission, launched on September 10, 2011, was the first mission to implement a low-energy lunar transfer as part of its primary mission. The *GRAIL* mission benefited from its low-energy route to the Moon in more ways than just saving fuel. It is fully expected that more missions will follow this lead, and low-energy transfers will become common among lunar missions.

1.5 THE LUNAR TRANSFER PROBLEM

Soon after the dawn of the Space Age, people were designing trajectories for spacecraft to travel to the Moon [20, 21]. In fact, not even a full year had elapsed since the launch of *Sputnik* (October 4, 1957) before the United States attempted to launch the *Pioneer 0* probe to the Moon (August 17, 1958). The first probes designed to explore the Moon were plagued with launch vehicle failures, including four Pioneer failures by the United States and three Luna failures by the Soviet Union. It was not until 1959 that *Luna 1* finally flew by the Moon. Later in 1959, *Luna 2* became the first probe to impact the Moon.

As technology improved, spacecraft were able to fly to the Moon using less fuel. Several general bounds exist that limit the movement of a spacecraft in the Earth–Moon system when other perturbations, such as the Sun's gravity, are ignored. Research in the circular restricted three-body problem (examined in Section 2.6.2) reveal that a spacecraft with enough energy to reach the Earth–Moon L_1 point has the minimum energy required to transfer to the Moon, without considering other perturbations. Sweetser computed that the theoretical minimum ΔV that a spacecraft would require to travel from a 167-km altitude circular orbit at the Earth to a 100-km altitude circular orbit at the Moon, just passing through L_1, is approximately 3.721 km/s [22]. Actual trajectories have since been computed that approach this theoretical minimum [23].

Early investigations concluded that it is impossible to launch from the Earth and arrive at the Moon such that the spacecraft becomes captured without performing a maneuver [21]; however, these analyses did not include the effects of the Sun's gravity. As early as 1968, Charles Conley began using dynamical systems methods to explore the construction of a theoretical trajectory that could become temporarily captured by the Moon without performing a capture maneuver [24]. A spacecraft with the proper energy could target the neck region near one of the collinear libration points in the Earth–Moon system (see Section 2.6.2). A planar periodic orbit exists in each of those regions that acts as a separatrix, separating the interior of the Moon's region from the rest of the Earth–Moon region. Conley's method implemented dynamical systems techniques to construct the transfer by targeting the gateway periodic orbit. His transfers were restricted to the Moon's orbital plane.

In the late 1980s and early 1990s, Belbruno and Miller began developing a method to construct lunar transfers using a new technique, which they have referred to as the weak stability boundary (WSB) theory [25–27]. The method involves targeting the region of space that is in gravitational balance between the Sun, Earth, and Moon, without involving any three-body periodic orbits or other dynamical structures. Ballistic capture occurs when the spacecraft's two-body energy becomes negative, as described by Yamakawa [28, 29]. In 1991, the Japanese mission *Hiten/MUSES-A* used the effects of the Earth, Moon, and Sun for its transfer to the Moon [30].

In the early 2000s, Ivashkin also developed a method to construct transfers between the Earth and Moon using the Sun's gravitational influence [31–34]. His methods involve beginning from a low lunar orbit, or from the surface of the Moon, and numerically targeting trajectories that depart from the Moon in the direction of the Earth's L_1 or L_2 points. A spacecraft on such a trajectory departs from the Moon with a negative two-body energy with respect to the Moon, but as it climbs away from the Moon, it gains energy from the effect of the Earth's and Sun's gravity. Eventually, it gains enough energy to escape the Moon's vicinity. The trajectory is then targeted such that it lingers near the chosen Lagrange point long enough to allow the Sun to lower the perigee radius of the next perigee passage down to an altitude of approximately 50 km.

In the mid 1990s, other methods were developed to construct a lunar transfer that takes advantage of the chaos in the Earth–Moon three-body system. Bollt and Meiss constructed a trajectory that sent a spacecraft into an orbit without sufficient energy to immediately reach the Moon, but with enough to get close enough to become substantially perturbed by the Moon [35]. Using a series of four very small maneuvers, the spacecraft could then hop between nearby trajectories in the chaotic sea of possible trajectories to become captured by the Moon using far less energy than standard direct transfers. In 1997, Schroer and Ott reduced the time of transfer for the chaotic lunar transfer by targeting specific three-body orbits near the Earth [36]. The total cost remained approximately the same as the transfer constructed by Bollt and Meiss, but the transfer duration was reduced from approximately 2.05 years to 0.8 years.

In 2000, Koon et al. [37, 38] constructed a planar lunar transfer that was almost entirely ballistic using the techniques involved in Conley's method [38]. Following Conley, Koon et al. [37] observed that the planar libration orbits act as a gateway between the interior and exterior regions of space about the Moon. Koon et al. [37, 38] constructed a trajectory that targets the interior of the stable invariant manifold of a planar libration orbit about the Earth–Moon L_2 point. Once inside the interior of the stable manifold, the spacecraft ballistically arrives at a temporarily captured orbit about the Moon. Many authors have debated what it means to be *temporarily captured* at the Moon; Koon et al., define a similar term, "ballistically captured" to be a trajectory that comes within the sphere of influence of the Moon and revolves about the Moon at least once [38].

Further advances have been made since 2004 to apply dynamical systems theory to the generation of three-dimensional low-energy lunar transfers [39–44]. Parker mapped out numerous families of low-energy transfers, illuminating different geometries that are available for spacecraft to travel to the Moon and arrive in lunar libration orbits without requiring any capture maneuver [2, 45–47]. Several authors have begun applying low-thrust techniques to further improve low-energy transfers, including transfers from the Earth to the Moon and transfers from one libration orbit to another [48–55]. In 60 years, research has advanced the knowledge of lunar transfers from the early spacecraft missions that implemented direct lunar transfers to modern analyses that reveal maps of entire families of low-energy transfers to the Moon.

1.6 HISTORICAL MISSIONS

Many historical missions have taken direct transfers from the Earth to the Moon, including a large number of spacecraft in the Luna, Zond, Ranger, Surveyor, Lunar Orbiter, and Apollo programs. A few of these missions implemented direct transfers back to the Earth again: most notably *Luna-16* and the nine Apollo missions that ventured to the Moon and returned. Several other missions have also flown direct transfers since the 1960s, and they are summarized below.

Low-energy lunar transfers are closely related to low-energy transfers in the Sun–Earth system, as is described later in this book. Since the 1970s, several spacecraft have been placed on three-body trajectories in the Sun–Earth system to conduct their scientific and technological missions, including *International Sun–Earth Explorer-3 (ISEE-3), Solar and Heliospheric Observatory (SOHO), Advanced Composition Explorer (ACE), Wind, Wilkinson Microwave Anisotropy Probe (WMAP)*, and *Genesis*, among others. Three spacecraft are known to have followed three-body trajectories in the Earth–Moon system, including *SMART-1* and the two *ARTEMIS* spacecraft. Between 1991 and 2011, five spacecraft have traversed low-energy transfers from the Earth to the Moon, including *Hiten/MUSES-A* in 1991, the two *ARTEMIS* spacecraft in 2010 and the two *GRAIL* spacecraft in 2011. A brief summary of each of these missions will be presented here.

1.6.1 Missions Implementing Direct Lunar Transfers

Table 1-2 summarizes many historical missions that have taken direct lunar transfers, noting their launch date and transfer duration, among other things. One notices that early missions implemented very quick transfers that required fewer than 1.5 days to reach the Moon. These involved lunar flybys or impacts, with no intention of inserting into orbit or landing softly. Indeed, their velocities at the Moon would be quite high. The first soft landing was performed by the Soviet Union's *Luna 9*, which took a 79-hour transfer to the Moon. The first robotic sample return attempt was performed by the Soviet Union's *Luna 15*, which took a 101.6-hour transfer to the Moon: longer to save fuel mass so that it would be capable of returning to the Earth. *Luna 16* was the first successful robotic sample return, taking a 105.1-hour lunar transfer. The first human landing, and first successful sample return was performed earlier, by *Apollo 11*. The direct transfer that *Apollo 11* took required about 73 hours, which was shorter in time and required more fuel, but required less total consumable mass than a longer transfer since the mission involved human occupants.

1.6.2 Low-Energy Missions to the Sun–Earth Lagrange Points

ISEE-3. On August 12, 1978, the *International Sun–Earth Explorer 3 (ISEE-3)* spacecraft was launched and placed in a halo orbit about the Sun–Earth L_1 point. It was the first spacecraft to be inserted into an orbit about a Lagrange point. On June 10, 1982, the spacecraft began performing 15 very small maneuvers to guide it on a series of lunar flybys. Its fifth and final lunar flyby was performed on December 22, 1983, coming within 120 km of the lunar surface. The lunar flyby ejected the spacecraft from the Earth–Moon system and it entered a heliocentric orbit. The spacecraft was renamed the *International Cometary Explorer (ICE)* as it readied for its encounter with the comet Giacobini-Zinner. On June 5, 1985, *ICE* entered the comet's tail and collected scientific information about the tail. *ICE* is expected to return to the vicinity of the Earth in 2014, when it may be captured and brought back to Earth, or repurposed for another comet observation mission. Figure 1-8 shows a plot of the trajectory of *ISEE-3/ICE* [60, 61].

Table 1-2 The transfer durations, among other information, of several historical missions that have implemented direct lunar transfers [56–59].

Launch Date	Spacecraft	Nation[a]	Transfer Duration	Notes
	There were 24 successful Soviet Luna missions; examples include:			
2 Jan. 1959	*Luna 1*	USSR	34 hr (1.42 days)	First lunar flyby (5995 km)
12 Sept. 1959	*Luna 2*	USSR	33.5 hr (1.40 days)	First lunar impact (29.10 N, 0.00 E)
4 Oct. 1959	*Luna 3*	USSR	60 hr (2.50 days)	Flyby (6200 km)
2 Apr. 1963	*Luna 4*	USSR	77.3 hr (3.22 days)	Flyby (8336.2 km)
9 May 1965	*Luna 5*	USSR	~83 hr (3.4 days)	First soft-landing attempt; impact (31 S, 8 W)
31 Jan. 1966	*Luna 9*	USSR	79 hr (3.29 days)	First soft landing (7.08 N, 64.37 W)
31 Mar. 1966	*Luna 10*	USSR	78.8 hr (3.29 days)	First orbiter
13 July 1969	*Luna 15*	USSR	101.6 hr (4.23 days)	First sample return attempt
12 Sep. 1970	*Luna 16*	USSR	105.1 hr (4.38 days)	First sample return (101 grams)
9 Aug. 1976	*Luna 24*	USSR	103.0 hr (4.29 days)	Sample return, landing within 1 km of *Luna 23* (170 grams returned)
	There were eight Soviet Zond missions; little accurate information is available.			
18 July 1965	*Zond 3*	USSR	33 hr (1.38 days)	Flyby (9200 km)
14 Sept. 1968	*Zond 5*	USSR	~3.4 days	First circumlunar return
	There were nine American Ranger missions; examples include:			
26 Jan. 1962	*Ranger 3*	USA	2–3 days	Flyby (~36,800 km)
23 Apr. 1962	*Ranger 4*	USA	64 hr (2.67 days)	Impact (15.5 S, 130.7 W)
18 Oct. 1962	*Ranger 5*	USA	2–3 days	Flyby (725 km)
30 Jan. 1964	*Ranger 6*	USA	65.5 hr (2.73 days)	Impact
28 July 1964	*Ranger 7*	USA	68.6 hr (2.86 days)	Impact (10.70 S, 20.67 W)
17 Feb. 1965	*Ranger 8*	USA	64.9 hr (2.70 days)	Impact (2.71 N, 24.81 E)
21 Mar. 1965	*Ranger 9*	USA	64.5 hr (2.69 days)	Impact (12.91 S, 2.38 W)
	There were seven American Surveyor missions, including:			
30 May 1966	*Surveyor 1*	USA	63 hr (2.63 days)	Landed (2.45 S, 43.21 W)
20 Sept. 1966	*Surveyor 2*	USA	~1.9 days	Impact (5.5 N, 12 W)
17 Apr. 1967	*Surveyor 3*	USA	64.5 hr (2.69 days)	Landed (3.01 S, 23.34 W)
14 July 1967	*Surveyor 4*	USA	~2.6 days	Impact (0.4 N, 1.33 W)
8 Sept. 1967	*Surveyor 5*	USA	64.8 hr (2.70 days)	Landed (1.41 N, 23.18 E)
7 Nov. 1967	*Surveyor 6*	USA	65.0 hr (2.71 days)	Landed (0.49 N, 1.4 W); First powered take-off
7 Jan. 1968	*Surveyor 7*	USA	66.0 hr (2.75 days)	Landed (40.86 S, 11.47 W)
	There were five American Lunar Orbiter missions; examples include:			
10 Aug. 1966	*Lunar Orbiter 1*	USA	91.6 hr (3.82 days)	Orbiter
6 Nov. 1966	*Lunar Orbiter 2*	USA	92.5 hr (3.85 days)	Orbiter
5 Feb. 1967	*Lunar Orbiter 3*	USA	92.6 hr (3.86 days)	Orbiter
	There were 9 American Apollo missions that orbited or orbited and landed on the Moon; examples include:			
21 Dec. 1968	*Apollo 8*	USA	66.3 hr (2.76 days)	First manned lunar orbiter
18 May 1969	*Apollo 10*	USA	73.3 hr (3.05 days)	Orbit and return
16 July 1969	*Apollo 11*	USA	73.1 hr (3.04 days)	First manned landing
7 Dec. 1972	*Apollo 17*	USA	83.0 hr (3.46 days)	Final manned landing 35-km traverse, 110.5 kg returned

[a] Union of Soviet Socialist Republic (USSR) and United States of America (USA)

Table 1-2 Continued.

Launch Date	Spacecraft	Nation	Transfer Duration	Notes
	Additional missions that have implemented direct transfers include:			
3 Mar. 1959	*Pioneer 4*	USA	29.3 hr (1.22 days)	Flyby, first USA spacecraft to reach escape velocity
19 July 1967	*Explorer 35*	USA	~2 days	Orbiter
10 June 1973	*Explorer 49*	USA	113.1 hr (4.71 days)	Orbiter
25 Jan. 1994	*Clementine*	USA	~4 days + 12 days phasing	Orbiter
24 Dec. 1997	*Asiasat 3* / HGS-1	China	~4.5 days	2 lunar flybys en route to GEO
7 Jan. 1998	*Lunar Prospector*	USA	105 hr (4.38 days)	Orbiter
26 Oct. 2006	*STEREO Ahead*	USA	85 hr (3.54 days) + 47 days phasing	1 lunar flyby
26 Oct. 2006	*STEREO Behind*	USA	83 hr (3.46 days) + 47 days phasing	2 lunar flybys
14 Sept. 2007	*Kaguya*/SELENE	Japan	127 hr (5.29 days)	Orbiter
24 Oct. 2007	*Chang'e 1*	China	~120 hr (~5 days) + 7 days phasing	Orbiter
22 Oct. 2008	*Chandrayaan-1*	India	107.9 hr (4.50 days) + 13 days phasing	Orbiter/impactor
18 June 2009	*LRO/LCROSS*	USA	108 hr (4.5 days)	Orbiter/impactor
1 Oct. 2010	*Chang'e 2*	China	112.1 hr (4.7 days)	Orbiter

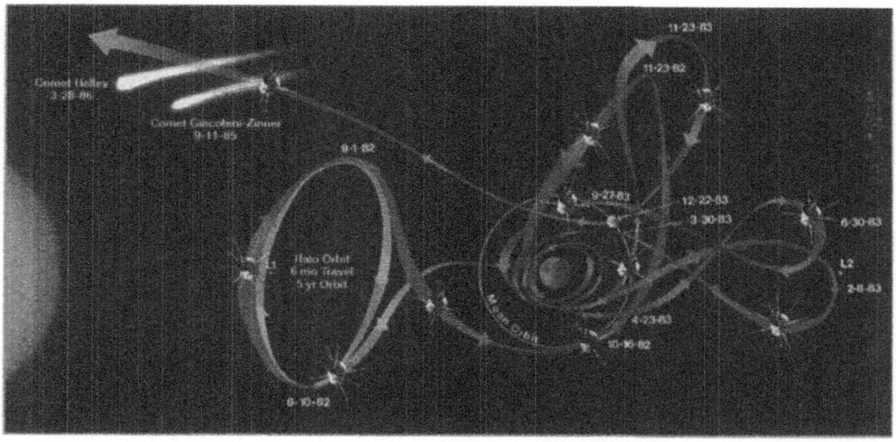

Figure 1-8 The trajectory of *ISEE-3 / ICE* [62]. (See color insert.)

Wind. The *Wind* mission was launched on November 1, 1994, and placed in a halo orbit about the Sun–Earth L_1 point. Its scientific objectives were to characterize the solar wind using a variety of particle and field measurements, all of which complemented several other spacecraft in a variety of other orbits, including the *Polar* and *Geotail* satellites, as part of the International Solar-Terrestrial Physics (ISTP) Science Initiative. After several years of measurements from the Sun–Earth L_1 environment, *Wind*'s orbit was altered to give it access to new areas in the near-Earth environment, including a campaign of "petal orbits" to send it out of the ecliptic plane (1998–1999), a lunar backflip (April, 1999), several revolutions about a distant prograde orbit (2001–2003), and a complex orbit that involved repeated lunar flybys and excursions out beyond the Sun–Earth L_1 and L_2 points (2003–2006). The first part of *Wind*'s trajectory resembles the first part of *ISEE-3*'s trajectory shown in Fig. 1-8. Figure 1-9 illustrates *Wind*'s orbits in the Sun–Earth system from 2003 through 2006 [63], illustrating a unique aspect of its low-energy mission design.

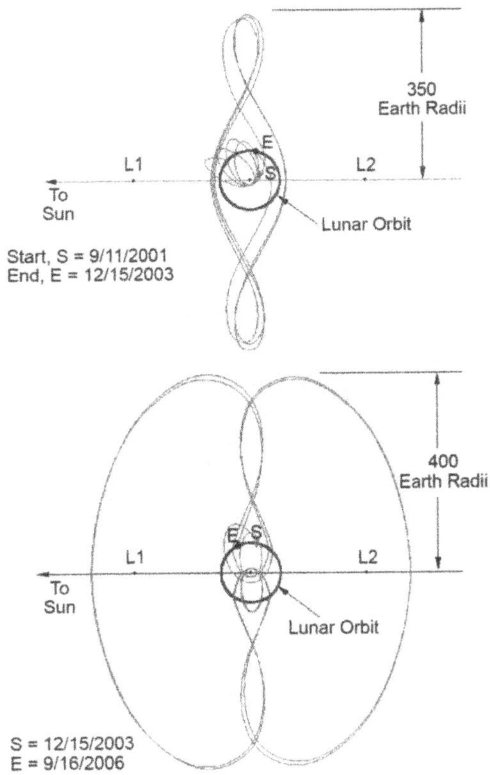

Figure 1-9 The trajectory of *Wind* from 2003 through 2006, viewed from above in the Sun–Earth rotating frame [63].

SOHO. The *Solar and Heliospheric Observatory (SOHO)* was launched on December 2, 1995, on a path taking it directly toward a libration orbit about the Sun–Earth L_1 point. On March 17, 1996, SOHO performed a small orbit insertion maneuver to formally enter the quasi-halo L_1 orbit 1.5 million kilometers away from the Earth. The L_1 halo orbit is ideal for the observatory because it provides an unobstructed view of the Sun on one side and a near-constant view of the Earth on the other side. Hence, it can collect scientific data about the Sun continuously, while being able to communicate with the Earth at any time. Figure 1-10 shows a plot of the trajectory that *SOHO* used to transfer to its halo orbit [64–67].

ACE. In 1997, the *Advanced Composition Explorer (ACE)* was launched and placed in a Lissajous orbit about the Sun–Earth L_1 point. Its mission, much like *SOHO*'s, is dedicated to collecting energetic particles to study the solar corona, interplanetary medium, solar wind, and cosmic rays. Its transfer appears very similar to *SOHO*'s transfer, shown in Fig. 1-10 [68, 69].

WMAP. Launched on June 30, 2001, the *Wilkinson Microwave Anisotropy Probe (WMAP)* is currently residing in a small-amplitude Lissajous orbit about the Sun–Earth L_2 point. From this orbit, *WMAP* continues to measure cosmic background radiation, unobstructed by the radiation originating from the Sun, Earth, or Moon. Figure 1-11 shows a plot of the trajectory that *WMAP* used to reach its libration orbit about L_2 [70].

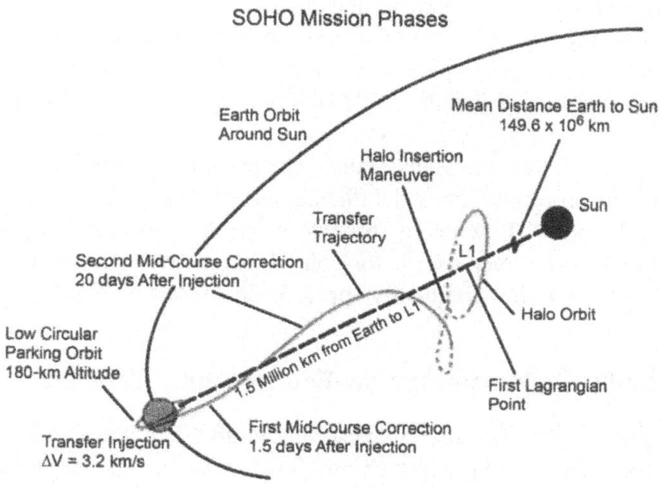

Figure 1-10 The transfer trajectories and mission phases of *SOHO* [68], used with permission of ESA.

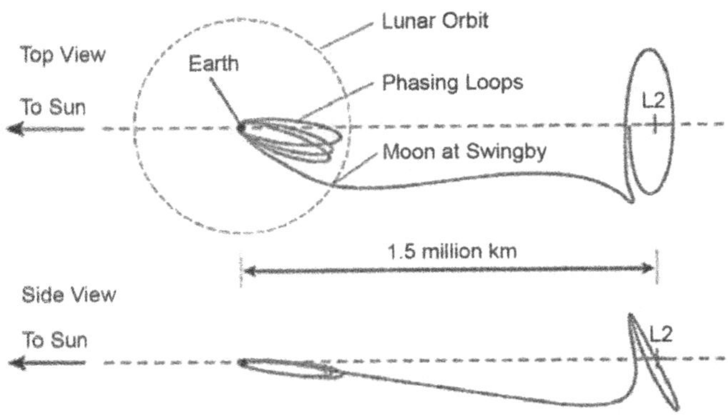

Figure 1-11 The transfer trajectory of *WMAP* [76].

Genesis. On August 8, 2001, *Genesis* launched and was quickly injected into a halo orbit about the Sun–Earth L_1 point. It traversed the halo orbit approximately five times, spending more than 2 years in the libration orbit collecting solar wind samples before turning back toward the Earth. Before returning to the Earth, however, it made a 3-million-mile (4.8×10^6 km) detour to visit the Sun–Earth L_2 point. The detour allowed it to deposit its science payload on the sunlit-side of the Earth. Figure 1-12 shows a plot of the trajectory that *Genesis* followed during its primary mission [71, 72].

Herschel and Planck. The Herschel and Planck space observatories were launched together on May 14, 2009 [73–75]. The two spacecraft separated soon after launch and traveled separately to Lissajous orbits about the Sun–Earth L_2 point. Their orbit transfers were heuristically similar to *WMAP*'s transfer to L_2, illustrated in Fig. 1-11.

Future Missions. There are plans to place the proposed James Webb Space Telescope [78] and the proposed Terrestrial Planet Finder [79] missions, among others, at the Sun–Earth L_2 point. Low-energy trajectories to the Sun–Earth Lagrange points have been shown to be very useful for solar observatories (L_1) and astrophysics observatories (L_2), and they frequently appear in spacecraft proposals.

1.6.3 Missions Implementing Low-Energy Lunar Transfers

Hiten/MUSES-A. In 1991, the Japanese spacecraft *Hiten* was the first spacecraft to transfer to the Moon using a low-energy lunar transfer. The spacecraft was not designed to go to the Moon, but rather to send a probe to the Moon. After the probe's communication system failed, mission designers scrambled to find a new mission for *Hiten*. Edward Belbruno and James Miller constructed a new trajectory—a "WSB transfer"—that required less fuel than traditional lunar transfers [80, 81]. The

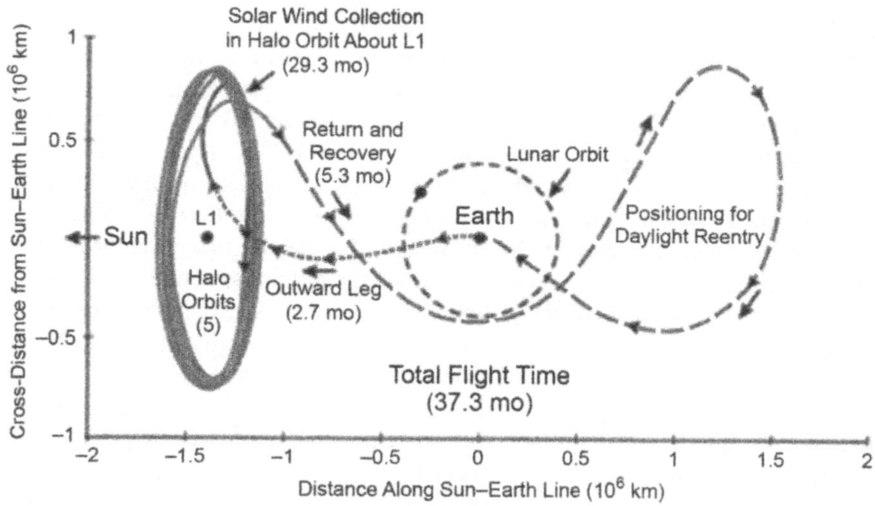

Figure 1-12 The low-energy trajectory that the *Genesis* spacecraft followed [77], viewed from above in the Sun–Earth rotating frame.

spacecraft *Hiten* did not have the fuel required for a conventional lunar transfer, but had the fuel to use this new lunar transfer to reach the Moon. *Hiten* became Japan's first lunar mission.

SMART-1. On September 27, 2003, the European Space Agency's *SMART-1* spacecraft followed a low-thrust 2-year trajectory to reach the Moon, becoming the first low-thrust spacecraft to transfer to the Moon [82].

ARTEMIS. The *Time History of Events and Macroscale Interactions during Substorms (THEMIS)* constellation was launched on February 17, 2007, to monitor the Earth's magnetic field from five different vantage points in high-altitude orbits, tracking the large-scale evolution of substorms. In 2009, two of those spacecraft were maneuvered to begin an extended mission called *ARTEMIS* [4]. The two spacecraft performed numerous maneuvers near their orbital perigees to gradually raise their orbits until they could take advantage of several lunar flybys to propel them onto two low-energy transfers. Both *ARTEMIS* spacecraft arrived at the Moon near the Earth–Moon L_2 point; one of them remained there and one immediately transferred to a libration orbit about the Earth–Moon L_1 point. After several months, the second spacecraft made the transfer and both orbited the L_1 point. After several more months, the two spacecraft departed their respective L_1 orbits, descended to the Moon, and entered smaller Keplerian orbits about the Moon. The two *ARTEMIS* spacecraft are the first two spacecraft to orbit either LL_1 or LL_2, and they each orbited both points.

GRAIL. The *GRAIL* mission (Fig. 1-13) [83–85] was launched on September 10, 2011, aboard a Delta II Heavy launch vehicle. Two vehicles, *GRAIL-A* (*Ebb*) and *GRAIL-B* (*Flow*), were separated soon after launch and flew independently to the Moon via two similar low-energy transfers. The two spacecraft arrived at the Moon approximately 25 hours apart, on December 31, 2011 and January 1, 2012. After a few months of orbit reductions and adjustments, the two spacecraft inserted into a formation, such that one spacecraft trailed the other in almost identical orbits about the Moon. By tracking each other, the two spacecraft were able to recover the Moon's gravity field to unprecedented precision and map the interior structure of the Moon. The two *GRAIL* spacecraft were the first ever to fly low-energy lunar transfers as part of their primary mission, and they were the first ever to arrive at the Moon and perform lunar orbit insertions directly from low-energy transfers.

GRAIL's trajectory design is illustrated in Fig. 1-13, including the first and last launch opportunity in a 26-day launch period. This is the launch period published in Ref. [83], however, it was actually extended by many days as the mission developed. As one can see in Fig. 1-13, *GRAIL*'s mission design includes two significant deterministic maneuvers executed per spacecraft during the cruise, performed primarily to separate their lunar orbit insertion dates.

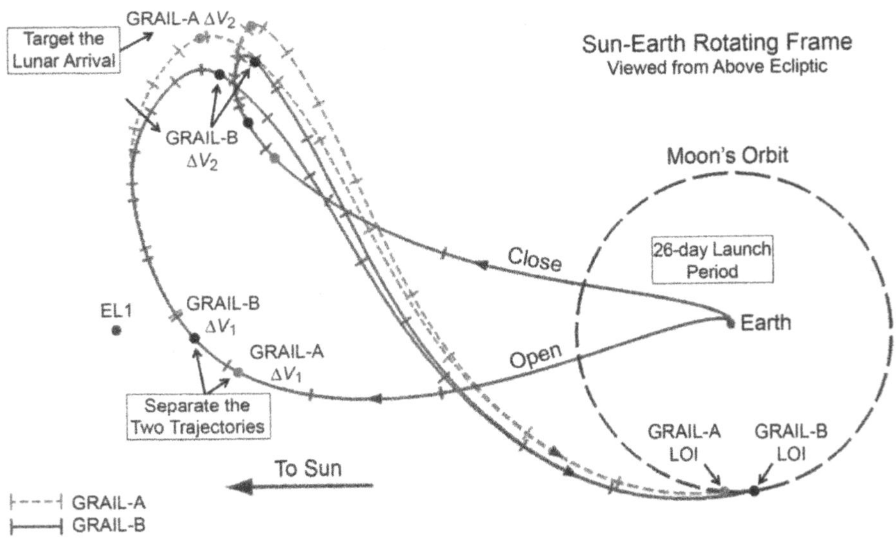

Figure 1-13 *GRAIL*'s mission design, including a 26-day launch period and two deterministic maneuvers for both *GRAIL*-A and *GRAIL*-B, designed to separate their lunar orbit insertion times by 25 hours (Ref. [83], originally published by AAS).

1.7 LOW-ENERGY LUNAR TRANSFERS

Low-energy transfers between the Earth and the Moon are the focus of this book; this section heuristically describes these transfers and how they are used.

A low-energy lunar transfer includes several segments and a wide variety of possible itineraries. The transfer may begin from a direct launch, a parking orbit, or some previous mission orbit. From the initial state, the spacecraft may depart immediately toward the low-energy transfer, or it may target an outbound lunar flyby. If the trajectory employs a lunar flyby, the mission may benefit by incorporating one or more Earth phasing orbits to target that flyby. The lunar transfer then spends 3–4 months before returning to the Moon. Upon arriving at the Moon, the spacecraft may immediately inject into a libration orbit or some other three-body orbit, a low lunar orbit, or it may immediately descend to the surface for a soft landing or a targeted impact. If the mission inserts into an orbit, it may later transfer to a different orbit and/or transfer to the surface. These itinerary choices and approximate performance parameters are illustrated in the flowchart shown in Fig. 1-14. This section describes each of these options in more detail.

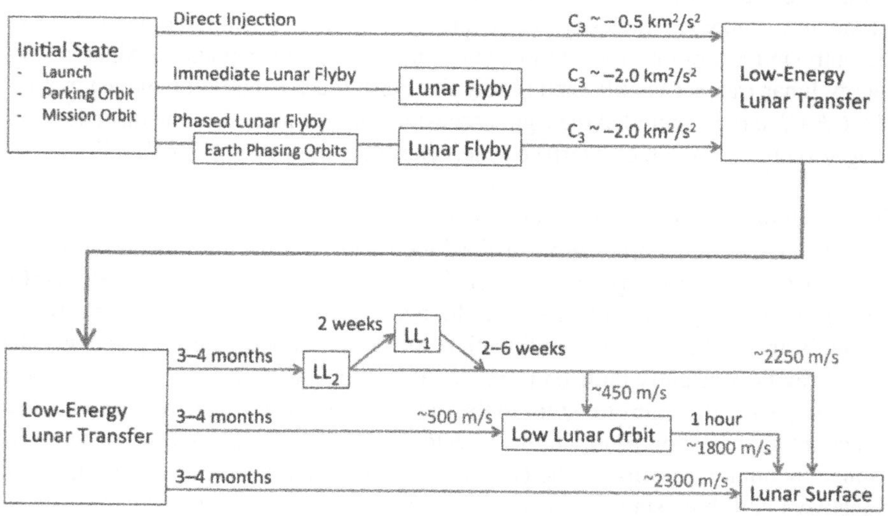

Figure 1-14 A flowchart illustrating different low-energy lunar transfer itineraries, with approximate C_3 values, transfer times, and ΔV values shown. For instance, a mission could use this flowchart to determine the approximate C_3 of taking a direct injection to a low-energy transfer (upper half), followed by the transfer duration and ΔV cost needed to transfer to a low lunar orbit (lower half). From there, one could transfer to the lunar surface, if desired (lower right).

Earth Parking Orbit. Low-energy lunar transfers may begin in any Earth parking orbit, including those compatible with a launch from any launch site around the world; they may also begin from nearly any preexisting mission orbit, which was the case for the *Hiten* and *ARTEMIS* missions [4, 81]. It is typically easier to tailor a mission to launch into a parking orbit and then depart that orbit onto a low-energy transfer than it is to adjust the orbit of a preexisting spacecraft to achieve a particular low-energy transfer. The surveys in this book assume that the mission begins in a 185-km circular low-Earth parking orbit, unless otherwise noted.

It will be shown that a given low-energy transfer has a natural Earth departure geometry—one that does not necessarily align with a desirable Earth parking orbit. Section 6.5 provides targeting procedures to connect a desirable Earth parking orbit, for example, one with an inclination of 28.5 deg, with a given low-energy transfer using 1–3 maneuvers and a minimal amount of fuel.

Trans-Lunar Injection. The trans-lunar injection (TLI) is modeled in this book as an impulsive ΔV tangent to the parking orbit. This maneuver is typically performed by the launch vehicle's upper stage. The launch vehicle's target C_3 value is typically in the range of -0.7 to -0.4 km^2/s^2, where C_3 is a parameter equal to twice the target specific energy. Since this target is negative, the resulting orbit is still captured by the Earth. If the trajectory is designed to implement a lunar gravity assist on the way out to the long cruise, then the launch target may be reduced to a C_3 of approximately -2 km^2/s^2. Launch vehicles typically target the right ascension and declination of the outbound asymptote for interplanetary missions to other planets. Since a low-energy lunar transfer is still captured by the Earth there is no outbound asymptote. The *GRAIL* targets included the right ascension and declination of the instantaneous apogee vector at the target interface time, referred to as RAV and DAV [83].

Trans-Lunar Cruise. A spacecraft's trans-lunar cruise on its low-energy lunar transfer takes it beyond the orbit of the Moon and typically in a direction toward either the second or fourth quadrant in the Sun–Earth synodic coordinate system [86]. The spacecraft typically ventures 1–2 million kilometers away from the Earth, where the Sun's gravity becomes very influential. As the spacecraft traverses its apogee the Sun's gravity constantly pulls on it, raising the spacecraft's perigee altitude. By the time the spacecraft begins to return to the Earth its perigee has risen high enough that it encounters the Moon. Further, the trajectory is designed to place the spacecraft on a lunar encounter trajectory. The *GRAIL* mission design involves two deterministic maneuvers and three statistical maneuvers for each spacecraft to navigate its trans-lunar cruise [84]. The transfers in this book may include up to two deterministic maneuvers performed during the trans-lunar cruise, and it is reasonable to assume that two or three statistical maneuvers are sufficient to implement a low-energy transfer unless a spacecraft has particularly challenging characteristics.

During this transfer, the spacecraft requires station-keeping to remain on its proper trajectory. The station-keeping cost is minimal and may be accounted for by trajectory correction maneuvers; the *Genesis* spacecraft followed a similar low-energy transfer and required only approximately 8.87 m/s of ΔV per year [71, 72, 87].

The low-energy transfer may include one or more Earth phasing orbits and/or one or more lunar flybys. These add complexity to the mission and may increase the number of maneuvers required to perform the mission, but may reduce the injection energy requirements or orbit insertion requirements upon arriving at the Moon.

Lunar Arrival. As the spacecraft approaches the Moon, it arrives on a trajectory that leads it to its initial lunar destination, be it a high-altitude three-body orbit, a low lunar orbit, or the surface of the Moon. If the spacecraft's destination is a three-body orbit, then the spacecraft often does not require any significant maneuver to enter the orbit (studied in Chapter 3); if the spacecraft's destination is a low lunar orbit, then the trajectory guides the spacecraft to its lunar orbit insertion state (studied in Chapter 4); finally, if the spacecraft's destination is the lunar surface, then the trajectory guides the spacecraft there at the designed flight path angle (studied in Chapter 5).

CHAPTER 2

METHODOLOGY

2.1 METHODOLOGY INTRODUCTION

This chapter introduces all of the models, coordinate frames, and methodology used in the analysis and construction of lunar transfers. The chapter begins by simply defining the physical constants used in these analyses, including the masses and radii of the Sun, the Moon, and the planets. It then defines the time systems used, coordinate frames, and models, including the circular restricted three-body problem and the Jet Propulsion Laboratory (JPL) developmental ephemerides used to model the motion of the planets and the Moon. A large portion of this chapter is then devoted to describing the dynamical systems methods employed in this work for the analysis and design of low-energy transfers in the Solar System. These methods may be used to design low-energy transfers from one orbit to another and/or one celestial body to another, such as low-energy transfers between the Earth and the Moon. Finally, this chapter discusses the tools used to generate the trajectories in this work.

2.2 PHYSICAL DATA

The trajectories generated in this work have been propagated using point masses for the Sun, the Moon, and the planets. Early analyses include just the Sun, Earth, and Moon, often in circular orbits that approximate the real orbits. Once early analyses are complete, high-fidelity trajectories are generated that include all of the planets, such that their positions are determined at each moment in time using accurate planetary ephemerides. Table 2-1 presents the masses, gravitational parameters, and average radii used to generate each trajectory, where it is assumed that the gravitational constant, G is equal to 6.673×10^{-20} cubic kilometers per second squared per kilogram ($km^3/s^2/kg$).

The values of gravitational constant times mass (GM) shown in cubic kilometers per second squared (km^3/s^2) in Table 2-1 are the best estimates of those values when modeling the entire Solar System as point masses. However, other GM values represent the best estimate for different cases, such as when one is modeling the gravity of the Moon using the spherical harmonic expansion. For instance, the LP150Q gravity field estimates the GM of the Moon to be approximately 4902.801076 km^3/s^2: slightly different than the value in the table [88].

Table 2-1 The masses, gravitational parameters, and average radii of the Sun, Moon, and planets used in this work [89, 90]. If the planet has natural satellites, the mass and gravitational parameter of the barycenter of the planetary system have been used.

Body	Mass (kg)	GM (km^3/s^2)	Radius (km)
Sun	$1.98879724 \times 10^{30}$	$1.32712440 \times 10^{11}$	696000.
Earth	$5.97333183 \times 10^{24}$	3.98600433×10^{5}	6378.14
Moon	$7.34722101 \times 10^{22}$	4.90280058×10^{3}	1737.4
Earth Barycenter	$6.04680404 \times 10^{24}$	4.03503233×10^{5}	–
Mercury	$3.30167548 \times 10^{23}$	2.20320805×10^{4}	2439.7
Venus	$4.86825414 \times 10^{24}$	3.24858599×10^{5}	6051.8
Mars	$6.41814926 \times 10^{23}$	4.28283100×10^{4}	3396.19
Mars Barycenter	$6.41814990 \times 10^{23}$	4.28283143×10^{4}	–
Jupiter	$1.89849445 \times 10^{27}$	1.26686534×10^{8}	71492.
Jupiter Barycenter	$1.89888757 \times 10^{27}$	1.26712768×10^{8}	–
Saturn	$5.68552375 \times 10^{26}$	3.79395000×10^{7}	60268.
Saturn Barycenter	$5.68569250 \times 10^{26}$	3.79406261×10^{7}	–
Uranus	$8.68269993 \times 10^{25}$	5.79396566×10^{6}	25559.
Uranus Barycenter	$8.68357412 \times 10^{25}$	5.79454901×10^{6}	–
Neptune	$1.02429180 \times 10^{26}$	6.83509920×10^{6}	24764.
Neptune Barycenter	$1.02450683 \times 10^{26}$	6.83653406×10^{6}	–
Pluto	$1.32300764 \times 10^{22}$	8.82843000×10^{2}	1195.
Pluto Barycenter	$1.47100388 \times 10^{22}$	9.81600888×10^{2}	–

The radius of the Earth at the Equator is equal to approximately 6378.14 km, according to the International Astronomical Union/International Association of Geodesy (IAU/IAG) 2000 Report [89]. The distance from the Earth's center to either pole is approximately 6356.75 km, shorter than at the Equator since the Earth has a significant oblateness about the Equator [89]. The radius that defines the atmospheric boundary at the Earth for sample return missions is equal to approximately 6503.14 km, approximately 125 km above the Earth's Equator [91].

2.3 TIME SYSTEMS

The passage of time may be represented in countless ways. One may define broad definitions of four types of time systems that are in common use in physics and astronomy. To varying degrees, each of these types of time systems, and the relationships between them, is important to the mission analyst [91, 92].

1. Dynamical time, in which the unit of duration is based on the orbital motion of the Earth, Moon, and planets.
2. Atomic time, in which the unit of duration corresponds to a defined number of wavelengths of radiation of a specified atomic transition of a chosen isotope.
3. Universal time, in which the unit of duration represents the solar day, defined to be as uniform as possible, despite variations in the rotation of the Earth.
4. Sidereal time, in which the unit of duration is the period of the Earth's rotation with respect to a point nearly fixed with respect to the stars.

It is very difficult to be both succinct and technically correct when defining the different types of time systems that exist. See Seidelmann, 1992, for more details [92].

2.3.1 Dynamical Time, ET

To a mission analyst, "ephemeris time" or "ET" refers to the independent variable in the equations of motion governing the motion of bodies in the Solar System. The time scale represents a smooth-flowing time coordinate that is used in the development of the numerically integrated Solar System ephemerides produced at JPL and distributed worldwide [91], as well as barycentric dynamical time (TDB). This time scale has also been referred to as T_{eph} in other studies [93]. Unfortunately, the label "ET" has a history of referring to a variety of slightly different time scales in previous studies.

2.3.2 International Atomic Time, TAI

As of 2012, the fundamental time period of a *second* is defined in the Système International (SI) system to be a specific number of oscillations of an undisturbed cesium atom. Specifically, the second is defined as the duration of time required for 9,192,631,770 periods of the radiation corresponding to the transition between the two hyperfine levels of the ground state of the cesium 133 atom. The Temps Atomique International (TAI), or international atomic time, is defined as a continuous time scale resulting from the statistical analysis of a large number of atomic clocks operating around the world, performed by the Bureau International des Poids et Mesures (BIPM). The difference between Terrestrial Time (TT) and TAI is approximately 32.184 seconds (s); that is, TT − TAI = 32.184 s. The difference between TAI and ET is: ET − TAI = 32.184 s + relativistic terms, where the relativistic terms contribute less than 2 milliseconds (ms) of variation [91].

2.3.3 Universal Time, UT

Universal Time (UT) is a time scale that is based upon the mean solar day. The time scale "UT1" represents the daily rotation of the Earth and is independent of the observing location, that is, it is independent of corrections for polar motion on the longitude of the observing site. The Earth's rotation rate changes continuously as its shape and mass distribution shifts; hence, this time scale is unpredictable. UT1 is computed using a combination of a variety of different types of observations, including very long baseline interferometry (VLBI) measurements of extragalactic radio sources (quasars), lunar laser ranging, satellite laser ranging, and Global Positioning System (GPS) measurements, to name a few.

2.3.4 Coordinated Universal Time, UTC

The Coordinated Universal Time (UTC) is the time scale that is used as the basis for the worldwide system of civil timekeeping and is available from radio broadcast signals. It is the time system used by flight operations teams and tracking stations. UTC was set equal to TAI in 1958; it was reset in 1972 such that the TAI time scale was 10 s ahead of UTC, corresponding to the approximate accumulation of drift by 1972. From then on it has been adjusted using leap seconds so that it remains within 0.9 s of UT1. As of early 2012, a total of 24 leap seconds had been added, such that the TAI time scale was 34 s ahead of UTC, that is, TAI − UTC = 34 leap seconds. The "ET" time scale was 66.184 s (excluding periodic relativistic terms) ahead of UTC, as it had been since January 1, 2009 [91].

2.3.5 Lunar Time

A "day" on the Moon is typically associated with a mean solar day, namely, the duration of time between sunrises and sunsets at a particular location on the surface. Put another way, a day on the Moon is equal to the mean interval of time between successive crossings of the Sun on a particular lunar longitude, that is, the lunar prime meridian. As a result, the period of one mean lunar day is equal to the period of a mean synodic lunar month, namely, approximately 29.53059 Earth days. The actual lunar month may vary from this mean value by nearly ± 2 hours due to the eccentricity of the Earth's orbit and small periodic variations in the Moon's rotation rate.

A lunar month may be defined in a variety of ways. Table 2-2 summarizes five ways that one may define a lunar month and their corresponding durations of time [92].

2.3.6 Local True Solar Time, LTST

The Local True Solar Time (LTST) represents the instantaneous time of day of an observer at a site on the Moon. It is a time system that does not flow constantly, but it is useful to the mission planner when measuring time for a lunar lander. The LTST on the Moon is defined as follows

$$\mathrm{LTST} = (\lambda_p - \lambda_{TS}\ \mathrm{deg})\,\frac{24\ \mathrm{hr}}{360\ \mathrm{deg}} + 12\ \mathrm{hr}$$

where λ_p is the east longitude of a point on the surface of the Moon and λ_{TS} is the east longitude of the true Sun. Using this relationship, 12 lunar hours corresponds to the time when the Sun is crossing the local meridian of the reference site, for example, local noon, and the lunar day includes 24 lunar hours.

2.3.7 Orbit Local Solar Time, OLST

During the development and operations of nearly all planetary and satellite orbiting missions, understanding how the geometry of the orbit plane changes relative to the Sun over time is extremely important, both from an engineering and a science perspective. A useful way to characterize the orbit geometry, particularly for high-inclination orbiters, is to report the local solar time of the ascending or descending node of the orbit, namely, the Orbit Local Solar Time (OLST). To be clear, this measurement describes the orientation of the orbit relative to the Moon's surface. The convention generally adopted is to report the local time of the orbit node relative to the true Sun. The Moon's gravity field will have an effect on the orbit's OLST

Table 2-2 Five ways to define a lunar month and their corresponding durations of Earth time [92].

Month	Duration (Earth days)	Duration			
		days	hr	min	s
Synodic (new moon to new moon)	29.53059	29	12	44	03
Anomalistic (perigee to perigee)	27.55455	27	13	18	33
Sidereal (fixed star to fixed star)	27.32166	27	07	43	12
Tropical (equinox to equinox)	27.32158	27	07	43	05
Nodical / Draconic (node to node)	27.21222	27	05	05	36

over time, but to first order the orbit remains essentially fixed in inertial space. The main reason that the orbit's OLST will change over time is due to the motion of the Earth–Moon system about the Sun. Thus, the following relationship describes the gross change in the OLST over time, derived from the mean synodic and sidereal periods of the Moon's orbit.

$$\text{OLST}_{\text{in LTST}} \;\text{ changes by }\; -3.94\frac{\text{minutes}}{\text{Earth day}} = -27.60\frac{\text{minutes}}{\text{week}} = -1.94\frac{\text{hours}}{\text{mean lunar day}}$$

Since the change in OLST over time is primarily a function of the rate at which the Earth–Moon system moves about the Sun, the partial will change slightly as a function of time due to the eccentricity of the Earth's orbit. For example, the partial will vary roughly within the following range each year during the 3-year period from 2009–2012

$$\text{OLST}_{\text{in LTST}} \;\text{ changes by }\; -4.1\frac{\text{minutes}}{\text{Earth day}} \;\text{ to }\; -3.8\frac{\text{minutes}}{\text{Earth day}} \;\text{ within 2009–2012}$$

2.4 COORDINATE FRAMES

This section describes several coordinate frames that are frequently used in lunar mission analysis. Each coordinate frame has its use: some are useful to describe states on the surface of the Earth, Moon, or other body; others are useful to describe the relative geometry between the Sun, Earth, and/or Moon.

Coordinate systems include a reference frame and an origin, and are often rotating or translating relative to other bodies. A coordinate system is inertial only when it is not accelerating. When referencing motion in the Solar System, the only truly "inertial" coordinate system is one that is not rotating and centered at the Solar System barycenter. Strictly speaking, no Earth-centered coordinate system can be inertial, even one that is not rotating, since the Earth is accelerating in its orbit as it revolves about the Sun. Although it is inaccurate, coordinate systems may be referred to in this book as "inertial" when they are merely nonrotating.

2.4.1 EME2000

The Earth Mean Equator and Equinox of J2000 (EME2000) coordinate frame is a nonrotating coordinate frame that is approximately aligned with the Earth's Equator. It is almost identical to the International Celestial Reference Frame (ICRF) [94]. The ICRF is defined by the IAU and is tied to the observations of a selection of quasars and other distant bright radio objects. It is a reference frame that is fixed as well as possible to the observable universe. Each of the quasars moves relative to the others, but very slowly; the motion of each of the quasars is averaged out in order to best approximate inertial space relative to the Earth's position in the universe. According to Feissel and Mignard [95], the pole of the EME2000 frame differs from the ICRF pole by ~18 milliarcseconds and the right ascension of the EME2000 x-axis differs from the right ascension of the ICRF x-axis by 78 milliarcseconds.

The coordinate axes are defined as follows:

- The z-axis of the EME2000 is defined as the pole vector of the Earth Mean Equator at the J2000 epoch, namely, at 1 January 2000 12:00:00 ET, or at Julian date 2451545.0 ET.
- The x-axis of the EME2000 is defined as the cross product of the z-axis and the Earth Mean Orbit pole of J2000, that is, the ecliptic pole of J2000. This axis defines the vernal equinox of J2000.
- The y-axis completes the right-handed coordinate frame.

This coordinate frame provides the fundamental reference for the definitions of other coordinate frames.

2.4.2 EMO2000

The Earth Mean Orbit of J2000 (EMO2000) coordinate frame is a nonrotating coordinate frame that is approximately aligned with the ecliptic. The frame shares the same x-axis as the EME2000 frame, but is rotated about that axis such that the EMO2000 z-axis is aligned with the mean ecliptic pole of J2000. This involves a rotation of approximately 23.4393 degrees (deg). The y-axis completes the right-handed coordinate frame.

2.4.3 Principal Axis Frame

The principal axis frame of a body is a body-fixed coordinate frame, that is, rotating frame, aligned with the principal axes of the body.

The coordinate axes are defined as follows:

- The z-axis points in the direction of the maximum moment of inertia; for the Earth and the Moon, this is the North Pole principal axis.
- The x-axis points in the direction of the minimum moment of inertial, that is, the prime meridian principal axis.
- The y-axis completes the right-handed coordinate frame.

It is common practice to define lunar gravity fields in the lunar principal-axis body-fixed frame (LPABF).

2.4.4 IAU Frames

The International Astronomical Union has developed definitions for coordinate frames that are tied to the surface of each planet, many satellites, and some other bodies in the Solar System. There are typically two variations of each coordinate frame: a fixed frame that rotates with the motion of the body about its primary spin axis and an "inertial" frame that shares the same z-axis but which does not rotate.

Detailed definitions of the IAU frames are described by Archinal et al. [96], and an overview of these frames is given here. The z-axis of the IAU body-fixed frame for a given body is aligned with the direction of the spin-axis of that body. The positive direction of the North Pole is defined to be on the north side of the invariable plane of the Solar System (defined by the angular momentum of the Solar System), and the pole's orientation is defined using measured values for the right ascension and declination [96]. Relatively simple low-degree polynomial approximations are used to compute the direction of this pole vector for most of the planets. Longitude is typically defined relative to a fixed surface feature for rigid bodies. In each case these quantities are specified relative to the ICRF, which varies slightly from the EME2000 coordinate frame as described above.

To give some idea for the variations between the Earth's IAU frame and EME2000, Fig. 2-1 illustrates the mapping of Greenwich, England, from the inertial IAU Earth frame to EME2000, where Greenwich has been defined in the IAU Earth frame to be at a latitude of 51.48 deg North and a longitude of 0.0 deg at the J2000 epoch.

2.4.5 Synodic Frames

It is often useful to describe a synodic frame that rotates with the motion of two massive bodies about their barycenter. Two synodic reference frames that are frequently used in this work are the Earth–Moon synodic frame, which rotates with the motion of the Earth and the Moon about their barycenter, and the Sun–Earth synodic frame, which rotates with the motion of the Earth–Moon barycenter about the Sun. The

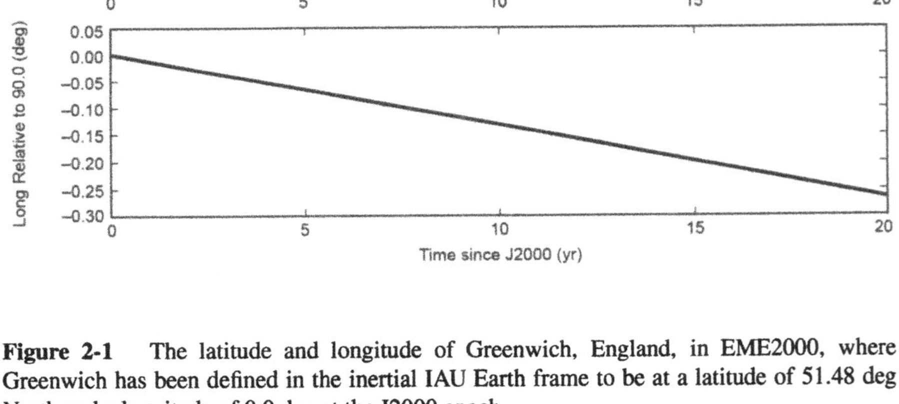

Figure 2-1 The latitude and longitude of Greenwich, England, in EME2000, where Greenwich has been defined in the inertial IAU Earth frame to be at a latitude of 51.48 deg North and a longitude of 0.0 deg at the J2000 epoch.

synodic frame may be constructed to rotate at a constant rate or at a rate that varies with the instantaneous motion of the bodies. In this book, the frames are always constructed such that the x-axis points from the larger body to the smaller body at each instant in time, the z-axis points in the direction of the angular momentum of the system, and the y-axis completes the right-handed coordinate system. This definition defines a frame that rotates at a rate that varies with the motion of the bodies in their orbits. Of course, if the bodies orbit their barycenter in circular orbits, then this frame rotates at a constant rate.

2.5 MODELS

This section describes the different models that have been used in this work to approximate the motion of spacecraft in the Solar System. Each model has a use in the analyses provided here.

The most basic model is the *two-body model*, which is used to approximate the motion of a spacecraft about a simple massive body without any other perturbations. This model is useful because one can use conic sections to approximate the spacecraft's motion, which are predictable and very quick to generate. This model is very well-known [97] and will not be further described here.

The next step up in complexity is a model that includes the gravitational attraction of two large bodies, namely, the model formulated by the circular restricted three-body problem (CRTBP). The CRTBP more closely approximates the motion of a spacecraft in the Earth–Moon and Sun–Earth three-body systems than the two-body model. Working within the CRTBP allows a mission designer to bring a wealth of techniques that have been developed over hundreds of years to a design problem. These techniques provide many qualitative insights that assist in the design of useful low-energy orbit transfers.

The patched three-body model gracefully introduces a fourth body into the design problem. The patched three-body model approximates the motion of a spacecraft using the Sun–Earth three-body model for all times, except when the spacecraft is within close proximity to the Moon, at which point the model approximates the motion of the spacecraft using the Earth–Moon three-body model. These features permit the design of four-body trajectories, such as low-energy lunar transfers, while retaining much of the useful structure found in the CRTBP.

Finally, the fourth and most complex model frequently used in this work is the JPL developmental ephemerides model, which approximates the motion of a spacecraft under the influence of the gravitational attraction of any or all of the planets and the Moon, using accurate ephemerides to model the motion of the planets and the Moon relative to the Sun.

Each of these models is described in detail in this section.

2.5.1 CRTBP

The CRTBP describes a dynamical model that is used to characterize the motion of a massless particle, for example, a spacecraft, in the presence of two massive bodies, such as the Earth and the Moon [86]. The model assumes the two massive bodies orbit their barycenter in circular orbits.

2.5.1.1 Equations of Motion It is convenient to characterize the motion of the third body, that is, the spacecraft, in a synodic reference frame that rotates at the same rate as the orbital motion of the two primary masses. The coordinate frame is centered at the barycenter of the system and oriented such that the x-axis extends from the barycenter toward the smaller primary, the z-axis extends toward the primary bodies' orbit normal, and the y-axis completes the right-handed coordinate frame. In that synodic frame, the two massive bodies are stationary, and the spacecraft moves about the system in non-Keplerian motion [46, 86]. It is convenient to normalize the units in the system such that the following measurements are equal to one: the distance between the two primaries, the sum of the mass of the two primaries, the rotation rate of the system, and the gravitational parameter. The three-body constant, μ, relates all of these normalized measurements and is easily computed by dividing the mass of the smaller primary by the total mass in the system. The equations of motion for the third body in the normalized rotating frame are equal to [86]

$$\ddot{x} = 2\dot{y} + x - (1-\mu)\frac{x+\mu}{r_1^3} - \mu\frac{x-1+\mu}{r_2^3} \tag{2.1}$$

$$\ddot{y} = -2\dot{x} + y - (1-\mu)\frac{y}{r_1^3} - \mu\frac{y}{r_2^3} \tag{2.2}$$

$$\ddot{z} = -(1-\mu)\frac{z}{r_1^3} - \mu\frac{z}{r_2^3} \tag{2.3}$$

where r_1 and r_2 are equal to the distance from the third body to the larger and smaller primary, respectively

$$r_1^2 = (x+\mu)^2 + y^2 + z^2 \tag{2.4}$$

$$r_2^2 = (x-1+\mu)^2 + y^2 + z^2 \tag{2.5}$$

The dynamics in the circular restricted three-body system depend only on the three-body constant, μ. Furthermore, as μ goes to zero, the dynamics approach two-body dynamics, although represented in a rotating frame.

2.5.1.2 Lagrange Points There are five well-known equilibrium solutions to the CRTBP, known as the five Lagrange points [86], or the five libration points. These points are referred to as L_1–L_5; this book adopts the nomenclature that L_1 lies between the two primary masses and L_2 lies on the far side of the smaller primary, relative to the barycenter of the system. The Lagrange points in the Earth–Moon system are abbreviated using the nomenclature LL_1–LL_5; the Sun–Earth Lagrange points are abbreviated EL_1–EL_5. The seven Lagrange points near the Earth are depicted in Fig. 2-2. More discussion about their locations and dynamics are provided in Sections 2.6.2 and 2.6.10.

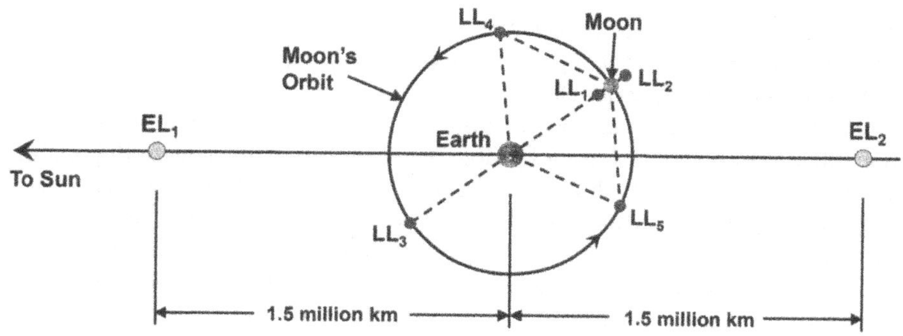

Figure 2-2 A plot depicting the relative proximity of the five Earth–Moon Lagrange points and the two nearby Sun–Earth points (first published in Ref. [97]; reproduced with kind permission from Springer Science+Business Media B.V.).

2.5.1.3 Jacobi Constant The dynamics of the CRTBP permit an integral of motion to exist in the synodic reference frame, known as the Jacobi integral or Jacobi constant [46, 97, 98]. The Jacobi constant of a spacecraft in the three-body system may be written simply as

$$C = 2U - V^2, \quad \text{where} \tag{2.6}$$

$$U = \frac{1}{2}\left(x^2 + y^2\right) + \frac{1-\mu}{r_1} + \frac{\mu}{r_2} \tag{2.7}$$

$$V^2 = \dot{x}^2 + \dot{y}^2 + \dot{z}^2 \tag{2.8}$$

The spacecraft's position and velocity coordinates in Equations 2.1–2.8 are given in nondimensional normalized synodic coordinates, relative to the barycenter of the three-body system. The Jacobi constant of a spacecraft moving in the CRTBP may not change unless the spacecraft is perturbed in some way other than by the gravitational attraction of the two primary bodies.

It is useful to consider the Jacobi constant of spacecraft in different practical orbits in order to place the value of the Jacobi constant in context. From two-body analyses, we know that spacecraft in orbits about the Earth below the geosynchronous Earth orbit (GEO) belt are only slightly perturbed by the gravity of the Moon. A spacecraft in a 185-km near-circular orbit about the Earth has a Jacobi constant of approximately 58.0, though there is some variation depending on the location of the Moon, the time of year, and the inclination of the orbit. If the spacecraft's orbital altitude is increased, its Jacobi constant decreases. A spacecraft in a 1000-km near-circular orbit has a Jacobi constant near 51.5, a GPS satellite has a Jacobi constant near 14.6, a GEO satellite has a Jacobi constant near 9.6, and so forth. The same trend exists for orbits about the Moon. A spacecraft in a near-circular lunar orbit at an altitude near 100 km has a Jacobi constant near 5.5, and a satellite in a lunar orbit at an altitude near 1000 km has a Jacobi constant near 4.7, and so forth. A spacecraft on a direct transfer to the Moon has a Jacobi constant in the vicinity of 2.3, depending on the particulars of the transfer. Likewise, a spacecraft on a low-energy transfer to the Moon departs with a Jacobi constant of about 0.8, though it changes significantly before it arrives at the Moon due to the gravity of the Sun.

One observes that a spacecraft with a smaller Jacobi constant can traverse further from either central body. A useful analysis is to identify the boundary of possible motion for a spacecraft with a particular Jacobi constant. These boundaries are computed by setting the velocity of the spacecraft equal to zero in Eq. (2.6); they are hence known as zero-velocity curves. Figure 2-3 illustrates the zero-velocity curves for several Jacobi constants for motion in the x–y plane in the Earth–Moon system.

2.5.1.4 Forbidden Regions A spacecraft traversing the Earth–Moon system with a Jacobi constant less than 2.988 (the approximate Jacobi constant of the L_4 and L_5 points) can theoretically reach any point in the entire system. Its velocity in the rotating frame will decrease to a minimum if it traverses through the L_4 or L_5 points, but no region is inaccessible. Any spacecraft that has a Jacobi constant greater than about 2.988 cannot physically reach all regions, but is bounded by the zero-velocity curves to regions of space that permit its Jacobi constant value. Those regions in space that the spacecraft cannot reach are known as *forbidden regions.*

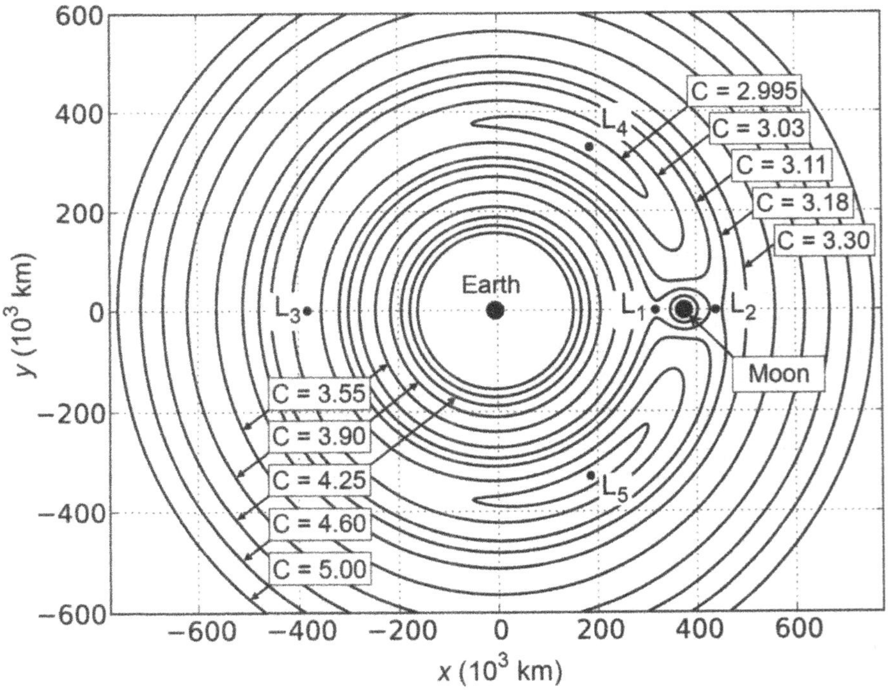

Figure 2-3 An illustration of zero-velocity curves for several Jacobi constant values in the planar Earth–Moon system.

Consider a spacecraft with a Jacobi constant of 3.18 in the Earth–Moon system (see Fig. 2-3). Its forbidden region encircles the Earth–Moon system, including the L_2–L_5 points. If the spacecraft begins at a point near the Earth or Moon, it can traverse anywhere between the Earth and Moon within the corresponding zero-velocity curve, including transferring through the gap at the L_1 point. If the spacecraft begins well outside of the system, then it must remain beyond the zero-velocity curve. It cannot match the angular velocity of the rotating frame any nearer than its zero-velocity curve.

2.5.1.5 Symmetries The existence of symmetries in the CRTBP is of particular interest for some of the analyses encountered in this book. One symmetry that is quite useful was demonstrated by Miele in his examination of image trajectories in the Earth–Moon space [99]. He showed that if $(x, y, z, \dot{x}, \dot{y}, \dot{z}, t)$ is a solution in the CRTBP, then $(x, -y, z, -\dot{x}, \dot{y}, -\dot{z}, -t)$ is also a solution. In other words, if a trajectory is reflected about the xz plane, a valid trajectory is obtained by traveling along the reflected trajectory in reverse. This property eliminates the need to compute approach and departure trajectories separately in the CRTBP. Another useful symmetry that exists in the CRTBP is that if $(x, y, z, \dot{x}, \dot{y}, \dot{z}, t)$ is a solution in the CRTBP, then $(x, y, -z, \dot{x}, \dot{y}, -\dot{z}, t)$ is also a solution. This permits trajectories to have a *Northern* and a *Southern* variety. Since the CRTBP approximates many aspects of the real Solar System, one may also frequently use CRTBP reflections as approximations for trajectories in the real Solar System.

2.5.2 Patched Three-Body Model

The patched three-body model [38–40, 45, 46] uses purely three-body dynamics to model the motion of a spacecraft in the presence of the Sun, Earth, and Moon. It retains many of the desirable characteristics of the CRTBP, while permitting a spacecraft in the near-Earth environment to be affected by all three massive bodies, albeit only two massive bodies at any given moment. When the spacecraft is near the Moon, the spacecraft's motion is modeled by the Earth–Moon three-body system. Otherwise, the spacecraft's motion is modeled by the Sun–Earth three-body system, where the secondary body is the barycenter of the Earth and Moon. For simplicity it is assumed that the Earth–Moon system is coplanar with the Sun–Earth system. The boundary of these two systems is referred to as the three-body sphere of influence (3BSOI); it is analogous to the two-body sphere of influence used in the patched conic method of interplanetary mission design.

Parker describes the 3BSOI as the boundary of a sphere centered at the Moon with a radius r_{SOI} computed using the following relationship [46]

$$r_{\text{SOI}} = a \left(\frac{m_{\text{Moon}}}{m_{\text{Sun}}} \right)^{2/5} \tag{2.9}$$

where m_{Moon} and m_{Sun} are the masses of the Moon and Sun, respectively, and a is the average distance between the Sun and Moon, equal to approximately 1 astronomical unit (AU). Thus, the 3BSOI has a radius of approximately 159,200 km, which is large enough to include LL_1 and LL_2.

2.5.3 JPL Ephemeris

The Jet Propulsion Laboratory and the California Institute of Technology have developed the DE421 Planetary and Lunar Ephemerides, which is the most accurate model of the Solar System used in this work. The model includes ephemerides of the positions and velocities of the Sun, the four terrestrial planets, the four gas-giant planets, the Pluto/Charon system, and the Moon [100]. The lunar orbit is accurate to within a meter; the orbits of Earth, Mars, and Venus are accurate to within a kilometer [100].

Low-energy lunar transfers modeled in the patched three-body model repeat perfectly from one synodic month to the next, since the dynamics and the Sun–Earth–Moon geometry are perfectly symmetric. It is often possible to build a very similar low-energy lunar transfer from one month to the next in the DE421 model of the Solar System, but its characteristics will vary in each month. This will be further discussed in later chapters.

2.5.3.1 Earth Orbit The Earth–Moon system orbits the Sun in a nearly circular orbit, but its nonzero eccentricity has an impact on the performance of a particular low-energy lunar transfer from one month to the next. Furthermore, its orbit changes over time due to the influence of Jupiter and the other planets. Figure 2-4 illustrates the Earth's osculating eccentricity, semi-major axis, and inclination over time in the DE421 model of the Solar System, relative to the Sun. One notices a nearly annual signal in the eccentricity and a bi-annual signal in the semi-major axis. This is predominantly due to the influence of Jupiter's gravity, which has a synodic period of about 399 days.

2.5.3.2 Lunar Orbit For the purposes of mission planning, the Moon's orbit about the Earth may be modeled as circular and coplanar with Earth's orbit about the Sun. In reality, the Moon's orbit about the Earth is inclined by about 5.15 deg relative to the ecliptic, and it has an average eccentricity of about 0.05490—quite a bit higher than the Earth's orbital eccentricity. Figure 2-5 illustrates the Moon's osculating eccentricity, semi-major axis, and inclination over time in the DE421 model of the Solar System, relative to the Earth. The Moon's orbit is strongly perturbed by the gravity of the Sun on several time scales. First, one can see a very clear signal in the time series of the Moon's orbital parameters that has a frequency of about 29.53 days, corresponding to the length of an average synodic month. Another very strong signal in the time series of the Moon's orbital parameters has a frequency of about 6 months, corresponding to the bi-annual impact of the Earth's orbit about the Sun. The relative orientation of the Moon's orbit to the Sun cycles over the course of a year, as well as the distance to the Sun. Both the orientation and the distance have a direct effect on the orbit. In addition to the solar perturbation, the planets Venus and Jupiter impact the lunar orbit, as does the Earth's asymmetric gravity field.

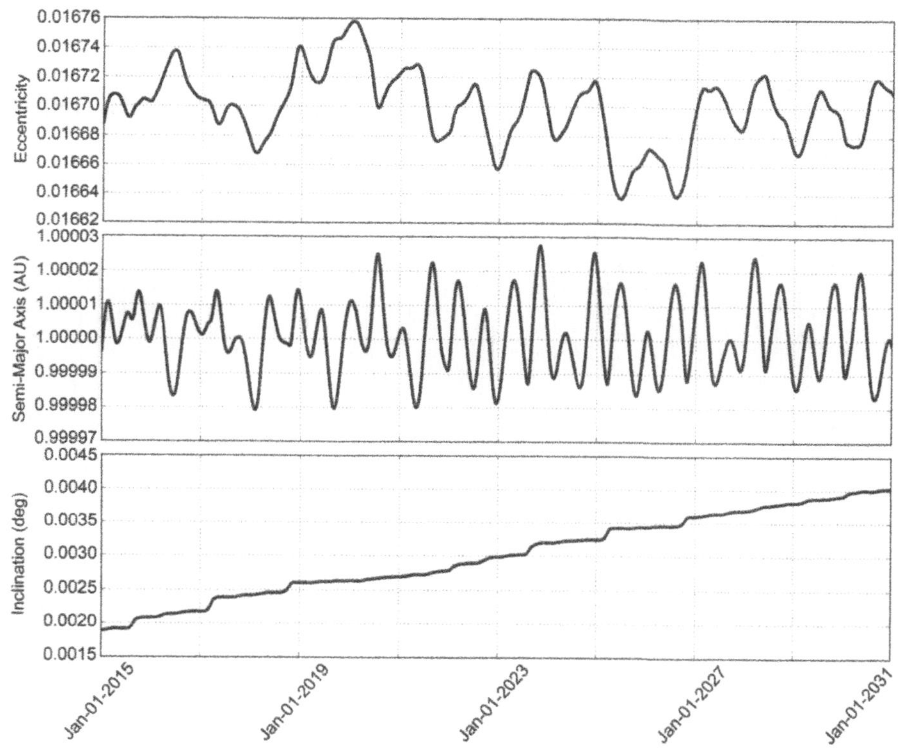

Figure 2-4 The instantaneous eccentricity (top), semi-major axis (middle), and inclination (bottom) of the Earth–Moon barycenter over time relative to the Sun in the EMO2000 coordinate frame.

In addition to the three orbital parameters illustrated in Fig. 2-5, the orientation of the Moon's orbit about the Earth undergoes both secular and periodic variations. Most notably, the Moon's orbit precesses about the ecliptic North Pole. The period of regression of the longitude of the lunar orbit's ascending node (Ω) is equal to about 18.6 years. The period of precession of the lunar orbit's argument of periapse (ω) is equal to about 6.0 years. Finally, the period of precession of the longitude of periapse ($\Omega + \omega$) is equal to about 8.85 years.

2.6 LOW-ENERGY MISSION DESIGN

The field of low-energy mission design relates to the study of trajectories that traverse unstable three-body orbits and take advantage of the dynamics to perform orbit transfers using very little fuel. This section will describe three-body orbits, their unstable manifolds, and how to construct low-energy transfers between them. Indeed, an ex-

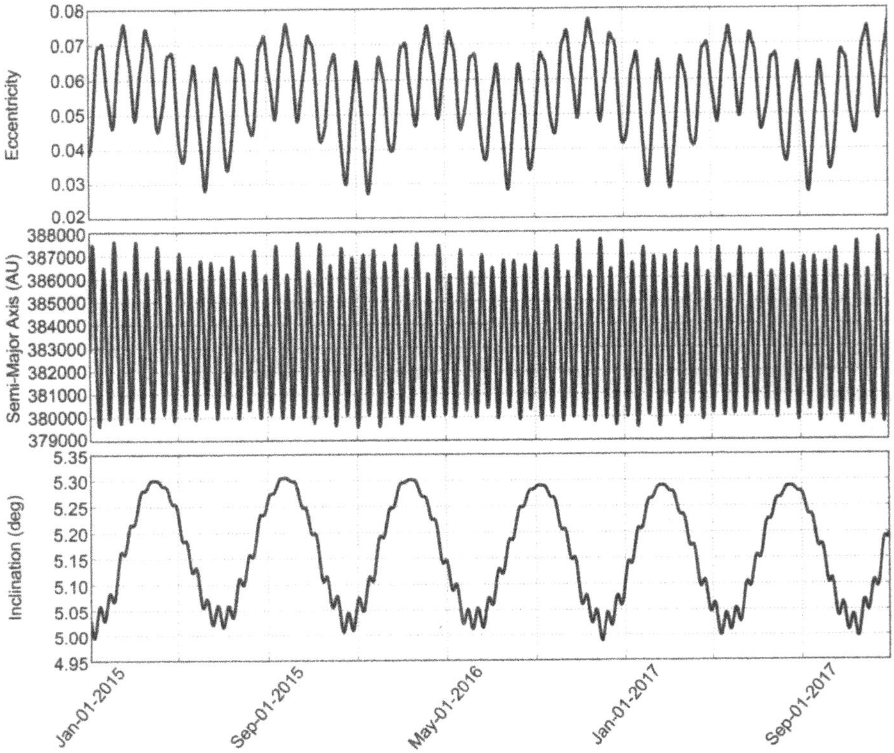

Figure 2-5 The instantaneous eccentricity (top), semi-major axis (middle), and inclination (bottom) of the Moon over time relative to the Earth in the EMO2000 coordinate frame.

ample low-energy lunar transfer is described later in dynamical systems terminology as a trajectory that first departs the Earth on the stable manifold of a Sun–Earth orbit, transfers from the stable manifold to an unstable manifold, and traverses that until it intersects the stable manifold of an orbit in the Earth–Moon system. This section describes dynamical systems analyses and how those methods may be applied to practical spacecraft mission design.

2.6.1 Dynamical Systems Theory

A dynamical system may be described as a state space with a set of rules, where the rules govern the evolution of objects' states through time within the system. The rules are deterministic; that is, the evolution of a state through a particular amount of time yields only one future state.

There are different types of dynamical systems depending on the mathematics involved and the allowable values of time. If time is continuous, capable of taking any value in the set of real numbers, then the dynamical system is smooth and is called a *flow*. If time may only take discrete values, then the dynamical system is a *map*. Models of the Solar System are generally described by flows. A spacecraft's trajectory in such dynamical systems is the set of states that the spacecraft will take as it moves through time, given its initial state. When integrating the equations of motion for a spacecraft through time using a machine, time cannot truly take on any value in the set of real numbers. The process of integration is a mapping of the spacecraft's state from one point in the state space to another point. A spacecraft's mapped trajectory is therefore only an approximation of the true trajectory.

There are many techniques that are commonly used to analyze dynamical systems. In this work, we begin our analysis of the CRTBP by identifying fixed points and periodic orbits that exist in the system. We continue by studying the stability of those solutions. These techniques provide an understanding of the motion of trajectories near those solutions. Further analysis gradually provides more information about the motion of trajectories throughout the dynamical system.

2.6.2 Solutions to the CRTBP

The CRTBP is a good example of a system in which dynamical systems methods of analysis work well. The CRTBP contains five fixed-point solutions and an infinite number of periodic orbit solutions. The characterization of these solutions helps to understand the flow of particles and spacecraft in the system. Useful trajectories may then be constructed that take advantage of the flow in the system, rather than forcing their way through the system. The following sections describe some of the simplest solutions to the CRTBP.

2.6.2.1 Fixed-Point Solutions: Five Lagrange Points The most basic solutions to the CRTBP are fixed-point solutions, that is, the trajectories in the CRTBP that particles may follow such that they stay at rest in the system indefinitely. There are five such fixed-point solutions in the CRTBP, namely, the five Lagrange points. These points were introduced in Section 2.5.1 and are displayed again in Fig. 2-6 for the case of the Earth–Moon CRTBP.

The locations of the five Lagrange points in the Sun–Earth and Earth–Moon circular three-body systems are given in Table 2-3, using the planetary masses and distances provided in the Constants, page 382. Appendix A provides an analytical derivation for the locations of the five Lagrange points for any three-body system, as well as algorithms to determine their locations. Table 2-4 summarizes the Jacobi constant of each of the five Lagrange points for both three-body systems.

2.6.2.2 Periodic and Quasiperiodic Orbit Solutions The CRTBP permits the existence of numerous families of periodic and quasiperiodic orbits. Authors have been studying such orbits since the late 1800s, though the introduction of modern computing capability dramatically improved the quantity and complexity of orbits that could be generated.

A periodic orbit in the three-body system may just be a two-body orbit about one of the bodies that is slightly perturbed by the other massive body and is in resonance

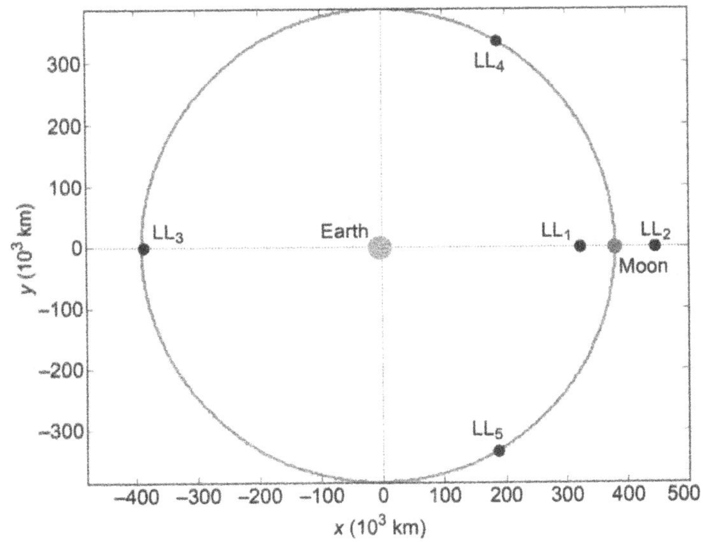

Figure 2-6 The locations of the five Lagrange points in the Earth–Moon CRTBP.

Table 2-3 The locations of the five Lagrange points in the Sun–Earth and Earth–Moon circular three-body systems. The positions are given in nondimensional normalized units and kilometers with respect to the barycenter of the system, assuming the masses and distances given in the Methodology Introduction in Section 2.1.

Lagrange Point		Position in normalized units			Position in kilometers		
		x	y	z	x	y	z
	L_1	0.9899859823	0	0	148,099,795.0	0	0
Sun–	L_2	1.0100752000	0	0	151,105,099.2	0	0
Earth	L_3	−1.0000012670	0	0	−149,598,060.2	0	0
	L_4	0.4999969596	08660254038	0	74,798,480.5	129,555,556.4	0
	L_5	0.4999969596	−08660254038	0	74,798,480.5	−129,555,556.4	0
	L_1	0.8369151324	0	0	321,710.177	0	0
Earth–	L_2	1.1556821603	0	0	444,244.222	0	0
Moon	L_3	−1.0050626453	0	0	−386,346.081	0	0
	L_4	0.4878494157	08660254038	0	187,529.315	332,900.165	0
	L_5	0.4878494157	−08660254038	0	187,529.315	−332,900.165	0

with the motion of the primaries, that is, a low Earth orbit with a period that is perfectly resonant with a sidereal month. Such an orbit has characteristics not unlike any other low Earth orbit, except that its orbital elements were carefully selected to be periodic with the Moon in the presence of the Moon's perturbations. Further, such a low Earth orbit is not quite periodic from one revolution to the next about the Earth,

Table 2-4 The Jacobi constant of each Lagrange point in the Earth–Moon and the Sun–Earth three-body systems, given in normalized coordinates.

Lagrange Point	Earth–Moon C	Sun–Earth C
L_1	3.18834129	3.00089794
L_2	3.17216060	3.00089388
L_3	3.01214717	3.00000304
L_4	2.98799703	2.99999696
L_5	2.98799703	2.99999696

due to the Moon's perturbations; it is only perfectly periodic over the course of its resonant cycle with the Moon.

Alternatively, one can construct a trajectory that carefully balances the three-body dynamics and can only exist in any form under the gravitational attraction of both bodies. Examples of three families of such periodic orbits are illustrated in Fig. 2-7. These orbits include *libration orbits* about the Earth–Moon L_1 and L_2 points and *distant prograde orbits* about the Moon. It should be noted that the smallest distant prograde orbits are very similar to two-body orbits about the Moon in resonance with the orbital motion of the Earth and Moon. Clearly the libration orbits about the Lagrange points only exist within a three-body system.

Between the 1890s and the 1930s, George Darwin [102, 103], George Hill [104], Henry Plummer [105], Forest Moulton [106], Elis Strömgren [107], and their col-

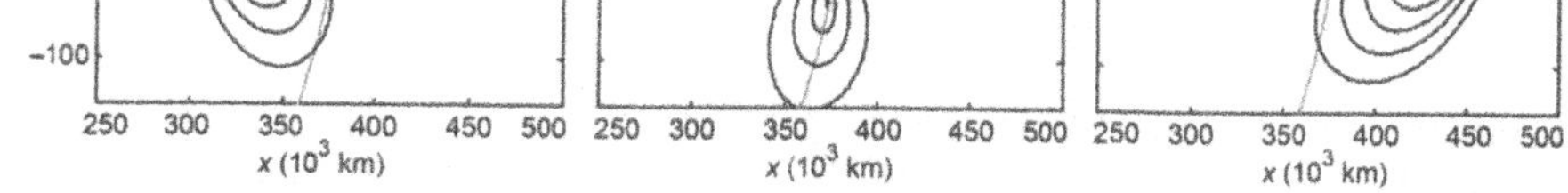

Figure 2-7 Several example orbits from three families of unstable periodic Earth–Moon three-body orbits, viewed from above in the Earth–Moon synodic reference frame. The orbits shown are from the family of Lyapunov orbits about L_1 (left), the family of distant prograde orbits about the Moon (center), and the family of Lyapunov orbits about L_2 (right). The arrows indicate the motion of objects traversing these orbits; the Moon's orbital radius about the Earth–Moon barycenter is shown in gray for reference [101] (*Acta Astronautica* by International Academy of Astronautics, reproduced with permission of Pergamon in the format reuse in a book/textbook via Copyright Clearance Center).

leagues contributed to the discovery of the first known periodic orbits in the circular restricted three-body problem. Over the course of 40 years, it is unlikely that more than 150 orbits were ever computed [108]; however, the general aspects of orbits in the three-body problem became well-understood.

In the 1960s, modern computers became accessible, and numerical techniques could be used to swiftly identify and compute periodic orbits. In 1968, Roger Broucke published a large catalog of families of planar periodic orbits that exist in the CRTBP with Earth–Moon masses [108]. Also in the 1960s, researchers computed and cataloged a large number of three-dimensional periodic orbits; significant contributors include Michel Hénon [109–113], Arenstorf [114], Goudas [115], Bray and Goudas [116, 117], and Kolenkiewicz and Carpenter [118], among others. Halo and quasi-halo orbits were discovered and analyzed beginning in the late 1960s (see, for example, Farquhar [119], Farquhar and Kamel [120], Breakwell and Brown [121], and Howell [122]). In 1980, David Richardson used the Lindstedt–Poincaré method to analytically produce periodic orbits, such as halo orbits, about the collinear libration points [123]. Additional work was accomplished toward the end of the 20th Century studying Lissajous and other quasi-halo orbits (see, for example, Farquhar and Kamel [120], Howell and Pernicka [124], and Gómez et al. [67, 125]). Many authors have studied how to take advantage of libration orbits for practical spacecraft missions, including scientific missions such as *WMAP* and *SOHO*, communication relays [5–7, 11], transportation nodes [14, 126], and navigation services [8, 10–13, 127, 128].

In this section, we demonstrate how to analytically construct one set of periodic and quasiperiodic orbits that exist about each of the collinear Lagrange points. This demonstration sheds light on why many periodic orbits exist [106, 123, 124].

We begin by translating the origin of the synodic frame to one of the collinear libration points, L_i. The parameter γ is defined to be equal to the distance from L_i to the smaller primary. The value of γ is positive when referring to L_2 and negative when referring to L_1 and L_3. The new position coordinates x', y', and z' are thus defined by the following

$$\begin{aligned} x' &= x - (1 - \mu + \gamma) \\ y' &= y \\ z' &= z \end{aligned}$$

If we now linearize the equations of motion of the CRTBP given in Eqs. (2.1)–(2.3) under this transformation, we find the following

$$\begin{aligned} \ddot{x}' - 2\dot{y}' - (1 + 2c)x' &= 0 \\ \ddot{y}' + 2\dot{x}' + (c - 1)y' &= 0 \\ \ddot{z}' + cz' &= 0 \end{aligned} \tag{2.10}$$

where c is a constant coefficient. The analytical solution to the out-of-plane z motion describes simple harmonic motion. The solution for the in-plane x–y motion involves a characteristic equation that has two real roots and two imaginary roots. The roots represent modes of motion, one divergent and one nondivergent. If the nondivergent mode is excited, then the solution is bounded and may be written as

$$\begin{aligned} x' &= -kA_y \cos(\lambda t + \phi) \\ y' &= A_y \sin(\lambda t + \phi) \\ z' &= A_z \sin(\nu t + \psi) \end{aligned} \tag{2.11}$$

This motion is described by six variables: the amplitudes of the in-plane and out-of-plane motion (A_y and A_z), the frequency of oscillation in the in-plane and out-of-plane motion (λ and ν), and the phase angles for the in-plane and out-of-plane motion (ϕ and ψ). The linearized approximation to the analytical solution for periodic motion about a Lagrange point may thus be characterized by oscillatory motion. If the two frequencies λ and ν are equal or otherwise commensurate, the resulting motion will be periodic; if the frequencies are incommensurate, the resulting motion will be quasiperiodic. The periodic orbits whose frequencies are *equal* are known as halo orbits, the more general quasiperiodic trajectories are known as Lissajous orbits or quasi-halo orbits. A portion of the family of halo orbits about the Earth–Moon L_2 is shown in Fig. 2-8, and characteristic views of several types of Lissajous orbits are shown in Fig. 2-9.

It should also be noted that there is a symmetry that exists in the CRTBP, as described in Section 2.5.1. If the CRTBP permits an orbit to exist, then it also permits a symmetric orbit to exist that is a reflection across the $y = 0$ plane. Hence, there are two families of halo orbits, a northern and a southern. By convention, if a spacecraft spends more than half of its time above the $y = 0$ plane in a halo orbit, then the spacecraft is following a northern halo orbit.

If A_z is set to zero in Eq. (2.11) the resulting orbits are planar and are known as Lyapunov orbits. Figure 2-7 shows a portion of the families of Lyapunov orbits about L_1 and L_2 in the Earth–Moon system.

These orbits may be constructed analytically since the linearization process near one of the Lagrange points produces a good approximation of the true dynamics found in the system. Other orbits do not have well-described linear approximations and must be constructed numerically. The process of numerically constructing simple periodic orbits is discussed in Section 2.6.6.

Periodic orbits in the three-body system exist that revolve about all five Lagrange points, the primary, the secondary, and also about the entire system. Periodic orbits exist that revolve about either body in a prograde sense and a retrograde sense. One may construct simple symmetric periodic orbits, such as those illustrated in this section, and one may construct asymmetric, complex orbits. A wide variety of periodic orbits exist that may be useful to the mission planner.

2.6.2.3 Orbit Parameters An orbit and a position in that orbit may be uniquely specified in the two-body problem using six parameters. Typical sets of two-body parameters include the Cartesian and Keplerian sets. Parameterization of orbits in the three-body problem has proven to be much more difficult, since there are no general analytical solutions to the system. Dynamical systems theory is very useful

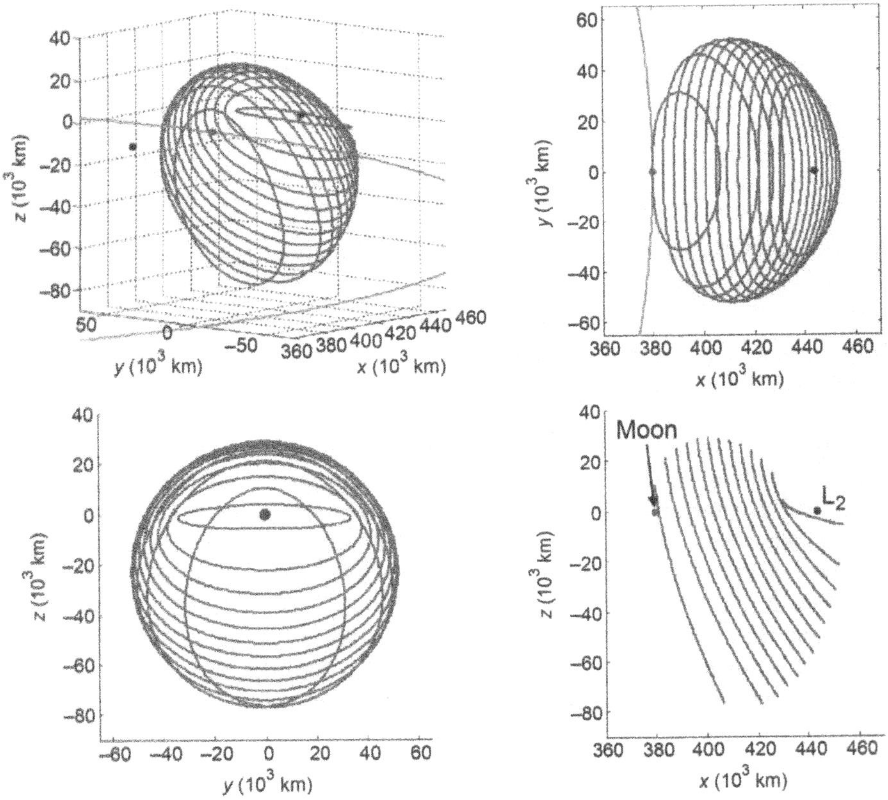

Figure 2-8 A portion of the family of halo orbits about L2 in the Earth–Moon system, shown from four perspectives.

in this regard because the methodology lends itself to many useful parameters. One such parameter, τ, is useful when describing periodic orbit solutions to the CRTBP. This parameter is described here; others are introduced in later chapters as their uses become apparent.

The parameter τ mimics the two-body mean anomaly. For the case of halo orbits, and other symmetric periodic orbits in the CRTBP, τ advances at a steady rate over time, beginning at one landmark (typically where the orbit pierces the $y = 0$ plane) and resetting when it completes an entire period. In some studies, τ takes on values in the range of 0–360 deg, much like the mean anomaly [11]. In other studies, τ is defined to take on values in the range of 0–1, indicating the periodic revolution number [46]. Most libration orbits, for example, halo and Lyapunov orbits, have a shape that resembles a conic section; in those cases it is intuitive to use an angular unit of measurement for τ. However, there are many classes of periodic orbits that

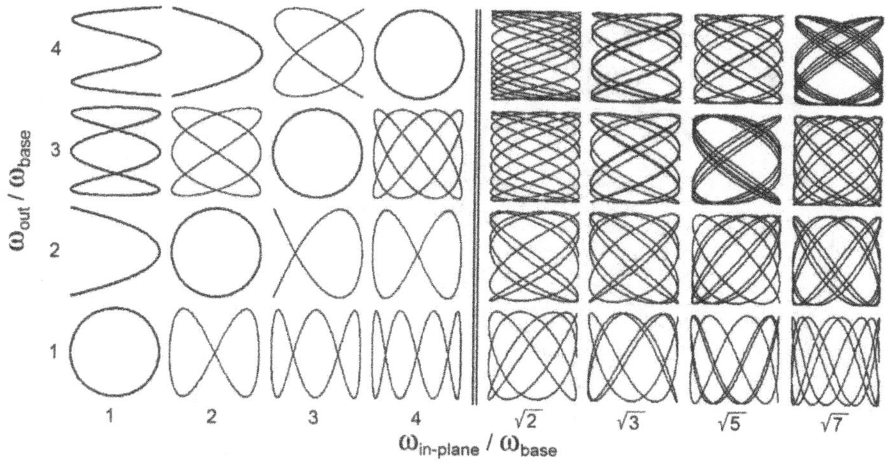

Figure 2-9 A sample of Lissajous curves representing the view of Lissajous orbits in the Earth–Moon system as viewed from an observer at the Earth looking toward the Moon; $\omega_{\text{in-plane}}$ and ω_{out} are multiples of some base frequency ω_{base}. The curves on the left are perfectly periodic; the curves on the right have incommensurate frequencies and have only been propagated for a short amount of time (first published in Ref. [97]; reproduced with kind permission from Springer Science+Business Media B.V.).

do not resemble any sort of conic section, and it may be confusing to refer to τ in angular units. Figure 2-10 shows two orbits, demonstrating how τ advances along each orbit, where τ has been represented as a revolution number for a complex orbit and as an angle for an L_2 libration orbit.

2.6.3 Poincaré Maps

A Poincaré map is a useful tool for analyzing dynamical systems and is often used to identify orbits and orbit transfers in a complex system. A Poincaré map is created by intersecting a trajectory in the n-dimensional flow $\dot{\mathbf{x}} = f(\mathbf{x})$ by an $(n-1)$-dimensional surface of section Σ. Thus, the Poincaré mapping replaces the flow of an nth order system with a discrete system of order $(n-1)$ [129]. A Poincaré mapping, P, may be described as a function that maps the state of a trajectory at the kth intersection with the surface of section, $\mathbf{x}_k$, to the next intersection, $\mathbf{x}_{k+1}$

$$\mathbf{x}_{k+1} = P(\mathbf{x}_k) \tag{2.12}$$

If a trajectory pierces Σ at the state $\mathbf{x}^*$ at time t and then returns to $\mathbf{x}^*$ at time $t + T$, then one may conclude that the trajectory is a periodic orbit with a period T [130].

There are three different types of Poincaré maps considered in this research, defined as follows [130]:

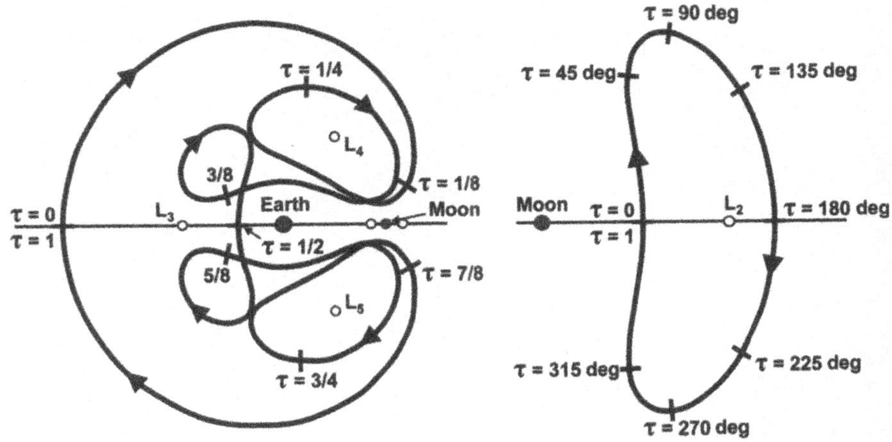

Figure 2-10 The two orbits shown demonstrate how the parameter τ advances from 0 to 1 about a complex orbit (left) or from 0 deg to 360 deg about a libration orbit (right). Both orbits are viewed from above in the Earth–Moon CRTBP synodic frame.

- P_+: The Poincaré map created from only the positive intersections of the trajectory with the surface of section. For instance, in the CRTBP, Σ may be defined as a y–z plane set to some x-value and P_+ includes only those intersections that have positive values of $\dot{x}$.
- P_-: The Poincaré map created from only the negative intersections of the trajectory with the surface of section.
- $P_\pm$: The Poincaré map created from all intersections of the trajectory with the surface of section.

The maps P_+ and P_- are called one-sided maps, while $P_\pm$ is called a two-sided map [130]. Figure 2-11 provides a simple illustration of a one-sided Poincaré mapping of two orbits, where one is periodic and one is not immediately periodic.

2.6.4 The State Transition and Monodromy Matrices

The state transition matrix $\Phi(t, t_0)$ monitors the divergent dynamics along a trajectory. Essentially, it approximates how a slight deviation in any of the state variables propagates along the trajectory. Its practical uses are twofold in this study:

1. to provide a means of adjusting the initial conditions of a trajectory to correct for unwanted motion, and
2. to provide information about the stability of an orbit, including the orientation of the eigenvectors along the orbit.

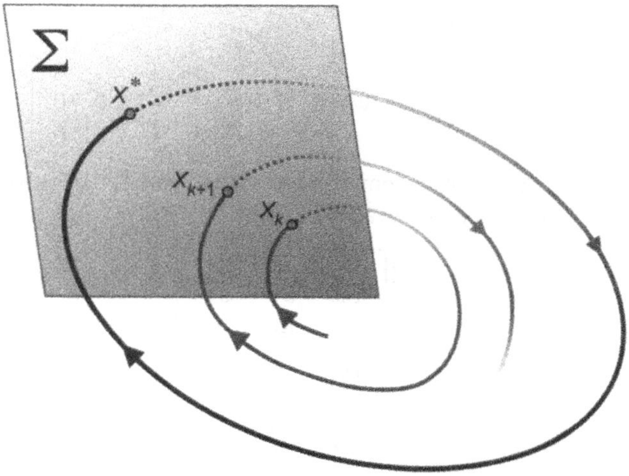

Figure 2-11 An illustration of a one-sided Poincaré mapping of two trajectories. The point x^* indicates a fixed point on the surface, corresponding to a periodic trajectory [101] (*Acta Astronautica* by International Academy of Astronautics, reproduced with permission of Pergamon in the format reuse in a book/textbook via Copyright Clearance Center).

The second use involves the *monodromy matrix*, a special case of the state transition matrix. We explore (1) in Sections 2.6.5 and 2.6.6 and (2) in Sections 2.6.8 and 2.6.10. This section discusses how to construct the state transition matrix and the monodromy matrix.

Let us define the state vector **X** to be a column vector that contains all of the state variables of interest in the system. In most cases, one is usually only interested in computing the six state variables of a particle or spacecraft in a system. Hence, **X** is defined as

$$\mathbf{X} = [x \quad y \quad z \quad \dot{x} \quad \dot{y} \quad \dot{z}]^T$$

Then the state transition matrix is a 6×6 matrix composed of the partial derivatives of the state

$$\Phi(t, t_0) = \frac{\partial \mathbf{X}(t)}{\partial \mathbf{X}(t_0)} \tag{2.13}$$

with initial conditions $\Phi(t_0, t_0) = I$. The state transition matrix is propagated using the following relationship

$$\dot{\Phi}(t, t_0) = A(t)\Phi(t, t_0) \tag{2.14}$$

where the matrix $A(t)$ is equal to

$$A(t) = \frac{\partial \dot{\mathbf{X}}(t)}{\partial \mathbf{X}(t)} \tag{2.15}$$

In the CRTBP, $A(t)$ is equal to

$$A(t) = \left[\begin{array}{c:c} 0 & I \\ \hdashline U_{XX} & 2\Omega \end{array}\right], \qquad \text{where} \qquad \Omega = \begin{bmatrix} 0 & 1 & 0 \\ -1 & 0 & 0 \\ 0 & 0 & 0 \end{bmatrix} \tag{2.16}$$

and U_{XX} is the symmetric matrix composed of second partial derivatives of U with respect to the third body's position evaluated along the orbit

$$U_{XX} = \begin{bmatrix} \frac{\partial \ddot{x}}{\partial x} & \frac{\partial \ddot{x}}{\partial y} & \frac{\partial \ddot{x}}{\partial z} \\ \frac{\partial \ddot{y}}{\partial x} & \frac{\partial \ddot{y}}{\partial y} & \frac{\partial \ddot{y}}{\partial z} \\ \frac{\partial \ddot{z}}{\partial x} & \frac{\partial \ddot{z}}{\partial y} & \frac{\partial \ddot{z}}{\partial z} \end{bmatrix} \tag{2.17}$$

The monodromy matrix, M, exists for periodic orbits and is computed by propagating the state transition matrix one entire orbit: $M = \Phi(t_0 + P, t_0)$ [131]. After being propagated for a full orbit, the matrix contains information about every region that a spacecraft would pass through along that orbit. This matrix's use is further explored in Sections 2.6.8 and 2.6.10.

2.6.5 Differential Correction

Differential correction, as it is implemented here, is a process by which the state transition matrix is used to change a set of initial conditions in order to better satisfy a set of criteria. It is a targeting scheme that converges on its constraints very swiftly within the basin of convergence. Two types of differential correction routines are used in this work: single-shooting and multiple-shooting correction routines.

2.6.5.1 Single-Shooting Differential Correction In the standard single-shooting differential correction routine used in this work, a spacecraft begins at some state $\mathbf{X}_0$, following a nominal trajectory $T(t)$, where $\mathbf{X}_0$ is composed of a position vector $\mathbf{R}_0$ and a velocity vector $\mathbf{V}_0$. It is desired that the spacecraft's trajectory be shifted such that at a later time, t_f, the trajectory encounters a desired state $\hat{\mathbf{X}}_f$ (including a desired position vector $\hat{\mathbf{R}}_f$ and a velocity vector $\hat{\mathbf{V}}_f$). There are usually two constraints to the problem: (1) the spacecraft's initial position may not change; and (2) the spacecraft's new trajectory leads it to a final specified position vector $\hat{\mathbf{R}}_f$. The routine is allowed to vary the initial velocity of the spacecraft (simulating a change in velocity (ΔV) in the mission design), and is oftentimes allowed to vary the time at which the spacecraft arrives at its final desired position. The velocity of the spacecraft at the final position is usually a free variable, and mission designers typically plan to perform an additional ΔV at that time. If the routine converges, a new ballistic trajectory is constructed, $\hat{T}(t)$, that satisfies the two conditions

$$\hat{T}(\hat{t}_0) = \hat{\mathbf{X}}_0 \qquad \text{with } \hat{t}_0 = t_0,\ \hat{\mathbf{R}}_0 = \mathbf{R}_0, \text{ and } \hat{\mathbf{V}}_0 \text{ free}$$

$$\hat{T}(\hat{t}_f) = \hat{\mathbf{X}}_f \qquad \text{with } \hat{t}_f \text{ constrained or fixed, } \hat{\mathbf{R}}_f \text{ constrained, and } \hat{\mathbf{V}}_f \text{ free}$$

This routine is diagrammed in Fig. 2-12.

The single-shooting method uses the state transition matrix $\Phi(t_f, t_0)$ to estimate what change to make in the initial velocity of the state, $\Delta\mathbf{V}_0$, in order to eliminate the deviation in position at the end of the trajectory $\delta\mathbf{R}_f$. The state transition matrix maps perturbations in the state over time using the following linearized equations

$$\begin{aligned} \delta\mathbf{X}_f &= \Phi(t_f, t_0)\Delta\mathbf{X}_0 \qquad \text{or} \\ \begin{bmatrix} \delta\mathbf{R}_f \\ \delta\mathbf{V}_f \end{bmatrix} &= \begin{bmatrix} \Phi_{RR}(t_f, t_0) & \Phi_{RV}(t_f, t_0) \\ \Phi_{VR}(t_f, t_0) & \Phi_{VV}(t_f, t_0) \end{bmatrix} \begin{bmatrix} \Delta\mathbf{R}_0 \\ \Delta\mathbf{V}_0 \end{bmatrix} \end{aligned} \tag{2.18}$$

Since $\Delta\mathbf{R}_0 = 0$ and $\delta\mathbf{V}_f$ is unconstrained, we may simplify Eq. (2.18) and solve for $\Delta\mathbf{V}_0$ to find

$$\Delta\mathbf{V}_0 = [\Phi_{RV}(t_f, t_0)]^{-1}\,\delta\mathbf{R}_f \tag{2.19}$$

Since the state transition matrix is propagated with linearized equations, the algorithm must be iterated until convergence. When the algorithm is converging, each iteration typically improves the solution by a factor of 10, although factors anywhere between 2 and 100 have been observed [46].

2.6.5.2 *Multiple-Shooting Differential Correction* Multiple-shooting differential correction takes a series of states and adjusts them all simultaneously to construct a complicated trajectory that satisfies a set of constraints. It is very useful when mission designers wish to construct a long trajectory in an unstable environment in the presence of machine precision. For example, the *Genesis* spacecraft departed the Earth, traversed a quasi-halo orbit about the Sun–Earth L_1 point several times, transferred to the vicinity of the Sun–Earth L_2 point, and then returned to Earth. This entire trajectory may in theory be constructed without a single maneuver. However, in this unstable environment, deviations even as small as round-off errors due to machine precision grow exponentially. A computer using finite-precision mathematics

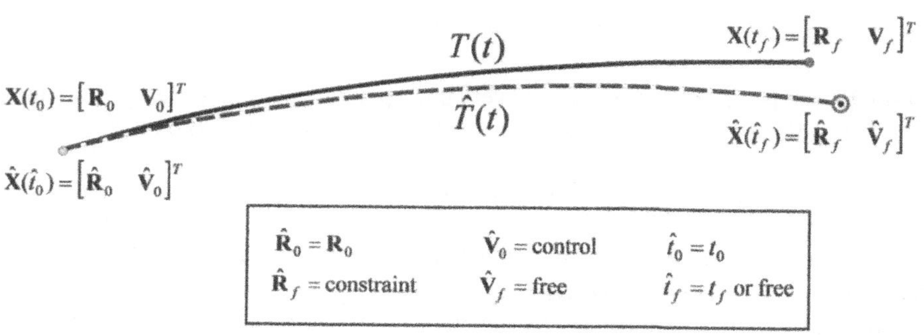

Figure 2-12 The single-shooting differential-correction routine. The solid-line trajectory, $T(t)$, is the initial trajectory; the dashed-line trajectory, $\hat{T}(t)$, is the corrected trajectory that encounters the target position, indicated by a bull's eye, at the target time.

does not normally have the precision required to propagate the spacecraft through its entire trajectory before the round-off errors grow large enough to create a large-scale deviation in the spacecraft's trajectory. To get around this numerical problem, mission designers break the trajectory into many segments and patch the segments together with very small maneuvers. The maneuvers counteract the build-up of propagated round-off errors. They may be as small as desired, depending on the length of each patched segment, and are typically much smaller than any expected station-keeping maneuver. Hence, they are not normally considered to be deterministic maneuvers in the mission.

The multiple-shooting differential correction method is described in detail by Pernicka [132, 133] and by Wilson [134, 135], among other authors [136–138]. This section provides a summary of the process. Section 2.6.5.3 then returns to the details and derives the tools needed to implement the multiple-shooting differential correction technique. The derivation follows that presented by Wilson [134].

The first step in the process of differentially correcting a complex trajectory is to define a series of patchpoints. Several things must be considered when setting up the patchpoints; a discussion of these considerations is given below. From then on, a two-level process is iterated until either the differential corrector fails or each constraint in the system is satisfied. The first level of the process adjusts the patchpoints' velocities; the second level of the process adjusts the patchpoints' positions and epochs. If the patchpoints fall into some basin of convergence for the differential corrector, then the process converges on a continuous trajectory swiftly. The following paragraphs provide more details about the two-level process:

Level 1. The velocities of every patchpoint along the trajectory except the last one are adjusted using the single-shooting differential corrector given in Section 2.6.5.1. The velocities are set such that the position of each segment ends at the following patchpoint. When this step has been completed, the trajectory is continuous, although a spacecraft must perform a ΔV at each patchpoint in order to follow the trajectory.

Level 2. The positions and epochs of every patchpoint, including the last patchpoint, are adjusted using a least-squares method that is designed to reduce the total ΔV cost of the trajectory. The result is a discontinuous trajectory that should require less total ΔV after the following iteration of Level 1.

This iteration process is repeated until the discontinuity at each patchpoint in position and velocity is below some tolerance.

The choice of patchpoints has a strong effect on the differential corrector's performance. First of all, since the single-shooting method is invoked between every adjacent patchpoint, the patchpoints must be close enough to permit the single-shooting method to converge within the desirable tolerance given the numerical precision of the machine. That is, if the patchpoints are too far apart, Level 1 will not converge. Secondly, Level 2 of the two-level process is generally designed with the assumption that each patchpoint is evenly spaced in time. If the patchpoints are not evenly spaced in time, then the time system should be normalized in some way. This improves the convergence characteristics of the algorithm, but it certainly does not guarantee that the system will converge on a desirable solution. Finally, it has been observed that the differential corrector converges more readily if there are more patchpoints where the dynamics become more unstable, for example, near massive bodies, though the time system often must be normalized for this to benefit the stability of the algorithm.

Section 2.6.5.1 provides the details of Level 1. In order to shed some light onto the functionality of Level 2, let us assume that we have a trajectory built from three patchpoints. The two segments meet in the middle at the second patchpoint, but are not continuous in velocity, that is, there is some velocity deviation $\Delta\mathbf{V}_2$. Let us assume that it is desirable to remove that discontinuity, and to do so we allow the positions and epochs of all three patchpoints to be adjusted. Section 2.6.5.3 provides details about how to numerically compute the variations of the target ($\Delta\mathbf{V}_2$) to the controls ($\mathbf{R}_1, t_1, \mathbf{R}_2, t_2, \mathbf{R}_3, t_3$). In this way, one can construct the following partial derivatives

$$\frac{\partial \Delta\mathbf{V}_2}{\partial \mathbf{R}_1}, \quad \frac{\partial \Delta\mathbf{V}_2}{\partial \mathbf{R}_2}, \quad \frac{\partial \Delta\mathbf{V}_2}{\partial \mathbf{R}_3},$$
$$\frac{\partial \Delta\mathbf{V}_2}{\partial t_1}, \quad \frac{\partial \Delta\mathbf{V}_2}{\partial t_2}, \quad \frac{\partial \Delta\mathbf{V}_2}{\partial t_3}$$

With these partial derivatives, one can build an approximation of the change in $\Delta\mathbf{V}_2$ when each of the control parameters are perturbed

$$[\delta\Delta\mathbf{V}_2] = \underbrace{\begin{bmatrix} \frac{\partial \Delta\mathbf{V}_2}{\partial \mathbf{R}_1} & \frac{\partial \Delta\mathbf{V}_2}{\partial t_1} & \frac{\partial \Delta\mathbf{V}_2}{\partial \mathbf{R}_2} & \frac{\partial \Delta\mathbf{V}_2}{\partial t_2} & \frac{\partial \Delta\mathbf{V}_2}{\partial \mathbf{R}_3} & \frac{\partial \Delta\mathbf{V}_2}{\partial t_3} \end{bmatrix}}_{[M]} \begin{bmatrix} \delta\mathbf{R}_1 \\ \delta t_1 \\ \delta\mathbf{R}_2 \\ \delta t_2 \\ \delta\mathbf{R}_3 \\ \delta t_3 \end{bmatrix} \tag{2.20}$$

In general, we wish to determine the appropriate changes to make to each of the control variables in order to reduce the value of $\Delta\mathbf{V}_2$ to zero. The linear system given in Eq. (2.20) is underdetermined; it is common practice to use the smallest Euclidean norm to produce a good solution [134]

$$\begin{bmatrix} \delta\mathbf{R}_1 \\ \delta t_1 \\ \delta\mathbf{R}_2 \\ \delta t_2 \\ \delta\mathbf{R}_3 \\ \delta t_3 \end{bmatrix} = M^T \left(MM^T\right)^{-1} [\delta\Delta\mathbf{V}_2] \tag{2.21}$$

These deviations in position and epoch are then added to the patchpoints' states to complete the Level 2 iteration. This example includes only three patchpoints; additional patchpoints may be added on indefinitely. With many patchpoints in the system, the majority of the matrix M is filled with zeros, since each velocity discontinuity is only dependent on the positions and epochs of the three nearest patchpoints.

In its simplest form, the Level 2 corrections are only constrained by the velocity discontinuities at each patchpoint. Wilson describes how to add many other types of constraints to the differential corrector [134]. Some examples of constraints that may be added include:

- Desirable Position Vector. One may target a particular position vector or position magnitude for any patchpoint in the trajectory. This may be with respect to a point in the coordinate axes or with respect to another body.
- Desirable Inclination. One may target many different orbital parameters, such as the inclination of one or more specified patchpoints.
- Maximum Change in Position. One may limit the differential corrector's capability to change one or more patchpoints' positions during each iteration of Level 2. This helps to keep a trajectory near some initial guess.

Many other types of constraints may be placed on the system. The inclusion of additional constraints is very useful for practical spacecraft missions, where the trajectory must be designed to begin from a particular state or to end at a particular state; however, it does often make it more difficult for the differential corrector to converge.

There are many practical applications of the multiple-shooting differential corrector. To demonstrate its use, we will examine its performance as it is used to differentially correct a periodic halo orbit from the CRTBP into a quasi-halo orbit in the DE421 model of the Solar System. Figure 2-13 provides several representative plots of the differential corrector in action. The plots are exaggerated to demonstrate the procedure clearly. The plot shown in (a) depicts the initial periodic halo orbit in the CRTBP. The trajectory is broken into four segments, separated by five patchpoints as shown in (b), where the fifth patchpoint is coincident with the first in the synodic frame. The first iteration of Level 1, shown in (c), forces the new trajectory to be continuous in position and time in the DE421 model, but permits velocity discontinuities at each interior patchpoint. The five patchpoints' positions and epochs are then adjusted in the first iteration of Level 2 as shown in (d). The plots shown in (e) and (f) give an exaggerated representation of the second iteration of Levels 1 and 2. The plot shown in (g) depicts the trajectory after the third iteration of Level 1; one can see that the trajectory is approaching a continuous trajectory. The plot shown in (h) depicts the final, converged trajectory that is continuous in the DE421 model within some tolerance limits.

Studying Figure 2-13, one can see that the differential corrector permits the ends of the trajectory to be altered substantially since there are no boundary conditions. If more revolutions of the halo orbit were originally sent into the differential corrector, then the final trajectory would resemble the original halo orbit more closely. This process is shown in more detail in Section 2.6.6.3.

The multiple-shooting differential corrector typically operates on a set of patchpoints that define a single trajectory, presumably to be followed by a single spacecraft.

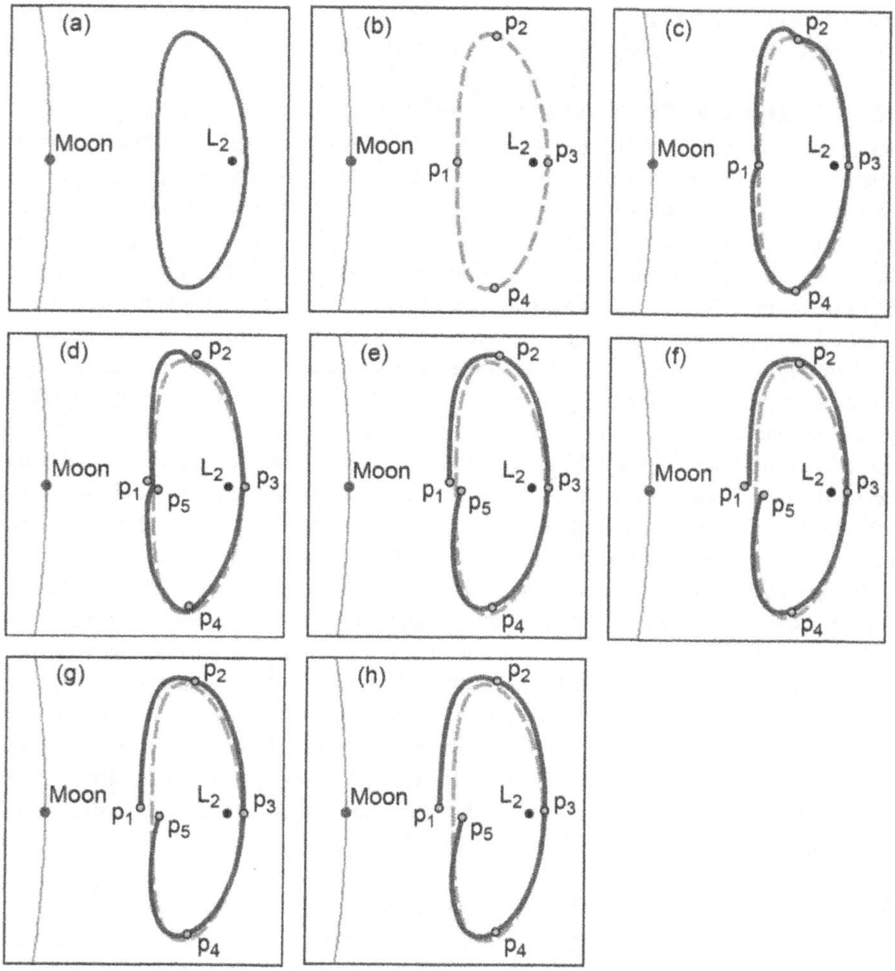

Figure 2-13 An exaggerated demonstration of the implementation of the multiple-shooting differential corrector used to convert a halo orbit from the CRTBP into the DE421 model.

(a) The initial CRTBP halo orbit.
(b) The initial placement of patchpoints in the DE421 model.
(c) Level 1: Differential correction to determine the ΔVs necessary to make the trajectory continuous in the DE421 model.
(d) Level 2: The adjustments of the patchpoints' positions and epochs to reduce the total ΔV.
(e) Level 1: The second adjustments of the patchpoints' velocities to make the trajectory continuous.
(f) Level 2: The second adjustments of the patchpoints' positions and epochs to reduce the total ΔV.
(g) Level 1: The third adjustments of the patchpoints' velocities to make the trajectory continuous.
(h) The final converged trajectory in the DE421 model after several additional iterations.

However, the differential corrector may certainly be designed to operate on segments that represent more than one spacecraft, including segments that branch, segments that rendezvous, and/or segments that define a formation.

2.6.5.3 Multiple-Shooting Implementation The multiple-shooting differential corrector is such a useful tool in the design of low-energy trajectories that further attention is given here to derive the algorithms needed to implement it. As described earlier, the multiple-shooting differential corrector involves a process that repeats two steps until a trajectory is generated that satisfies all given constraints. Level 1 is fully described in Section 2.6.5.1, including everything needed to generate software to implement it. Level 2 is introduced in Section 2.6.5.2, but the details have been omitted in order to demonstrate its operation. Those details are provided here.

The engine of the most basic implementation of Level 2 is given by Eq. (2.21), which computes a linear approximation of the changes that must be made to the positions and/or times of the three patchpoints in the scenario in order to reduce the ΔV at the interior patchpoint. The multiple-shooting differential corrector may certainly be extended to include many trajectory segments and a wide variety of constraints. Further, the trajectory segments are not restricted to a single trajectory, but may define multiple trajectories that are simultaneously optimized.

This section begins by describing a basic formulation of Level 2 that involves a single trajectory defined by at least three patchpoints such that the only goal is to reduce the total ΔV required to traverse that trajectory. Next, the section describes the algorithms required to add constraints to the patchpoints in the trajectory. Finally, the section includes a discussion about how to implement the multiple-shooting technique such that it operates on several codependent trajectories simultaneously. In each case, it is always assumed that a ΔV or constraint applied to a particular patchpoint is only affected by the position and/or time of that patchpoint and its neighbors, which is an important feature in the formulation of Level 2.

Basic Level 2. The basic Level 2 formulation is one that operates on a single trajectory and works only to reduce the ΔV of each interior patchpoint. It is assumed that the position and/or time of each patchpoint may be changed to accomplish this goal. Hence, the ΔV at the second patchpoint, ΔV_2, may be reduced by changing the position, the time, or both of the first, second, and third patchpoints. Any other patchpoints do not directly influence ΔV_2, though their influences are transmitted through the connecting patchpoints.

Equation (2.21) captures the linear estimate of the change in the positions and times of three patchpoints needed to reduce ΔV_2, the velocity discontinuity at the interior patchpoint. This expression may be extended to include multiple patchpoints as follows

$$\begin{bmatrix} \delta\mathbf{R}_1 \\ \delta t_1 \\ \delta\mathbf{R}_2 \\ \delta t_2 \\ \delta\mathbf{R}_3 \\ \delta t_3 \\ \vdots \\ \delta\mathbf{R}_n \\ \delta t_n \end{bmatrix} = M^T \left(MM^T\right)^{-1} \begin{bmatrix} \delta\Delta\mathbf{V}_2 \\ \delta\Delta\mathbf{V}_3 \\ \vdots \\ \delta\Delta\mathbf{V}_{n-1} \end{bmatrix} \tag{2.22}$$

where the matrix M is constructed using the relationship

$$\begin{bmatrix} \delta\Delta\mathbf{V}_2 \\ \delta\Delta\mathbf{V}_3 \\ \vdots \\ \delta\Delta\mathbf{V}_{n-1} \end{bmatrix} = M \begin{bmatrix} \delta\mathbf{R}_1 \\ \delta t_1 \\ \delta\mathbf{R}_2 \\ \delta t_2 \\ \delta\mathbf{R}_3 \\ \delta t_3 \\ \vdots \\ \delta\mathbf{R}_n \\ \delta t_n \end{bmatrix} \tag{2.23}$$

Thus, M is equal to

$$\underbrace{\begin{bmatrix} \frac{\partial \Delta \mathbf{V}_2}{\partial \mathbf{R}_1} & \frac{\partial \Delta \mathbf{V}_2}{\partial t_1} & \frac{\partial \Delta \mathbf{V}_2}{\partial \mathbf{R}_2} & \frac{\partial \Delta \mathbf{V}_2}{\partial t_2} & \frac{\partial \Delta \mathbf{V}_2}{\partial \mathbf{R}_3} & \frac{\partial \Delta \mathbf{V}_2}{\partial t_3} & 0 & 0 & \dots & 0 & 0 \\ 0 & 0 & \frac{\partial \Delta \mathbf{V}_3}{\partial \mathbf{R}_2} & \frac{\partial \Delta \mathbf{V}_3}{\partial t_2} & \frac{\partial \Delta \mathbf{V}_3}{\partial \mathbf{R}_3} & \frac{\partial \Delta \mathbf{V}_3}{\partial t_3} & \frac{\partial \Delta \mathbf{V}_3}{\partial \mathbf{R}_4} & \frac{\partial \Delta \mathbf{V}_3}{\partial t_4} & \dots & 0 & 0 \\ \vdots & & \vdots & & \vdots & & \vdots & & & & \vdots \\ 0 & 0 & 0 & 0 & \dots & \frac{\partial \Delta \mathbf{V}_{n-1}}{\partial \mathbf{R}_{n-2}} & \frac{\partial \Delta \mathbf{V}_{n-1}}{\partial t_{n-2}} & \frac{\partial \Delta \mathbf{V}_{n-1}}{\partial \mathbf{R}_{n-1}} & \frac{\partial \Delta \mathbf{V}_{n-1}}{\partial t_{n-1}} & \frac{\partial \Delta \mathbf{V}_{n-1}}{\partial \mathbf{R}_{n}} & \frac{\partial \Delta \mathbf{V}_{n-1}}{\partial t_{n}} \end{bmatrix}}_{[M]} \tag{2.24}$$

In order to generate M, one requires knowledge of each of the partials given in Eq. (2.24). Fortunately, each of these partials may be constructed using the state transition matrix, provided that the linear approximations are acceptable. In order to derive the formulae needed to represent each of these partials, we must examine the problem definition more closely.

We again consider the first two segments, defined by the first three patchpoints: P_1, P_2, and P_3. Each of these patchpoints is characterized by its position $\mathbf{R}$, velocity $\mathbf{V}$, and time t. After the application of the Level 1 correction, Segment 1 traverses from P_1 to P_2 and Segment 2 traverses from P_2 to P_3. The resulting trajectory is continuous in position over time (within some small tolerance at P_2) and continuous in velocity over time except at P_2, where $\Delta \mathbf{V}_2$ defines the difference between $\mathbf{V}_2^+$ (the velocity at the start of Segment 2) and $\mathbf{V}_2^-$ (the velocity at the end of Segment 1)

$$\Delta \mathbf{V}_2 = \mathbf{V}_2^+ - \mathbf{V}_2^-$$

The superscripts "$-$" and "$+$" differentiate between the incoming and outgoing parameters, respectively, at a particular patchpoint. The position, velocity, and time of the end of Segment 1 are indicated as $\mathbf{R}_2^-$, $\mathbf{V}_2^-$, and t_2^-, respectively. Likewise, the position, velocity, and time of the initial state of Segment 2 are indicated as $\mathbf{R}_2^+$, $\mathbf{V}_2^+$, and t_2^+, respectively. After applying Level 1 to P_1, $\mathbf{R}_2^- = \mathbf{R}_2^+$ and $t_2^- = t_2^+$. These are fixed constraints and assumed in the formulation of Level 2.

The state transition matrix, Φ, may be mapped from P_1 to P_2 and from P_3 to P_2 to approximate the response of $\mathbf{V}_2^-$ and $\mathbf{V}_2^+$, respectively, given a change in the states of P_1 and P_3. The basic Level 2 formulation defines the state transition matrix to be a 6×6 matrix as given in Eq. (2.18)

$$\begin{bmatrix} \delta \mathbf{R}_f \\ \delta \mathbf{V}_f \end{bmatrix} = \Phi \begin{bmatrix} \Delta \mathbf{R}_0 \\ \Delta \mathbf{V}_0 \end{bmatrix}$$

where we indicate a change in parameters performed by the user by a "Δ" and the response by a "δ". The 6×6 state transition matrix may be broken up into four 3×3 submatrices as illustrated previously in Eq. (2.18) and repeated here

$$\begin{bmatrix} \delta \mathbf{R}_f \\ \delta \mathbf{V}_f \end{bmatrix} = \begin{bmatrix} \Phi_{RR}(t_f, t_0) & \Phi_{RV}(t_f, t_0) \\ \Phi_{VR}(t_f, t_0) & \Phi_{VV}(t_f, t_0) \end{bmatrix} \begin{bmatrix} \Delta \mathbf{R}_0 \\ \Delta \mathbf{V}_0 \end{bmatrix} \tag{2.25}$$

In order to simplify the nomenclature, we abbreviate the pieces of Eq. (2.25) as follows, where the state transition matrix is now demonstrating a mapping of the deviations from patchpoint P_1 to patchpoint P_2

$$\begin{bmatrix} \delta\mathbf{R}_2^- \\ \delta\mathbf{V}_2^- \end{bmatrix} = \begin{bmatrix} A_{21} & B_{21} \\ C_{21} & D_{21} \end{bmatrix} \begin{bmatrix} \Delta\mathbf{R}_1^+ \\ \Delta\mathbf{V}_1^+ \end{bmatrix} \tag{2.26}$$

This simplified notation is commonly found in literature [134]. Using this simplification, the matrix C_{23} would describe the change in $\mathbf{V}_2^+$ caused by a perturbation in the position of P_3, namely, $\mathbf{R}_3^-$.

The Level 1 differential corrector process given in Eq. (2.19) may be written using this simplified notation as follows, where we have again applied it to describe the linear approximation of the change in velocity of P_1 needed to achieve a position difference at the time t_2

$$\Delta\mathbf{V}_1^+ = B_{21}^{-1}\delta\mathbf{R}_2^-$$

The following two linear systems represent approximations of the changes in P_2 that are caused by deviations in the patchpoints P_1 and P_3 [134], where it is assumed that the deviations are all small enough to be in the linear regime of the dynamics along each trajectory

$$\begin{bmatrix} \delta\mathbf{R}_2^- - \mathbf{V}_2^-\delta t_2^- \\ \delta\mathbf{V}_2^- - \mathbf{a}_2^-\delta t_2^- \end{bmatrix} = \begin{bmatrix} A_{21} & B_{21} \\ C_{21} & D_{21} \end{bmatrix} \begin{bmatrix} \Delta\mathbf{R}_1^+ - \mathbf{V}_1^+\Delta t_1^+ \\ \Delta\mathbf{V}_1^+ - \mathbf{a}_1^+\Delta t_1^+ \end{bmatrix} \tag{2.27}$$

$$\begin{bmatrix} \delta\mathbf{R}_2^+ - \mathbf{V}_2^+\delta t_2^+ \\ \delta\mathbf{V}_2^+ - \mathbf{a}_2^+\delta t_2^+ \end{bmatrix} = \begin{bmatrix} A_{23} & B_{23} \\ C_{23} & D_{23} \end{bmatrix} \begin{bmatrix} \Delta\mathbf{R}_3^- - \mathbf{V}_3^-\Delta t_3^- \\ \Delta\mathbf{V}_3^- - \mathbf{a}_3^-\Delta t_3^- \end{bmatrix} \tag{2.28}$$

The formulation for this particular Level 2 differential corrector includes the fixed constraints that the trajectory be continuous in position and time across each patchpoint. Hence, $\mathbf{R}_2^- = \mathbf{R}_2^+ = \mathbf{R}_2$ and $t_2^- = t_2^+ = t_2$. For most applications, this also implies that $a_2^- = a_2^+ = a_2$, though that may not be the case in the presence of dynamics that are velocity-dependent, such as atmospheric drag. These fixed constraints will be applied to each and every patchpoint in turn as the matrix M is constructed.

The targets for this Level 2 are that $\mathbf{V}_2^- = \mathbf{V}_2^+$ in order that the trajectory require no ΔV. The formulation is nearly identical for the case when a mission designer wishes to specify that a particular ΔV be performed at a patchpoint. Hence, the more general target is given by

$$\Delta\hat{\mathbf{V}}_2 - (\mathbf{V}_2^+ - \mathbf{V}_2^-) = 0 \tag{2.29}$$

where the vector $\Delta\hat{\mathbf{V}}_2$ is specified by the designer.

As described earlier, the controls available to achieve the target $\Delta\hat{\mathbf{V}}_2$ include the position vectors and times of P_1, P_2, and P_3. The controls and constraints applied to achieve the target $\Delta\hat{\mathbf{V}}_2$ permit $\mathbf{V}_1^+$ and $\mathbf{V}_3^-$ to be free variables, though those may be targeted by neighboring constraints as the matrix M is constructed.

In summary, there are 28 parameters involved with the goal of achieving a desirable ΔV across P_2, including the position, velocity, and time at the beginning and end of each trajectory segment, organized as follows

$$\begin{aligned}
\textit{Fixed Constraints:}\quad & \mathbf{R}_2^- = \mathbf{R}_2^+ = \mathbf{R}_2, \quad t_2^- = t_2^+ = t_2 \\
\textit{Controls:}\quad & \Delta\mathbf{R}_1,\ \Delta t_1,\ \Delta\mathbf{R}_2,\ \Delta t_2,\ \Delta\mathbf{R}_3,\ \Delta t_3 \\
\textit{Free Variables:}\quad & \Delta\mathbf{V}_1^+, \quad \Delta\mathbf{V}_3^- \\
\textit{Targets:}\quad & \delta\mathbf{V}_2^+ - \delta\mathbf{V}_2^- = \Delta\hat{\mathbf{V}}_2 - \left(\mathbf{V}_2^+ - \mathbf{V}_2^-\right)
\end{aligned}$$

A similar set of parameters is defined for each patchpoint that is included in the differential correction process.

The first row of the M matrix requires six partial derivatives. These in turn require other partial derivatives, as follows

$$\frac{\partial \Delta\mathbf{V}_2}{\partial \mathbf{R}_1} = -\frac{\partial \mathbf{V}_2^-}{\partial \mathbf{R}_1^+} \qquad \frac{\partial \Delta\mathbf{V}_2}{\partial \mathbf{R}_2} = \frac{\partial \mathbf{V}_2^+}{\partial \mathbf{R}_2^-} - \frac{\partial \mathbf{V}_2^-}{\partial \mathbf{R}_2^+} \qquad \frac{\partial \Delta\mathbf{V}_2}{\partial \mathbf{R}_3} = \frac{\partial \mathbf{V}_2^+}{\partial \mathbf{R}_3^-}$$
$$\frac{\partial \Delta\mathbf{V}_2}{\partial t_1} = -\frac{\partial \mathbf{V}_2^-}{\partial t_1^+} \qquad \frac{\partial \Delta\mathbf{V}_2}{\partial t_2} = \frac{\partial \mathbf{V}_2^+}{\partial t_2^-} - \frac{\partial \mathbf{V}_2^-}{\partial t_2^+} \qquad \frac{\partial \Delta\mathbf{V}_2}{\partial t_3} = \frac{\partial \mathbf{V}_2^+}{\partial t_3^-}$$

Wilson provides details to construct each of these partials [134]; we will demonstrate the process and illustrate the construction of $\frac{\partial \mathbf{V}_2^-}{\partial \mathbf{R}_1^+}$; the process may be applied in the same manner to construct each of these partials.

In order to construct $\frac{\partial \mathbf{V}_2^-}{\partial \mathbf{R}_1^+}$, we first set the perturbation of every other independent control to zero, namely

$$\begin{aligned}
& & \delta\mathbf{R}_2 &= \delta\mathbf{R}_3 = \mathbf{0} \\
\delta t_1 &= & \delta t_2 &= \delta t_3 = 0
\end{aligned}$$

These values may then be inserted into Eqs. (2.27) and (2.28) or their inverses, whichever generates the most practical result. There are often many ways to describe the partials, and we are interested in the simplest relationships. For this particular case, the simplest relationship comes from substituting these values into the inverse of Eq. (2.27)

$$\begin{bmatrix} \delta\mathbf{R}_1^+ - \mathbf{V}_1^+\delta t_1^+ \\ \delta\mathbf{V}_1^+ - \mathbf{a}_1^+\delta t_1^+ \end{bmatrix} = \begin{bmatrix} A_{12} & B_{12} \\ C_{12} & D_{12} \end{bmatrix} \begin{bmatrix} \delta\mathbf{R}_2^- - \mathbf{V}_2^-\delta t_2^- \\ \delta\mathbf{V}_2^- - \mathbf{a}_2^-\delta t_2^- \end{bmatrix} \tag{2.30}$$

$$\begin{bmatrix} \delta\mathbf{R}_1^+ \\ \delta\mathbf{V}_1^+ \end{bmatrix} = \begin{bmatrix} A_{12} & B_{12} \\ C_{12} & D_{12} \end{bmatrix} \begin{bmatrix} \mathbf{0} \\ \delta\mathbf{V}_2^- \end{bmatrix} \tag{2.31}$$

This yields a system of two equations

$$\delta\mathbf{R}_1^+ = B_{12}\delta\mathbf{V}_2^- \quad \text{and} \tag{2.32}$$
$$\delta\mathbf{V}_1^+ = D_{12}\delta\mathbf{V}_2^- \tag{2.33}$$

The first equation provides the relationship we are interested in, namely

$$\frac{\delta \mathbf{V}_2^-}{\delta \mathbf{R}_1^+} = B_{12}^{-1} \tag{2.34}$$

Be aware that although $\Phi_{21} = \Phi_{12}^{-1}$, the submatrices do not typically follow such inverse relationships; that is, $B_{12}^{-1} \neq B_{21}$.

This procedure may be followed to generate relationships for each partial required for the production of the matrix M. The result is the following

$$\begin{aligned}
\frac{\partial \Delta \mathbf{V}_2}{\partial \mathbf{R}_1} &= -\frac{\partial \mathbf{V}_2^-}{\partial \mathbf{R}_1^+} &&= -B_{12}^{-1} \\
\frac{\partial \Delta \mathbf{V}_2}{\partial t_1} &= -\frac{\partial \mathbf{V}_2^-}{\partial t_1^+} &&= B_{12}^{-1}\mathbf{V}_1^+ \\
\frac{\partial \Delta \mathbf{V}_2}{\partial \mathbf{R}_2} &= \frac{\partial \mathbf{V}_2^+}{\partial \mathbf{R}_2^-} - \frac{\partial \mathbf{V}_2^-}{\partial \mathbf{R}_2^+} &&= -B_{32}^{-1}A_{32} + B_{12}^{-1}A_{12} \\
\frac{\partial \Delta \mathbf{V}_2}{\partial t_2} &= \frac{\partial \mathbf{V}_2^+}{\partial t_2^-} - \frac{\partial \mathbf{V}_2^-}{\partial t_2^+} &&= \left[\mathbf{a}_2^+ - \mathbf{a}_2^-\right] + \left[B_{32}^{-1}A_{32}\mathbf{V}_2^+ - B_{12}^{-1}A_{12}\mathbf{V}_2^-\right] \\
\frac{\partial \Delta \mathbf{V}_2}{\partial \mathbf{R}_3} &= \frac{\partial \mathbf{V}_2^+}{\partial \mathbf{R}_3^-} &&= B_{32}^{-1} \\
\frac{\partial \Delta \mathbf{V}_2}{\partial t_3} &= \frac{\partial \mathbf{V}_2^+}{\partial t_3^-} &&= -B_{32}^{-1}\mathbf{V}_3^-
\end{aligned}$$

Finally, we have all of the pieces to use Eq. (2.22) to determine an approximation of the adjustments that must be made in the positions and times of each patchpoint as a function of the unwanted velocity changes in each patchpoint.

Level 2 with Constraints. The Level 2 differential corrector can be modified to place a wide variety of constraints on the patchpoints in the system. For instance, we already observed in the derivation of the partials above that it is quite arbitrary to enforce the ΔV at each patchpoint to zero; rather, one can specify a list of ΔV values to perform at particular times and drive the trajectory to that solution instead. Before doing that, we must have a way of preventing the Level 2 corrector from adjusting a patchpoint's time. This is one example of a constraint that may be placed on the system. Other examples include constraining a patchpoint to have a particular inclination relative to some body, or to be located at a particular position or distance relative to a body. These constraints are very important when designing a practical trajectory for a spacecraft mission. For instance, the trajectory being designed may be an extension to a spacecraft's mission that is already in orbit, such that the trajectory must originate from the spacecraft's current trajectory. Or perhaps the trajectory being designed must land on the Moon at a particular landing site. The multiple-shooting differential corrector can accommodate any of these scenarios.

Any constraint may be added to the Level 2 architecture as long as it may be described in the form

$$\alpha_{ij} = f(\mathbf{R}_i, \mathbf{V}_i, t_i) \tag{2.35}$$

where the subscript i represents the patchpoint that the constraint is placed upon and the subscript j indicates the constraint number applied to that patchpoint. This nomenclature is consistent with that used by previous authors [134]. In this form, a constraint may be treated precisely the same as the ΔV targets described in the previous section. The constraint will be added to the list of targets for the differential corrector. It will be assumed, once again, that the only controls that may influence the constraint are the position and time of the patchpoint that the constraint is applied to as well as the positions and times of the two neighboring patchpoints (or the single neighboring patchpoint in the case of a constraint placed on the first or last patchpoint of a trajectory). The differential corrector may certainly be re-derived to operate with constraints that act upon many patchpoints, but this discussion is limited to constraints that act upon a single patchpoint.

It is straightforward to add a constraint to the list of targets in the differential corrector. The relationship given in Eq. (2.23) is augmented as follows

$$\begin{bmatrix} \delta\Delta\mathbf{V}_i \\ \delta\alpha_{ij} \end{bmatrix} = \underbrace{\begin{bmatrix} \dfrac{\partial\Delta\mathbf{V}_i}{\partial\mathbf{R}_k} & \dfrac{\partial\Delta\mathbf{V}_i}{\partial t_k} \\ \dfrac{\partial\alpha_{ij}}{\partial\mathbf{R}_k} & \dfrac{\partial\alpha_{ij}}{\partial t_k} \end{bmatrix}}_{[P]} \begin{bmatrix} \delta\mathbf{R}_k \\ \delta t_k \end{bmatrix} \tag{2.36}$$

where the matrix P is known as the augmented state relationship matrix (SRM). Equation (2.36) is highly compressed: P is typically sparsely populated roughly along the diagonal, such that each constraint and each ΔV may only be influenced by the patchpoint it is assigned to and that patchpoint's nearest neighbors. Much like the ΔV constraints described in the previous section, each constraint requires the definition of the following six partials

$$\begin{matrix} \dfrac{\partial\alpha_{ij}}{\partial\mathbf{R}_{i-1}} & \dfrac{\partial\alpha_{ij}}{\partial\mathbf{R}_{i}} & \dfrac{\partial\alpha_{ij}}{\partial\mathbf{R}_{i+1}} \\ \dfrac{\partial\alpha_{ij}}{\partial t_{i-1}} & \dfrac{\partial\alpha_{ij}}{\partial t_{i}} & \dfrac{\partial\alpha_{ij}}{\partial t_{i+1}} \end{matrix} \tag{2.37}$$

A quick observation shows that the ΔV constraints described in the previous section are a specific case of a constraint, where $\alpha_{ij} = \Delta\mathbf{V}_i$.

Any constraint that is a function of the position and/or time of one of the control patchpoints, and not a direct function of the velocity of any patchpoint, may be easily defined. For instance, if one wishes to constrain the time of patchpoint P_i, one simply characterizes that constraint as

$$\alpha_{ij} = t_i - \hat{t}_i$$

where $\hat{t}_i$ is the desired time. One then computes the partials given in Eq. (2.37) and finds that the only non-zero partial is

$$\frac{\partial \alpha_{ij}}{\partial t_i} = 1$$

Similarly, if one wishes to constrain the position vector of patchpoint P_i, one characterizes that constraint as

$$\alpha_{ij} = \mathbf{R}_i - \hat{\mathbf{R}}_i$$

where $\hat{\mathbf{R}}_i$ is the desired position vector. One then finds that the only non-zero partial is

$$\frac{\partial \alpha_{ij}}{\partial \mathbf{R}_i} = I_{3\times 3}$$

Constraints that depend on velocity are more complex, as demonstrated by the ΔV constraints given above. In order to compute the partials given in Eq. (2.37), one must perform the chain rule and compute additional partial derivatives. Fortunately, many of these were computed in the previous section, and many go to zero for numerous constraint formulations. The relationships are

$$\begin{aligned}
\frac{\partial \alpha_{ij}}{\partial \mathbf{R}_{i-1}} &= \frac{\partial \alpha_{ij}}{\partial \mathbf{R}_{i-1}} + \frac{\partial \alpha_{ij}}{\partial \mathbf{V}_i^-}\frac{\partial \mathbf{V}_i^-}{\partial \mathbf{R}_{i-1}} \\
\frac{\partial \alpha_{ij}}{\partial t_{i-1}} &= \frac{\partial \alpha_{ij}}{\partial t_{i-1}} + \frac{\partial \alpha_{ij}}{\partial \mathbf{V}_i^-}\frac{\partial \mathbf{V}_i^-}{\partial t_{i-1}} \\
\frac{\partial \alpha_{ij}}{\partial \mathbf{R}_i} &= \frac{\partial \alpha_{ij}}{\partial \mathbf{R}_i} + \frac{\partial \alpha_{ij}}{\partial \mathbf{V}_i^-}\frac{\partial \mathbf{V}_i^-}{\partial \mathbf{R}_i} + \frac{\partial \alpha_{ij}}{\partial \mathbf{V}_i^+}\frac{\partial \mathbf{V}_i^+}{\partial \mathbf{R}_i} \\
\frac{\partial \alpha_{ij}}{\partial t_i} &= \frac{\partial \alpha_{ij}}{\partial t_i} + \frac{\partial \alpha_{ij}}{\partial \mathbf{V}_i^-}\frac{\partial \mathbf{V}_i^-}{\partial t_i} + \frac{\partial \alpha_{ij}}{\partial \mathbf{V}_i^+}\frac{\partial \mathbf{V}_i^+}{\partial t_i} \\
\frac{\partial \alpha_{ij}}{\partial \mathbf{R}_{i+1}} &= \frac{\partial \alpha_{ij}}{\partial \mathbf{R}_{i+1}} + \frac{\partial \alpha_{ij}}{\partial \mathbf{V}_i^+}\frac{\partial \mathbf{V}_i^+}{\partial \mathbf{R}_{i+1}} \\
\frac{\partial \alpha_{ij}}{\partial t_{i+1}} &= \frac{\partial \alpha_{ij}}{\partial t_{i+1}} + \frac{\partial \alpha_{ij}}{\partial \mathbf{V}_i^+}\frac{\partial \mathbf{V}_i^+}{\partial t_{i+1}}
\end{aligned}$$

Wilson derives the formulae that may be used to constrain a patchpoint's velocity, velocity magnitude, inclination, apse location, flight path angle, declination, right ascension, and conic energy [134]. For example, the conic energy relative to a massive body may be described as

$$\begin{aligned}
\alpha_{ij} &= \frac{|\mathbf{V}_i|^2}{2} - \frac{\mu}{|\mathbf{R}_i|} \\
&= \frac{\mathbf{V}_i \cdot \mathbf{V}_i}{2} - \frac{\mu}{(\mathbf{R}_i \cdot \mathbf{R}_i)^{1/2}}
\end{aligned}$$

where μ is the gravitational parameter for the central body. The majority of the partials given above are either zero or already known. The remaining partials may be computed as follows

$$\begin{aligned}\frac{\partial \alpha_{ij}}{\partial \mathbf{R}_i} &= \frac{\mu \mathbf{R}_i^T}{|\mathbf{R}_i|^3} \\ \frac{\partial \alpha_{ij}}{\partial \mathbf{V}_i^{\pm}} &= \mathbf{V}_i^{\pm T} \\ \frac{\partial \alpha_{ij}}{\partial t_i} &= \mathbf{0}\end{aligned}$$

The implementation of additional constraints is left to the designer.

A practical constraint that is not formulated in the same way is to restrict the size of the steps that the Level 2 differential corrector may take between iterations. The differential corrector estimates the change in each patchpoint's position, time, or both in order to achieve the given targets, and it does so using a large system of linearized equations. It is often the case that small perturbations drive the realized deviations in the trajectory into highly nonlinear regimes. In practice, it is often the case that the application of a full adjustment in the controls will push the trajectory further from the desired solution than it started. If a designer observes the trajectory diverging from the desired target, one common solution is to limit the maximum deviation that the patchpoints may shift in position or time per iteration of the differential corrector. If implemented properly, the smaller steps should keep the trajectory within the basin of convergence of the solution.

Level 2 with Multiple Trajectories. The Level 2 differential corrector formulated here operates on a large system of controls, targets, and constraints, where ultimately each patchpoint in the system contributes to the satisfaction of all goals, though each patchpoint is only directly influenced by its neighboring patchpoints at any given iteration. This system may be applied to multiple trajectories simultaneously in much the same way as it is applied to a single trajectory. This has clear practical applications for many spacecraft missions that involve deployments, separations, and/or formation-flying activities.

One may formulate the Level 2 differential corrector with multiple trajectories by augmenting the SRM, P, once again, such that it includes the patchpoints, targets, and constraints of every trajectory. One must be sure to permit the system some ΔV leverage to allow any given pair of trajectories to separate (either forward in time or backward in time, as appropriate).

For example, let us assume we have a scenario that involves one spacecraft deploying a secondary payload via a spring mechanism, which imparts a specified ΔV between the two spacecraft. Let us also assume that the differential corrector is permitted to vary the trajectory of the joined system prior to the deployment, as well as both trajectories after the deployment, and the deployment ΔV may occur in any direction. One way to model this scenario is to set up two series of patchpoints that define each spacecraft and then carefully lock the two spacecraft together. A practical way to lock the two trajectories together is to define the first patchpoint of the deployed payload to be prior to the deployment, such that its position, velocity, and time are all constrained to be equal to the relevant parameters of the host spacecraft. That is, it is entirely constrained to match the corresponding patchpoint on the host spacecraft's trajectory. The second patchpoint in the deployed payload's trajectory is then defined to be the deployment event, such that its position and time are constrained to be equal to the position and time of the host spacecraft at the deployment, and its outgoing velocity is constrained to have the appropriate ΔV magnitude applied. From there the trajectory departs in the same way as any other trajectory. In this scenario, one would also have to take care to model the appropriate reaction to the host spacecraft's trajectory via constraints.

The augmented SRM for the case of multiple trajectories is very sparse, and it may be beneficial to implement numerical algorithms that take advantage of this feature. The simple Step 2 SRM includes nonzero elements only on a diagonal swath six elements wide. The SRM shifts further away from diagonal each time it is augmented by an additional constraint or an additional trajectory, though it remains approximately diagonal.

2.6.6 Constructing Periodic Orbits

Periodic orbits are important when analyzing and constructing trajectories using dynamical systems methods, since they help to characterize the flow in the system. There are many methods that are frequently used to identify and construct periodic orbits. Three categories of methods are described here:

1. Analytical Expansion Techniques. The discussion given in Section 2.6.2.2 demonstrates how to use basic analytical techniques to identify planar and three-dimensional periodic orbits in the CRTBP. Many authors have constructed analytical expansions that may be used to approximate periodic orbits in the CRTBP or in more complex systems [67, 123, 139].

2. Shooting Techniques. One may numerically construct a periodic orbit by targeting a single state as both the initial and final states in a trajectory using either a single- or multiple-shooting technique. This technique is difficult without any constraints, but it has proven to be very useful when numerically constructing certain types of periodic orbits, such as simple symmetric periodic orbits [46, 107, 108, 122].

3. The Poincaré Method. The Poincaré Method is a notable method that has proven to be very successful at identifying periodic and quasiperiodic orbits, especially stable orbits. Poincaré's technique involves numerically integrating many trajectories for a large amount of time. Trajectories that are close to periodic tend to linger near the same regions of the state space. One can readily identify stable periodic orbits or trajectories near such orbits if one places a plane in the state space, that is, a *Poincaré Surface of Section*. Then one records the state of each trajectory as the trajectory pierces the plane.

A periodic orbit appears as a fixed-point in the plane; a quasiperiodic orbit appears as a closed loop in the plane. Regions that are unstable in the state space appear as a chaotic sea of points, since unstable trajectories are very sensitive to their initial conditions.

Many other types of methods certainly exist, but these three categories provide a good overview of the variety of methods that are frequently used.

2.6.6.1 Periodic Orbits in the CRTBP If the Lagrange points represent the five simplest solutions to the CRTBP, it may be argued that the next set of solutions to introduce is the set of simple periodic symmetric orbits in the CRTBP. Simple periodic symmetric orbits are orbits that are symmetric about the $y = 0$ plane, pierce the $y = 0$ plane exactly twice per orbit, and pierce the plane orthogonally each time. Libration orbits, such as halo and Lyapunov orbits, are good examples of such orbits. A simple single-shooting differential correction scheme may be used to construct these orbits by taking advantage of their well-defined structure. Section 2.6.6.2 provides more information about this differential correction scheme. It should be noted that although this class of orbits does include what might be argued to be the simplest periodic orbits in the CRTBP, this class of orbits also includes families of very complex orbits.

Many other types of periodic orbits exist in the CRTBP, including orbits that pierce the $y = 0$ plane multiple times per orbit and orbits that are not symmetric, such as orbits about the triangular Lagrange points. One may also construct arbitrarily complex periodic orbits by chaining simple unstable orbits together, as is discussed in Section 2.6.11.

2.6.6.2 Single-Shooting Method for Constructing Simple Periodic Symmetric Orbits in the CRTBP One may formulate many types of shooting techniques to identify periodic orbits using the techniques introduced in Section 2.6.5. Howell identified a simple procedure that has been used by many researchers in the field [122]. The technique is easily applied to the families of halo orbits, Lyapunov orbits, distant prograde orbits, distant retrograde orbits, symmetric resonant orbits, and a variety of other classes of symmetric periodic orbits [46]. Since it is a very common and straightforward procedure, and since it has been used repeatedly in relevant research to construct halo orbits and other similar orbits, it is reviewed here.

As mentioned earlier, simple periodic symmetric orbits are orbits that are symmetric about the $y = 0$ plane, pierce the $y = 0$ plane exactly twice per orbit, and pierce the plane orthogonally each time. Let us define $\mathbf{X}(t_0)$ to be the state of a simple periodic symmetric orbit at the $y = 0$ plane-crossing with a positive $\dot{y}$ and $\mathbf{X}(t_{T/2})$ to be the state of the orbit half of its orbital period later at the $y = 0$ plane-crossing with a negative $\dot{y}$. For this orbit to be periodic and symmetric, these states must have the following form

$$\begin{aligned} \mathbf{X}(t_0) &= [\, x_0 \quad 0 \quad z_0 \quad 0 \quad \dot{y}_0 \quad 0 \,]^T \\ \mathbf{X}(t_{T/2}) &= [\, x_{T/2} \quad 0 \quad z_{T/2} \quad 0 \quad \dot{y}_{T/2} \quad 0 \,]^T \end{aligned} \tag{2.38}$$

Let us assume that we have an initial guess, $\hat{\mathbf{X}}(t_0)$, that is near the initial state of a desirable orbit. When we integrate this state forward in time until the next $y = 0$ plane, we obtain the state $\hat{\mathbf{X}}(t_{\hat{T}/2})$

$$\hat{\mathbf{X}}(t_{\hat{T}/2}) = \begin{bmatrix} x_{\hat{T}/2} & 0 & z_{\hat{T}/2} & \dot{x}_{\hat{T}/2} & \dot{y}_{\hat{T}/2} & \dot{z}_{\hat{T}/2} \end{bmatrix}^T$$

We now wish to adjust the initial state of the trajectory in such a way as to drive the values of $\dot{x}_{\hat{T}/2}$ and $\dot{z}_{\hat{T}/2}$ to zero. One notices that by adjusting the initial state, not only do the values of $\dot{x}$ and $\dot{z}$ change, but the propagation time, $\hat{T}/2$, needed to pierce the $y = 0$ plane also changes. In order to target a proper state $\mathbf{X}(t_{T/2})$, one may vary the initial values of x, z, and/or $\dot{y}$.

The linearized system of equations relating the final state to the initial state may be written as

$$\delta\mathbf{X}(t_{T/2}) \approx \Phi\left(t_{T/2}, t_0\right)\delta\mathbf{X}(t_0) + \frac{\partial\mathbf{X}}{\partial t}\delta(T/2) \tag{2.39}$$

where $\delta\mathbf{X}(t_{T/2})$ is the deviation in the final state due to a deviation in the initial state, $\delta\mathbf{X}(t_0)$, and a corresponding deviation in the orbit's period, $\delta(T/2)$. The time-derivative of the state, $\partial\mathbf{X}/\partial t$, may be computed at the second plane-crossing, namely, at time $t = T/2$. Equation (2.39) may be used as the driver for a differential corrector by setting $\delta\mathbf{X}(t_{T/2})$ to be the desired change in the final state's components and solving for $\delta\mathbf{X}(t_0)$, the approximate correction to the initial state needed to produce such a change.

We now consider what the desired change in the final state's components should be. For our purposes, the only desired change in the final state is a change in the values of $\dot{x}$ and $\dot{z}$, but it is not important if the other components of the final state change. However, we know that the deviation in the final value of y will always be equal to zero since the trajectory is always propagated to that point. Thus we set $\delta\mathbf{X}(t_{T/2})$ to

$$\delta\mathbf{X}(t_{T/2}) = \begin{bmatrix} \delta x_{T/2} & 0 & \delta z_{T/2} & -\dot{x}_{T/2} & \delta\dot{y}_{T/2} & -\dot{z}_{T/2} \end{bmatrix}^T$$

Furthermore, in order to restrict our search to simple periodic symmetrical orbits, we restrict the allowed correction in the initial conditions to

$$\delta\mathbf{X}(t_0) = \begin{bmatrix} \delta x_0 & 0 & \delta z_0 & 0 & \delta\dot{y}_0 & 0 \end{bmatrix}^T$$

Now Eq. (2.39) simplifies to

$$\begin{bmatrix} \delta x_{T/2} \\ 0 \\ \delta z_{T/2} \\ -\dot{x}_{T/2} \\ \delta\dot{y}_{T/2} \\ -\dot{z}_{T/2} \end{bmatrix} \approx \begin{bmatrix} \phi_{11} & \phi_{12} & \phi_{13} & \phi_{14} & \phi_{15} & \phi_{16} \\ \phi_{21} & \phi_{22} & \phi_{23} & \phi_{24} & \phi_{25} & \phi_{26} \\ \phi_{31} & \phi_{32} & \phi_{33} & \phi_{34} & \phi_{35} & \phi_{36} \\ \phi_{41} & \phi_{42} & \phi_{43} & \phi_{44} & \phi_{45} & \phi_{46} \\ \phi_{51} & \phi_{52} & \phi_{53} & \phi_{54} & \phi_{55} & \phi_{56} \\ \phi_{61} & \phi_{62} & \phi_{63} & \phi_{64} & \phi_{65} & \phi_{66} \end{bmatrix} (t_{T/2}, t_0) \begin{bmatrix} \delta x_0 \\ 0 \\ \delta z_0 \\ 0 \\ \delta\dot{y}_0 \\ 0 \end{bmatrix} + \begin{bmatrix} \dot{x} \\ \dot{y} \\ \dot{z} \\ \ddot{x} \\ \ddot{y} \\ \ddot{z} \end{bmatrix} \delta(T/2) \tag{2.40}$$

The value of $\delta(T/2)$ may be determined from the second line of Eq. (2.40) to be

$$\delta(T/2) = \frac{-\phi_{21}\delta x_0 - \phi_{23}\delta z_0 - \phi_{25}\delta\dot{y}_0}{\dot{y}} \tag{2.41}$$

Substituting this value into the fourth and sixth lines of Eq. (2.40) yields

$$-\dot{x}_{T/2} \approx \left(\phi_{41} - \phi_{21}\frac{\ddot{x}}{\dot{y}}\right)\delta x_0 + \left(\phi_{43} - \phi_{23}\frac{\ddot{x}}{\dot{y}}\right)\delta z_0 + \left(\phi_{45} - \phi_{25}\frac{\ddot{x}}{\dot{y}}\right)\delta \dot{y}_0 \quad (2.42)$$

$$-\dot{z}_{T/2} \approx \left(\phi_{61} - \phi_{21}\frac{\ddot{z}}{\dot{y}}\right)\delta x_0 + \left(\phi_{63} - \phi_{23}\frac{\ddot{z}}{\dot{y}}\right)\delta z_0 + \left(\phi_{65} - \phi_{25}\frac{\ddot{z}}{\dot{y}}\right)\delta \dot{y}_0 \quad (2.43)$$

Equations (2.42) and (2.43) give expressions for the approximate deviation in the final x- and z-velocities as functions of the deviation in all three initial conditions x_0, z_0, and $\dot{y}_0$. It is sufficient to change only two of the initial conditions, if that is desirable, or a combination of all three. For the purpose of this description, the value of x_0 will be kept constant, and the values of z_0 and $\dot{y}_0$ will be permitted to vary. The following expression summarizes the approximate changes that must be made to z_0 and $\dot{y}_0$ to produce a desirable change in the final state (while keeping the other initial conditions constant)

$$\begin{bmatrix} \delta z_0 \\ \delta \dot{y}_0 \end{bmatrix} \approx \begin{bmatrix} \phi_{43} - \phi_{23}\frac{\ddot{x}}{\dot{y}} & \phi_{45} - \phi_{25}\frac{\ddot{x}}{\dot{y}} \\ \phi_{63} - \phi_{23}\frac{\ddot{z}}{\dot{y}} & \phi_{65} - \phi_{25}\frac{\ddot{z}}{\dot{y}} \end{bmatrix}^{-1} \begin{bmatrix} -\dot{x}_{T/2} \\ -\dot{z}_{T/2} \end{bmatrix} \quad (2.44)$$

Since the system was linearized in order to produce this procedure, the adjustments will not correct the unwanted motion perfectly; this procedure must be iterated until it converges on an orbit.

When all is said and done, a simple, symmetric periodic orbit has three nonzero states at its orthogonal $y = 0$ plane crossing: x_0, z_0, and $\dot{y}$ (see Eq. (2.38)). The procedure outlined here is used to generate the periodic orbit given one of those parameters and estimates of the other two. Because of this, a family of periodic orbits may be well represented by plotting its initial $\dot{y}$ values or its Jacobi constant values as a function of its initial x values. Figure 2-14 illustrates these curves using the family of Lyapunov orbits about LL_1 as an example.

2.6.6.3 Differentially Correcting Orbits into the DE421 Model An orbit that is perfectly periodic in the CRTBP is not perfectly periodic in the real Solar System since the planets and moons in the real Solar System do not move in circular, coplanar orbits. Various perturbations lead the orbit to diverge from being periodic; the most notable of which is the nonzero eccentricity of the orbits of the primary bodies in the system [100].

To produce a quasiperiodic orbit in the real Solar System, one can use a multiple-shooting differential corrector with the periodic CRTBP orbit as the initial guess of the real trajectory. This technique was demonstrated in Section 2.6.5.2. The differential corrector takes the CRTBP orbit and perturbs it to keep it near its initial guess while eliminating the need to perform large maneuvers. In the case of generating a quasi-periodic halo or Lissajous orbit in the DE421 model of the Solar System, one may use an analytical approximation of the orbit as the initial guess to the differential corrector [123, 139]. This has been demonstrated on many occasions and has been shown to work well [47].

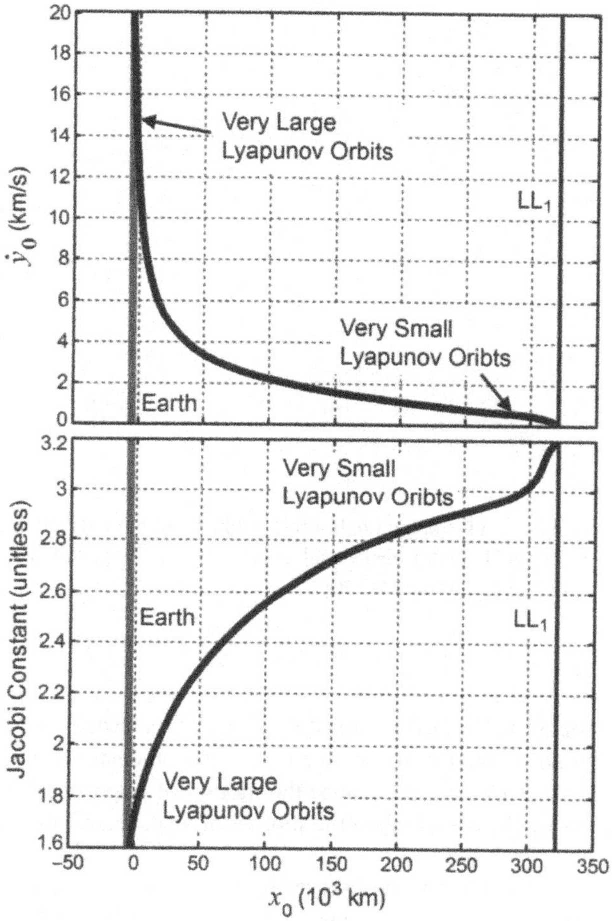

Figure 2-14 Plots of $\mathbf{x_0}$ vs. $\mathbf{\dot{y}_0}$ (top) and $\mathbf{x_0}$ vs. **C** (bottom) for the family of Lyapunov orbits about the Earth–Moon L_1 point. The initial values of the other Cartesian coordinates in the Earth–Moon synodic frame are all equal to zero for each orbit in this family [101] (*Acta Astronautica* by International Academy of Astronautics, reproduced with permission of Pergamon in the format reuse in a book/textbook via Copyright Clearance Center).

Figure 2-15 shows the difference between a halo orbit about the lunar L_2 point produced in the CRTBP compared with the same halo orbit differentially corrected into the DE421 model of the real Solar System. One can see that the real halo orbit is quasiperiodic, tracing out the same vicinity of space on each orbit, but never truly retracing itself. For this illustration, the realistic quasi-halo orbit is plotted in a coordinate frame that is normalized over time based on the instantaneous distance between the Earth and the Moon, and then re-scaled to the average distance between the Earth and the Moon.

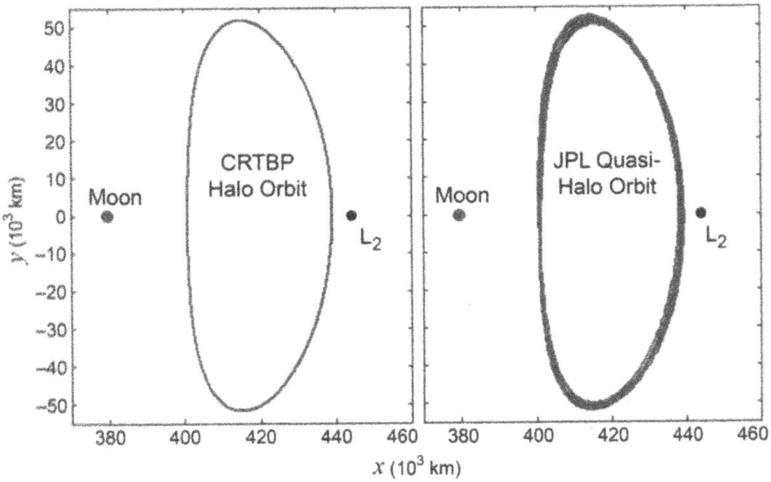

Figure 2-15 A comparison between a halo orbit produced in the CRTBP and a quasi-halo orbit produced in the DE421 model of the real Solar System. The orbits are shown in the Earth–Moon synodic reference frame [44, 46].

The perfectly periodic CRTBP orbit is typically a very close approximation of the real quasiperiodic orbit, enough so that the early mission design may be developed in the CRTBP. This is convenient because the motion of a spacecraft in a perfectly periodic orbit is more predictable than the motion of a spacecraft in a quasiperiodic orbit.

On several occasions, it has been observed that some of the structure of a periodic orbit in the CRTBP becomes lost or significantly altered as the orbit is differentially corrected into the DE421 model. This is often seen when a single revolution of a periodic orbit is sent into the differential corrector. Ordinarily, a differential corrector converges on a continuous trajectory more readily if the trajectory's endpoints are not constrained. Without the boundary values constrained, it is often the case that the differential corrector significantly alters the states of the trajectory's endpoints. The resulting trajectory, although continuous, may not resemble the original orbit much at all. This effect may be observed in Fig. 2-16.

One way to combat this effect is to differentially correct several orbits of the periodic orbit together. For the purpose of this discussion, let us say that four periodic orbits are differentially corrected together. Then, two of the orbits are "outer" orbits (the first and last orbits) that are vulnerable to substantial changes in the differential correction process, and two of the orbits are "inner" orbits (the second and third orbits) that are more protected from significant alteration in the process. Normally, the differential corrector converges on a continuous trajectory before the inner orbits are substantially altered. Once the differential corrector has converged on the final trajectory, then the outer orbits may be pruned off in order to observe the structure of

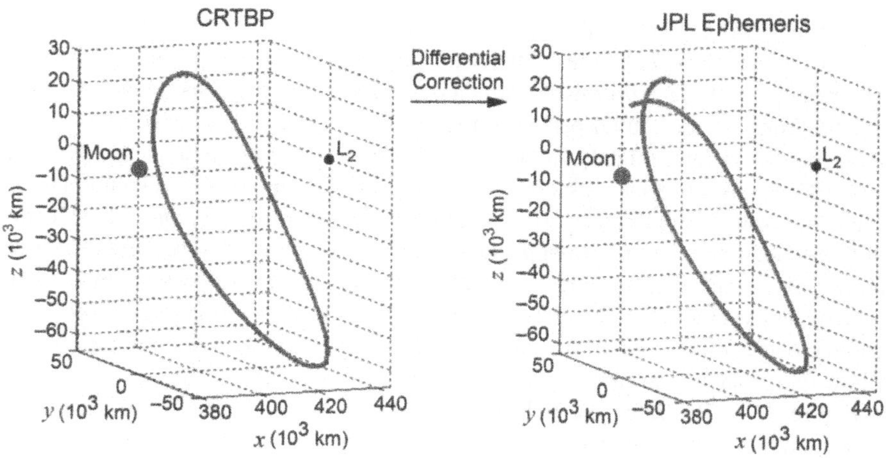

Figure 2-16 A single L_2 halo orbit in the Earth–Moon CRTBP (left) is differentially-corrected into the DE421 model (right) [46].

the resulting quasiperiodic orbit. Ordinarily, this procedure results in quasiperiodic orbits that exist in the DE421 model that retain the same structure as the periodic orbits that exist in the CRTBP. Figure 2-17 shows an example of this process.

Since halo orbits are used frequently in later chapters of this book, some discussion is given here regarding the largest observable deviations between the perfectly periodic halo orbit in the CRTBP and the quasi-halo orbit in the real Solar System. Arguably the most substantial deviation between the CRTBP and the real Solar Sys-

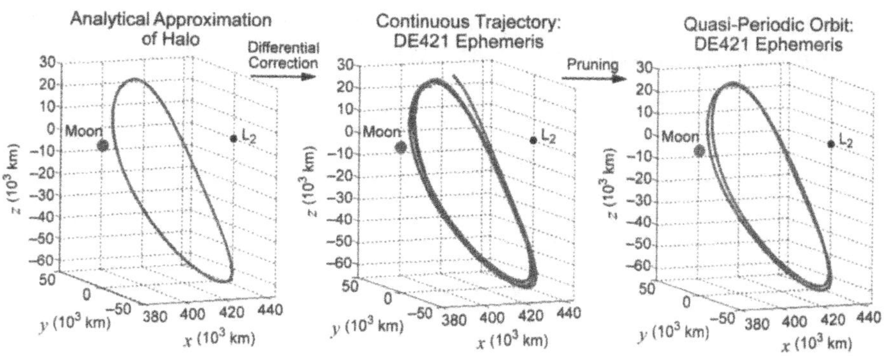

Figure 2-17 An example of the process of differentially correcting and pruning a halo orbit from the Earth–Moon CRTBP into the DE421 model. Left: the nominal periodic halo orbit in the Earth–Moon CRTBP; center: the differentially corrected trajectory in the DE421 model; right: the pruned quasiperiodic halo trajectory in the DE421 model [44, 46].

tem, at least in the case of the Sun–Earth and Earth–Moon three-body systems, is the nonzero eccentricity of the orbits of the primary bodies in the system. The real, eccentric orbit of the primaries imparts a deviation in the quasi-halo orbits that has a period equal to the orbital period of the primaries. Since most halo orbits have a period equal to approximately half of the orbital period of the primaries, this dynamic perturbation tends to appear as a resonant pulsation. One quasi-halo revolution tends to deviate from the perfectly periodic halo orbit in one direction, and the next revolution tends to deviate in the opposite direction. The result is that a spacecraft on a quasi-halo orbit tends to retrace its path very closely every other revolution. This effect is less visible when the reference frame is centered on a Lagrange point rather than the barycenter, since the Lagrange point pulses in and out as the primary bodies follow their noncircular orbits.

Figure 2-18 illustrates the pulsation that exists in the Earth–Moon system by showing a plot of the distance between the Moon and a spacecraft traversing an orbit much like that one shown in Figs. 2-16 and 2-17. One can see that every other revolution retraces a similar path. The moments in time when the Moon reaches its perigee and apogee are indicated for reference. Figure 2-19 illustrates how this same quasi-halo orbit appears in the DE421 ephemeris when viewed in different synodic coordinate systems, including an Earth-centered synodic frame, out to an LL_2-centered synodic frame.

2.6.7 The Continuation Method

Periodic orbits in the CRTBP may be grouped into families, where a family consists of an infinite number of periodic orbits whose properties vary continuously from one end of the family to the other. All orbits in the same family may be uniquely identified by a single parameter of that family, for example, their position on a perpendicular $y = 0$ plane crossing, their velocity at that crossing, or some other specified parameter. This property of the CRTBP is due to the existence of the Jacobi

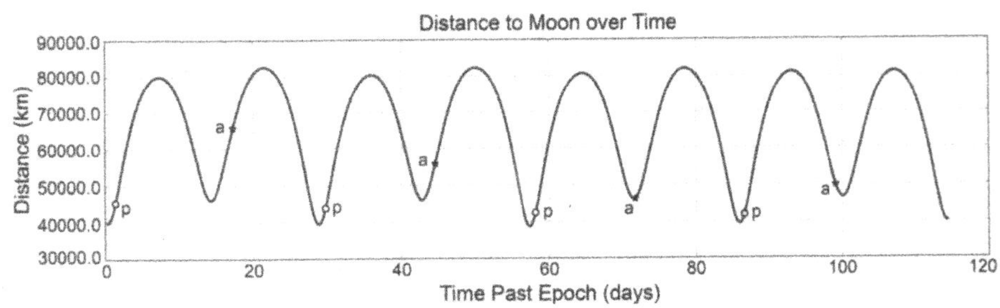

Figure 2-18 The distance between the orbit and the Moon over time for a realistic quasi-halo orbit. The moments in time when the Moon reaches its perigee and apogee are indicated by the symbols "p" and "a", respectively.

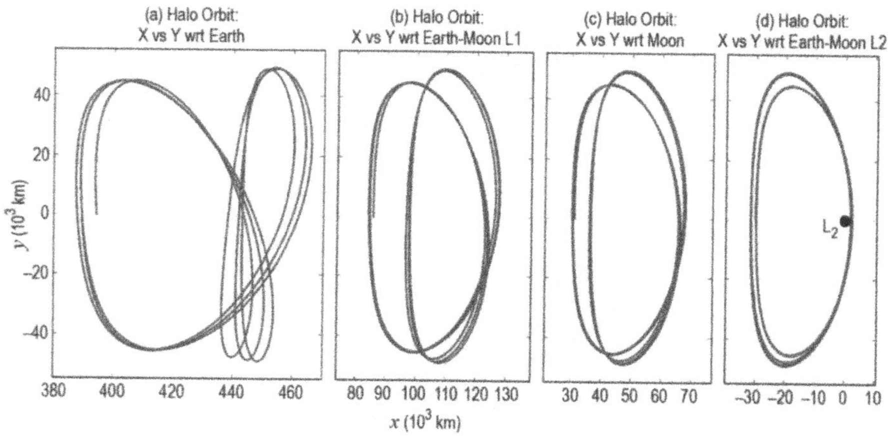

Figure 2-19 An illustration of how the same quasi-halo orbit appears in different synodic coordinate systems. From left to right, the systems include an Earth-centered synodic frame, an L_1-centered frame, a Moon-centered frame, and an L_2-centered frame.

constant, the CRTBP's unique integral of motion. Hénon provides more discussion about the existence of families of solutions in the CRTBP [113].

Once a single periodic orbit is known in the CRTBP, then the continuation method may be used to traverse that orbit's family. The method starts by perturbing some parameter of the known periodic orbit and then differentially correcting the new conditions to find that periodic orbit's neighbor in its family. The differential corrector presented in Section 2.6.6 is well-suited to this method for simple periodic symmetrical orbits because one may vary the initial position and correct for the initial velocity that corresponds to the next periodic orbit in the family (or vice versa, if desired).

To demonstrate this method, the continuation method has been applied to the family of Lyapunov orbits that exist about the Earth–Moon L_2 point. First, a single Lyapunov orbit is identified, for example, the gray orbit in Figs. 2-20 and 2-21. The orbit's initial position, x_0, is then systematically varied while a differential corrector fills out the curve shown in Fig. 2-20. The initial conditions in the curve correspond to the family of orbits shown in Fig. 2-21.

The continuation method works well when the perturbations are small; in practice it is beneficial to predict the differential corrector's adjustment to the perturbation because this allows larger jumps in the varying parameter. Furthermore, if the perturbations are too large, the differential corrector may converge on a solution of a different family. Thus smaller steps or better prediction methods may be required to make the continuation method more reliable. The work for this study has implemented a quadratic prediction method that uses the three previous data points of the family to predict the next data point. This has been sufficient to allow the differential corrector to converge quickly while allowing the curve of the family to evolve naturally over the state space. Two-dimensional curve tracking algorithms may also work well since

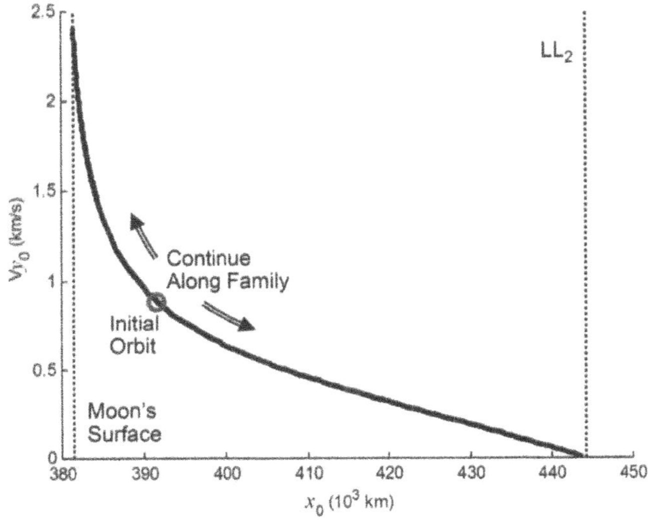

Figure 2-20 A plot of the initial conditions of the family of Lyapunov orbits about the Earth–Moon L_2 point (LL_2) [140] (Copyright ©2006 by American Astronautical Society Publications Office, San Diego, California [website http://www.univelt.com], all rights reserved; reprinted with permission of the AAS).

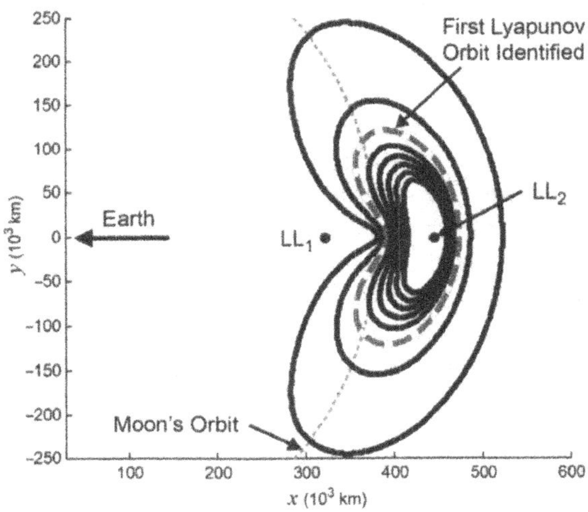

Figure 2-21 Plots of the orbits in the family of LL_2 Lyapunov orbits corresponding to those initial conditions shown in Fig. 2-20 [140] (Copyright ©2006 by American Astronautical Society Publications Office, San Diego, California [website http://www.univelt.com], all rights reserved; reprinted with permission of the AAS).

state space curves are not necessarily well-modeled by polynomials. For instance, one may extrapolate curves using a constant arc-length of two parameters [141].

2.6.8 Orbit Stability

The stability of a periodic orbit may be determined by analyzing the eigenvalues of the orbit's monodromy matrix. A random perturbation in the state of a spacecraft on an unstable orbit will cause the spacecraft's state to exponentially diverge from that of the original orbit over time; hence, the monodromy matrix of an unstable orbit includes at least one eigenvalue for which the real component is outside of the unit circle. This section explores the stability characteristics of periodic orbits via the eigenvalues of their monodromy matrices.

2.6.8.1 Eigenvalues of an Orbit's Monodromy Matrix The monodromy matrices of orbits in the CRTBP have six eigenvalues, λ_i for $i = 1, 2, \ldots, 6$, corresponding to the eigenvectors $\mathbf{v}_i$. The eigenvalues of the monodromy matrix occur in reciprocal pairs [142], which is a direct consequence of the symplectic nature of the monodromy matrix, and of the state transition matrix in the CRTBP in general [143]. Additionally, a pair of eigenvalues of the monodromy matrix will be equal to unity because of the Jacobi integral of motion in the CRTBP [131, 142]. The eigenvalues are thus related in the following way

$$\lambda_2 = \frac{1}{\lambda_1} \qquad \lambda_4 = \frac{1}{\lambda_3} \qquad \lambda_5 = \lambda_6 = 1 \tag{2.45}$$

The monodromy matrices of periodic orbits in the planar CRTBP only have four eigenvalues: $(\lambda_1, 1/\lambda_1, 1, 1)$. Since those orbits may be computed in the spatial CRTBP by setting their z- and $\dot{z}$-components to zero, the remainder of this section only considers orbits in the full three-dimensional system.

The eigenvalues of the monodromy matrix of a periodic orbit in the CRTBP are the roots of a characteristic equation; furthermore, each has a characteristic exponent, α, where $\lambda = e^{\alpha T}$ and T is the period of the orbit. Then, the reciprocal of that eigenvalue is equal to: $1/\lambda = e^{-\alpha T}$. The characteristic exponents are sometimes referred to as Lyapunov characteristic exponents [144].

The monodromy matrices of Keplerian orbits, such as low Earth orbits (LEOs), have three pairs of eigenvalues that are all equal to 1, indicating that after a full orbit any given perturbation neither grows nor decays exponentially. The monodromy matrices of periodic orbits in the CRTBP may have other eigenvalue pairs, including real values not equal to 1 and pairs of complex numbers. Table 2-5 provides a summary of the resulting motion of a spacecraft in a periodic orbit, whose state is perturbed along the eigenvector corresponding to any type of given eigenvalue.

The stability of a periodic orbit may be identified by observing the resulting motion of a perturbed particle in that orbit or by computing the eigenvalues of the orbit's monodromy matrix and comparing those eigenvalues to the results given in Table 2-5. The following classification scheme for an orbit's stability is used in this work:

Table 2-5 A summary of the resulting motion of a spacecraft in a periodic orbit, whose state is perturbed along the eigenvector corresponding to a given eigenvalue. The result of a perturbation along the eigenvector corresponding to a complex eigenvalue includes a combination of the imaginary result and one of the real results listed.

Eigenvalue	Result of the perturbation
Real, within the range [−1,1]	The perturbation exponentially decays.
Real, equal to 1 or −1	The perturbation neither exponentially decays nor grows.
Real, outside of the range [−1,1]	The perturbation exponentially grows.
Imaginary	After each orbital period, the perturbation oscillates about the spacecraft's original state.

- If an eigenvalue exists whose real component is outside of the range [−1,1], then the periodic orbit is asymptotically unstable, referred to here as *unstable*, along the corresponding eigenvector.
- If the real component of each and every eigenvalue of the monodromy matrix is between −1 and 1, then the orbit is *stable*. Given the relationships in Eq. (2.45), an orbit may only be stable in the CRTBP if each and every eigenvalue pair is complex with real components in the range [−1,1].
- If the orbit is not unstable and there is at least one eigenvalue pair whose real component is equal to 1, then the periodic orbit is *neutrally stable*, or a *center* [130].

Since every periodic orbit in the CRTBP has at least one pair of eigenvalues with values are equal to unity, then it is customary to ignore that pair of eigenvalues when classifying the stability of the orbit [108, 145].

To determine the eigenvalues of the monodromy matrix, it is useful to consider the characteristic equation, since many of the roots of this equation are already known

$$\begin{aligned}\det(M-\lambda I) &= (\lambda-\lambda_1)(\lambda-\lambda_2)(\lambda-\lambda_3)(\lambda-\lambda_4)(\lambda-\lambda_5)(\lambda-\lambda_6)=0\\ &= (\lambda-1)^2(\lambda-\lambda_1)(\lambda-1/\lambda_1)(\lambda-\lambda_3)(\lambda-1/\lambda_3)=0\end{aligned} \tag{2.46}$$

The relationship given in Eq. (2.46) may be re-written in terms of the new parameters p and q, keeping consistent with the nomenclature found in the literature [116]

$$(\lambda-1)^2\left(\lambda^2+p\lambda+1\right)\left(\lambda^2+q\lambda+1\right)=0 \tag{2.47}$$

Thus, $p=-(\lambda_1+1/\lambda_1)$ and $q=-(\lambda_3+1/\lambda_3)$. Equation (2.47) may also be factored in the following manner

$$(\lambda-1)^2\left(\lambda^4+(p+q)\lambda^3+(pq+2)\lambda^2+(p+q)\lambda+1\right)=0 \tag{2.48}$$

Equation (2.48) may be re-written using the new parameters α, β, and γ, once again to keep consistent with the nomenclature found in the literature [116] (where α should not be confused with the characteristic exponent that corresponds to each eigenvalue)

$$(\lambda - 1)^2 \left(\lambda^4 + \alpha\lambda^3 + \beta\lambda^2 + \alpha\lambda + \gamma\right) = 0 \tag{2.49}$$

In this form it is clear that $\alpha = p + q$, $\beta = pq + 2$, and $\gamma = 1$. The benefits of factoring the characteristic equation into the parameters α, β, and γ arises at this point. Bray and Goudas derive a fast and simple method to compute α and β using the monodromy matrix [116, 145]

$$\alpha = 2 - \text{trace}(M) \tag{2.50}$$

$$\beta = \frac{\alpha^2 - \text{trace}(M^2)}{2} + 1 \tag{2.51}$$

It is then simple to determine the parameters p and q using knowledge of α and β

$$\left.\begin{matrix} p \\ q \end{matrix}\right\} = \frac{\alpha \pm \sqrt{\alpha^2 - 4\beta + 8}}{2} \tag{2.52}$$

It then follows that with knowledge of p and q one may determine the corresponding eigenvalues

$$\left.\begin{matrix} \lambda_1 \\ 1/\lambda_1 \end{matrix}\right\} = \frac{-p \pm \sqrt{p^2 - 4}}{2} \tag{2.53}$$

$$\left.\begin{matrix} \lambda_3 \\ 1/\lambda_3 \end{matrix}\right\} = \frac{-q \pm \sqrt{q^2 - 4}}{2} \tag{2.54}$$

The final two eigenvalues have already been predetermined and are given in Eq. (2.45) as $\lambda_5 = \lambda_6 = 1$. Thus, Eqs. (2.50)–(2.54) provide a fast and simple method to compute the six eigenvalues of the monodromy matrix. The corresponding eigenvectors may be computed in any standard way using the equation $M\mathbf{v}_i = \lambda_i \mathbf{v}_i$. It should be noted that the stable and unstable eigenvalues, λ^S and λ^U, of an orbit's monodromy matrix, are equal to the pair of real eigenvalues with the smallest and largest values, respectively, if they exist.

2.6.8.2 The Stability Index An orbit's *stability index* is defined in various ways in the literature depending on the author. Several authors, for example, Broucke [108], define the stability of a periodic orbit on the value of k, where k is equal to the sum of the real eigenvalues of the orbit. If $|k| > n$, where n is equal to the number of real eigenvalues in the orbit's monodromy matrix, then the orbit is unstable; if $|k| < n$, the orbit is stable; otherwise $|k| = n$ and the orbit is neutrally stable. One problem with such a definition is that the value of n may change depending on the orbit.

Another definition of the stability index is defined by Howell, among others, as follows [122]. If one considers the definition $k_i = \lambda_i + 1/\lambda_i$, one notices several things. First, the values of k_i may be easily computed using the parameters p and

q that were introduced above, namely: $k_1 = -p$, $k_2 = -q$, and $k_3 = 2$. Next, the value of k_i is always real and in the range $-2 \leq k \leq 2$ for stable orbits since the sum of a real pair is real and the sum of a complex conjugate pair is also real. Furthermore, if $|k_i| > 2$, then the real component of at least one of the eigenvalues summed is greater than 1 and the orbit is unstable. Since two of the eigenvalues of the orbit's monodromy matrix are equal to unity and it is conventional to ignore them, the stability index, k, may then be given by

$$k = \sup\{|\text{Re}(k_1)|, |\text{Re}(k_2)|\} \tag{2.55}$$

where the operator Re() only observes the real component of the operand. We have the final test: if $k > 2$ the orbit is unstable, if $k = 2$ the orbit is neutrally stable, and if $k < 2$ the orbit is stable.

2.6.8.3 The Perturbation Doubling Time The stability index defined in Eq. (2.55) certainly provides information about the stability of the orbit in question. However, it only provides limited information about the relative stability of different orbits. A highly unstable orbit may appear to be more stable than a weakly unstable orbit if the weakly unstable orbit's period is much greater than the highly unstable orbit's period. It is now of interest to find a parameter that may be used to directly compare the stability of two orbits regardless of their relative orbital periods.

The eigenvalues of the monodromy matrix of a periodic orbit are a function of the orbit's period, T, and a characteristic exponent, α, as follows

$$\lambda = e^{\alpha T} \tag{2.56}$$

To compare the stability of several orbits directly, one may either normalize the eigenvalues of the monodromy matrices or, equivalently, compare the characteristic exponents in some way.

An intuitive measure for comparison is the orbit's *perturbation doubling time* (for unstable orbits) or the orbit's *perturbation half-life* (for stable orbits). In this work, we refer to this time measurement as $\hat{\tau}$ for two reasons: first, to indicate that it is a normalized measurement and second, to distinguish it from the parameter τ that is used to identify points along an orbit (see Section 2.6.2.3). Given a spacecraft in an unstable orbit, the perturbation doubling time characterizes the length of time that is required for a perturbation in the spacecraft's state to double in magnitude. Similarly, given a spacecraft in a stable orbit, the perturbation half-life characterizes the length of time that is required for a perturbation in the spacecraft's state to be reduced by one half. For simplicity, we refer to this time measurement only as the perturbation doubling time, since it is generally more useful when designing real missions to compare this time measurement for unstable orbits.

After determining the eigenvalues of the orbit's monodromy matrix, one may use Eq. (2.56) to determine the corresponding characteristic exponents. If a spacecraft's state is perturbed at time $t = t_0$ from its nominal state by a perturbation with magnitude $\delta(t_0)$ along the eigendirection corresponding to the characteristic exponent α, the perturbation magnitude grows over time by the following expression

$$\delta(t) = \delta(t_0)e^{\alpha(t-t_0)} \tag{2.57}$$

Given a random perturbation in the spacecraft's state, the spacecraft's deviation over time is dominated by the component of that deviation that exists in the most unstable eigendirection, namely, by the direction indicated by the unstable eigenvalue λ^U. The perturbation doubling time may be computed by identifying the time, $t = t_0 + \hat{\tau}$, when the spacecraft's perturbed state is twice as far from its nominal position compared to its perturbed state at time $t = t_0$. One can find the perturbation doubling time by solving for $\hat{\tau}$ in Eq. (2.58), derived as follows

$$\begin{aligned} \delta(t) &= \delta(t_0)e^{\alpha(t-t_0)} \\ 2\delta(t_0) &= \delta(t_0)e^{\alpha\hat{\tau}} \\ 2 &= e^{\alpha\hat{\tau}} \end{aligned} \tag{2.58}$$

where α is the characteristic exponent that corresponds to the unstable eigenvalue, λ^U, of the orbit's monodromy matrix. The value of α may be computed using the simple relationship $\alpha T = \ln \lambda^U$, derived from Eq. (2.56). Hence, the time duration $\hat{\tau}$ may be computed using the expression

$$\hat{\tau} = \frac{\ln 2}{\ln \lambda^U} T \tag{2.59}$$

2.6.9 Examples of Practical Three-Body Orbits

The three-body problem contains a wide variety of interesting and potentially useful periodic and quasiperiodic orbits. Numerous authors have catalogued families of orbits and a brief history of these efforts is given in Section 2.6.2.2. This section illustrates several example families of three-body orbits, all of which appear frequently in the literature, and often in spacecraft mission proposals.

2.6.9.1 Lyapunov Orbits Lyapunov orbits were introduced in Section 2.6.2.2; they are two-dimensional periodic solutions to the circular restricted three-body problem. Lyapunov orbits exist about all three of the collinear Lagrange points, as illustrated in Fig. 2-22. The LL_1 and LL_2 families include orbits with orbital periods between two and four weeks—closer to two weeks for orbits closer to the Lagrange point; the LL_3 family includes orbits with orbital periods of approximately four weeks [140]. These orbits are all unstable.

2.6.9.2 Distant Prograde Orbits Periodic three-body orbits certainly exist about the Earth and the Moon as well as the Lagrange points. Figure 2-23 illustrates the family of planar distant prograde orbits and shows how that family of orbits fits in between the family of L_1 and L_2 Lyapunov orbits. A spacecraft only needs to adjust its state slightly to transfer from a Lyapunov orbit to a distant prograde orbit and vice versa. This is explored in Section 2.6.11.3. Most distant prograde orbits are unstable; their orbital periods vary from two weeks to four weeks, much like the Lyapunov orbits.

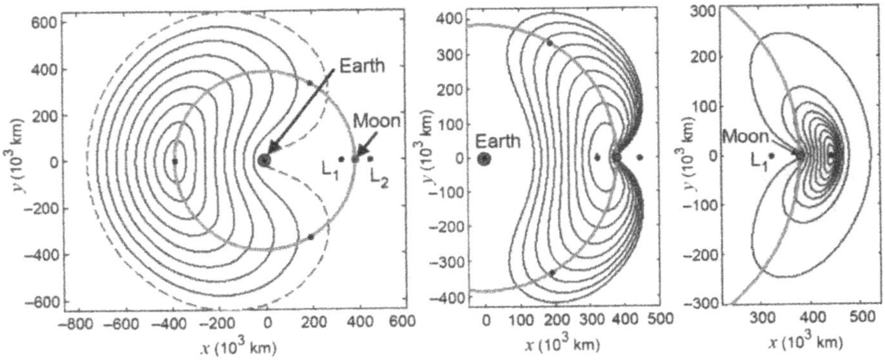

Figure 2-22 Example orbits in the families of Lyapunov orbits about the Earth–Moon L_3 point (left), L_1 point (middle), and L_2 point (right), viewed in the Earth–Moon rotating frame from above [46].

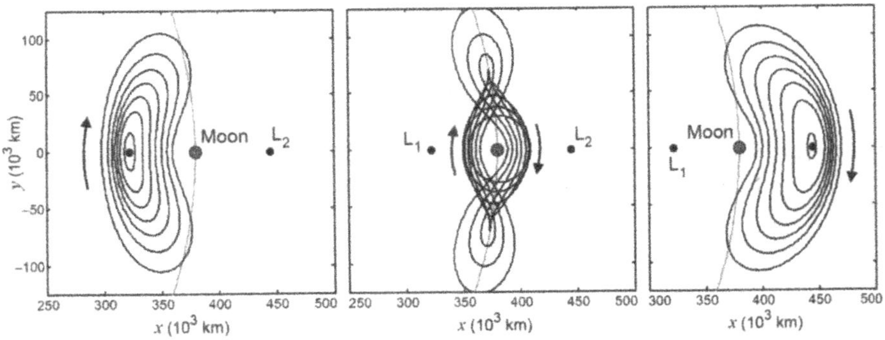

Figure 2-23 Example orbits in the families of Lyapunov orbits about the Earth–Moon L_1 point (left), distant prograde orbits about the Moon (middle), and Lyapunov orbits about the Earth–Moon L_2 point (right), viewed in the Earth–Moon rotating frame from above [46].

2.6.9.3 Distant Retrograde Orbits Distant retrograde orbits (DROs) are periodic three-body orbits that exist about the smaller primary, for example, the Moon in the Earth–Moon system, such that a spacecraft revolves about the body in a retrograde fashion. They are commonly found in the literature and in proposed spacecraft missions because they are frequently stable. They behave just like a normal two-body orbit, but occur in resonance with the motion of the three-body system. Figure 2-24 illustrates several examples of Earth–Moon DROs of varying radii from the Moon.

2.6.9.4 Halo Orbits Halo orbits are very well-known three-dimensional periodic solutions [119, 121, 122] to the circular restricted three-body problem. Fig-

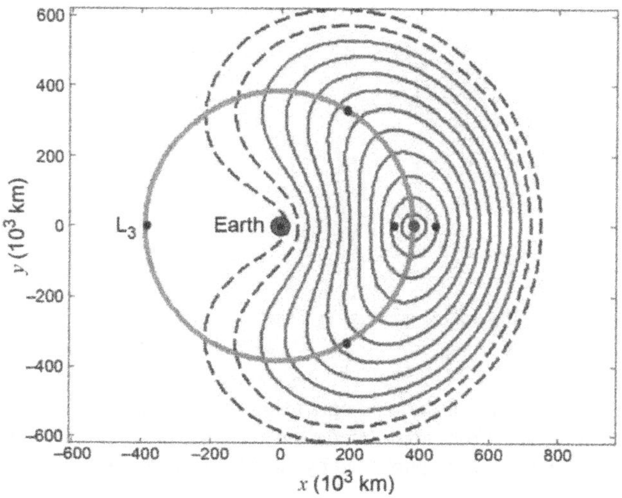

Figure 2-24 Example orbits in the family of distant retrograde orbits about the Moon, viewed in the Earth–Moon rotating frame from above.

ure 2-25 shows a plot of several example halo orbits about the lunar L_1 and L_2 points. Many authors have studied how to take advantage of halo orbits for practical missions to the Moon [5–7]. Halo orbits are of particular use for lunar communication and navigation satellites [11]: a satellite in a halo orbit has an unimpeded view of both the Earth and either the near-side of the Moon or the far-side of the Moon, for lunar L_1 and L_2 halo orbits, respectively. Furthermore, a satellite may be placed in a halo orbit such that its view of the Sun is also never impeded, simplifying the satellite's power and thermal systems.

Since the force field in the CRTBP is symmetric about the xy plane (see Section 2.5.1), and since halo orbits are assymetric about this plane, each halo orbit solution to the CRTBP comes in a symmetric pair with a northern and a southern variety [121]. As one can see in Fig. 2-25, a satellite in a southern orbit spends more than half of its time below the Moon's orbital plane, which gives that satellite benefits for communicating with objects in the southern hemisphere of the Moon.

It is convenient to specify a halo orbit by its z-axis amplitude, A_z, since one may formulate an analytical approximation to a halo orbit using that parameter as an input [106, 123, 124]. Other studies have specified a halo orbit using its Jacobi constant or its x_0-value, namely, the x-value of the location where the orbit has a y-position of 0 km and a positive y-velocity in the synodic reference frame [46, 122]. Figure 2-26 shows several northern LL_1 and LL_2 halo orbits from the side in the synodic frame to illustrate the relationship between a halo orbit's shape and its z-axis amplitude. Figure 2-27 shows the relationship between a halo orbit's z-axis amplitude and its period for reference.

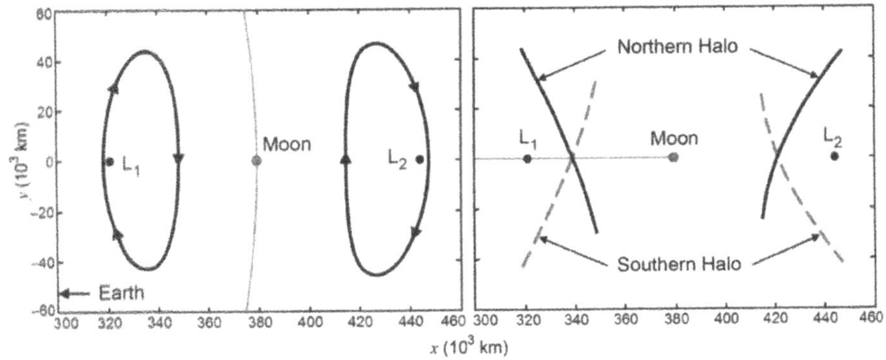

Figure 2-25 An illustration of four example halo orbits about the lunar L_1 and L_2 points. The halo orbits are viewed from above (left) and from the side (right) in the Earth–Moon synodic reference frame [47] (first published by the American Astronautical Society).

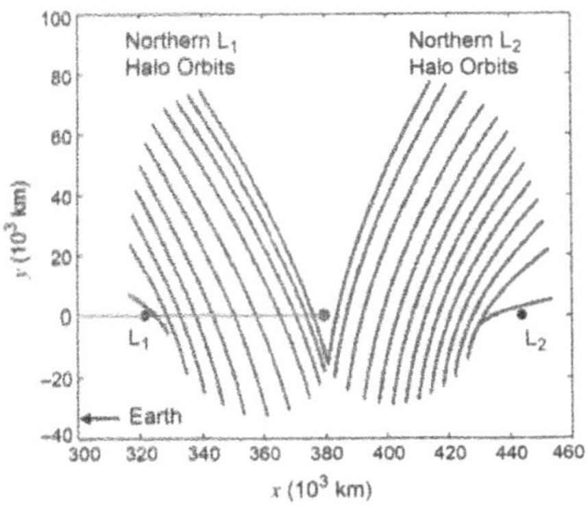

Figure 2-26 Northern Earth–Moon halo orbits [146] (Copyright © 2008 by American Astronautical Society Publications Office, San Diego, California [Web Site: http://www.univelt.com], all rights reserved; reprinted with permission of the AAS).

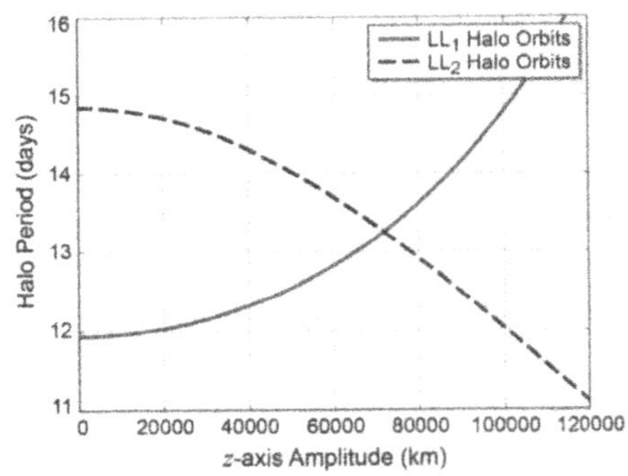

Figure 2-27 Earth–Moon halo orbit periods [44] (Copyright © 2009 by American Astronautical Society Publications Office, all rights reserved, reprinted with permission of the AAS).

2.6.9.5 Vertical Lyapunov Orbits Another family of libration orbits that exist about each of the collinear Lagrange points is the family of vertical Lyapunov orbits, also known as *vertical* orbits for short. Vertical orbits oscillate out of the xy plane, piercing the plane at the Lagrange point itself. They are symmetric orbits, traversing the same route above the plane as below it. Figure 2-28 provides several views of example orbits in the family of LL_1 vertical Lyapunov orbits.

2.6.9.6 Resonant Orbits Although there are numerous other interesting families of periodic orbits in the three-body system, the last type of orbit that will be

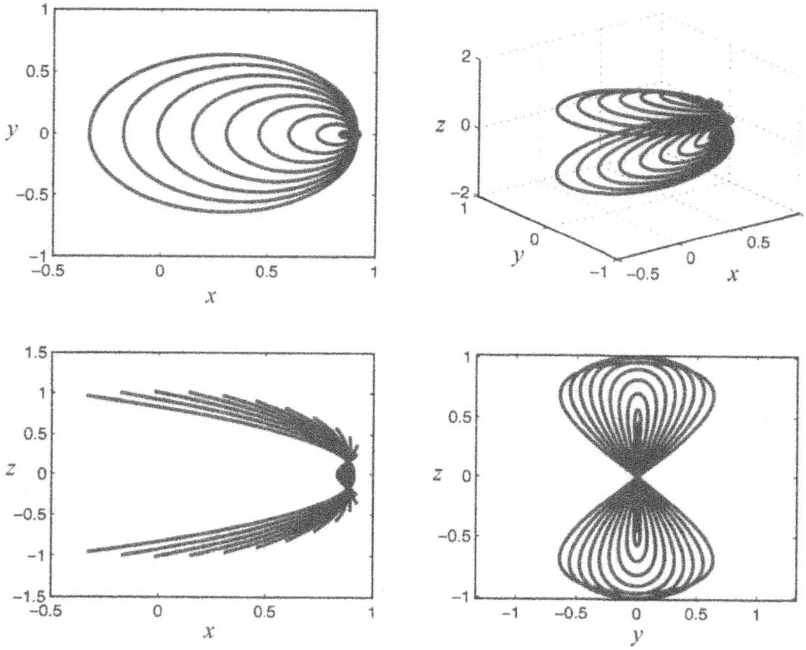

Figure 2-28 Four perspectives of example orbits within the family of L_1 vertical Lyapunov orbits computed in the Earth–Moon CRTBP.

described here is the resonant orbit. Resonant orbits in the Earth–Moon three-body problem are essentially two-body orbits about the Earth that are in resonance with the Moon, and which have been significantly perturbed by the Moon. As one may expect, there are different families of resonant orbits for each resonant period, namely, 3:1, 3:2, 5:1, 5:2, and so forth, where an $m{:}n$ resonant orbit is one where the spacecraft traverses the resonant orbit n times while the primaries orbit their barycenter m times. Figure 2-29 illustrates four families of resonant orbits in the Earth–Moon system, shown in the synodic reference frame.

The resonant orbits shown in Fig. 2-29 are particularly unstable as they pass by the Moon, but they are generally stable elsewhere. It is possible to transition a spacecraft off of one three-body orbit, such as a Lyapunov orbit, and onto a resonant orbit for very little fuel, if the transition is performed near the Moon. A spacecraft that arrives onto a resonant orbit may then sit in it without requiring any significant station-keeping fuel, until the spacecraft returns to the Moon. In that way, resonant orbits may play a useful role as a staging orbit, quarantine orbit, or a destination for a spacecraft to remain to avoid performing station-keeping maneuvers. One may also select how much time should pass between lunar swingbys, based on the resonance; for instance, a spacecraft traversing a 7:3 resonant orbit will spend far longer between lunar swingbys than a spacecraft traversing a 3:2 resonant orbit.

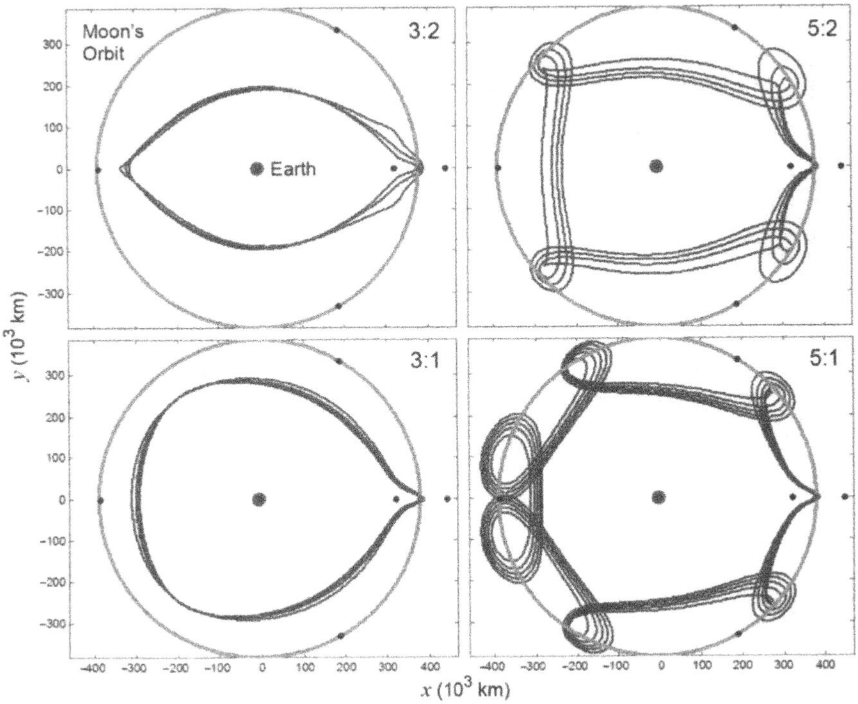

Figure 2-29 Four example families of resonant orbits in the Earth–Moon system, viewed from above in the Earth–Moon rotating frame.

2.6.10 Invariant Manifolds

The dynamics in the circular restricted three-body system permit the existence of five fixed points (Section 2.6.2.1) and numerous periodic orbits (Section 2.6.6.1). The three collinear libration points and many of the periodic orbit solutions in the Earth–Moon three-body system are unstable (Section 2.6.8). An unstable orbit has at least one stable and one unstable eigenvalue with corresponding eigenvectors. A spacecraft traveling along an unstable orbit that experiences a perturbation even slightly in the unstable direction will exponentially fall away from its nominal position on that orbit, tracing out a smooth trajectory away from the orbit. In a similar sense, a spacecraft that has the right initial conditions will follow a smooth trajectory that exponentially approaches an unstable orbit and eventually arrives on that orbit from the orbit's stable direction. These two trajectories describe what is known as an orbit's stable and unstable invariant manifolds.

An orbit's *unstable invariant manifold* (W^U) contains the set of all trajectories that a spacecraft may take if it was perturbed anywhere on that orbit in the direction of the orbit's unstable eigenvector. Similarly, an orbit's *stable invariant manifold* (W^S) contains the set of all trajectories that a spacecraft may take to asymptotically arrive onto that orbit along the orbit's local stable eigenvector. Put another way, the orbit's stable invariant manifold is the set of all trajectories that a spacecraft may take backward through time after a perturbation in the direction of the orbit's stable eigenvector.

Mathematically, the invariant manifolds are defined as follows. First, the CRTBP may be defined as a vector field bound in $\mathbb{R}^6$. One and only one vector is bound to every point in the vector field. Thus, the integration of any point p in the vector field with respect to time generates only one trajectory. Let us define T_p as the trajectory generated by the point p. The α- and ω-limits are defined to be the set of points in $\mathbb{R}^6$ as T_p tends toward $-\infty$ and $+\infty$, respectively. The α- and ω-limits may include a single point, a periodic orbit, or, if T_p has no asymptotic behavior, they may include a large portion of the state space. The set of all points defining trajectories that have the same α-limit set is called the *unstable manifold* of that limit set. Similarly, the set of all points defining trajectories that have the same ω-limit set is called the *stable manifold* of that limit set.

2.6.10.1 *Invariant Manifolds of the Unstable Lagrange Points* The three collinear Lagrange points are unstable in both the Sun–Earth and Earth–Moon three-body systems; hence, they have associated invariant manifolds. Since the Lagrange points are single points in space, their invariant manifolds are one-dimensional structures. To produce them, one first computes the eigenvalues of the Jacobian of their states. If $\mathbf{X}$ is the state of one of the collinear Lagrange points, equal to $[x\ y\ z\ \dot{x}\ \dot{y}\ \dot{z}]^T$, then its Jacobian is equal to

$$J = \frac{\partial \dot{\mathbf{X}}}{\partial \mathbf{X}} = \begin{bmatrix} \frac{\partial \dot{x}}{\partial x} & \frac{\partial \dot{x}}{\partial y} & \cdots & \frac{\partial \dot{x}}{\partial \dot{z}} \\ \frac{\partial \dot{y}}{\partial x} & \frac{\partial \dot{y}}{\partial y} & \cdots & \frac{\partial \dot{y}}{\partial \dot{z}} \\ \vdots & \vdots & \ddots & \vdots \\ \frac{\partial \ddot{z}}{\partial x} & \frac{\partial \ddot{z}}{\partial y} & \cdots & \frac{\partial \ddot{z}}{\partial \dot{z}} \end{bmatrix} \tag{2.60}$$

After plugging in the equations of motion of the CRTBP given in Eqs. (2.1)–(2.3) in Section 2.5.1, Eq. (2.60) simplifies to

$$J = \begin{bmatrix} 0 & 0 & 0 & 1 & 0 & 0 \\ 0 & 0 & 0 & 0 & 1 & 0 \\ 0 & 0 & 0 & 0 & 0 & 1 \\ \frac{\partial \ddot{x}}{\partial x} & \frac{\partial \ddot{x}}{\partial y} & \frac{\partial \ddot{x}}{\partial z} & 0 & 2 & 0 \\ \frac{\partial \ddot{y}}{\partial x} & \frac{\partial \ddot{y}}{\partial y} & \frac{\partial \ddot{y}}{\partial z} & -2 & 0 & 0 \\ \frac{\partial \ddot{z}}{\partial x} & \frac{\partial \ddot{z}}{\partial y} & \frac{\partial \ddot{z}}{\partial z} & 0 & 0 & 0 \end{bmatrix} \tag{2.61}$$

It is apparent that the Jacobian is the same as the A-matrix given in Eq. (2.16).

The eigenvalues of the Jacobian for each of the three collinear Lagrange points include two pairs of imaginary numbers and one pair of real numbers. Tables 2-6 and 2-7 summarize the six eigenvalues for the Jacobian of each of the five Lagrange points for the Earth–Moon system and for the Sun–Earth system, respectively. The eigenvector corresponding to the larger real eigenvalue indicates the unstable direction: $\mathbf{v}^U$; the eigenvector corresponding to the other real eigenvalue indicates the stable direction: $\mathbf{v}^S$. The unstable manifold of the Lagrange point, W^U, may be mapped by propagating the state $\mathbf{X}^U$ forward in time, where $\mathbf{X}^U = \mathbf{X} \pm \epsilon \mathbf{v}^U$ and ϵ is some small perturbation. Similarly, the stable manifold, W^S, may be mapped by propagating the state $\mathbf{X}^S$ backward in time, where $\mathbf{X}^S = \mathbf{X} \pm \epsilon \mathbf{v}^S$.

The perturbation $\epsilon \mathbf{v}$ may be applied to the state $\mathbf{X}$ in either a positive or a negative sense, corresponding to two halves of each manifold. One perturbation will result in motion that departs the Lagrange point toward the smaller body (for example, toward the Moon in the Earth–Moon system), and one will result in motion that departs the Lagrange point away from the smaller body. It is conventional to refer to the half of the manifold that moves toward the smaller body as the *interior* manifold, since it remains in the interior of the smaller body's influence, at least for a short while, and

Table 2-6 A summary of the eigenvalues of the Jacobian of each Lagrange point in the Earth–Moon CRTBP.

Component	LL_1	LL_2	LL_3	LL_4	LL_5
λ_1	-2.932056	-2.158674	-0.177875	1i	1i
λ_2	2.932056	2.158674	0.177875	-1i	-1i
λ_3	2.334386i	1.862646i	1.01041991i	0.95450078i	0.95450078i
λ_4	-2.334386i	-1.862646i	-1.01041991i	-0.95450078i	-0.95450078i
λ_5	2.268831i	1.786176i	1.00533144i	0.29820842i	0.29820842i
λ_6	-2.268831i	-1.786176i	-1.00533144i	-0.29820842i	-0.29820842i

Table 2-7 A summary of the eigenvalues of the Jacobian of each Lagrange point in the Sun–Earth CRTBP.

Component	EL_1	EL_2	EL_3	EL_4	EL_5
λ_1	-2.532659	-2.484317	-0.002825	1i	1i
λ_2	2.532659	2.484317	0.002825	-1i	-1i
λ_3	2.0864535i	2.057014i	1.00000266i	0.99998974i	0.99998974i
λ_4	-2.0864535i	-2.057014i	-1.00000266i	-0.99998974i	-0.99998974i
λ_5	2.0152106i	1.985075i	1.00000133i	0.00453024i	0.00453024i
λ_6	-2.0152106i	-1.985075i	-1.00000133i	-0.00453024i	-0.00453024i

to refer to the half that moves away from the smaller body as the *exterior* manifold [37].

The process of analyzing and constructing the invariant manifolds of the unstable Lagrange points may be visualized by considering that each unstable Lagrange point is a dynamical saddle point, as illustrated by the plot shown in Fig. 2-30. One can see that a spacecraft's motion will follow the unstable manifold when propagated forward in time after a perturbation, and it will follow the point's stable manifold when propagated backward in time. Figure 2-30 also demonstrates how there are two halves of each manifold.

Figures 2-31–2-33 show plots of the stable and unstable manifolds of the first three Lagrange points in the Earth–Moon three-body system. The forbidden region is shown shaded in gray in each plot.

The eigenvalues of the Jacobian of the triangular Lagrange points include three imaginary pairs for the Sun–Earth and Earth–Moon three-body systems; hence, they do not have interesting associated invariant manifolds. A spacecraft following a trajectory near one of these Lagrange points will oscillate about the point. If the spacecraft is perturbed, its motion will change but it will not exponentially deviate from its nominal path.

2.6.10.2 Invariant Manifolds of Unstable Periodic Orbits Every unstable periodic orbit in the CRTBP has a set of invariant manifolds, much like the Lagrange points. The only substantial difference between the invariant manifolds of periodic

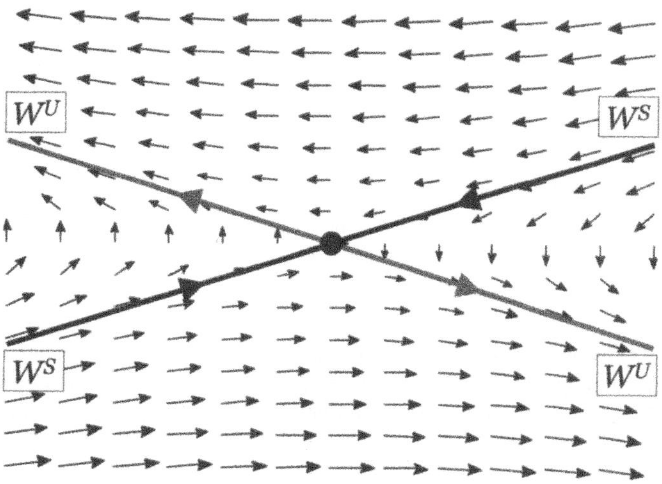

Figure 2-30 A dynamical saddle point, such as that of the unstable Lagrange points in the CRTBP, with a vector field shown that indicates the motion of a spacecraft near the point. There are two lines of stable (W^S) and unstable (W^U) manifolds of the saddle point (first published in Ref. [97]; reproduced with kind permission from Springer Science+Business Media B. V.).

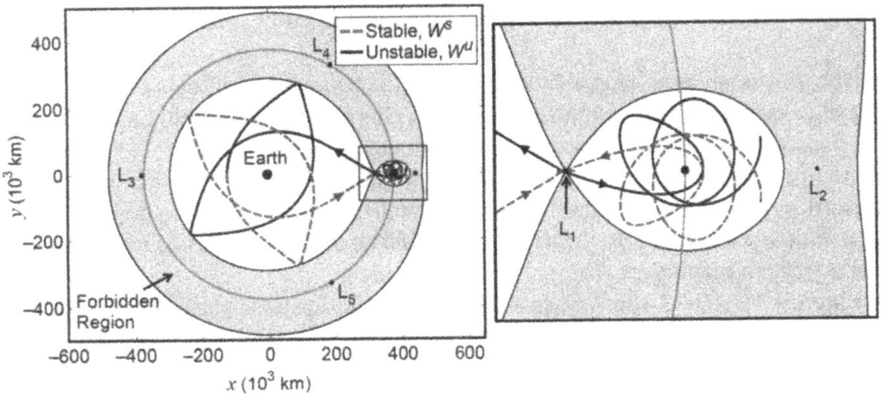

Figure 2-31 The stable and unstable invariant manifolds of the first Lagrange point of the Earth–Moon three-body system. *See inset at right for expanded view of the lunar vicinity* (First published in Ref. [97]; reproduced with kind permission from Springer Science+Business Media B. V.).

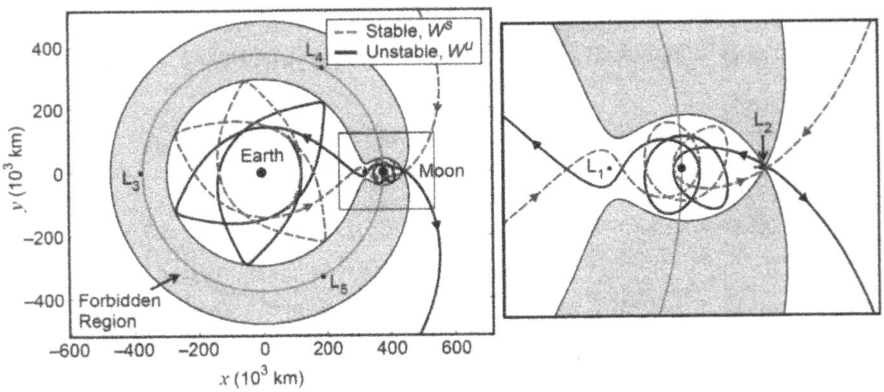

Figure 2-32 The stable and unstable invariant manifolds of the second Lagrange point of the Earth–Moon three-body system. *See inset at right for expanded view of the lunar vicinity* (First published in Ref. [97]; reproduced with kind permission from Springer Science+Business Media B. V.).

orbits and of the Lagrange points is that an additional dimension is added when considering periodic orbits: periodic orbits are one-dimensional structures where the Lagrange points are zero-dimensional structures. Consequently, the invariant manifolds of unstable periodic orbits are two-dimensional structures. They are constructed of a set of trajectories, where each trajectory corresponds to a point along

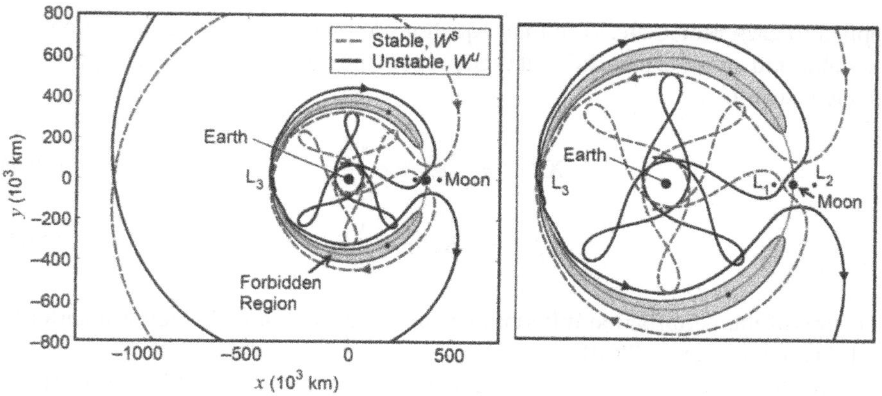

Figure 2-33 The stable and unstable invariant manifolds of the third Lagrange point of the Earth–Moon three-body system. *See inset at right for an expanded view in the vicinity of Earth* (First published in Ref. [97]; reproduced with kind permission from Springer Science+Business Media B. V.).

the periodic orbit. The set of trajectories wraps about itself, forming a topological tube. This is further explained below.

To produce the invariant manifolds of an unstable periodic orbit, one requires information about the local stability characteristics of each point along the orbit. In theory, one may evaluate the eigenvalues and eigenvectors of the Jacobian at each and every state along the orbit, and use that information to produce the orbit's invariant manifold. However, evaluating so many eigenvalues requires a great deal of computation. A more efficient manner of producing the invariant manifolds uses the eigenvalues and eigenvectors of the monodromy matrix [147, 148].

Since the monodromy matrix is produced by propagating the state transition matrix all the way around the orbit, from time $t = t_0$ to time $t = t_0 + T$, it contains information about the stability of the entire orbit. To determine the stable and unstable directions at each point along the orbit, one only has to propagate the stable and unstable eigenvectors of the monodromy matrix about the orbit using the state transition matrix. That is, the stable and unstable vectors at time t_i about the orbit, $\mathbf{v}_i^S$ and $\mathbf{v}_i^U$, respectively, may be determined using the stable and unstable eigenvectors of the monodromy matrix, $\mathbf{v}^S$ and $\mathbf{v}^U$, respectively, using the following equations

$$\mathbf{v}_i^S = \Phi(t_i, t_0)\mathbf{v}^S \tag{2.62}$$
$$\mathbf{v}_i^U = \Phi(t_i, t_0)\mathbf{v}^U \tag{2.63}$$

A small perturbation, ϵ, is then applied to the state of the orbit at that time, $\mathbf{X}_i$, and the result is propagated in time. Since the state transition matrix grows exponentially along an unstable orbit, the magnitudes of the vectors $\mathbf{v}_i^S$ and $\mathbf{v}_i^U$ grow along the orbit. It is therefore important to normalize the vectors so that a consistent perturbation is applied to each orbit state. The final equations to produce the initial conditions for the stable and unstable manifolds at time t_i about the orbit, $\mathbf{X}_i^S$ and $\mathbf{X}_i^U$, respectively, are then equal to

$$\mathbf{X}_i^S = \mathbf{X}_i \pm \epsilon \frac{\mathbf{v}_i^S}{|\mathbf{v}_i^S|} \tag{2.64}$$

$$\mathbf{X}_i^U = \mathbf{X}_i \pm \epsilon \frac{\mathbf{v}_i^U}{|\mathbf{v}_i^U|} \tag{2.65}$$

The sign of the perturbation differentiates between interior and exterior manifolds, as discussed in Section 2.6.10.1.

Some discussion should be provided regarding the magnitude of the perturbations applied to the state to produce the manifolds. The theoretical invariant manifolds of the orbit include the set of all trajectories that asymptotically approach the orbit as time goes either forward or backward. In fact, they never truly arrive on the orbit in finite time, but just come arbitrarily close to the orbit. To map them, one approximates the manifolds by perturbing a state slightly off of the orbit and then propagating that state in time. The smaller the perturbation, the closer the approximation comes to mapping the true manifolds; however, small perturbations require more time to depart from the orbit than larger perturbations. When designing practical missions, one is less interested in precisely mapping the invariant manifolds of the orbits, and generally more interested in computationally-swift algorithms. Additionally, the dynamics of the trajectories depend the greatest on the largest eigenvalues since motion in those directions grows exponentially faster than motion in any other direction. Hence, somewhat large perturbations may be used to map out the motion of the trajectories in the manifolds, for example, on the order of 100 km in the Earth–Moon system and 1000 km in the Sun–Earth system. In practice, the perturbation magnitudes are given in either units of position or units of velocity, but the perturbation is applied proportionally to all six components. A 100-km perturbation means that the magnitude of the perturbation applied to the position coordinates is equal to 100 km, and the resulting proportionality is used to apply the perturbation to the velocity components, that is

$$\epsilon = \frac{100 \text{ km}}{\sqrt{v_x^2 + v_y^2 + v_z^2}} \tag{2.66}$$

The structure of the manifolds of an orbit greatly depends on the stability characteristics of each portion of the orbit. Orbits such as libration orbits are fairly uniformly unstable; that is, the local Lyapunov exponent does not vary much along the orbit (Anderson, among others, provides a detailed exploration about the local Lyapunov exponent of libration orbits [149]). Consequently, their manifolds are fairly smooth as they extend from the orbit. Various other orbits are unstable due to a close flyby of one of the primary bodies. The local stability of these orbits changes drastically, becoming very unstable as the orbit approaches one of the massive bodies. Hence, their manifolds spread out quickly near the body and remain fairly close together elsewhere.

Figure 2-34 shows the stable and unstable manifolds of a two-dimensional Lyapunov orbit about the Earth–Moon L_2 point. One can see that the manifolds are smooth and form a tube-like structure. They remain well-defined until they encounter the Moon, at which time they spread out very rapidly, and the tube-like structure becomes less obvious. One can also see that the stable and unstable manifolds are

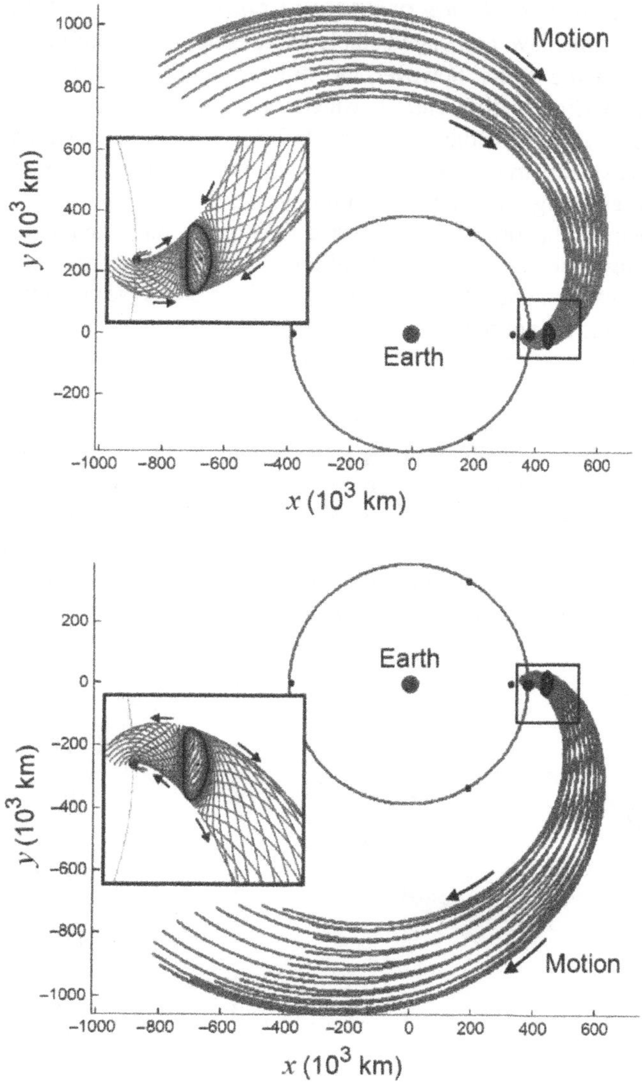

Figure 2-34 The stable (left) and unstable (right) manifolds of a Lyapunov orbit about the Earth–Moon L_2 point.

symmetric about the x-axis, due to the symmetry in the CRTBP. For comparison, Fig. 2-35 shows the stable manifold of a resonant flyby orbit, which shows how the structure of the manifolds depends on the local stability of the orbit. One notices that the trajectories in the manifold diverge quickly near the Moon but remain near the orbit elsewhere. The unstable manifold is not shown but is symmetric to the stable manifold.

2.6.10.3 Invariant Manifolds of Unstable Quasiperiodic Orbits Unstable quasi-periodic orbits have associated stable and unstable invariant manifolds, much like unstable periodic orbits; however, the structure and the procedures required to produce them are slightly different. Quasiperiodic Lissajous and quasi-halo orbits in the CRTBP are two-dimensional structures [125]. Hence, their invariant manifolds are three-dimensional structures. The additional dimension adds benefits as well as complexity when using them in practical mission designs.

Since quasiperiodic orbits never retrace their path, one cannot produce them entirely, although one can use a variety of numerical tools to represent them and to produce desirable segments of them [150]. Since these orbits are not periodic, they do not have associated monodromy matrices. Hence, one cannot use the same simplified procedures to produce their invariant manifolds as those procedures discussed in Section 2.6.10.2.

To produce a quasiperiodic orbit's invariant manifolds, one can always compute the eigenvectors of the Jacobian of the states at each point along sample segments of the orbit, and follow the same procedures as given in the previous sections. However, that procedure is numerically intensive and slow. Alternatively, to reduce the computational load, one may approximate the manifolds by producing an analog to the monodromy matrix. One may propagate the state transition matrix from one $y = 0$ plane crossing in the synodic frame to the next (or to any later crossing) and

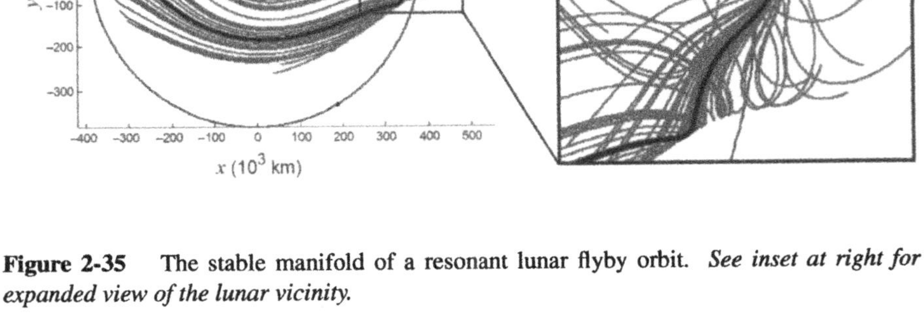

Figure 2-35 The stable manifold of a resonant lunar flyby orbit. *See inset at right for expanded view of the lunar vicinity.*

use the resulting matrix as a pseudo-monodromy matrix. When one propagates this matrix's stable and unstable eigenvectors about the orbit segment, and then follows the process outlined in Section 2.6.10.2, one produces approximations of the invariant manifolds of the quasiperiodic orbit. These approximations are often good enough to be used for preliminary spacecraft mission design, such as that used for the *Genesis* spacecraft mission [137].

2.6.11 Orbit Transfers

Dynamical systems theory provides the tools needed to systematically produce transfers to/from unstable orbits in the CRTBP. This section discusses several example orbit transfers as demonstrations of the application of dynamical systems theory. Section 2.6.11.1 discusses the construction of a transfer from the Earth to a halo orbit about the Sun–Earth L_2 point, and that transfer is very similar to that used by the *WMAP* mission [70]. Section 2.6.11.3 discusses the construction of a chain of periodic orbits in the CRTBP, which is relevant to missions like *Genesis* [71, 72] and *Wind* [63]. That is, trajectories are constructed that transfer a spacecraft back and forth between several periodic orbits in the CRTBP. These examples demonstrate the procedures that may be followed to construct any type of orbit transfer in the CRTBP using dynamical systems theory.

2.6.11.1 Surface to Orbit Transfers Several missions (including *WMAP*, *Herschel*, and *Planck*) have demonstrated the benefits of operating in a libration orbit about the Sun–Earth L_2 point; many other missions have been proposed to operate in similar orbits, including the *James Webb Space Telescope* and the *Terrestrial Planet Finder*. In this section, we demonstrate how to construct a ballistic transfer from the Earth to a halo orbit about EL_2, a transfer that might prove to be very useful for missions such as these proposed missions. The transfers produced here do not require any orbit insertion maneuvers; after their LEO departures, each transfer is thereafter entirely free of any deterministic maneuvers. The process used here may be generalized to compute a transfer from the surface of the secondary body in most three-body systems into many unstable three-body periodic orbits, or vice-versa.

We first consider the family of halo orbits about the EL_2 point, illustrated in Section 2.6.9.4. The family begins as a bifurcation of the family of planar Lyapunov orbits about EL_2. The orbits in the family gradually move farther out of the plane until they eventually make close approaches with the Earth. Example orbits in the family of northern EL_2 halo orbits are shown in Fig. 2-36.

We next consider a single unstable halo orbit and produce its stable invariant manifold. This manifold includes all the trajectories that a spacecraft may take to arrive onto the orbit. A plot of the example halo orbit and its stable manifold is shown in Fig. 2-37. The trajectories shown in blue have a perigee altitude below 500 km. The halo orbit chosen here has a Jacobi constant equal to approximately 3.00077207. The CRTBP is a good model of the real Solar System for trajectories propagated for a reasonably short amount of time, namely, for one orbital period of the two primary masses about their barycenter, or about a year in the Sun–Earth system. Beyond

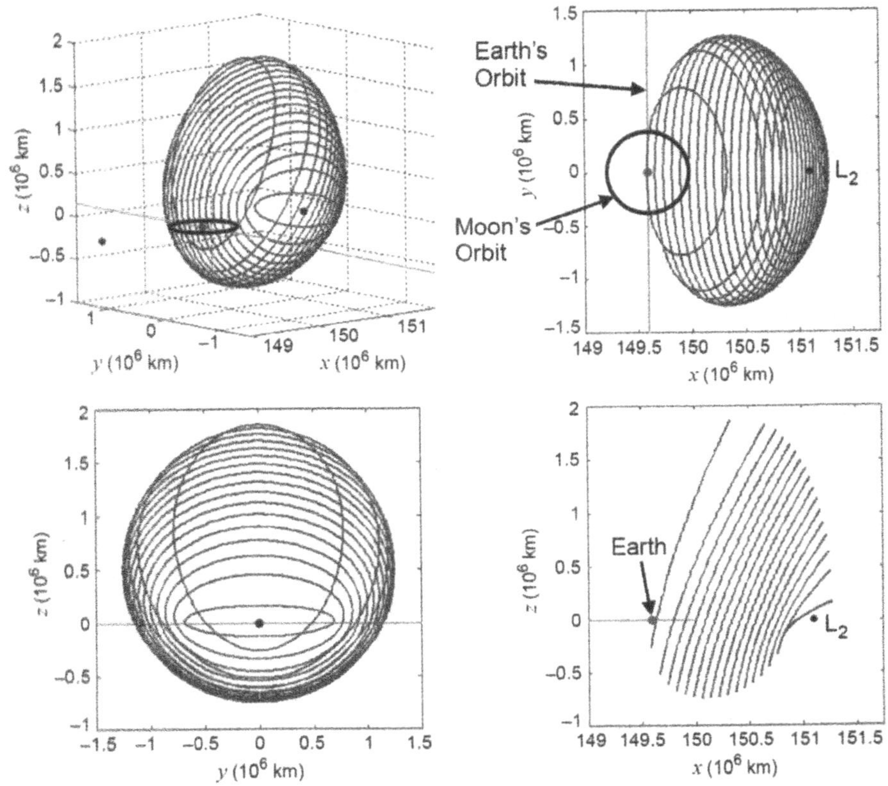

Figure 2-36 Example halo orbits in the family of northern halo orbits about the Sun–Earth L_2 point. The orbits are shown from four perspectives.

that, the accumulation of errors due to perturbations in the real Solar System causes the CRTBP approximation to break down. The trajectories shown in Fig. 2-37 have been propagated for at most 365 days—they are only plotted in the figure until they cross the plane of the Earth for clarity. As the propagation time is increased, the trajectories may make additional close approaches to the Earth. In some cases the second or third perigee passes closer to the Earth than the first. These features will be explored below.

Each trajectory shown in Fig. 2-37 may be characterized using several parameters. The parameter τ, defined in Section 2.6.2.3, indicates the point where the trajectory arrives at the halo orbit. The closest approach of each trajectory with the Earth is identified to compute the perigee altitude and ecliptic inclination with respect to the Earth. These two parameters are useful because they indicate the altitude and inclination of a low Earth orbit that may be used as a staging orbit to transfer a spacecraft to the halo orbit. Figure 2-38 shows two plots: one of the perigee altitude and one of the corresponding ecliptic inclination as functions of τ, where the vertical

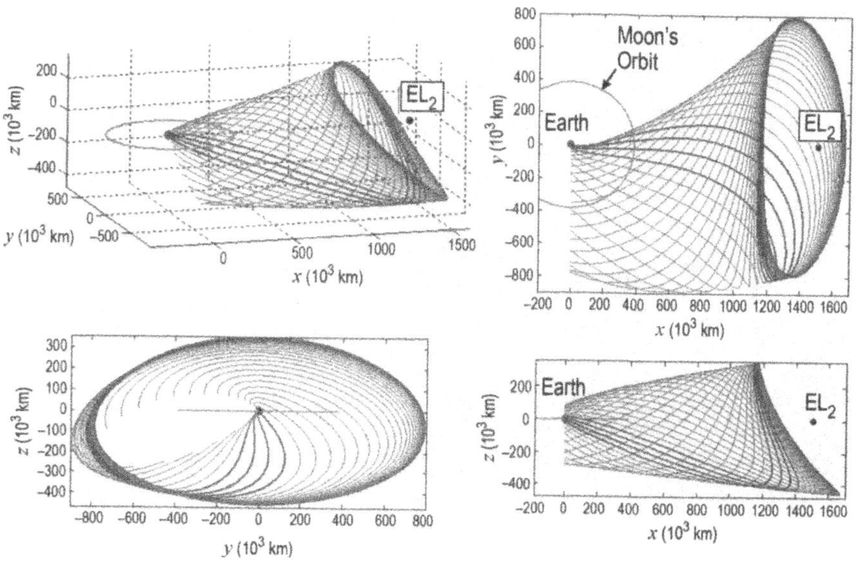

Figure 2-37 An example unstable halo orbit (red) about the Sun–Earth L_2 point and its stable invariant manifold (green/blue). Trajectories shown in blue intersect the Earth. *(See insert for color representation of this figure.)*

bars indicate the locations in the manifold that have perigee altitudes below 500 km. For example, one can see that the trajectory with a τ-value of 0.751 encounters a closest approach with the Earth with a perigee altitude of approximately 185 km and an ecliptic inclination of approximately 34.8 deg. Hence, a spacecraft in a circular low Earth parking orbit with an altitude of 185 km and an ecliptic inclination of 34.8 deg may perform a tangential ΔV to transfer onto the manifold; once on the manifold, the spacecraft ballistically follows it and asymptotically arrives on the halo orbit.

There are two statements in these results that need to be addressed. The first is that the inclination values displayed in Fig. 2-38 are the inclination values computed in the axes of the CRTBP: namely, in a plane that is very similar to the ecliptic. The equatorial inclination values of these perigee points depend on which date a spacecraft launches. Since the Earth's rotational axis is tilted by approximately 23.45 deg with respect to the ecliptic [97], many equatorial inclinations may be used to inject onto a desired trajectory, depending on the date. The second statement that should be addressed is that the results shown in Fig. 2-38 depend greatly on the perturbation magnitude, ϵ, described in Section 2.6.10. Implementing a different perturbation magnitude results in a change in the τ-values required to obtain a certain trajectory. For example, if ϵ were reduced, the trajectories modeling the orbit's manifold would spend more time asymptotically approaching/departing the orbit.

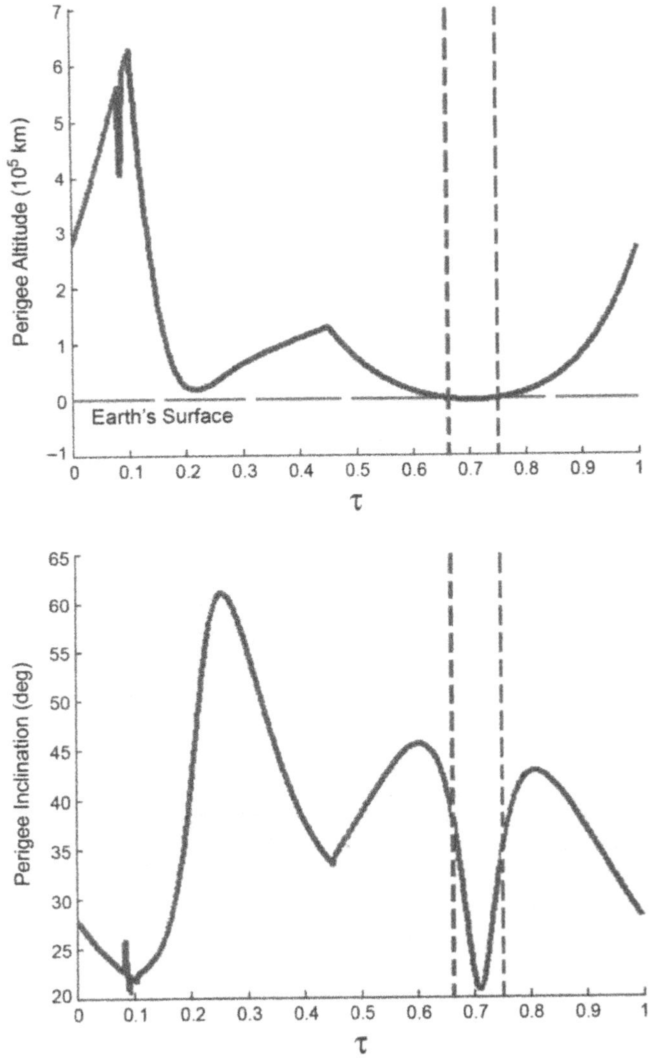

Figure 2-38 The perigee altitude (top) and corresponding ecliptic inclination (bottom) of the trajectories in the stable manifold shown in Fig. 2-37 as functions of τ.

Once the trajectories are sufficiently far from the orbit, their characteristics are nearly unchanged. The result is that the τ-value for a desired trajectory is strongly related to the value of ϵ. This has no significant effect for practical spacecraft mission designs; a spacecraft following a trajectory in the halo orbit's stable manifold will asymptotically approach the halo orbit—the value of ϵ is only used for modeling the stable manifold.

It is now of interest to identify how the manifolds change and how the plots shown in Fig. 2-38 change as the manifold's propagation time is varied. Figure 2-39 shows plots of the stable manifold of the same halo orbit propagated for successively longer amounts of time. One can see that the trajectories on the manifold spend some amount of time near the halo orbit (where, once again, the amount of time depends on the value of ϵ), and then depart. It may be seen that many of the trajectories in the manifold make closer approaches with the Earth after their first perigee. Figure 2-40 shows many plots of the closest approach each manifold makes with the Earth for varying amounts of propagation time. It is clear that the longer propagation times yield closer perigee passages.

The procedures given in this section may be repeated for each halo orbit in the entire family of halo orbits, and maps may be produced showing the range of perigee altitudes and the range of inclination values obtainable for each halo orbit. These are useful for identifying the approximate location of desirable trajectories in the real Solar System.

2.6.11.2 Homoclinic and Heteroclinic Connections Many unstable periodic orbits in the CRTBP contain *homoclinic* connections with themselves and/or *heteroclinic* connections with other unstable periodic orbits [71, 98, 151]. If a trajectory in an orbit's unstable manifold departs that orbit, traverses the three-body system for some time, and then later arrives back onto the same orbit, it makes what is known as a homoclinic connection with the host orbit [151]. This trajectory is contained within both the orbit's unstable and stable manifolds. McGehee proved the existence of homoclinic connections in both the interior and exterior regions of the three-body system [152]. In a similar sense, a different trajectory within the unstable manifold of one orbit may depart that orbit and eventually arrive onto a second orbit. The trajectory is thus contained within both the unstable manifold of the first orbit and the stable manifold of the second orbit. Such a trajectory forms what is known as a heteroclinic connection between the two unstable orbits [151].

In theory, heteroclinic connections asymptotically depart one orbit and asymptotically approach another orbit. In practice, the spacecraft is never truly on any host periodic orbit, but is instead within some small distance from the orbit. For the purpose of the discussions provided here, it is assumed that a spacecraft departs an orbit when its state is perturbed off of that orbit, and it arrives on the new orbit when it arrives at the state that corresponds to the perturbation that generated the stable manifold. For the case of orbit transfers in the Earth–Moon system, this means that the duration of an orbit transfer includes all time that the spacecraft is further than 100 km from a host orbit.

Many authors have explored homoclinic and heteroclinic transfers between three-body orbits as transport mechanisms for spacecraft and comets [98, 147, 149, 151, 153–160]. Using dynamical systems theory, Lo and Ross noted that the orbit of the comet Oterma appeared to shadow the invariant manifolds of libration orbits about the L_1 and L_2 points in the Sun–Jupiter three-body system [161]. Koon et al. later showed that the comet closely followed a homoclinic-heteroclinic chain [151]. Gómez et al. began exploring the numerical construction of orbits with prescribed

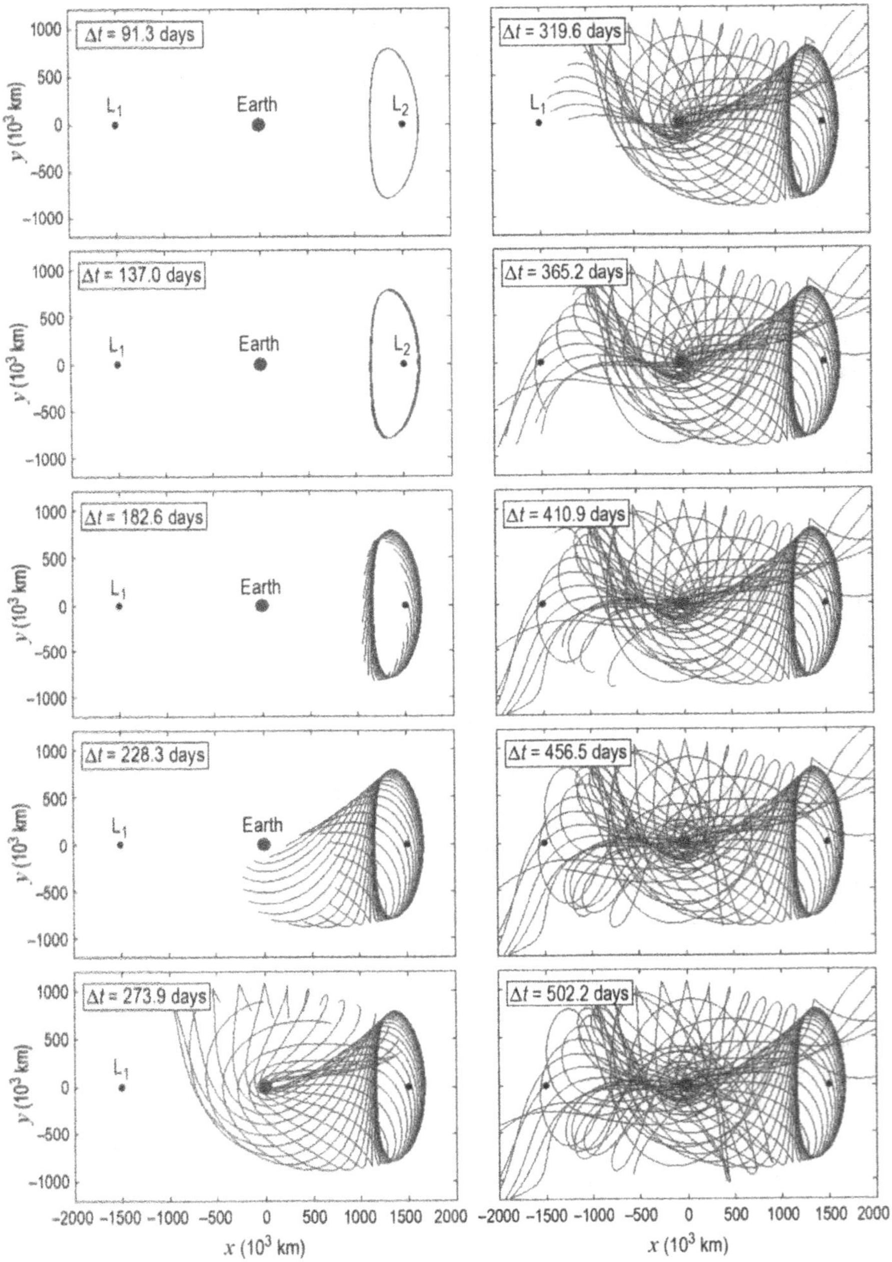

Figure 2-39 The stable manifold of a Sun–Earth L_2 halo orbit propagated for successively longer amounts of time. The duration of each propagation is indicated in each plot by the value Δt.

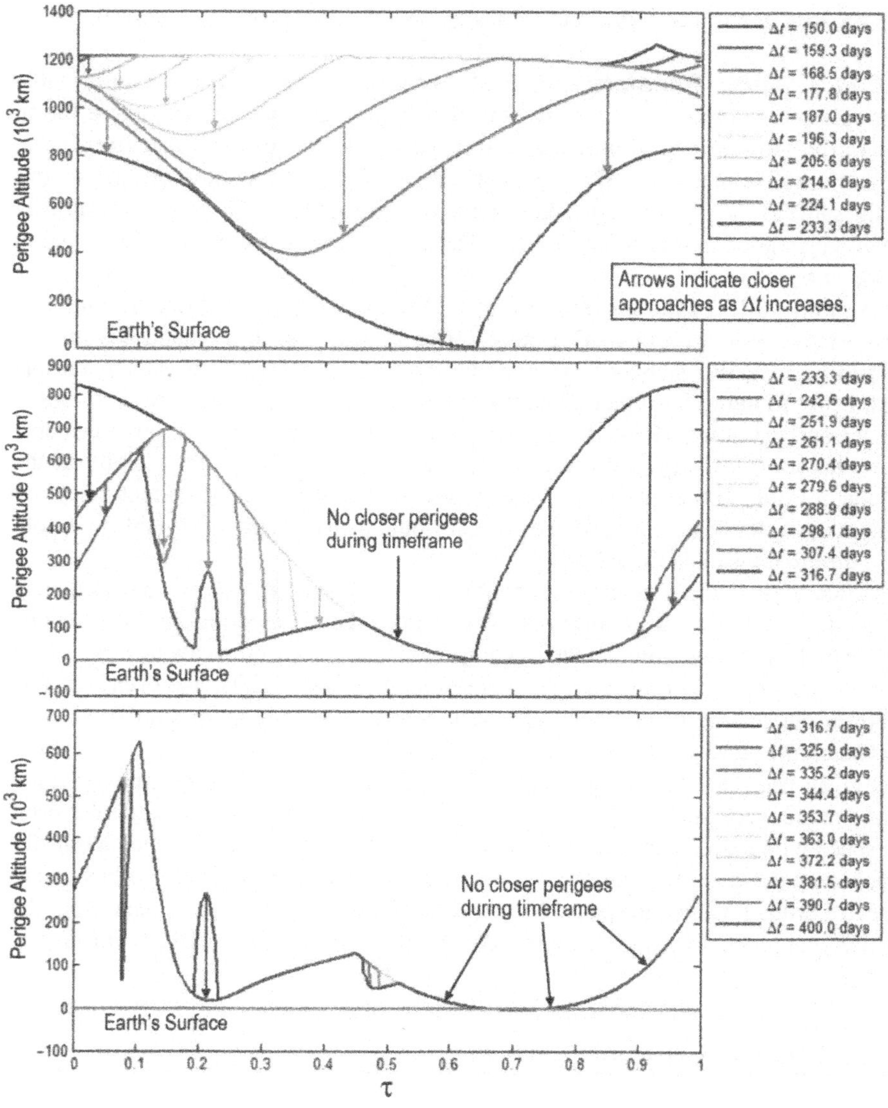

Figure 2-40 The altitude of closest approach of each trajectory in the stable manifolds shown in Fig. 2-39 with respect to the Earth. The propagation times of each manifold are shown in the legend. Longer propagation times yield closer perigee passages. (*See insert for color representation of this figure.*)

itineraries to describe the resonant transitions exhibited by the comet Oterma [98]. The material presented in Section 2.6.11.3 extends their work, applying a method for the construction of prescribed orbit transfers in the Earth–Moon system [101].

2.6.11.3 Orbit Transfers and Chains Once a spacecraft is on an unstable periodic orbit in the three-body system, then it may theoretically stay there for an arbitrarily long time, or it may depart that orbit by following any trajectory on that orbit's unstable manifold. The practical ΔV cost for a spacecraft to depart an orbit is the same as the cost of station-keeping to remain on that orbit: both are arbitrarily small given good navigational support. These considerations are further explored in Chapter 6.

Section 2.6.10.2 shows several examples of stable and unstable invariant manifolds of unstable periodic orbits. One may notice by studying these manifolds that by controlling exactly when the spacecraft departs from its periodic orbit, it may be able to transfer to numerous other locations in the state space, including, but not limited to, the surface of the Moon, any of the five lunar Lagrange points, another unstable periodic orbit in the system, or an escape trajectory away from the vicinity of the Moon or Earth. If the spacecraft were carefully navigated onto the correct trajectory within the unstable manifold of one orbit, it would then encounter the stable manifold of a different unstable three-body orbit.

After considering a spacecraft's options, several categories of orbit transfers may be identified. Table 2-8 summarizes a few characteristic categories of orbit transfers. In the table, a "stable orbit" includes conventional two-body orbits about either of the two primaries in the system, as well as stable three-body orbits, and even transfers to/from the surface of one of the primary bodies. The minimum number of ΔVs indicates the fewest number of maneuvers that may typically be used to perform the given transfer. There are many cases when a particular transfer might require more maneuvers, such as a transfer from the surface of a body to a particular orbit in space with a time constraint. There are also certain special cases when a transfer might require fewer maneuvers, such as a transfer between two stable orbits where the two orbits intersect in space. Nonetheless, Table 2-8 gives a good idea about the minimum number of required maneuvers for orbit transfers in several circumstances.

Table 2-8 A summary of several categories of orbit transfers in the CRTBP.

Orbit 1	Orbit 2	Constraints	Minimum Number of ΔVs
Stable Orbit	Stable Orbit	None	2
Stable Orbit	Unstable Orbit	None	1
Unstable Orbit	Stable Orbit	None	1
Unstable Orbit	Unstable Orbit	None	1
Any Orbit	Any Orbit	Transfer Time	2
Unstable Orbit	Unstable Orbit	Same Jacobi Constant	0

In this section, low-energy orbit transfers are introduced that may be useful for practical mission design, and which are useful background for the discussions of low-energy lunar transfers in later chapters. These orbit transfers are in the category represented by the last row of Table 2-8, transfers that are free of deterministic maneuvers.

Low-energy transfers between unstable orbits may be located in the CRTBP by analyzing Poincaré maps (Section 2.6.3) [156]. Suppose there are two unstable Lyapunov orbits in the Earth–Moon three-body system: one about LL_1 and the other about LL_2. Both of these orbits have a set of stable and unstable invariant manifolds. In the planar CRTBP, each point along a manifold may be characterized by a four-dimensional state $[x, y, \dot{x}, \dot{y}]$. If a surface of section is placed in $\mathbb{R}^4$ at some x-position, the resulting intersection is a surface in $\mathbb{R}^3$. If it is further specified that the two Lyapunov orbits have the same Jacobi constant, then each point along any trajectory within both orbits' manifolds will have the same Jacobi constant and the phase space of the problem is reduced to $\mathbb{R}^2$. The state at any intersection in the surface may only be reconstructed if the Poincaré map is one-sided, since the Jacobi constant has a sign ambiguity. The stable and unstable manifolds of both orbits appear as curves in the two-dimensional Poincaré map. Any intersection of these curves corresponds to a free transfer between the two orbits.

Figure 2-41 illustrates the process of identifying free transfers from a Lyapunov orbit about LL_1 to a Lyapunov orbit about LL_2. In this case, the value of the Jacobi constant of both orbits has been selected to be 3.13443929. A P_+ Poincaré map has been constructed, where the surface, Σ, has been placed at the x-coordinate of the Moon, namely, at a value of approximately 379,730 km with respect to the barycenter of the Earth–Moon system. The top of Fig. 2-41 illustrates the unstable and stable manifolds integrated to the first intersection with the surface of section. The intersection of both manifolds with the surface of section is shown on the bottom of Fig. 2-41. One can see that there are two intersections that correspond to the two free transfers indicated in the figure.

The simple illustration shown in Fig. 2-41 may be extended by propagating the manifolds longer and identifying intersections in the manifolds that correspond to longer, more complicated heteroclinic connections. The Poincaré map shown in Fig. 2-42 is produced by propagating the unstable manifold of the LL_1 Lyapunov orbit and the stable manifold of the LL_2 Lyapunov orbit for 60 days each. In addition, the map shown in Fig. 2-42 is a $P_\pm$ map, displaying all intersections of both manifolds with the surface of section. In this particular mapping, the majority of the points shown below the $y = 0$ line are members of the P_+ map (including the points shown in Fig. 2-41), the majority of the points shown above it are members of the P_- map, and all observed intersections of the two manifolds do indeed intersect, even accounting for the sign ambiguity of $\dot{x}$.

Figure 2-42 includes eight example orbit transfers to illustrate what sort of heteroclinic connections exist between these two libration orbits. Certain types of motion appear in more than one heteroclinic connection. For example, the trajectories labeled (1), (2), and (7) appear to graze a distant prograde orbit, whereas the trajectories labeled (1), (3), and (4) appear to traverse a figure-eight type orbit. The appearance of

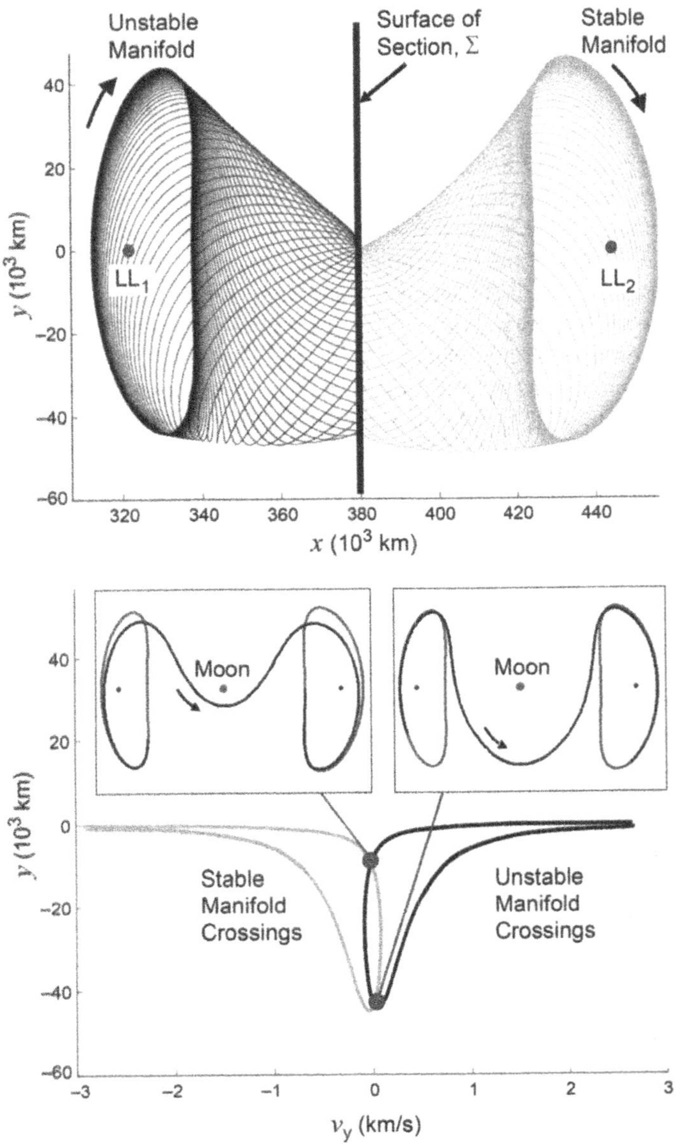

Figure 2-41 An illustration of the process of using a Poincaré map to identify free transfers between two Lyapunov orbits. Both orbits have a Jacobi constant of 3.13443929. Top: the unstable manifold of an LL_1 Lyapunov orbit and the stable manifold of an LL_2 Lyapunov integrated to the surface of section. Bottom: the corresponding P_+ Poincaré map and two free transfers [101] (*Acta Astronautica* by International Academy of Astronautics, reproduced with permission of Pergamon in the format reuse in a book/textbook via Copyright Clearance Center).

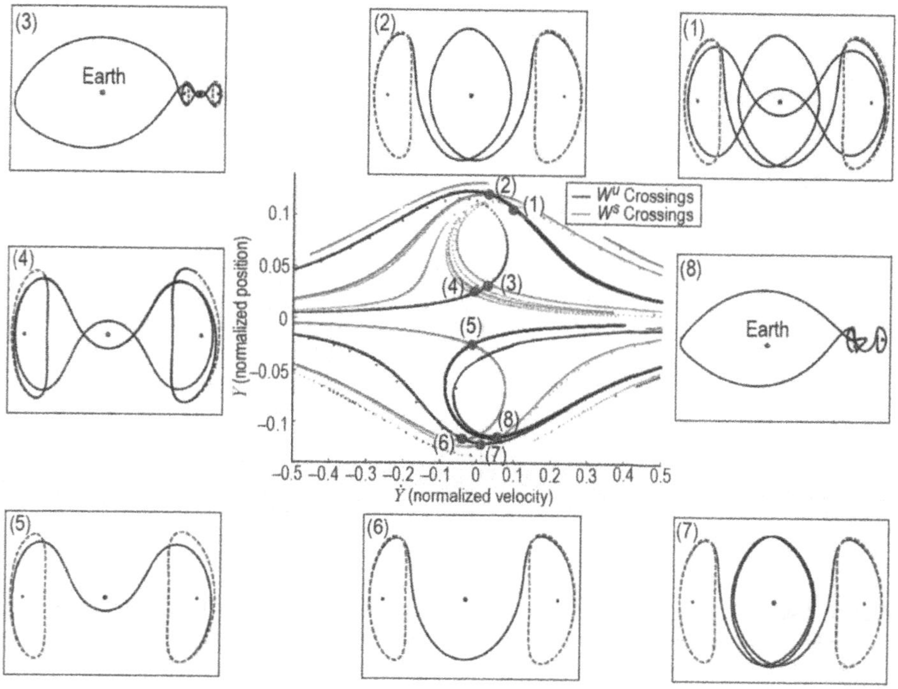

Figure 2-42 The P± Poincaré map produced from the same system and surface of section shown in Fig. 2-41, now with an extended manifold propagation duration of 60 days. The plots shown surrounding the Poincaré map illustrate several example free transfers that have been identified in the central map [101] (*Acta Astronautica* by International Academy of Astronautics, reproduced with permission of Pergamon in the format reuse in a book/textbook via Copyright Clearance Center).

such orbits in the Poincaré maps reinforces the idea that one may construct a specific chain of simple orbits to construct a complicated itinerary of orbit transfers.

The Poincaré map is a useful tool to identify what sorts of orbit transfers exist, but it does not immediately reveal the shape or geometry of the transfers. For instance, the transfer labeled (8) in Fig. 2-42 includes a lunar flyby, which may or may not be desirable. Section 2.6.12 introduces a method that may be used to construct a desirable sequence of orbit transfers after identifying that such orbit transfers exist.

Free transfers only exist in the CRTBP between two unstable orbits that have the same Jacobi constant. Figure 2-43 shows a plot of several families of three-body orbits in the Earth–Moon CRTBP, where the orbits' Jacobi constant values are plotted as functions of their x_0-values. The curves shaded in black correspond to unstable three-body orbits; the curves shaded in gray correspond to orbits that are neutrally stable [130]. The horizontal line indicates the Jacobi constant value used to produce

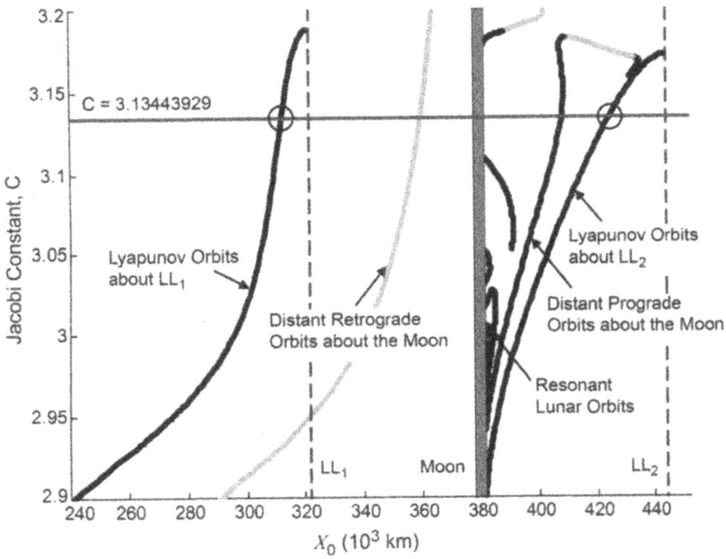

Figure 2-43 A plot of several families of three-body orbits in the Earth–Moon CRTBP, where the orbits' Jacobi constant values are plotted as functions of their x_0-values. The curves shaded in black correspond to unstable three-body orbits; the curves shaded in gray correspond to orbits that are neutrally stable [162] (Copyright ©2006 by American Astronautical Society Publications Office (web site: http://www.univelt.com), all rights reserved, reprinted with permission of the AAS).

the heteroclinic connections observed in Figs. 2-41 and 2-42. The figure verifies that the families of Lyapunov orbits about LL_1 and LL_2 both include unstable orbits at the same indicated Jacobi constant value, along with the family of distant prograde orbits, which helps to explain the appearance of such an orbit in the transfers labeled (1), (2), and (7) in Fig. 2-42.

2.6.12 Building Complex Orbit Chains

In the previous section, a technique was presented that may be used to identify the heteroclinic connections between two unstable periodic orbits. Previous papers have theorized using symbolic dynamics that if a heteroclinic connection exists, it is possible to find a trajectory that transfers back and forth arbitrarily between those orbits. Robinson provides a thorough review of the background of symbolic dynamics [163]. Canalias et al. [155], provide a methodology to search for a combination of homoclinic transfers that may be used to change the phase of a spacecraft traversing an unstable periodic orbit. In this section we study a practical method to construct a complex orbit chain given a desired sequence of homoclinic and/or heteroclinic transfers.

2.6.12.1 *Constructing a Complex Orbit Chain* One may describe a spacecraft's itinerary between simple periodic orbits in the CRTBP by considering its state at each x-axis crossing. A spacecraft traversing any simple periodic orbit pierces the x-axis twice: once with positive and once with negative values of $\dot{y}$. One may model a simplified orbit transfer by considering that the spacecraft departs the initial orbit at one x-axis crossing, is midway through the transfer at the next x-axis crossing, and completes the transfer at a later x-axis crossing. Using this conceptualization, one may construct a set of x-axis states to describe a given itinerary between two orbits. A set of eight states are summarized in Fig. 2-44 and Table 2-9 for transfers between an example LL_1 Lyapunov orbit and an example distant prograde orbit (DPO) that have the same Jacobi constant.

The states given in Fig. 2-44 and Table 2-9 have been collected from two sources. The states corresponding to the simple periodic orbits (A, B, E, and F) have been taken directly from those orbits; the algorithm described in Section 2.6.6.2 is well-suited to generate the states of a periodic orbit at their orthogonal x-axis crossings. The states that correspond to the orbit transfers (C, D, G, and H) have been taken from their heteroclinic connections identified using the Poincaré analysis described in Section 2.6.3. A theoretical heteroclinic connection between these orbits asymp-

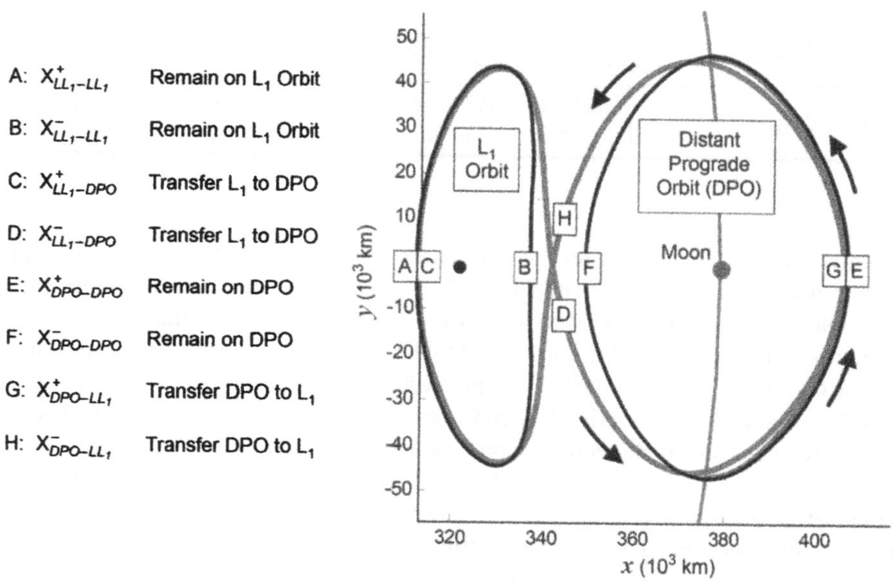

Figure 2-44 A summary of the states needed to produce complex itineraries between two orbits. In this case, the two orbits are a Lyapunov orbit about L_1 and a DPO about the Moon. The states "D" and "H" are on the x-axis, although the labels are offset [101] (*Acta Astronautica* by International Academy of Astronautics, reproduced with permission of Pergamon in the format reuse in a book/textbook via Copyright Clearance Center).

Table 2-9 The eight states shown in Fig. 2-44. The state coordinates are given in the Earth–Moon synodic reference frame, relative to the Earth–Moon barycenter, in both nondimensional normalized units and SI units [101] (*Acta Astronautica* by International Academy of Astronautics, reproduced with permission of Pergamon in the format reuse in a book/textbook via Copyright Clearance Center).

State		Units	x	y	$\dot{x}$	$\dot{y}$
A:	$\mathbf{X}^{+}_{LL_1-LL_1}$	Normalized	0.812255	0.0	0.0	0.248312
		SI (km, m/s)	312,230	0.0	0.0	254.418
B:	$\mathbf{X}^{-}_{LL_1-LL_1}$	Normalized	0.878585	0.0	0.0	−v0.281719
		SI (km, m/s)	337,728	0.0	0.0	−288.647
C:	$\mathbf{X}^{+}_{LL_1-DPO}$	Normalized	0.813049	0.0	0.0	0.247532
		SI (km, m/s)	312,536	0.0	0.0	253.618
D:	$\mathbf{X}^{-}_{LL_1-DPO}$	Normalized	0.890940	0.0	0.049050	−0.311179
		SI (km, m/s)	342,477	0.0	50.256	−318.830
E:	$\mathbf{X}^{+}_{DPO-DPO}$	Normalized	1.061692	0.0	0.0	0.403877
		SI (km, m/s)	408,115	0.0	0.0	413.809
F:	$\mathbf{X}^{-}_{DPO-DPO}$	Normalized	0.909845	0.0	0.0	−0.386264
		SI (km, m/s)	349,745	0.0	0.0	−395.762
G:	$\mathbf{X}^{+}_{DPO-LL_1}$	Normalized	1.056340	0.0	0.0	0.432104
		SI (km, m/s)	406,057	0.0	0.0	442.729
H:	$\mathbf{X}^{-}_{DPO-LL_1}$	Normalized	0.890940	0.0	−0.049050	−0.311179
		SI (km, m/s)	342,477	0.0	−50.256	−318.830

totically wraps off one orbit and onto the next as ϵ in Eq. (2.65) approaches 0; an infinite number of x-axis crossings precede the theoretical heteroclinic connection. The states D and H correspond to the x-axis crossings that are furthest from either host orbit. The states C and G correspond to the previous respective x-axis crossing. As one can see in Table 2-9, state C is approximately 306 km and 0.8 m/s away from state A, and state G is approximately 2058 km and 28.9 m/s away from state E. These state differences are small enough to proceed without difficulty.

The states summarized in Fig. 2-44 and Table 2-9 may be used to construct a sequence of states that represent any itinerary between the two given orbits. This sequence may then be converted into a series of patchpoints that may be inputted into a differential corrector in order to produce a continuous trajectory. For instance, the trajectory of a spacecraft in orbit about the LL_1 Lyapunov orbit may be represented by the sequence

$$\{\ldots, A, B, A, B, \ldots\}$$

A differential corrector may be used to convert this sequence into a continuous trajectory. If a mission designer wishes to transfer the spacecraft from the LL_1 orbit to the distant prograde orbit, the designer would construct the sequence

$$\{\ldots, A, B, A, B, C, D, E, F, E, F, \ldots\}$$

and input that sequence into the differential corrector. The differential correction process adjusts every state in the sequence to accommodate the slight differences between the states A and C to make the transfer continuous.

Table 2-10 provides two example sequences that may be used as inputs to a differential corrector in order to produce continuous trajectories with different itineraries. To demonstrate this process, the first sequence in Table 2-10 has been converted into patchpoints and processed by the multiple-shooting differential corrector described in Section 2.6.5.2. Table 2-11 displays the results of the differential correction process, comparing the states of the patchpoints before and after the process. One can see that the differential corrector adjusted each patchpoint away from the x-axis in order to produce a continuous trajectory, however, none of the patchpoints moved far. In this example, the differential corrector achieved a trajectory that met the requested continuity tolerances: the largest position and velocity discontinuities that were observed in any of the patchpoints along the final trajectory were less than 0.4 mm and 3.1×10^{-9} m/s, respectively.

2.6.12.2 Complex Periodic Orbits A complex periodic orbit may be constructed by repeating a given sequence of states *ad infinitum* and inputting that theoretical sequence into the differential corrector. For instance, the following sequence may be used to represent a periodic orbit that consists of two revolutions

Table 2-10 Two sequences that may be used as inputs to a differential corrector in order to produce continuous trajectories with different example itineraries. The letters correspond to the states summarized in Fig. 2-44 [101] (*Acta Astronautica* by International Academy of Astronautics, reproduced with permission of Pergamon in the format reuse in a book/textbook via Copyright Clearance Center).

Example 1		Example 2	
Sequence	Objective	Sequence	Objective
A, B	Traverse LL_1	A, B	Traverse LL_1
C, D	Transfer to DPO	C, D	Transfer to DPO
E, F	Traverse DPO (1)	G, H	Transfer to LL_1
E, F	Traverse DPO (2)	C, D	Transfer to DPO
G, H	Transfer to LL_1	E, F	Traverse DPO

Table 2-11 The results of a differential correction process that converted the first sequence presented in Table 2-10 into a continuous trajectory. The coordinates shown here are in the Earth–Moon synodic reference frame, relative to the Earth–Moon barycenter [101] (*Acta Astronautica* by International Academy of Astronautics, reproduced with permission of Pergamon in the format reuse in a book/textbook via Copyright Clearance Center).

	Initial State				Final State			
Patchpoint	x (km)	y (km)	$\dot{x}$ (m/s)	$\dot{y}$ (m/s)	x (km)	y (km)	$\dot{x}$ (m/s)	$\dot{y}$ (m/s)
A: $\mathbf{X}^+_{LL_1-LL_1}$	312,230	0.0	0.0	254.418	312,242	0.000	0.000	254.092
B: $\mathbf{X}^-_{LL_1-LL_1}$	337,728	0.0	0.0	−288.647	337,704	−0.003	0.034	−288.236
C: $\mathbf{X}^+_{LL_1-DPO}$	312,536	0.0	0.0	253.618	312,399	0.640	1.180	253.675
D: $\mathbf{X}^-_{LL_1-DPO}$	342,477	0.0	50.256	−318.830	342,553	12.782	51.463	−319.032
E: $\mathbf{X}^+_{DPO-DPO}$	408,115	0.0	0.0	413.809	407,079	19.004	6.862	427.761
F: $\mathbf{X}^-_{DPO-DPO}$	349,745	0.0	0.0	−395.762	349,595	−0.175	1.079	−393.771
E: $\mathbf{X}^+_{DPO-DPO}$	408,115	0.0	0.0	413.809	408,081	0.000	0.000	414.071
F: $\mathbf{X}^-_{DPO-DPO}$	349,745	0.0	0.0	−395.762	349,595	2.644	−1.094	−393.772
G: $\mathbf{X}^+_{DPO-LL_1}$	406,057	0.0	0.0	442.729	407,079	16.916	−7.171	427.755
H: $\mathbf{X}^-_{DPO-LL_1}$	342,477	0.0	−50.256	−318.830	342,557	−12.750	−51.431	−319.076
A: $\mathbf{X}^+_{LL_1-LL_1}$	312,230	0.0	0.0	254.418	312,555	0.000	0.000	254.418

about the LL_1 Lyapunov orbit, followed by one revolution about the distant prograde orbit, repeating itself indefinitely, where the orbits and letters are defined in Fig. 2-44

$$\{\ldots, \underbrace{A, B, A, B, C, D, G, H}, \underbrace{A, B, A, B, C, D, G, H}, \ldots\}$$

Figure 2-45 shows a plot of such a periodic orbit. One can see that a trajectory following a complex itinerary gets very close to one of its generating three-body orbits even with as few as two revolutions about the orbit.

Since each unstable three-body orbit exists in a family, where the characteristics of each orbit in the family vary continuously from one end of the family to the other, it is hypothesized that a complex periodic orbit also exists in a family. The family of any given periodic orbit is limited in extent to some range of parameters [46]. The extent of the family of complex orbits is also limited in extent, and it is hypothesized that the family may only extend through a range where each of its fundamental orbits and orbit transfers exists. Figure 2-46 shows several example complex periodic orbits that exist in the same family as the orbit shown in Fig. 2-45. Each of these orbits has a different Jacobi constant, but the same morphology.

2.6.12.3 Generalization The method demonstrated here has been illustrated by a very straightforward example, namely, the construction of orbit transfers between an LL_1 Lyapunov orbit, and a distant prograde orbit, two simple periodic three-body orbits. These orbits have been used because they are easily visualized and may be characterized using only a handful of states. Each state is placed at an x-axis crossing, although one can see in Table 2-11 that the states may be displaced during the differential correction process.

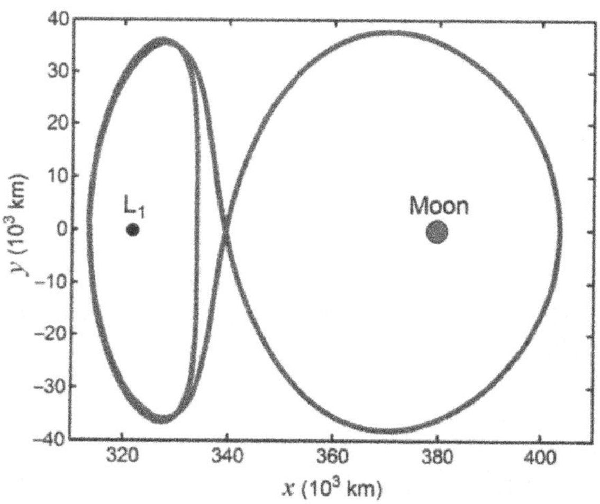

Figure 2-45 A complex periodic orbit that consists of two revolutions about the LL_1 Lyapunov orbit, followed by one revolution about the distant prograde orbit, repeating itself indefinitely. This orbit is viewed from above in the Earth–Moon synodic reference frame.

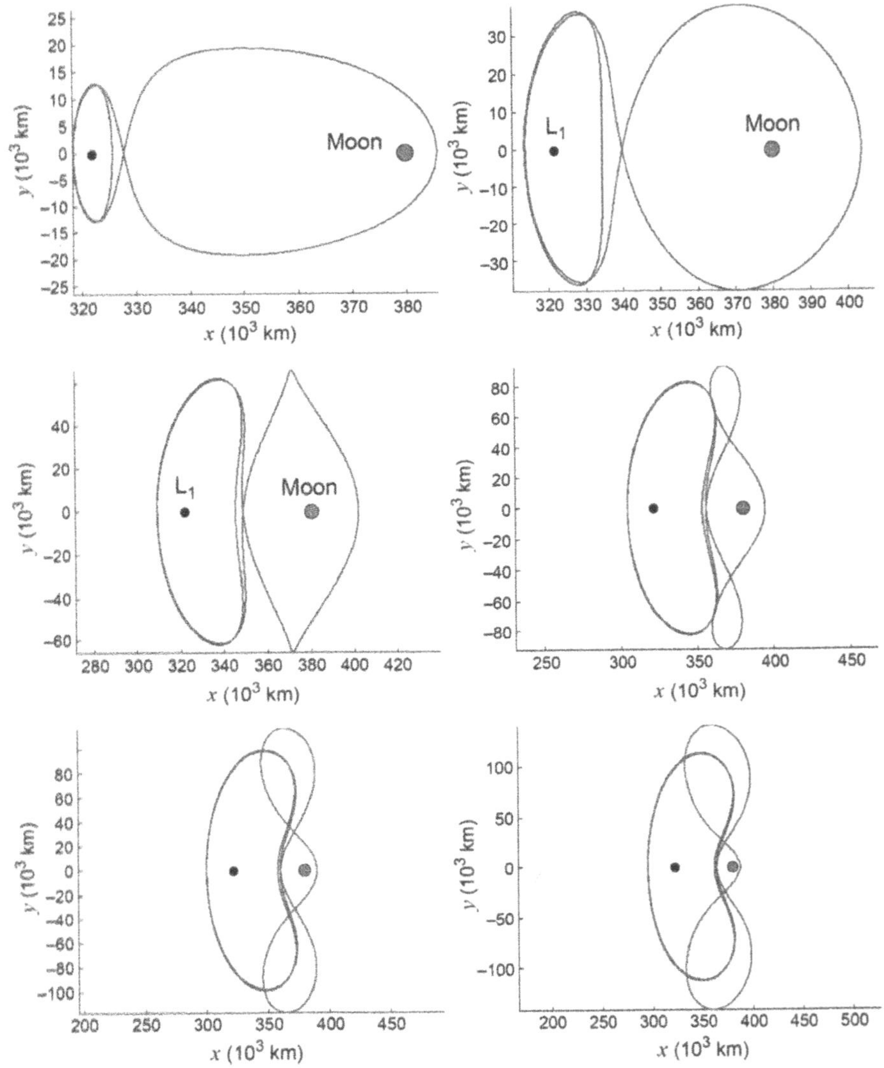

Figure 2-46 Several example complex periodic orbits that exist in the same family as the orbit shown in Fig. 2-45.

This method may certainly be applied to orbit transfers between other unstable three-body orbits, including non-symmetric orbits. In addition, a chain of orbits may certainly contain more than two different three-body orbits. Longer orbits and orbit transfers will likely require more states per segment for the differential corrector to converge. In that case, it is easier to visualize the problem by defining a sequence of states per segment and using symbols that represent sequences rather than individual states. Table 2-12 provides an example where the states A–H given above have been mapped to four such sequences.

If one refers to Fig. 2-41, one notices that there are two low-energy transfers between the example Lyapunov orbits about LL_1 and LL_2. One may construct a different sequence of states for each of those transfers, for example, $S^1_{LL1-LL2}$ and $S^2_{LL1-LL2}$, which may be constructed from three or more states, including an initial state and two intermediate states in order to keep the trajectory segment lengths short enough to permit the differential corrector to converge.

Figure 2-42 shows several low-energy transfers that exist from an orbit about LL_1 to an orbit about LL_2 that were generated using a Poincaré map. The transfer labeled (7) may be described as a complex chain that starts in an orbit about LL_1, transfers to a DPO, remains in that orbit for a revolution, and then transfers from there to an orbit about LL_2. This complex chain was identified using a Poincaré map, but it may be quickly generated by differentially correcting the series of states represented by the following sequence of

$$\{\ldots, S_{LL1}, S_{LL1}, S_{LL1-DPO}, S_{DPO}, S_{DPO-LL2}, S_{LL2}, S_{LL2}, \ldots\}$$

2.6.13 Discussion

This section introduced the tools that may be used to construct interplanetary transfers in the CRTBP using dynamical systems theory. It introduced the basic solutions of the CRTBP, including the Lagrange points and simple periodic orbits. It discussed several methods that may be used to build periodic orbits in the CRTBP. The stability of a trajectory or an orbit may be evaluated using the eigenvalues of the state transition or monodromy matrices. The state transition matrix is also very useful when implementing targeting tools such as the differential corrector. The unstable nature of many trajectories in the CRTBP leads to divergent behavior and chaos, but it also permits a mission designer to build low-energy transfers from one orbit to another. Mission designers trace structure in an orbit's stable and unstable manifolds and use that information to target a transfer to/from that orbit. Such transfers require

Table 2-12 The mapping of the states A–H to sequences [101] (*Acta Astronautica* by International Academy of Astronautics, reproduced with permission of Pergamon in the format reuse in a book/textbook via Copyright Clearance Center).

Sequence	States	Purpose
S_{LL1}	$\{A, B\}$	Traverse the LL_1 Orbit
$S_{LL1-DPO}$	$\{C, D\}$	Transfer from LL_1 to DPO
S_{DPO}	$\{E, F\}$	Traverse the DPO
$S_{DPO-LL1}$	$\{G, H\}$	Transfer from DPO to LL_1

very little energy and may be used to move a spacecraft a great distance around the three-body system without expending much fuel. These transfers are the basis for building ballistic transfers between the Earth and the Moon, which is the topic of the next few chapters.

2.7 TOOLS

Many tools are used in the design of low-energy lunar transfers. Dynamical systems methods and the corresponding tools, such as the differential corrector and Poincaré sections, are described earlier in this chapter. Other tools include numerical integrators and optimizers. These will be briefly described here.

2.7.1 Numerical Integrators

The two primary integrators used in the analyses contained in this work are the DIVA integrator [164–166] and a Runge-Kutta-Fehlberg seventh-order (RKF78) integrator with step-size control [167]. The DIVA integrator is currently implemented in both the Mission-analysis, Operations, and Navigation Toolkit Environment (MONTE) and libration point mission design tool (LTool) software (see Section 2.7.3) and has a rich heritage spanning more than three decades as an integrator for interplanetary missions at the Jet Propulsion Laboratory. It uses a variable-order Adams method for solving ordinary differential equations that has been written specifically for integrating trajectories. The RKF78 integrator is implemented in JPL's LTool. It allows for a variable step size as described by Fehlberg [167], and it is also widely used for astrodynamics and mission design.

2.7.2 Optimizers

Many mission designs presented in this book take advantage of the software package SNOPT (Sparse Nonlinear OPTimizer) [168, 169] to adjust the various parameters in order to identify solutions that require minimal amounts of fuel. SNOPT is written to use a particular sequential quadratic programming (SQP) method, one that takes advantage of the sparsity of the Jacobian matrix of the constraints of the system while maintaining a quasi-Newtonian approximation of the Hessian of the Lagrangian of the system. The details of the algorithms are beyond the scope of this discussion, except to say that they are written to be highly effective when applied to a system that has smooth nonlinear objective functions [169].

The objective functions and constraints studied here are indeed nearly smooth functions. There are two reasons why the functions studied in this paper are not perfectly smooth. First, the unstable nature of low-energy lunar transfers combined with finite-precision computers yields functions that have discontinuities. In general, these discontinuities are several orders of magnitude smaller than the trends being studied in this paper and are therefore ignored. Second, several objective functions studied in this paper involve iterative algorithms; there are discontinuities between a set of parameters that require n iterations to generate a solution and a neighboring set of parameters that require $n+1$ iterations to converge. The discontinuities observed are small relative to the topography in the state space; thus, SNOPT tends to work well in these studies.

2.7.3 Software

JPL's MONTE software [170] has been used for the majority of the analyses contained in this book. It provides an interface with JPL's DE421 Planetary and Lunar Ephemerides as well as integration using the DIVA propagator. JPL's LTool has been used for many of the computations involving libration orbits and their invariant manifolds. Targeting and optimization algorithms have been implemented in both sets of software for analyses in the CRTBP and in the ephemeris model. All of the coordinate frames described in Section 2.4 are accessed through the SPICE Toolkit in both software suites [171].

CHAPTER 3

TRANSFERS TO LUNAR LIBRATION ORBITS

3.1 EXECUTIVE SUMMARY

This chapter focuses on the performance of low-energy transfers to lunar libration orbits and other three-body orbits in the Earth–Moon system. This chapter presents surveys of direct transfers as well as low-energy transfers to libration orbits, and provides details about how to construct a desirable transfer, be it a short-duration direct transfer or a longer-duration low-energy transfer. The work presented here uses lunar halo orbits as destinations, but any unstable three-body orbit may certainly be used in place of those example destinations.

For illustration, Figs. 3-1 and 3-2 show some example direct and low-energy transfers to lunar halo orbits, respectively. One can see that these transfers are ballistic in nature: they require a standard trans-lunar injection maneuver, a few trajectory correction maneuvers, and a halo orbit insertion maneuver. One may also add Earth phasing orbits and/or lunar flybys to the trajectories, which change their performance characteristics.

Many thousands of direct and low-energy trajectories are surveyed in this chapter. Table 3-1 provides a quick guide for several types of transfers that are presented here, comparing their launch energy costs, the breadth of their launch period, that is, the

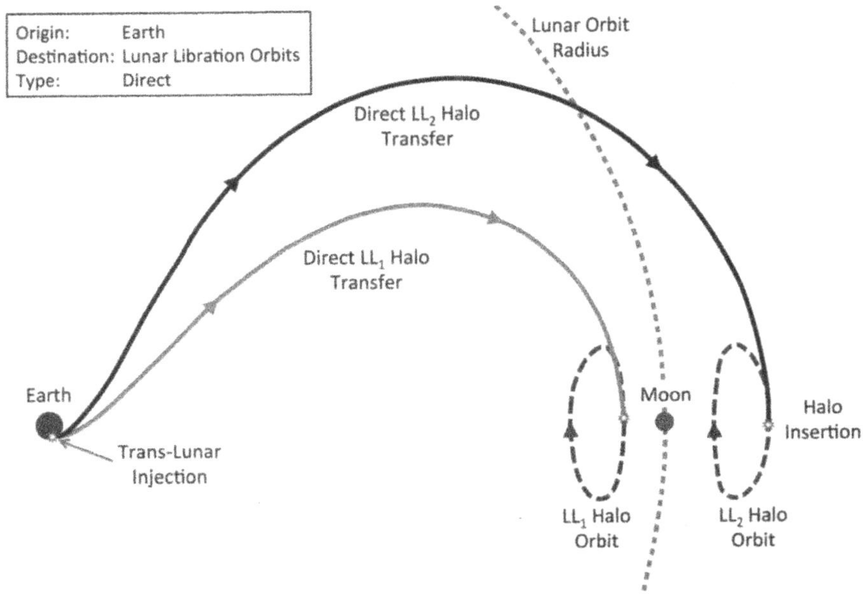

Figure 3-1 The profile for a simple direct transfer from the Earth to a lunar libration orbit about either the Earth–Moon L_1 or L_2 point.

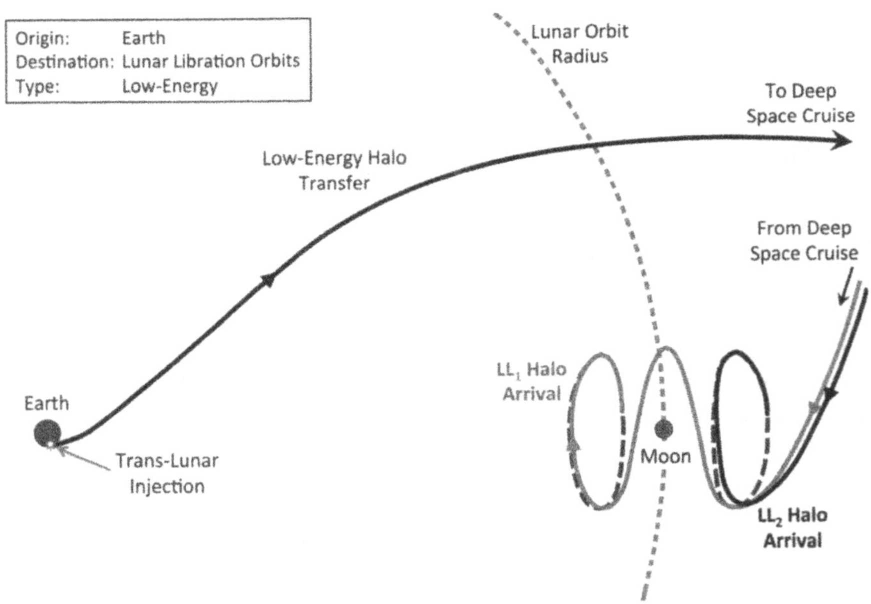

Figure 3-2 The profile for a simple, low-energy transfer from the Earth to a lunar libration orbit about either the Earth–Moon L_1 or L_2 point.

Table 3-1 A summary of several parameters that are typical for different mission scenarios to libration orbits about either the Earth–Moon L_1 or L_2 points. EPOs = Earth phasing orbits, BLT = low-energy ballistic lunar transfer.

Mission Element	Direct Transfer	Direct w/EPOs	Simple BLT	BLT w/Outbound Lunar Flyby	BLT w/EPOs
Launch C_3 (km^2/s^2)	-2.2 to -1.5	< -1.5	-0.7 to -0.4	-2.1 to -0.7	< -1.5
Launch Period	Short	Extended	Extended	Short	Extended
Transfer Duration (days)	3–6	13+	70–120+	70–120+	80–130+
Outbound Lunar Flyby	No	No	No	Yes	Yes
Libration Orbit Insertion ΔV (m/s)	~500	~500	~0	~0	~0

number of consecutive days they may be launched, their transfer duration, and the relative magnitude of the orbit insertion change in velocity (ΔV) upon arriving at the lunar libration orbit. These are representative and may be used for high-level mission design judgements, though the details will likely vary from mission to mission.

Direct transfers to lunar libration orbits are presented in Section 3.3. That section surveys thousands of transfers to libration orbits about both the Earth–Moon L_1 and L_2 points and presents methods to construct them. The trajectories minimize the halo orbit insertion ΔV cost while keeping the total transfer duration low, between 5 days and 2 months. The trajectories include no maneuvers other than the trans-lunar injection maneuver and the halo orbit insertion maneuver. Hence, there are no high-risk maneuvers, such as powered lunar flybys, though such maneuvers may indeed reduce the total transfer ΔV cost [172].

The surveys show that one may depart the Earth from any parking orbit, certainly including low-altitude parking orbits with an inclination of 28.5 degrees (deg). The transfers involve trans-lunar injections with launch injection energy (C_3) requirements as low as -2.6 kilometers squared per second squared (km^2/s^2) and as high as -2.0 km^2/s^2 for transfers to LL_1 or as high as -1.0 km^2/s^2 for transfers to LL_2. The halo orbit insertion maneuver may be as low as 430 meters per second (m/s) or as high as 950 m/s, depending on the mission's requirements, though most are in the range of 500–600 m/s. The quickest transfers arrive at their libration orbit destinations within 5 or 6 days. Some missions can reduce the total transfer ΔV by ~50 m/s by implementing a longer, 30-day transfer. In some cases it is beneficial to extend the duration to 40 or 50 days. Finally, direct lunar transfers exist in families, such that very similar transfers exist to neighboring libration orbits. That is, if a mission's requirements change slightly and a new libration orbit is required, one can usually build a very similar transfer to that orbit as to the original orbit.

Low-energy transfers to lunar libration orbits are presented in Section 3.4. Much like the analyses of direct transfers, Section 3.4 surveys thousands of transfers to libration orbits about both the Earth–Moon L_1 and L_2 points and presents methods to construct them. The trajectories are always entirely ballistic, except for the trans-lunar injection maneuver. None of the transfers studied requires an orbit insertion maneuver; every trajectory asymptotically arrives at the target orbit and inserts automatically. Trajectories are studied with a wide variety of geometry characteristics, but all require less ΔV than direct transfers.

Much like the analyses of direct transfers, the surveys in Section 3.4 show that one may depart the Earth from any given low Earth parking orbit, or any higher orbit as needed. The transfers involve trans-lunar injections with C_3 requirements as low as -0.75 km^2/s^2 and as high as -0.35 km^2/s^2. This C_3 requirement may be reduced to about -2.1 km^2/s^2 if a lunar flyby is implemented at an altitude of about 2000 km. The quickest transfers identified require about 83 days between the trans-lunar injection and the point when the trajectory has arrived within 100 km of the lunar libration orbit. Many transfer options exist that require 90–140 days between the injection point and the orbit arrival point. Since the transfers asymptotically approach the target libration orbit, they are essentially at the target orbit as many as 10 days prior to the "arrival" time. Finally, much like direct transfers to lunar libration orbits, low-energy transfers exist in families, such that very similar transfers exist to neighboring libration orbits. Very similar transfers also exist to the same orbit when the arrival time or arrival location is adjusted.

This chapter summarizes nearly ballistic transfers between the Earth and lunar libration orbits. Techniques to use these transfers in practical spacecraft mission design (for example, building launch periods, and budgeting station keeping ΔV) are studied in Chapter 6.

3.2 INTRODUCTION

This chapter describes methods to construct both direct and low-energy transfers between the Earth and libration orbits near the Moon. The focus of this book is on the analysis and construction of low-energy transfers, but it is helpful to have a good understanding of the costs and benefits of direct transfers as well. In addition, this chapter provides some transfers that one may take after arriving at a lunar libration orbit; transfers are presented from those libration orbits to other libration orbits, to low lunar orbits, and to the lunar surface.

Direct transfers include any sort of high-energy conventional trajectories using chemical propulsion systems. Low-energy transfers use the same propulsion systems but travel well beyond the orbit of the Moon, taking advantage of the Sun's gravity to reduce the ΔV cost of the transfer. Direct transfers to lunar libration orbits (and other three-body orbits) typically require 3–6 days, though there are benefits to increasing the transfer duration as long as 1 or 2 months. Low-energy transfers typically require 3–4 months of transfer time or more in some circumstances.

Figure 3-1 illustrates two example direct transfers between the Earth and libration orbits about the Earth–Moon L_1 and L_2 points. Figure 3-2 illustrates two low-energy transfers to the same two libration orbits, viewed in the same reference frame. One can see that the trajectories traverse beyond the orbit of the Moon and return after 2–3 months to arrive at the Moon in such a way that they insert into the target orbits without requiring any insertion maneuver. The lack of a large orbit insertion maneuver is the primary reason why these transfers save so much fuel (the direct transfers require an orbit insertion maneuver near 500 m/s).

Figure 3-3 illustrates two different low-energy transfers viewed in the Sun–Earth rotating frame to show that spacecraft may fly either toward the Sun or away from it during their transfers.

This chapter describes techniques to build direct and low-energy transfers to lunar libration orbits and surveys the performance of both types of transfers. Section 3.3 describes the techniques and provides performance data for direct transfers to lunar libration orbits. Section 3.4 does the same for low-energy transfers to the same orbits. Section 3.5 provides information about orbit transfers from the libration orbits to other libration orbits, to low lunar orbits, and to the lunar surface. Finally, Section 3.5 discusses transfers that a spacecraft could take to depart its lunar libration orbit and travel to another three-body orbit, a low lunar orbit, the lunar surface, or back to the Earth.

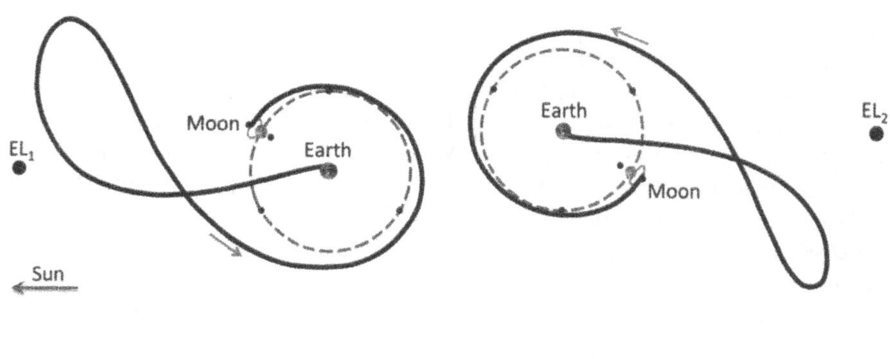

Figure 3-3 Two example low-energy transfers between the Earth and an LL_2 libration orbit. The transfers are viewed from above in the Sun–Earth rotating coordinate frame [44] (Copyright © 2009 by American Astronautical Society Publications Office, all rights reseved, reprinted with permission of the AAS.).

3.3 DIRECT TRANSFERS BETWEEN EARTH AND LUNAR LIBRATION ORBITS

As of 2012 no missions have flown a direct transfer from the Earth to a lunar libration orbit. Many researchers have considered the problem, dating back to 1970 when Edelbaum studied the case of transferring a spacecraft from the Earth to the L_1 point itself via a direct transfer [173]. Certainly NASA has considered the problem as it considers destinations for future missions [126]. The work presented here is based upon the work of Parker and Born [174, 175], who performed a robust survey of direct transfers to lunar halo orbits about both L_1 and L_2. Several other authors have also studied this problem, including Rausch [176], Gordon [177], and Alessi et al. [178].

The trajectories generated here are constructed by intersecting a low Earth orbit (LEO) parking orbit with a trajectory within the stable invariant manifold of the target libration orbit. Hence, the trajectories include two maneuvers: a maneuver to depart the Earth and a maneuver to inject onto the target orbit's stable manifold. Once on the stable manifold, the spacecraft asymptotically arrives at the target orbit.

3.3.1 Methodology

Direct transfers are constructed here by targeting states within the stable manifold of a desirable halo orbit or other libration orbit. This strategy has been implemented before for transfers to many types of Sun–Earth libration orbits, yielding trajectories for missions such as *Genesis* [72], *Wilkinson Microwave Anisotropy Probe* (*WMAP*) [70], and *Solar and Heliospheric Observatory* (*SOHO*) [66]. The technique has been highly successful for missions in the Sun–Earth system because the stable manifolds of many Sun–Earth halo orbits intersect the Earth. Unfortunately, as one can begin to see in Fig. 3-4, the stable manifolds of libration orbits near the Earth–Moon L_1 and L_2 points do not intersect the Earth within as much as two months of time. Consequently, at least two maneuvers must be performed to directly transfer onto the lunar halo orbit's stable manifold from an initial LEO parking orbit, rather than the single maneuver required to inject onto the stable manifold of a Sun–Earth halo orbit.

In theory, a direct transfer to a lunar halo orbit could involve many burns, each performed in some arbitrary direction. We have chosen to survey the simplest type of direct lunar halo orbit transfers, namely, transfers with only two burns that are each performed in a direction tangential to the spacecraft's velocity vector. These transfers are not guaranteed to have the lowest ΔV cost of any type of direct lunar halo transfer, but they should provide a good estimate for the ΔV requirement of such transfers. Even with this simplification, this design problem yields a very rich design space and is a useful foundation for future studies.

Figure 3-5 shows two perspectives of a scenario that illustrates the strategy used here to transfer a spacecraft from a 185-km LEO parking orbit to a lunar L_1 halo orbit. The scenario requires a large maneuver at the LEO injection point (ΔV_{LEO}; also known as the trans-lunar injection maneuver) and a second large maneuver at the manifold injection point (ΔV_{MI}). The two ballistic mission segments are referred to

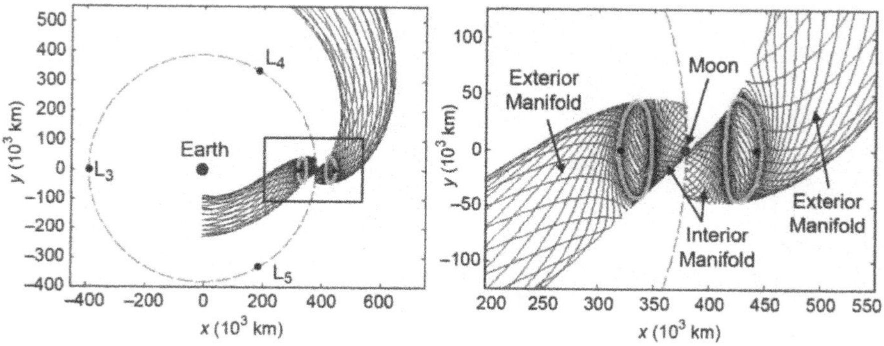

Figure 3-4 Plots of the stable manifolds of example L_1 and L_2 halo orbits, viewed from above in the Earth–Moon synodic reference frame. A spacecraft that travels along any one of these trajectories will asymptotically arrive onto the corresponding halo orbit [174] (Copyright © 2008 by American Astronautical Society Publications Office, all rights reserved, reprinted with permission of the AAS).

as the *bridge* segment and the *manifold* segment. Once the spacecraft arrives onto the manifold segment, after performing the ΔV_{MI} maneuver, it asymptotically transfers onto the lunar halo orbit.

It is assumed that each transfer constructed here begins in a 185-km circular prograde Earth parking orbit. In this way, the performance of each transfer may be directly compared. In reality, the same sorts of transfers that are constructed here may begin from a LEO parking orbit at any altitude and with any eccentricity, or even from the surface of the Earth, provided that the vehicle is at the correct position at the correct time to perform the ΔV_{LEO} maneuver successfully.

The following strategy has been followed to construct direct transfers to lunar halo orbits:

Step 1. Construct the desired halo orbit.

Step 2. Construct the *manifold* segment:

1. Choose a τ-value, that is, a point along the halo orbit as illustrated in Fig. 2-10 (page 50); choose a direction, that is, either "interior" or "exterior" as shown in Fig. 3-4; and choose a manifold propagation duration, Δt_m.

2. The manifold segment is constructed by propagating the specific trajectory in the halo orbit's stable manifold that corresponds to the given τ-value. The trajectory departs the halo orbit either in the interior or exterior direction, as indicated. It is propagated in the Earth–Moon three-body system backward in time for the given duration.

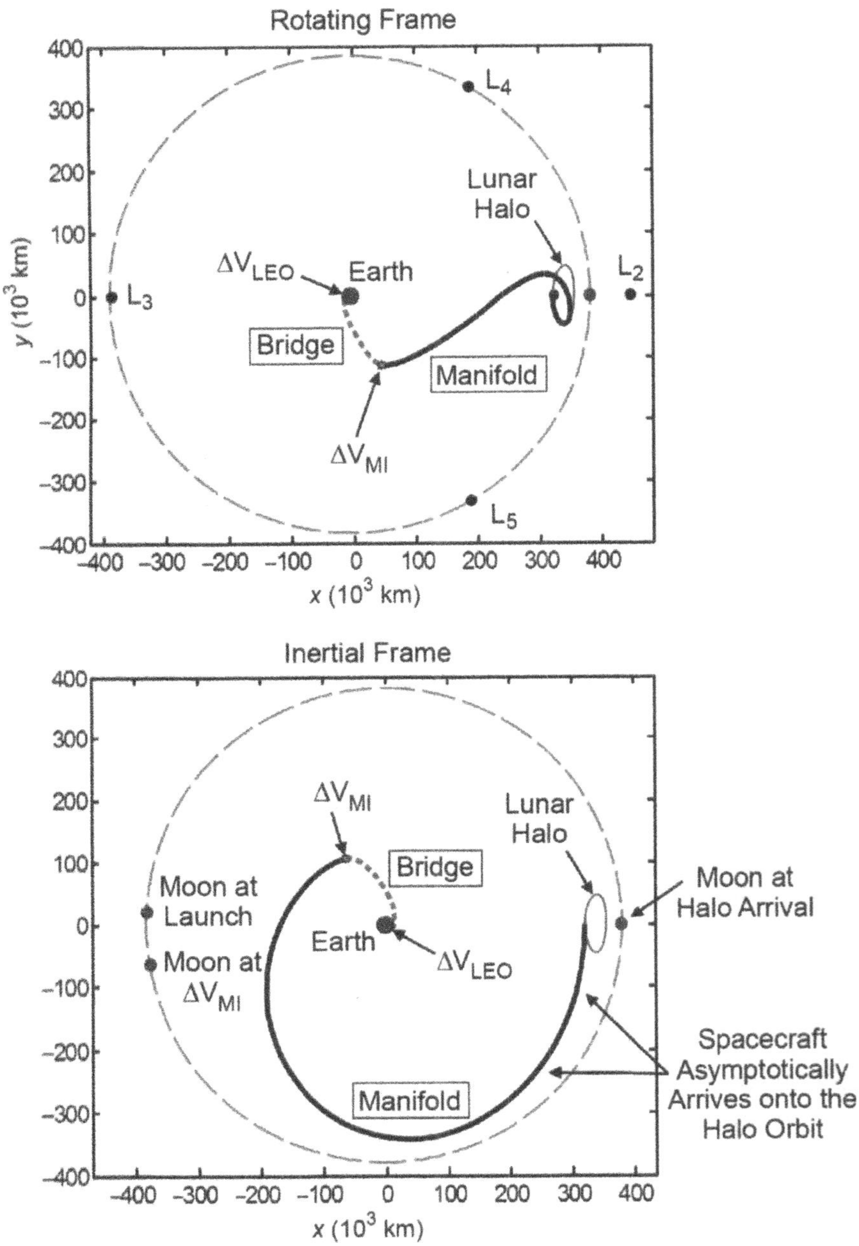

Figure 3-5 Two perspectives of an example scenario that may be used to transfer a spacecraft from a 185-km LEO parking orbit to a lunar L_1 halo orbit. The transfer is shown in the Earth–Moon rotating frame (top) and the corresponding inertial frame of reference (bottom). The halo orbit is shown in the inertial frame only for reference.

Step 3. Define X_{MI} to be the final state of the manifold segment. This is the state that a spacecraft would need to obtain in order to inject onto the manifold segment.

Step 4. Construct ΔV_{MI} and the *bridge* segment:

1. Define ΔV_{MI} to be the tangential ΔV that may be applied to X_{MI} in order to construct the bridge segment.
2. When propagated further backward in time, the bridge segment will encounter the prograde 185-km LEO orbit at the bridge's first perigee point. The bridge segment is propagated in the Earth–Moon three-body system.

Step 5. Construct ΔV_{LEO}, the tangential ΔV that may be applied to transfer the spacecraft from its LEO orbit onto the bridge segment.

This procedure is used here to produce a direct, two-burn transfer to a lunar halo orbit given an arbitrary lunar halo orbit and any given value for those parameters specified in Step 2 above. A significant benefit of this procedure is that it requires no knowledge of what a transfer should look like, except that the bridge segment is only propagated backward in time to its first perigee passage.

This process generates three-dimensional transfers in the idealized Earth–Moon circular restricted three-body problem (CRTBP). The inclination of the Earth departure is a free variable; it is computed and reported, but not targeted in any way. Furthermore, since no date is specified, the inclination is presented relative to the orbital plane of the Moon. The performance of actual transfers to real halo orbits will vary based on the date and orientations of each body and its orbit in the Solar System. Nevertheless, this exploration sheds light on what sorts of transfers exist and their approximate performance.

Several scenarios have been explored to identify optimal transfers, given the confines of this survey. The first suspicion is that the optimal transfer may be constructed by building a bridge segment that connects the LEO departure with the manifold segment's perigee point. Since energy-change maneuvers are more efficient when a spacecraft is traveling faster [97], the perigee of the manifold segment seems like a good location to perform the ΔV_{MI} maneuver. The best transfer for a specific halo orbit would then be the one that requires the least total ΔV over all τ-values. This *perigee-point* scenario is presented first. It turns out that this strategy does not produce the most efficient transfers—the next strategy generates better transfers—but the *perigee-point* scheme will still be presented because it illuminates the problem very well.

3.3.2 The Perigee-Point Scenario

Figure 3-6 shows two perspectives of several example trajectories that may be used to transfer a spacecraft onto a single lunar L_1 halo orbit using the perigee-point scheme. Each transfer implements a different τ-value about the same halo orbit. For

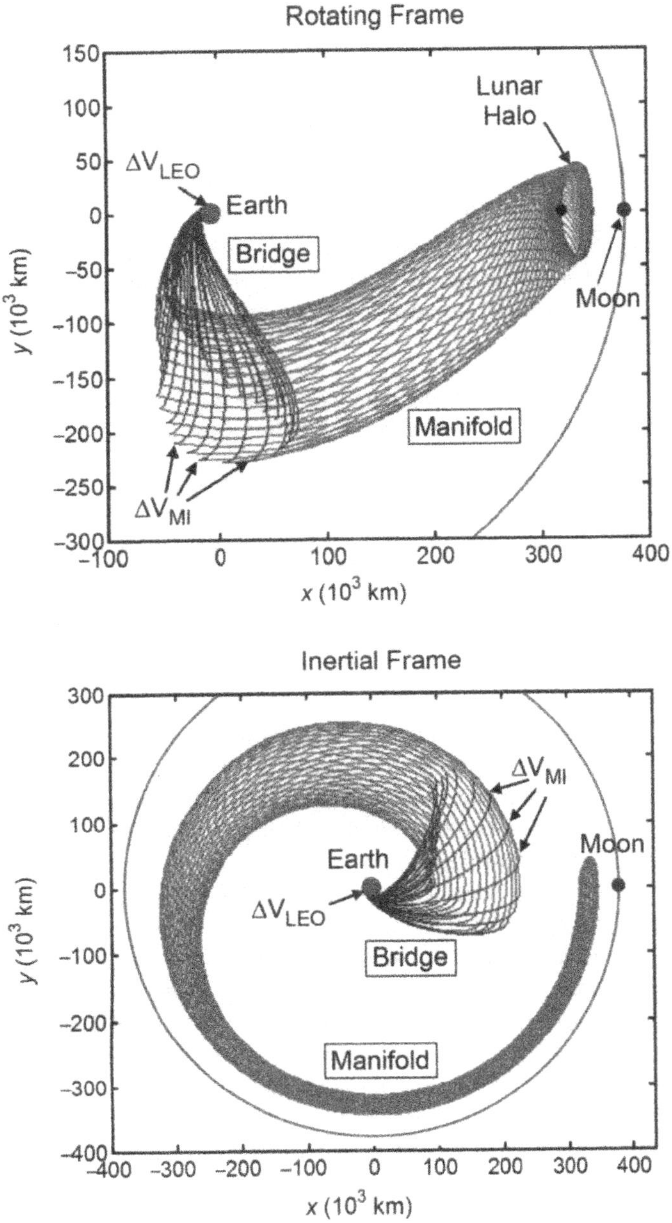

Figure 3-6 Example trajectories that implement the *perigee-point* scheme to directly transfer from LEO to a lunar L_1 halo orbit. The transfers are shown in the Earth–Moon rotating frame (top) and the corresponding inertial frame of reference (bottom). *(See insert for color representation of this figure.)*

reference, the halo orbit is a northern L_1 halo orbit with an x_0-value of approximately 319,052 kilometers (km). The manifold segment in each case has been propagated to its perigee point, and the corresponding bridge segment has been constructed to transfer from a 185-km prograde LEO orbit to that perigee point. The trajectories are shown in both the Earth–Moon rotating frame and the corresponding inertial frame of reference.

Several of the trajectories shown in the left plot of Fig. 3-6 appear to have non-tangential ΔV_{MI} maneuvers; this is only a visual effect caused by the rotating frame of reference. As the spacecraft departs the Earth on the bridge segment, it quickly crosses a point where the frame of reference rotates about the Earth faster than the spacecraft. After that point, the spacecraft appears to travel in a retrograde fashion about the Earth, seemingly in conflict with its inertially prograde orbit. If the spacecraft then performs a large enough ΔV_{MI} maneuver, the spacecraft's rotational velocity will once again exceed the rotational velocity of the frame of reference. The spacecraft will appear to have switched directions when it actually just increased its inertial velocity.

Figure 3-7 shows plots of the magnitudes of the two required maneuvers, ΔV_{LEO} and ΔV_{MI}, as well as the total maneuver cost as functions of the parameter τ. One can see that the minimum ΔV cost to transfer from the 185-km LEO orbit to this halo orbit using the perigee-point scheme is approximately 4.14 kilometers per second (km/s). One can also see that this minimum occurs at the point where ΔV_{LEO} is at its maximum. Figure 3-8 shows plots of the minimum- and maximum-ΔV transfers and verifies that the minimum-ΔV transfer involves the largest bridge segment observed in Fig. 3-6. The total transfer duration from the point where the spacecraft performs its ΔV_{LEO} maneuver to the point where it is within 100 km of the given halo orbit ranges between approximately 17.7 days ($\tau \approx 0.30$) and 22.9 days ($\tau \approx 0.83$).

3.3.3 The Open-Point Scenario

Although it may be intuitive to perform ΔV_{MI} at the manifold segment's perigee point because of the energy considerations, it is actually better to perform a larger ΔV_{LEO} and a smaller, although less-efficient, ΔV_{MI}. This is because the maneuver at LEO can take advantage of its close proximity to the Earth to make the total energy change required in the transfer as efficient as possible. That is, it is most efficient to change as much of the spacecraft's energy at LEO as possible, since that is the location where the spacecraft will be traveling the fastest during the lunar transfer. This result is evident by studying the results of the perigee-point scheme.

An alternate scheme is presented here where the second maneuver, ΔV_{MI}, may be placed anywhere along the stable manifold of the halo orbit. The manifold segment may be propagated well beyond its perigee point, although it has an imposed maximum propagation time of 1 or 2 months: 1 month for exterior manifolds since they depart the Moon's vicinity quickly and 2 months for interior manifolds since they linger near the Moon for longer amounts of time. The transfers have an additional degree of freedom compared with the perigee-point scheme, but they are otherwise constructed in exactly the same manner as listed above. This new scheme will be

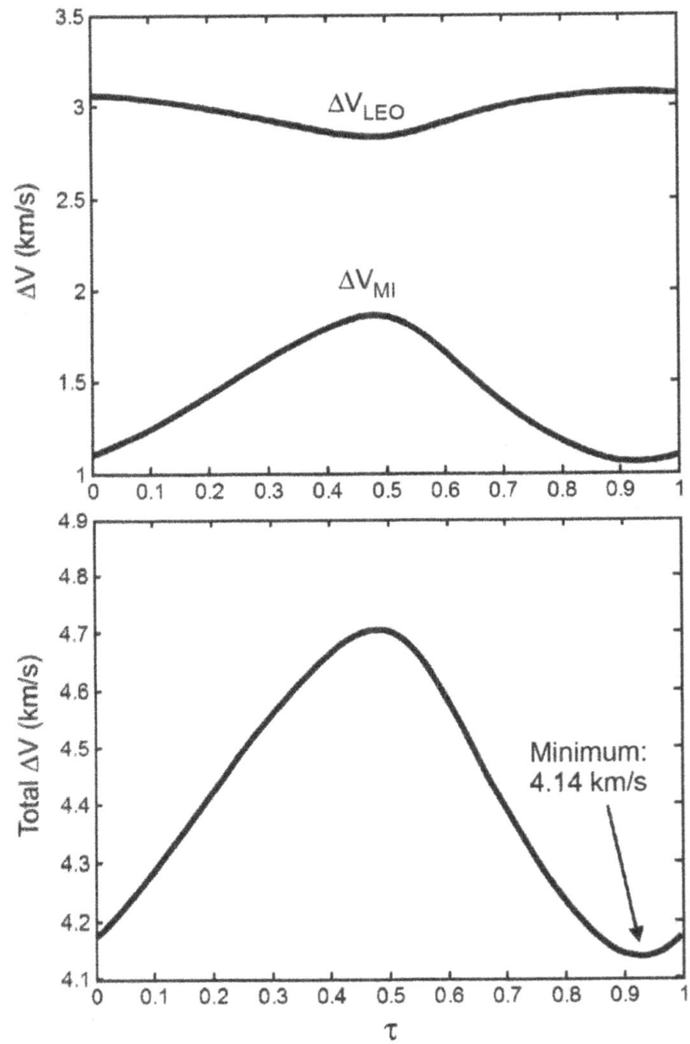

Figure 3-7 Plots of the maneuver requirements to transfer onto a lunar L_1 halo orbit using the perigee-point scheme. Top: the magnitudes of the two maneuvers ΔV_{LEO} and ΔV_{MI} as functions of τ; bottom: the total ΔV cost as a function of τ.

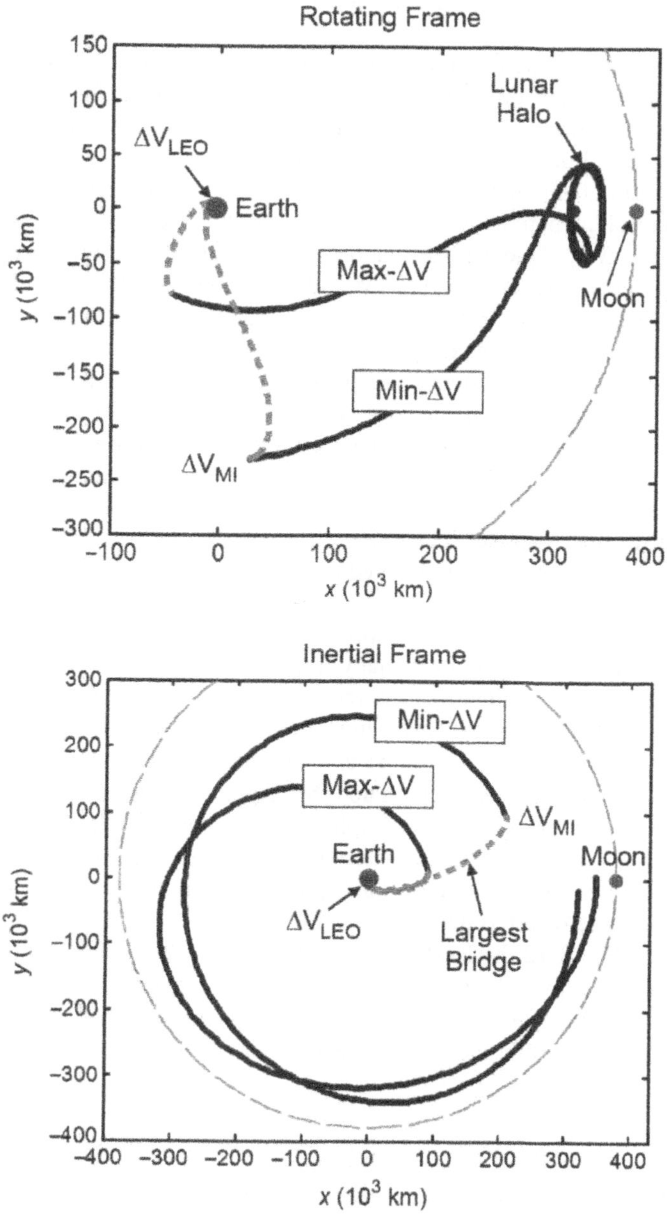

Figure 3-8 The minimum- and maximum-ΔV transfers produced using the perigee-point scheme. One can see that the minimum-ΔV transfer contains the largest bridge segment observed in Fig. 3-6 [174] (Copyright © 2008 by American Astronautical Society Publications Office, all rights reserved, reprinted with permission of the AAS).

referred to as the *open-point* scheme, since the manifold insertion point has had its position constraint opened.

To demonstrate the open-point transfer strategy, Fig. 3-9 shows several transfers that may be constructed from LEO to an arbitrary trajectory along the stable manifold of a particular halo orbit. The halo orbit shown in Fig. 3-9 is the same northern L_1 halo orbit presented in Section 3.3.2, and the manifold shown has a τ-value of 0.3. Figure 3-10 shows the maneuver cost associated with transferring to various points along the manifold, where the location of ΔV_{MI} is specified by the manifold propagation duration, Δt_m. One can see that there are two local minima that correspond to low-energy lunar transfers: one at a Δt_m of approximately 10.0 days and the next at a Δt_m of approximately 22.7 days, neither of which corresponds to a mission that transfers to the manifold segment's perigee point, which has a Δt_m of approximately 16.86 days. In fact, these transfers correspond to missions where the bridge segment connects the spacecraft to a point very near the apogee of the manifold segment. Figure 3-11 shows plots of the extreme cases, namely, the four transfers indicated by the labels (1)–(4) in Fig. 3-10. One can see that the two local minima observed in Fig. 3-10, that is, the trajectories marked with a (2) and a (4), coincide very near to the manifold segment's apogee locations.

Figures 3-9 to 3-11 have demonstrated the open-point scheme applied to a single trajectory (where $\tau = 0.3$) on the stable manifold of a single halo orbit (the lunar L_1 halo orbit with an x_0-value of approximately 319,052 km). The open-point scheme is easily extended to cover many trajectories along the halo orbit's stable manifold. Figure 3-12 summarizes the required maneuvers and the total maneuver cost associated with the least expensive lunar transfer for each trajectory on the stable manifold of the same halo orbit. One can see that the lowest-energy open-point transfer constructed to this particular halo orbit requires a total ΔV of approximately 3.62 km/s. This low-energy transfer implements the trajectory in the orbit's stable manifold with a τ-value of approximately 0.48. For verification, Fig. 3-12 shows that the trajectory with a τ-value of 0.3 requires a minimum ΔV of approximately 3.67 km/s: the same result as that shown in Fig. 3-10.

Note that in Fig. 3-12 the least-expensive transfers to this halo orbit use the first maneuver, ΔV_{LEO}, to perform the vast majority of the spacecraft's energy change. This is consistent with the notion that the most efficient transfer performs as much ΔV as possible deep within the Earth's gravity well where the spacecraft is traveling fastest.

3.3.4 Surveying Direct Lunar Halo Orbit Transfers

The previous section illustrates the open-point scheme applied to a single halo orbit about the Earth–Moon L_1 point using the halo orbit's exterior stable manifold. The process results in a low-energy, two-maneuver, direct lunar transfer to that halo orbit, following the exterior stable manifold. This section surveys low-energy direct lunar transfers to a large number of orbits within the families of halo orbits about both the Earth–Moon L_1 and L_2 points, taking advantage of both the exterior and interior

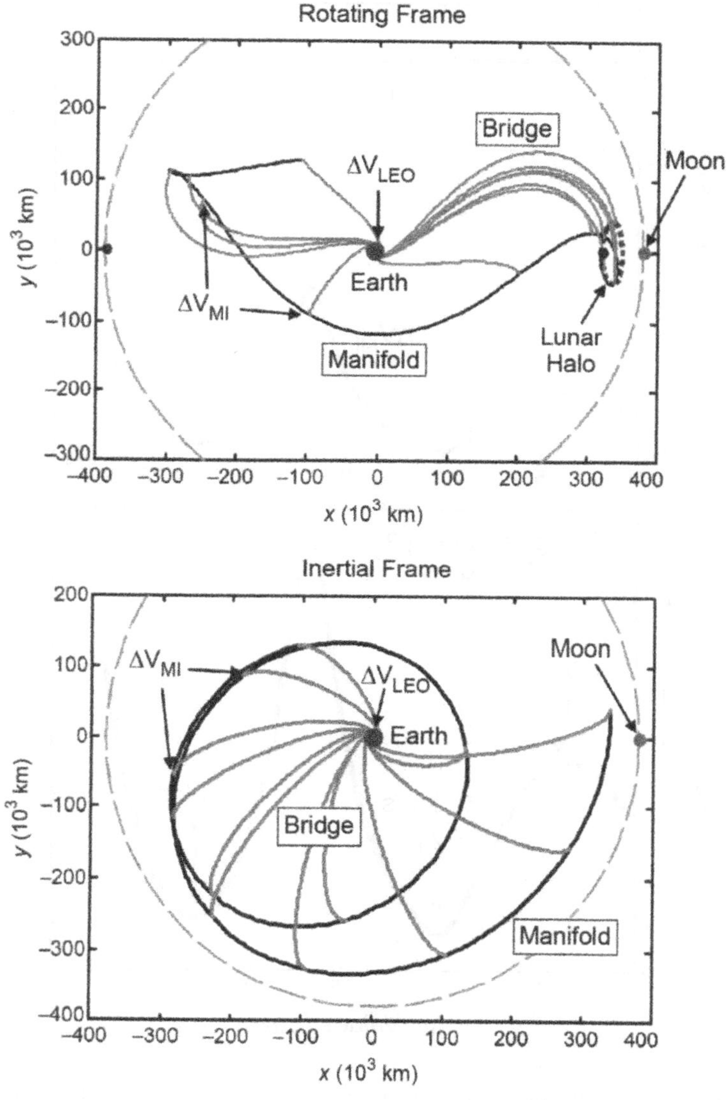

Figure 3-9 Example trajectories that implement the open-point scheme to directly transfer from LEO to a specific manifold of a particular lunar L_1 halo orbit.

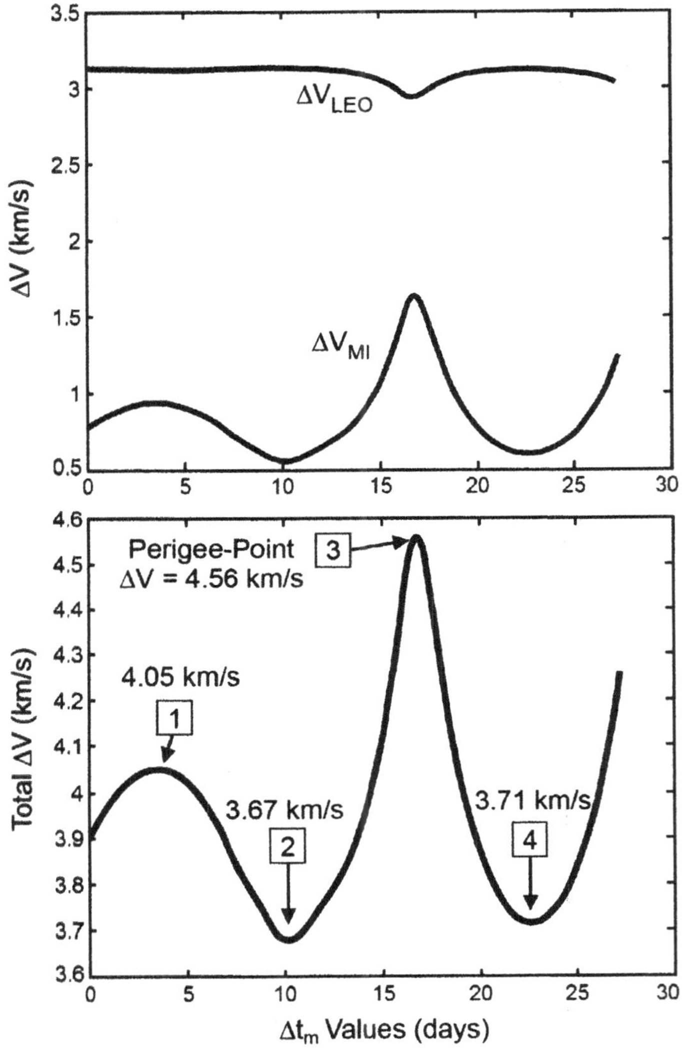

Figure 3-10 Plots of the maneuver requirements to transfer onto a specific manifold of a specific lunar L_1 halo orbit using the open-point scheme. Top: the magnitudes of the two maneuvers ΔV_{LEO} and ΔV_{MI} as functions of Δt_m; bottom: the total ΔV cost as a function of Δt_m.

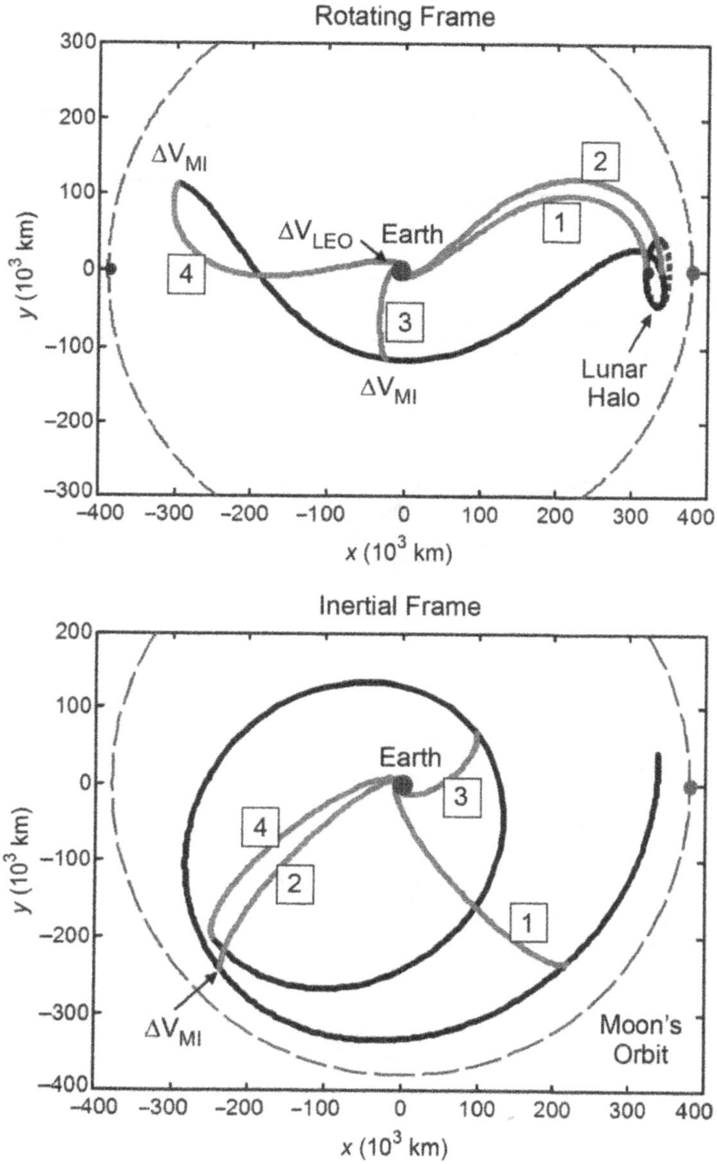

Figure 3-11 The four transfers with locally extreme ΔV requirements as indicated by the labels (1)–(4) in Fig. 3-10. The transfers are shown in the Earth–Moon rotating frame (top) and the corresponding inertial frame of reference (bottom) [174] (Copyright © 2008 by American Astronautical Society Publications Office, all rights reserved, reprinted with permission of the AAS).

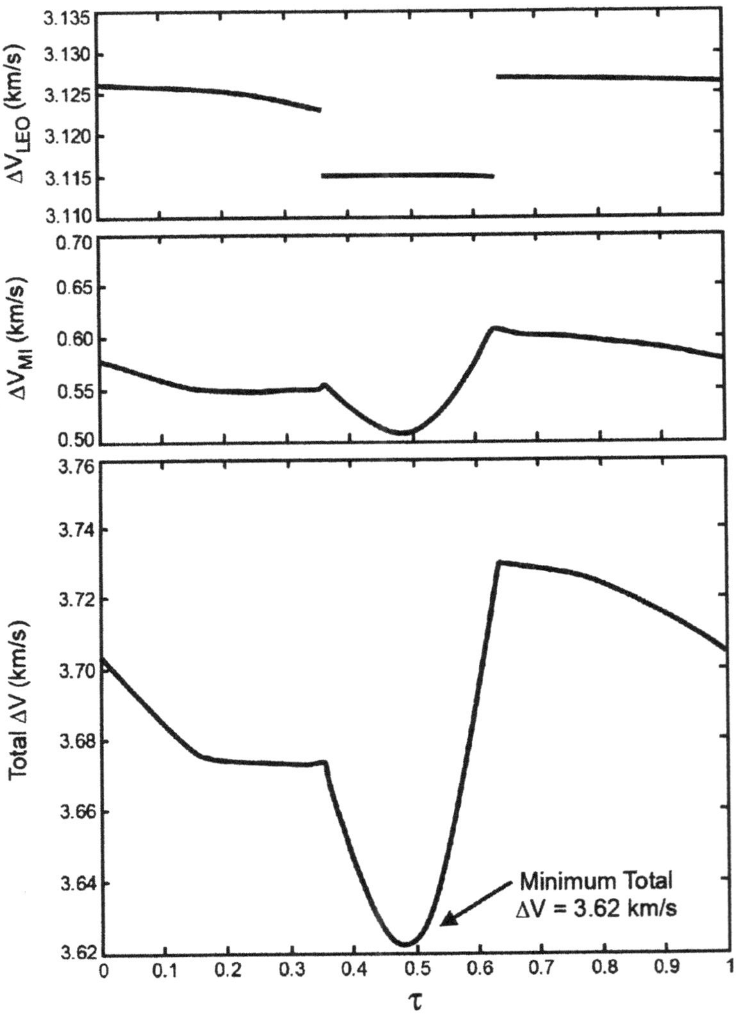

Figure 3-12 Plots of the magnitudes of the two required maneuvers ΔV_{LEO} and ΔV_{MI} (top) and the total ΔV cost (bottom) associated with the least-expensive lunar transfer for each trajectory on the stable manifold of a single lunar L_1 halo orbit.

stable manifolds. These results should be representative of other three-body orbits as well, such as Lissajous and vertical Lyapunov orbits.

The following sections summarize the results of four surveys performed here:

Section	Halo Family	Interior/Exterior Stable Manifold	Page Number
Section 3.3.4.1	L_1	Exterior	136
Section 3.3.4.2	L_1	Interior	140
Section 3.3.4.3	L_2	Exterior	142
Section 3.3.4.4	L_2	Interior	146

In each of these four cases, it would be ideal to perform an exhaustive search for the very best transfer to each halo orbit implementing the given stable manifold. However, it is very time-consuming to construct a transfer to each point along each trajectory in each halo orbit's stable manifold. The corresponding phase space is three-dimensional, and every combination of parameters takes a significant amount of computation time. To reduce the computation load, while still performing a survey of a large portion of the phase space, several numerical optimization routines have been implemented.

It has been found that a combination of hill-climbing and genetic algorithms performs very well at identifying the least-expensive transfers to a given halo orbit very swiftly [46]. The numerical algorithms use the state $X = [x_0, \tau, \Delta t_m]^T$ to define a direct two-maneuver lunar transfer, given the procedure outlined in Section 3.3.1. The numerical optimization process begins by implementing a genetic algorithm to identify a local ΔV-minimum in the phase space. The implementation of the genetic algorithm will not be discussed here for brevity, but may be found in many sources in literature [179]. After several iterations of the genetic algorithm, the state that corresponds to the least-expensive lunar transfer is refined using a dynamic hill-climbing algorithm, also known as the steepest-descent algorithm [180]. In this way, the local minima of the three-dimensional phase space are quickly explored. In order to survey specific orbits within a family of halo orbits, the parameter x_0 is held constant and the remaining two parameters are varied.

The majority of the locally-optimal transfers found in this work were identified by specifying a value for x_0 and varying the values of τ and Δt_m using ten iterations of a genetic algorithm with a population of twenty states. The least-expensive state resulting from the genetic algorithm was then iterated in the dynamic hill-climbing algorithm until a solution was found whose ΔV cost could not be improved by varying τ by more than 1×10^{-5} or by varying Δt_m by more than 4 seconds.

The numerical optimization routine is not guaranteed to converge on the most efficient transfer, but it easily converges on relatively efficient transfers. The results given in the following sections include the most efficient transfers identified, as well as somewhat less efficient transfers. The results then trace out a Pareto front of optimal solutions [181]. Other nonoptimal points have been added to the results to give an impression of the range of costs of transfers that exist. Each result is discussed in more detail in the following sections.

3.3.4.1 Survey of Exterior Transfers to L_1 Halo Orbits This section presents the results of open-point transfers constructed between 185-km LEO parking orbits and the exterior stable manifold of halo orbits in the family of lunar L_1 halo orbits. Figure 3-13 shows the cost of many such example transfers to halo orbits in the family. One can see that there are several types of efficient transfers. To help identify the trends and differences between each type of transfer, Fig. 3-14 shows plots of several example transfers. Finally, Tables 3-2 through 3-5 provide additional information about sample transfers of several varieties observed in the figures. Table 3-2 summarizes the characteristics of the numbered transfers shown in Fig. 3-14; Table 3-3 provides details about the shortest-duration transfers identified; Tables 3-4 and 3-5 summarize the transfers labeled "efficient" and "complex" in Fig. 3-13, respectively.

Figures 3-13 and 3-14 show many interesting patterns. After studying the transfers presented in these figures, as well as the corresponding data presented in Tables 3-2 through 3-5, the following observations have been made:

- The majority of the least-expensive transfers of this type are very fast transfers, requiring only five days to transfer to a close proximity of each corresponding halo orbit. Table 3-3 provides details about examples of such fast transfers. Their bridge segments take the spacecraft nearly directly to the halo orbit. These transfers compose the majority of the Pareto front observed in the figures.

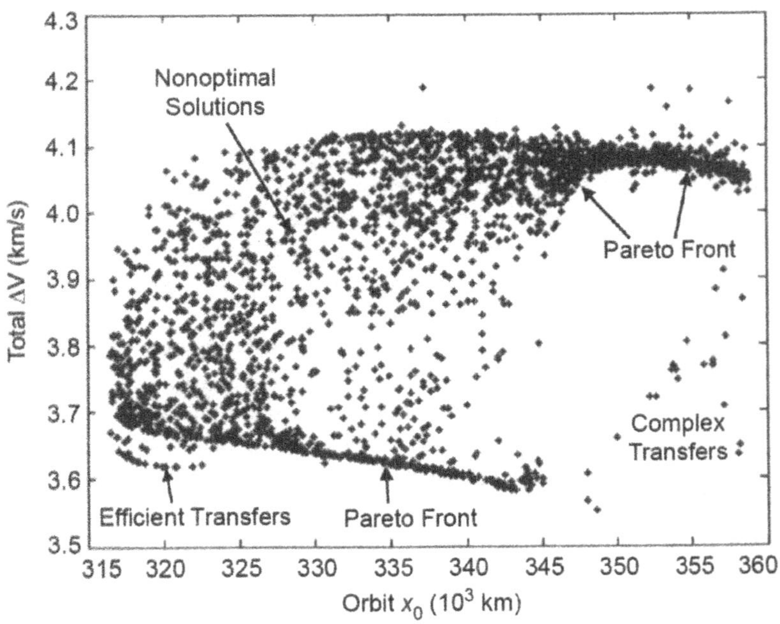

Figure 3-13 The total ΔV cost of many surveyed transfers to the exterior stable manifold of orbits in the family of lunar L_1 halo orbits.

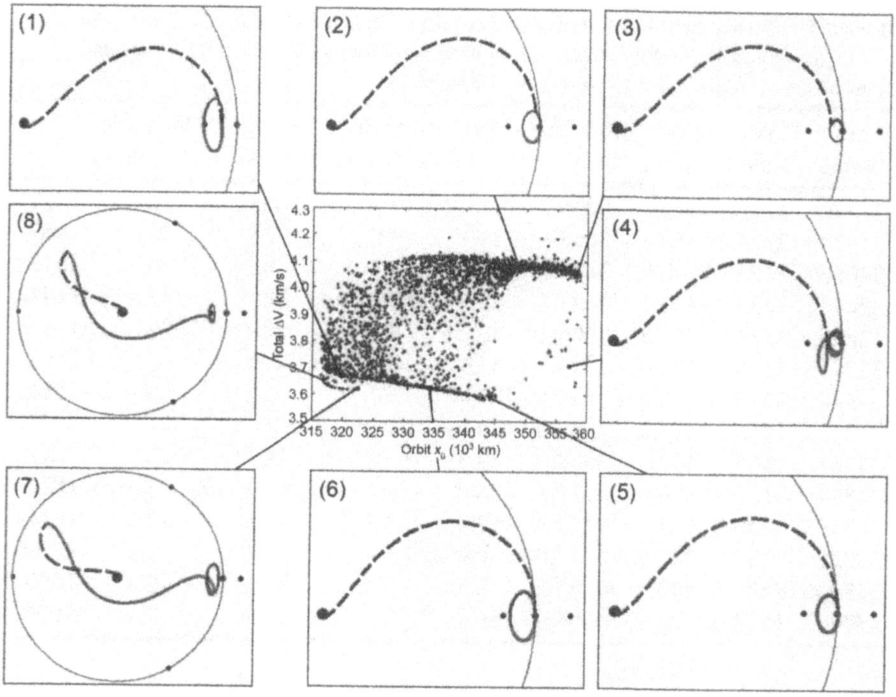

Figure 3-14 Several example transfers between 185-km LEO parking orbits and the exterior stable manifold of lunar L_1 halo orbits. The parameters of the numbered transfers are summarized in Table 3-2.

Table 3-2 Characteristics of the numbered transfers identified in Fig. 3-14 [174] (Copyright © 2008 by American Astronautical Society Publications Office, all rights reserved, reprinted with permission of the AAS).

#	x_0 (km)	ΔV_{LEO} (m/s)	ΔV_{MI} (m/s)	Total ΔV (m/s)	Inc* (deg)	Transfer Δt (days)	Bridge Δt (days)	Manifold Δt (days)	τ
1	320265	3128.0	539.0	3667.0	26.3	13.1	4.5	8.5	0.179
2	348963	3134.1	934.1	4068.2	8.2	8.8	4.7	4.1	0.888
3	357643	3132.9	923.1	4056.0	16.1	7.5	4.6	2.9	0.500
4	357177	3129.0	579.0	3708.0	25.7	30.6	4.6	26.0	0.501
5	342539	3136.2	453.5	3589.7	48.1	22.1	4.9	17.2	0.461
6	334016	3135.9	493.6	3629.5	46.9	28.8	4.9	23.9	0.911
7	322568	3119.2	503.1	3622.2	9.8	31.3	4.0	27.3	0.800
8	317035	3111.3	531.7	3643.0	5.0	23.5	3.4	20.1	0.281

*The inclination of the LEO parking orbit in the CRTBP.

Table 3-3 Characteristics of example fast transfers identified in Figs. 3-13 and 3-14 [174] (Copyright © 2008 by American Astronautical Society Publications Office, all rights reserved, reprinted with permission of the AAS).

x_0 (km)	ΔV_{LEO} (m/s)	ΔV_{MI} (m/s)	Total ΔV (m/s)	Inc (deg)	Transfer Δt (days)	Bridge Δt (days)	Manifold Δt (days)	τ
317406	3123.6	572.1	3695.7	17.0	7.6	4.4	3.1	0.757
318240	3125.3	558.3	3683.6	21.1	9.0	4.4	4.5	0.863
320569	3129.4	540.9	3670.3	30.0	5.4	4.6	0.7	0.543
324912	3133.4	522.2	3655.6	38.3	6.6	4.8	1.8	0.638
328382	3134.8	511.3	3646.2	42.2	6.8	4.8	2.0	0.664
332715	3135.7	497.9	3633.5	45.9	5.7	4.9	0.7	0.562
335440	3135.9	488.6	3624.5	47.7	5.7	4.9	0.8	0.566
339191	3136.1	471.9	3608.0	49.3	6.5	4.8	1.7	0.673
341814	3136.1	457.9	3594.0	49.5	8.2	4.8	3.5	0.878
345948	3135.8	876.0	4011.8	26.0	8.3	4.9	3.4	0.874
347333	3135.2	915.6	4050.8	13.4	7.3	4.8	2.5	0.745
350325	3133.4	940.5	4073.9	8.6	5.9	4.7	1.3	0.500
353906	3132.5	940.7	4073.2	12.7	6.7	4.6	2.1	0.500
357643	3132.9	923.1	4056.0	16.1	7.5	4.6	2.9	0.500

Table 3-4 Characteristics of example transfers from the family labeled "Efficient Transfers" in Fig. 3-13 [174] (Copyright © 2008 by American Astronautical Society Publications Office, all rights reserved, reprinted with permission of the AAS).

x_0 (km)	ΔV_{LEO} (m/s)	ΔV_{MI} (m/s)	Total ΔV (m/s)	Inc (deg)	Transfer Δt (days)	Bridge Δt (days)	Manifold Δt (days)	τ
316507	3108.9	561.7	3670.6	1.4	22.3	3.4	18.9	0.204
317035	3111.3	531.7	3643.0	5.0	23.5	3.4	20.1	0.281
317353	3112.1	524.8	3636.9	5.7	24.0	3.5	20.5	0.312
317721	3112.9	519.0	3631.9	6.3	24.6	3.5	21.1	0.348
318219	3113.8	513.2	3627.0	6.9	25.4	3.6	21.8	0.398
318745	3114.7	508.7	3623.4	7.5	26.2	3.6	22.6	0.453
319497	3115.8	504.6	3620.3	8.1	27.4	3.7	23.7	0.525
320179	3116.7	502.3	3619.0	8.6	28.3	3.7	24.6	0.583
320899	3117.6	501.2	3618.8	9.1	29.2	3.8	25.5	0.645
321932	3118.7	501.1	3619.8	9.6	30.7	3.8	26.8	0.743

- The bridge segments that do connect the spacecraft nearly directly with the halo orbit appear to do so in an organized manner. For halo orbits with x_0-values below a value of approximately 345,000 km, the bridge segments connect the spacecraft with the far side of the halo orbit. Beyond x_0-values of 345,000 km,

Table 3-5 Characteristics of example transfers from those labeled "Complex Transfers" in Fig. 3-13 [174] (Copyright © 2008 by American Astronautical Society Publications Office, all rights reserved, reprinted with permission of the AAS).

x_0 (km)	ΔV_{LEO} (m/s)	ΔV_{MI} (m/s)	Total ΔV (m/s)	Inc (deg)	Transfer Δt (days)	Bridge Δt (days)	Manifold Δt (days)	τ
348011	3137.0	430.7	3567.6	31.9	28.4	4.2	24.2	0.476
352619	3135.9	587.1	3723.0	32.7	31.7	4.5	27.3	0.904
354615	3136.4	669.3	3805.7	19.7	29.9	4.3	25.6	0.703
358106	3134.2	501.1	3635.3	24.8	31.2	4.9	26.3	0.499
358150	3131.1	519.1	3650.2	15.7	31.8	4.6	27.2	0.499

that is, for very large z-amplitude halo orbits, the optimal direct transfers tend to connect closer to the near-side of the halo orbit. This pattern may be observed in the plots shown around the perimeter of Fig. 3-14.

- A family of very efficient direct transfers of this kind appears for transfers to halo orbits with x_0-values between approximately 316,000 km and approximately 323,000 km. The bridge segments of these transfers connect the spacecraft with the first apogee of the manifold segments after the manifold segments traverse to the opposite side of the Earth–Moon system. This family of transfers may be seen on the left side of the figures and corresponds to halo orbits that have small z-amplitudes. Table 3-4 summarizes additional details about these transfers.

- A few transfers have been found that require less total ΔV than the vast majority of locally optimal transfers. These transfers appear toward the lower right portion of the plot shown in Fig. 3-13 and are labeled as complex transfers. These transfers tend to involve several close flybys of the Moon. This study has not fully explored these transfers, since they are much more complicated by nature, but Table 3-5 provides details about several example transfers of this type.

- The transfers shown in Figs. 3-13 and 3-14 implement LEO parking orbits with ecliptic inclinations anywhere between 0 deg and 50 deg. The equatorial inclination, by comparison, depends on the specific launch date and varies from the ecliptic inclination by as much as ± 23.45 deg.

- The duration of time required to transfer within 100 km of the halo orbit may be anywhere between 5–30 days. Transfers may certainly be constructed that require more time; however, these transfers are not considered in this study since they may be more influenced by the Sun's gravity.

- The least-expensive transfers to lunar L_1 halo orbits following their exterior stable manifolds generally require a total ΔV no smaller than approximately

3.60 km/s, depending on the halo orbit of choice. Halo orbits with x_0-values greater than approximately 345,000 km tend to require more total ΔV: in the range of 4.05 km/s $\leq \Delta V \leq$ 4.08 km/s.

In many practical missions, the launch vehicle provides a set amount of ΔV, given a payload mass, and mission designers must optimize their transfer trajectories around that performance. Hence, many times it is useful to consider the two transfer maneuvers separately as well as the total cost of the transfer. Figure 3-15 shows the magnitudes of the two maneuvers separately, which combine to produce the total ΔV cost of the transfers shown in Figs. 3-13 and 3-14. One can see that nearly all of the transfers require the magnitude of the trans-lunar injection maneuver (ΔV_{LEO}) to be between 3.120 and 3.136 km/s. This suggests that the same launch vehicle can perform the trans-lunar injection maneuver for nearly all of these transfers given the same payload mass. Although it is difficult to see in these plots, the least-expensive transfers require the most-expensive ΔV_{LEO}-magnitudes. The second maneuver, ΔV_{MI}, contributes most of the variations seen in the total cost of the lunar transfer.

3.3.4.2 Survey of Interior Transfers to L_1 Halo Orbits This section presents the survey of transfers constructed between 185-km LEO parking orbits and the interior stable manifold of halo orbits in the family of lunar L_1 halo orbits. Figure 3-16 shows the cost of many such example transfers, where several families of locally optimal transfers have been plotted in a more prominent shade. Other nonoptimal transfers have been scattered about the plot to demonstrate that an entire field of options are available. To help identify the trends and differences between each type of transfer, Fig. 3-17 shows plots of several example transfers and Tables 3-6 through 3-9 summarize the characteristics of many of these transfer types.

The following observations may be made after studying the plots shown in Figs. 3-16 and 3-17 and the data displayed in Tables 3-6 through 3-9:

- The same types of *fast* transfers exist to L_1 halo orbits via their interior stable manifolds as via their exterior stable manifolds, because the manifold segments of those transfers do not extend far beyond the halo orbits. Hence, the cost and performance of such fast transfers closely resemble the cost of the fast transfers explored in Section 3.3.4.1. This is apparent when comparing the data shown in Tables 3-3 and 3-7.

- Many families of longer-duration transfers exist that often require less total ΔV than the faster transfers. Examples of these transfers may be seen in the lower left and lower right regions of Figs. 3-16 and 3-17, as well as in Tables 3-8 and 3-9. In general, each of these transfers involves at least one close lunar encounter, and many are constructed by intersecting the transfer's bridge segment with a point very near apogee of the transfer's manifold segment.

- The transfers shown in Figs. 3-16 and 3-17 implement LEO parking orbits with ecliptic inclinations anywhere between 0 deg and 60 deg. Again, the equatorial inclinations of the LEO parking orbits depend on the launch date.

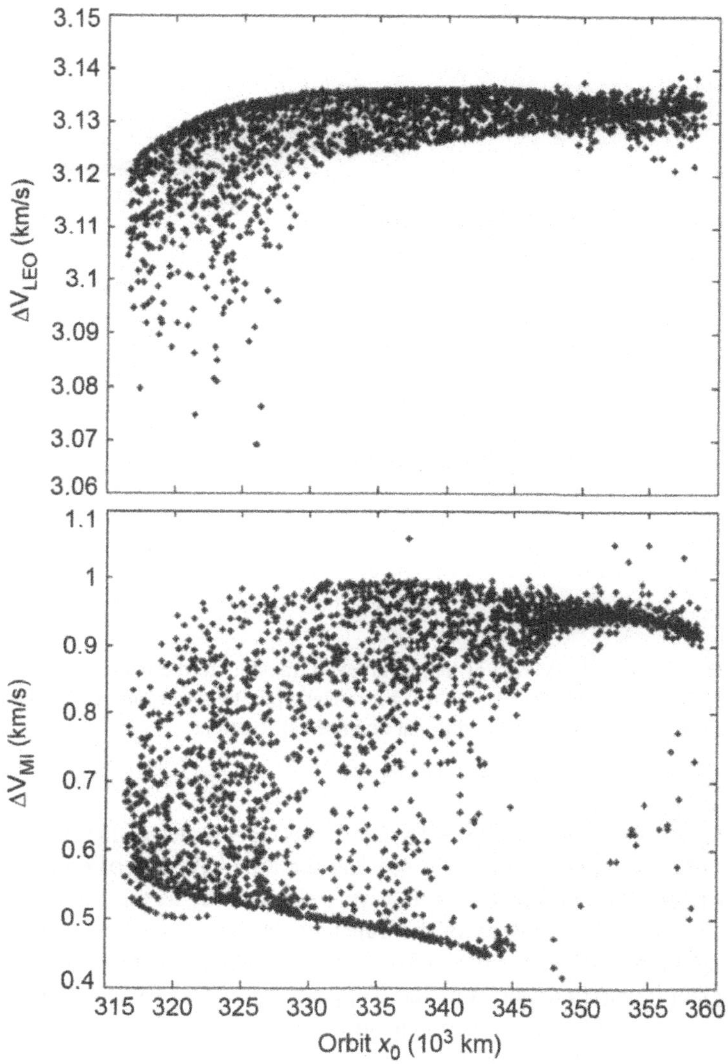

Figure 3-15 The two transfer maneuver magnitudes that combine to produce the total ΔV cost of the transfers shown in Figs. 3-13 and 3-14. Top: The magnitudes of the trans-lunar injection maneuvers (ΔV_{LEO}) in each transfer; bottom: The magnitudes of the manifold-insertion maneuvers (ΔV_{MI}) in each transfer.

- The least-expensive transfers to lunar L_1 halo orbits following their interior stable manifolds generally require a total ΔV no smaller than approximately 3.60 km/s, depending on the halo orbit of choice. The trend is very similar to that presented in Section 3.3.4.1 for short-duration lunar halo transfers.

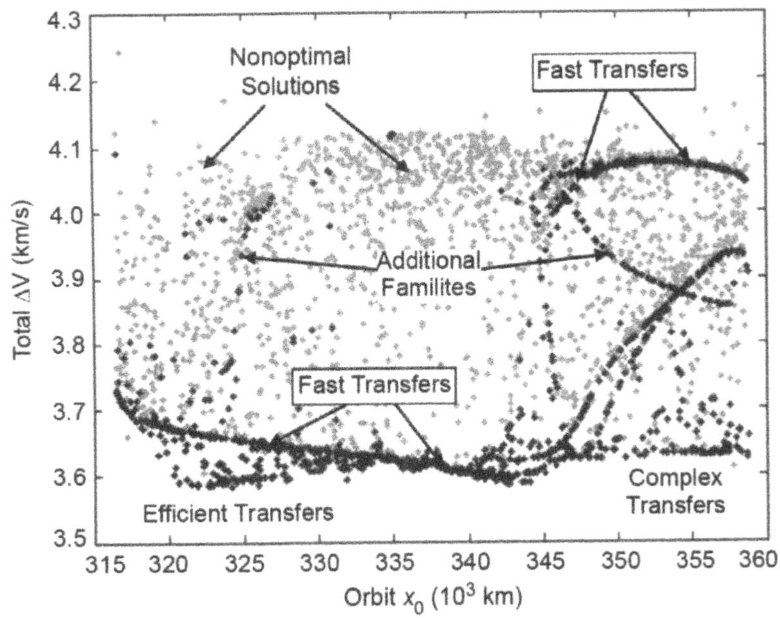

Figure 3-16 The total ΔV cost of many surveyed transfers to the interior stable manifold of orbits in the family of lunar L_1 halo orbits. Dark points correspond to locally optimal trajectories; faint points represent additional nonoptimal solutions.

To continue this analysis, Fig. 3-18 shows the magnitudes of the two deterministic maneuvers separately. One can see that the total ΔV cost of each transfer is divided between the two maneuvers in a very similar way as the exterior transfers shown in Section 3.3.4.1. Many of the transfers require a trans-lunar injection maneuver magnitude (ΔV_{LEO}) between 3.120 and 3.136 km/s. Some of the families of more-efficient transfers require smaller ΔV_{LEO}-magnitudes. Even with these slight reductions, the second maneuver, ΔV_{MI}, still contributes most of the variations seen in the total cost of the lunar transfer.

3.3.4.3 Survey of Exterior Transfers to L_2 Halo Orbits This section presents the survey of transfers constructed between 185-km LEO parking orbits and the exterior stable manifold of halo orbits in the family of lunar L_2 halo orbits. Figure 3-19 shows the cost of many such example transfers to halo orbits in the family, including a Pareto front of optimal transfers. To help identify the trends and differences between each type of transfer, Fig. 3-20 shows plots of several example transfers, and Tables 3-10 through 3-13 summarize the characteristics of many sample transfers of this type.

Figures 3-19 and 3-20 show many interesting patterns. After studying the transfers presented in these figures, and the data summarized in Tables 3-10 through 3-13, the following observations have been made:

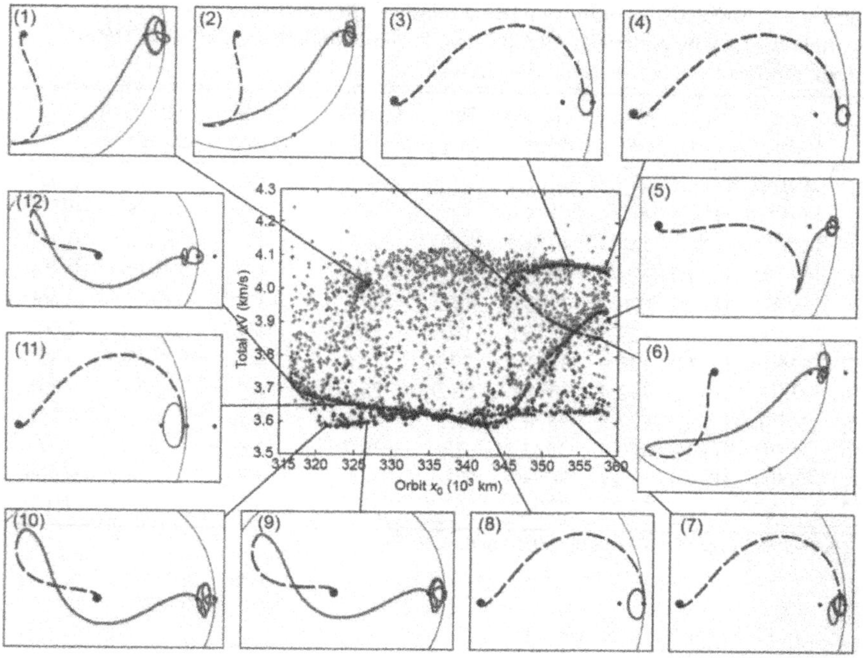

Figure 3-17 Several example transfers between 185-km LEO parking orbits and the interior stable manifold of lunar L_1 halo orbits. The parameters of the numbered transfers are summarized in Table 3-6. Dark points correspond to locally optimal trajectories; faint points represent additional nonoptimal solutions.

- Two dominant types of efficient transfers exist that transfer to the halo orbits' exterior stable manifold. The first one, indicated by the upper prominent curve in Fig. 3-19, includes transfers whose bridge segments connect the spacecraft directly with the far side of the L_2 halo orbit. These are short-duration transfers, characterized by data shown in Table 3-11, and they are similar to the short-duration transfers explored in Sections 3.3.4.1 and 3.3.4.2. The second dominant type of transfer, indicated by the lower prominent curve in Fig. 3-19, includes trajectories whose bridge segments send the spacecraft well beyond the Moon, where they intersect the corresponding manifold segments near the segments' apogee points. The first type of transfer requires only 5–6 days to accomplish, whereas the second type requires as many as 35–50 days before the spacecraft is within 100 km of the lunar halo orbit.
- Additional benefit may be obtained for transfers to L_2 halos with x_0-values greater than approximately 425,000 km by flying near the Moon en route to the ΔV_{MI} maneuver. The lunar flyby reduces the total required ΔV, albeit at the expense of more sensitive navigation requirements near that lunar flyby.

Table 3-6 Characteristics of the numbered transfers identified in Fig. 3-17 [174] (Copyright © 2008 by American Astronautical Society Publications Office, all rights reserved, reprinted with permission of the AAS).

#	x_0 (km)	ΔV_{LEO} (m/s)	ΔV_{MI} (m/s)	Total ΔV (m/s)	Inc* (deg)	Transfer Δt (days)	Bridge Δt (days)	Manifold Δt (days)	τ
1	326808	3111.3	902.6	4013.9	12.2	48.8	3.5	45.4	0.447
2	348529	3118.2	837.7	3955.9	12.1	58.4	3.9	54.6	0.193
3	353325	3132.6	941.9	4074.5	12.0	8.7	4.6	4.1	0.772
4	358234	3129.9	920.8	4050.8	18.3	24.1	4.5	19.6	0.499
5	358745	3116.7	796.4	3913.1	17.5	39.4	3.9	35.5	0.660
6	357400	3127.1	729.5	3856.6	10.6	57.3	4.3	53.0	0.008
7	353001	3133.0	498.9	3631.9	27.0	37.6	4.7	32.9	0.456
8	341601	3136.2	462.4	3598.6	51.9	25.7	4.9	20.8	0.848
9	326786	3121.0	477.5	3598.5	9.3	48.5	3.9	44.6	0.756
10	321441	3117.8	469.6	3587.4	7.9	39.7	3.8	35.9	0.121
11	325594	3133.8	520.1	3653.9	39.5	6.2	4.8	1.4	0.609
12	317083	3111.4	593.1	3704.5	6.7	31.5	3.5	28.0	0.079

*The inclination of the LEO parking orbit in the CRTBP.

Table 3-7 Characteristics of example fast transfers identified in Figs. 3-16 and 3-17 [174] (Copyright © 2008 by American Astronautical Society Publications Office, all rights reserved, reprinted with permission of the AAS).

x_0 (km)	ΔV_{LEO} (m/s)	ΔV_{MI} (m/s)	Total ΔV (m/s)	Inc (deg)	Transfer Δt (days)	Bridge Δt (days)	Manifold Δt (days)	τ
316536	3121.0	607.9	3728.9	7.0	8.2	4.3	3.9	0.834
318562	3126.3	556.5	3682.9	23.3	7.5	4.5	3.0	0.738
320977	3130.0	538.2	3668.2	31.3	6.5	4.7	1.8	0.635
324263	3133.1	523.2	3656.4	37.5	9.5	4.7	4.7	0.892
328038	3134.8	513.0	3647.8	41.7	4.9	4.8	0.1	0.487
331309	3135.5	500.8	3636.2	44.2	9.2	4.8	4.3	0.905
335684	3135.9	485.5	3621.5	47.6	9.2	4.8	4.4	0.946
339602	3136.1	469.3	3605.4	50.0	7.7	4.8	2.9	0.805
341979	3136.1	456.2	3592.3	49.9	9.5	4.8	4.7	0.025
345722	3135.9	859.6	3995.5	30.4	6.6	4.9	1.7	0.671
347918	3134.8	924.3	4059.2	10.1	7.1	4.8	2.4	0.707
349968	3133.6	939.0	4072.6	8.4	8.5	4.6	3.8	0.828
351974	3132.9	942.7	4075.6	10.4	8.3	4.6	3.7	0.764
354725	3132.5	939.9	4072.5	13.7	5.0	4.7	0.3	0.256
358661	3133.3	910.4	4043.7	16.7	7.8	4.6	3.2	0.500

Table 3-8 Characteristics of example transfers from the region labeled "Efficient Transfers" in Fig. 3-16. The rows of the table are organized in groups, where each group describes example transfers in a different family [174] (Copyright © 2008 by American Astronautical Society Publications Office, all rights reserved, reprinted with permission of the AAS).

x_0 (km)	ΔV_{LEO} (m/s)	ΔV_{MI} (m/s)	Total ΔV (m/s)	Inc (deg)	Transfer Δt (days)	Bridge Δt (days)	Manifold Δt (days)	τ
320688	3117.0	473.3	3590.3	7.9	38.3	3.8	34.6	0.057
321925	3118.2	468.3	3586.6	7.9	40.6	3.8	36.8	0.173
323219	3119.4	470.2	3589.6	8.0	44.6	3.8	40.8	0.456
324345	3120.2	476.2	3596.4	8.3	46.2	3.9	42.4	0.556
326087	3120.9	477.5	3598.4	8.8	47.8	3.9	43.9	0.686
327737	3121.3	480.6	3601.8	9.5	49.6	3.9	45.7	0.859
327189	3127.3	497.7	3625.0	15.3	45.4	4.3	41.1	0.658
328326	3129.4	491.6	3621.0	20.1	46.3	4.5	41.8	0.745
329353	3131.5	486.5	3618.0	25.9	47.1	4.7	42.5	0.830
330278	3133.7	480.9	3614.6	33.9	47.9	4.9	43.1	0.933
322265	3126.8	495.6	3622.4	18.1	31.2	4.4	26.8	0.212
325061	3128.2	498.4	3626.6	20.1	36.2	4.4	31.8	0.599
326012	3129.2	496.2	3625.4	22.5	36.9	4.5	32.5	0.669
328613	3130.8	496.3	3627.1	26.6	39.5	4.6	34.9	0.947
329737	3136.3	486.6	3622.8	52.5	29.0	4.9	24.1	0.861
329778	3136.5	486.8	3623.3	52.9	30.0	5.0	25.0	0.972
330195	3136.2	484.3	3620.6	51.8	33.7	4.8	28.8	0.326
330545	3136.2	487.4	3623.6	51.5	30.4	4.9	25.5	0.018

These transfers may be seen in the lower right portions of the plots shown in Figs. 3-19 and 3-20; Tables 3-12 and 3-13 compare the characteristics of transfers with and without the lunar flyby.

- The transfers shown in Figs. 3-19 and 3-20 implement LEO parking orbits with different ranges of ecliptic inclinations. The transfers indicated by the upper prominent curve in Fig. 3-19 may be launched from LEO parking orbits with ecliptic inclination values anywhere in the range of 0 deg–25 deg. Those transfers indicated by the lower prominent curve have a narrower range of 0 deg–19 deg. Finally, the lowest ΔV transfers shown in the lower right portion of the figures may implement LEO parking orbits with a much more broad range of ecliptic inclinations: anywhere in the range of 20 deg–120 deg and possibly beyond.

Table 3-9 Characteristics of example transfers in the region labeled "Complex Transfers" in Fig. 3-16. The examples summarized here belong to many different families, demonstrating the variety of transfers that exist [174] (Copyright © 2008 by American Astronautical Society Publications Office, all rights reserved, reprinted with permission of the AAS).

x_0 (km)	ΔV_{LEO} (m/s)	ΔV_{MI} (m/s)	Total ΔV (m/s)	Inc (deg)	Transfer Δt (days)	Bridge Δt (days)	Manifold Δt (days)	τ
351166	3134.5	501.3	3635.8	34.8	37.4	4.8	32.6	0.448
351444	3121.5	506.1	3627.5	11.6	49.9	3.9	46.0	0.515
352138	3135.3	528.4	3663.7	40.2	35.2	4.7	30.5	0.260
353639	3120.0	522.0	3642.0	12.7	50.0	3.8	46.3	0.508
355251	3135.1	529.1	3664.2	35.3	35.3	4.7	30.5	0.250
355550	3131.2	504.5	3635.7	19.5	37.5	4.6	32.9	0.405
355848	3121.8	508.6	3630.4	12.1	49.2	4.0	45.3	0.454
358221	3135.3	532.3	3667.6	31.6	35.1	4.7	30.4	0.108
358332	3130.8	501.3	3632.1	15.9	37.9	4.5	33.4	0.317
358677	3122.8	502.7	3625.5	13.0	50.2	4.0	46.2	0.374
358837	3135.6	523.4	3659.0	28.1	36.6	4.7	31.8	0.142

- The total ΔV cost of the least-expensive transfers to lunar L_2 halo orbits following their exterior stable manifolds greatly depend on which halo orbit is being targeted. Halo orbits with x_0-values less than 385,000 km, that is, very large z-amplitude halo orbits, require no less than approximately 3.95 km/s to reach in this way. The cost steadily decreases for halo orbits with x_0-values between 385,000 km and 415,000 km. Halo orbits with x_0-values greater than approximately 415,000 km, that is, very low z-amplitude halo orbits, require no less than approximately 3.77 km/s to reach in this way. Finally, those halo orbits that may be reached using an additional lunar flyby en route have a total ΔV requirement that may be reduced to as low as approximately 3.69 km/s.

Once again, to continue this analysis, Fig. 3-21 shows the magnitudes of the two transfer maneuvers separately. One can see that the total ΔV cost of each transfer is divided between the two maneuvers in a similar way as the transfers shown in Sections 3.3.4.1 and 3.3.4.2. However, in these exterior transfers to the L_2 halo orbits, the first maneuver, ΔV_{LEO}, must perform somewhat larger ΔVs than it did for transfers to L_1 halo orbits: between 3.145 and 3.185 km/s. The second maneuver, ΔV_{MI}, still contributes most of the variations seen in the total cost of the lunar transfer.

3.3.4.4 Survey of Interior Transfers to L_2 Halo Orbits This section presents the survey of transfers constructed between 185-km LEO parking orbits and the interior stable manifold of halo orbits in the family of lunar L_2 halo orbits. Figure 3-22 shows the cost of many such example transfers to halo orbits in the family. Several families of locally optimal transfers have been highlighted in a more prominent

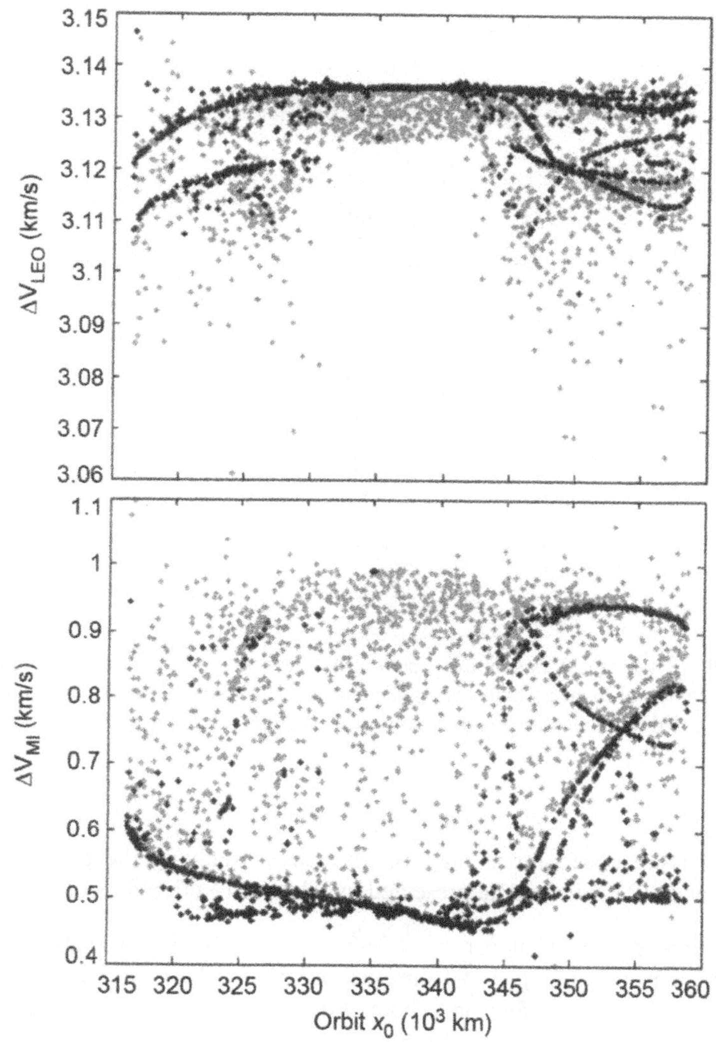

Figure 3-18 The two transfer maneuver magnitudes that combine to produce the total ΔV cost of the transfers shown in Figs. 3-16 and 3-17. Dark points correspond to locally optimal trajectories; faint points represent additional nonoptimal solutions. Top: The magnitudes of the trans-lunar injection maneuvers (ΔV_{LEO}) in each transfer; Bottom: The magnitudes of the manifold-insertion maneuvers (ΔV_{MI}) in each transfer.

shade to be distinguished from the scattered nonoptimal transfers. To help identify the trends and differences between each type of transfer, Fig. 3-23 shows plots of several example transfers and Tables 3-14 through 3-16 summarize the characteristics of many sample transfers of this type.

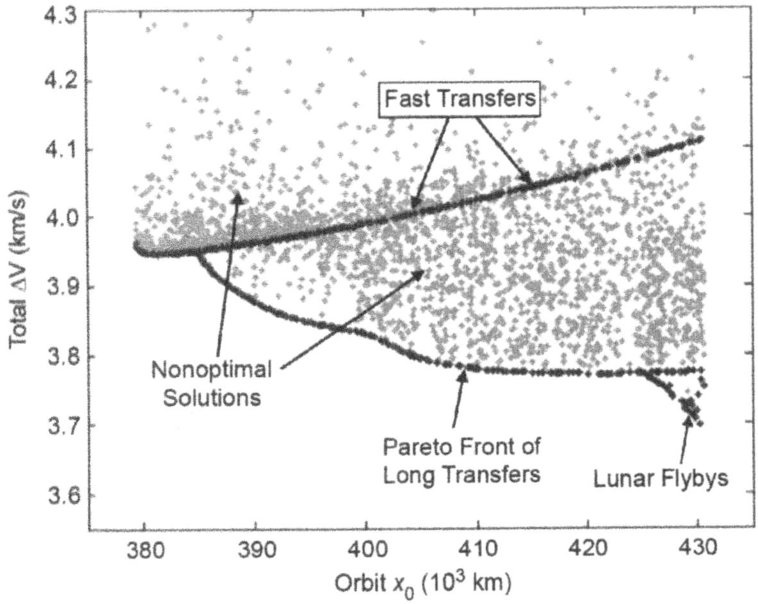

Figure 3-19 The total ΔV cost of many surveyed transfers to the exterior stable manifold of orbits in the family of lunar L_2 halo orbits, including a Pareto front of optimal solutions. Dark points correspond to locally optimal trajectories; faint points represent additional nonoptimal solutions.

The following observations may be made after studying the plots shown in Figs. 3-22 and 3-23 and the data presented in Tables 3-14 through 3-16:

- The most prominent upper curve in Fig. 3-22 is nearly identical to the most prominent curve in Fig. 3-19 from Section 3.3.4.3. This is because the manifold segments of the transfers along both of those curves do not depart far from the corresponding halo orbits. Both of these curves correspond to the shortest-duration transfers to lunar L_2 halo orbits, although they are certainly not the least-expensive in most cases.

- Many transfers exist that may be modeled as a transfer from LEO to an orbit about the Moon's L_1 point, followed by a transfer from L_1 to L_2. It makes sense, then, that many transfers to L_2 require no more ΔV than transfers to L_1. These transfers require more transfer time than the shortest-duration transfers previously described.

- The transfers shown in the lower left plots in Fig. 3-23 include manifold segments that extend well beyond the lunar vicinity. The bridge segments in those transfers connect with a point near one of the apogee points of the corresponding manifold segments. Several such families exist; in fact, a

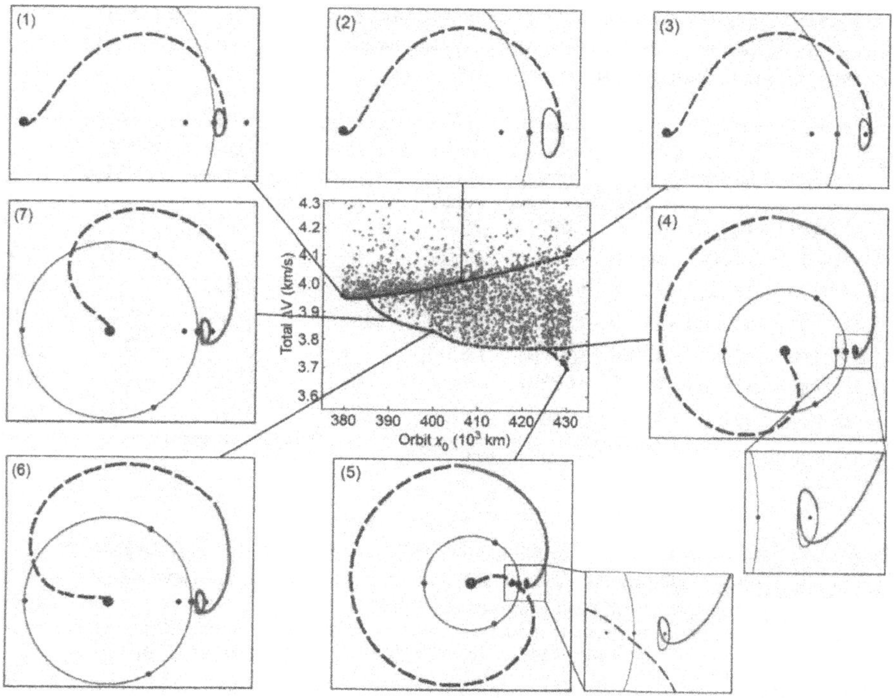

Figure 3-20 Several example transfers between 185-km low Earth orbits and the exterior stable manifold of lunar L_2 halo orbits. Dark points correspond to locally optimal trajectories; faint points represent additional nonoptimal solutions. Parameters of the transfers shown are summarized in Table 3-10.

different family may be produced for transfers that connect with any given apogee of the corresponding manifold segments. Figure 3-23 shows two plots of transfers that connect with the manifold segment's first apogee point opposite of the Moon, as well as one plot of a transfer that connects with the manifold segment's second apogee point. Families of transfers that intersect with later apogee points have not been produced here because they require longer transfer durations. The characteristics of example transfers from several of these families are shown in Table 3-16.

- There exist many types of transfers that make at least one close lunar passage en route to the L_2 halo orbit. It is apparent when studying the figures that the total required ΔV of a transfer is very dependent on the distance between the Moon and the manifold-insertion maneuver. That is, as the proximity of ΔV_{MI} with the Moon is reduced the total required ΔV in the transfer is reduced. This makes sense because more of the energy change in the transfer is performed

Table 3-10 Characteristics of the numbered transfers identified in Fig. 3-20 [174] (Copyright © 2008 by American Astronautical Society Publications Office, all rights reserved, reprinted with permission of the AAS).

#	x_0 (km)	ΔV_{LEO} (m/s)	ΔV_{MI} (m/s)	Total ΔV (m/s)	Inc* (deg)	Transfer Δt (days)	Bridge Δt (days)	Manifold Δt (days)	τ
1	379441	3142.6	820.4	3963.0	22.9	25.6	5.3	20.3	0.046
2	406016	3150.5	860.0	4010.4	13.4	9.4	5.9	3.5	0.740
3	430307	3152.4	957.0	4109.4	1.7	13.3	6.0	7.3	0.970
4	427287	3185.5	588.5	3774.1	3.2	44.2	16.9	27.3	0.465
5	430167	3162.2	536.4	3698.6	85.4	51.5	24.2	27.3	0.593
6	399548	3173.9	659.5	3833.4	9.1	39.2	11.9	27.3	0.004
7	391748	3169.2	696.2	3865.3	9.8	37.4	10.1	27.3	0.203

*The inclination of the LEO parking orbit in the CRTBP.

Table 3-11 Characteristics of example fast transfers observed in Figs. 3-19 and 3-20 [174] (Copyright © 2008 by American Astronautical Society Publications Office, all rights reserved, reprinted with permission of the AAS).

x_0 (km)	ΔV_{LEO} (m/s)	ΔV_{MI} (m/s)	Total ΔV (m/s)	Inc (deg)	Transfer Δt (days)	Bridge Δt (days)	Manifold Δt (days)	τ
382074	3146.3	801.5	3947.7	19.0	6.2	5.6	0.6	0.547
388716	3148.2	811.5	3959.7	17.1	7.5	5.7	1.8	0.636
400469	3149.9	842.4	3992.3	14.8	7.3	5.9	1.5	0.594
407319	3150.6	864.3	4014.8	13.4	6.2	6.0	0.2	0.506
412311	3151.0	881.8	4032.8	12.1	6.2	6.0	0.3	0.506
418589	3151.4	905.9	4057.3	10.0	6.2	6.0	0.3	0.506
423782	3151.8	927.8	4079.6	7.7	6.3	6.0	0.3	0.506
430202	3152.1	957.8	4109.9	2.0	6.3	6.0	0.2	0.506

deeper in a gravity well, where the spacecraft is traveling faster. The transfer shown in the lower right plot of Fig. 3-23 is a good example of this effect: its ΔV_{MI} is performed very close to the Moon; hence, its total ΔV cost is lower.

- Several of the nonoptimal transfers (plotted in a lighter shade in Fig. 3-22) appear to require less total ΔV than other locally optimal transfers. It is likely that those nonoptimal transfers are in a different class of transfer, that is, they require a different combination of lunar flybys en route to the L_2 halo orbit, such that the optimized transfers of that class require a longer transfer time. Only transfers requiring fewer than 60 days are plotted in the figures; the locally optimal transfers that require more than 60 days, and perhaps less total ΔV, are not displayed.

Table 3-12 Characteristics of example long-duration transfers observed in Figs. 3-19 and 3-20 that do not include a lunar flyby [174] (Copyright © 2008 by American Astronautical Society Publications Office, all rights reserved, reprinted with permission of the AAS).

x_0 (km)	ΔV_{LEO} (m/s)	ΔV_{MI} (m/s)	Total ΔV (m/s)	Inc (deg)	Transfer Δt (days)	Bridge Δt (days)	Manifold Δt (days)	τ
384950	3152.7	795.4	3948.1	12.3	33.2	6.4	26.8	0.650
392617	3169.9	689.9	3859.8	9.7	37.6	10.4	27.3	0.172
399992	3173.8	657.8	3831.5	9.1	39.1	11.8	27.3	0.988
407598	3178.5	607.7	3786.1	8.6	41.0	13.7	27.3	0.725
415027	3181.8	593.0	3774.7	7.2	42.4	15.1	27.3	0.600
422804	3184.3	588.5	3772.7	5.0	43.6	16.3	27.3	0.508
430370	3186.4	589.4	3775.8	0.9	44.7	17.4	27.3	0.440

Table 3-13 Characteristics of example long-duration transfers seen in Figs. 3-19 and 3-20 that do include a lunar flyby in their corresponding bridge segments.

x_0 (km)	ΔV_{LEO} (m/s)	ΔV_{MI} (m/s)	Total ΔV (m/s)	Inc (deg)	Transfer Δt (days)	Bridge Δt (days)	Manifold Δt (days)	τ
424719	3183.8	592.0	3775.9	22.1	50.9	23.6	27.3	0.684
426208	3182.5	578.6	3761.1	23.5	50.8	23.6	27.3	0.655
427590	3179.4	562.6	3742.0	30.9	51.0	23.7	27.3	0.631
428819	3174.2	549.2	3723.4	48.4	51.2	23.9	27.3	0.612
430167	3162.2	536.4	3698.6	85.4	51.5	24.2	27.3	0.593

- The transfers shown in Figs. 3-22 and 3-23 implement LEO parking orbits with ecliptic inclinations generally in the range of 0–55 deg.
- The duration of time required to transfer within 100 km of the halo orbit may be anywhere between 5 and 60 days.
- The least-expensive transfers to lunar L_2 halo orbits following their exterior stable manifolds generally require a total ΔV no smaller than approximately 3.60 or 3.65 km/s, depending on the halo orbit of choice.

The final analysis in this section is to study the performance of the two maneuvers separately for each interior lunar L_2 halo transfer. Figure 3-24 shows the magnitudes of the two transfer maneuvers. One can see that the majority of each transfer's ΔV cost is performed in the first maneuver, ΔV_{LEO}, but the variations in the magnitude of ΔV_{LEO} between transfers is very small, ranging between approximately 3.11 and 3.15 km/s. The second maneuver, ΔV_{MI}, although much smaller, has a great deal more variability and therefore determines the total cost of the transfer.

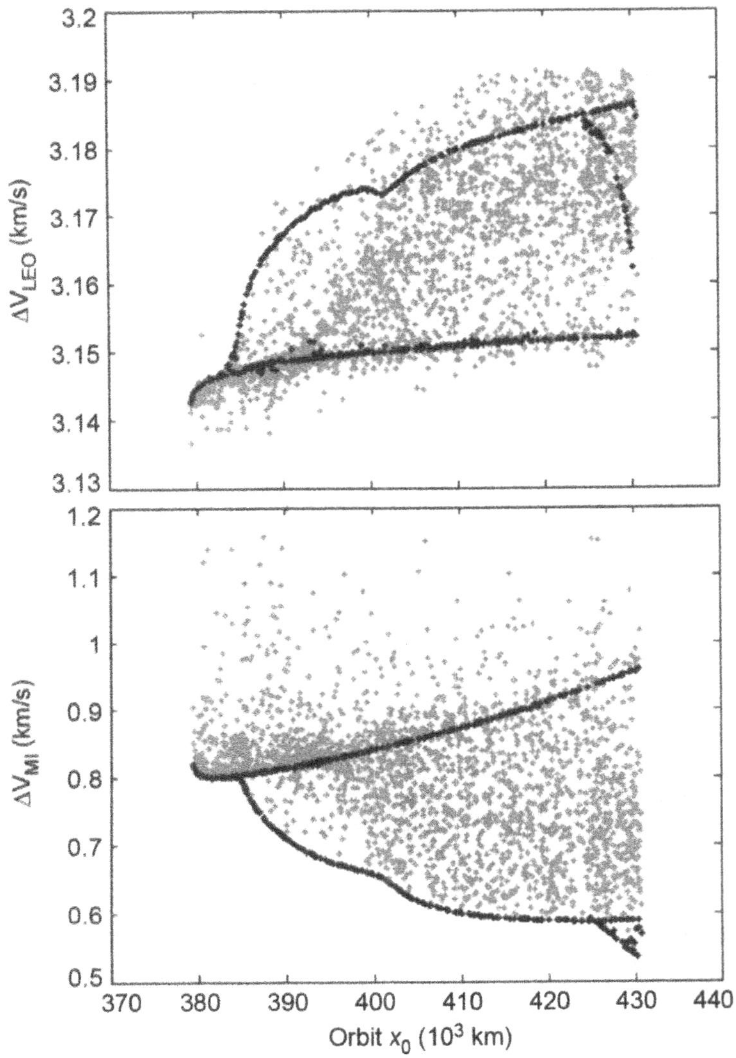

Figure 3-21 The two transfer maneuver magnitudes that combine to produce the total ΔV cost of the transfers shown in Figs. 3-19 and 3-20. Dark points correspond to locally optimal trajectories; faint points represent additional nonoptimal solutions. Top: The magnitudes of the trans-lunar injection maneuvers (ΔV_{LEO}) in each transfer. Bottom: The magnitudes of the manifold-insertion maneuvers (ΔV_{MI}) in each transfer.

3.3.5 Discussion of Results

The previous four sections surveyed four different types of direct lunar halo transfers; this section studies them together to draw several overall conclusions.

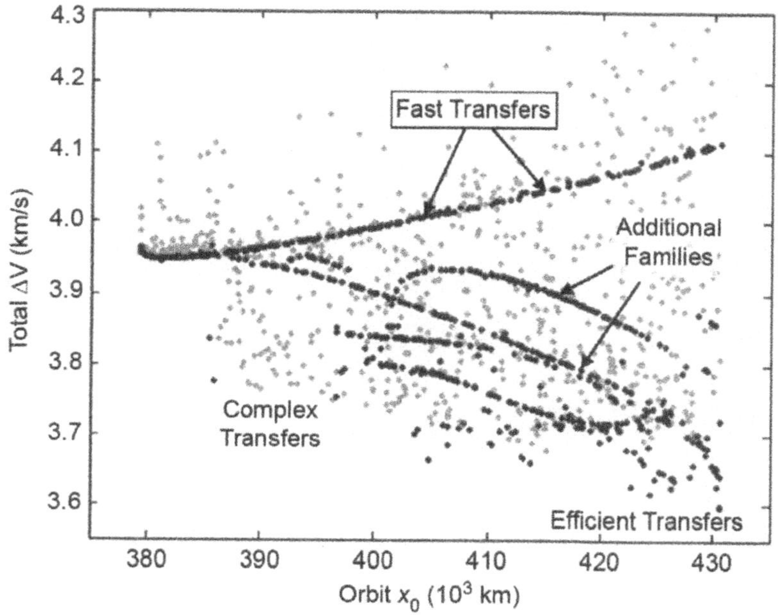

Figure 3-22 The total ΔV cost of many surveyed transfers to the interior stable manifold of orbits in the family of lunar L_2 halo orbits. Dark points correspond to locally optimal trajectories; faint points represent additional nonoptimal solutions.

Each of the results presented above implemented direct lunar transfers found by searching through only one half of the stable manifold of the targeted halo orbits. In reality, it most likely doesn't matter whether a particular trajectory implements an interior or an exterior transfer—just that the spacecraft arrives at the halo orbit in some way. Figure 3-25 shows a summary of the ΔV requirements for both interior and exterior transfers to lunar L_1 halo orbits, plotted in the same axes. Hence, Fig. 3-25 may be used to identify the least-expensive transfers to any lunar L_1 halo orbit no matter which type of manifold is taken. Figure 3-26 shows the same ΔV summary for transfers to lunar L_2 halo orbits.

Theoretically, it is possible to transfer to any given lunar L_2 halo orbit from a lunar L_1 halo orbit with the same Jacobi constant, and vice versa. The dynamical systems methodology presented in this work has been used in previous studies to construct low-energy orbit transfers and orbit chains [162]. To explore this concept further, Fig. 3-27 shows a plot of the Jacobi constant, C, of the lunar halo orbits surveyed in this work as a function of the halo orbits' x_0-values. One can see that there is a lunar L_1 halo orbit with the same Jacobi constant as each and every lunar L_2 halo orbit in this study. The family of lunar L_2 halo orbits includes orbits with Jacobi constants in the approximate range $3.015 < C < 3.152$; the family of lunar L_1 halo orbits spans that entire range and then extends a bit further in each direction. In theory, it is thus

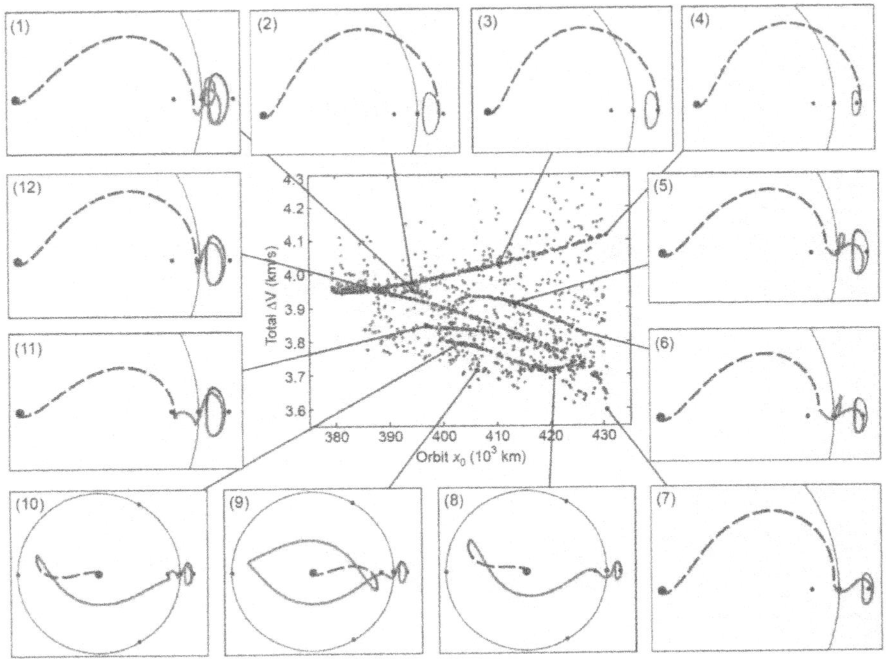

Figure 3-23 Several example transfers between 185-km LEO parking orbits and the interior stable manifold of lunar L_2 halo orbits. Dark points correspond to locally optimal trajectories; faint points represent additional nonoptimal solutions. The parameters of the numbered transfers are summarized in Table 3-14.

possible to transfer to any lunar L_2 halo orbit from the corresponding lunar L_1 halo orbit for very little energy.

Figure 3-28 shows the same results shown in Figs. 3-25 and 3-26, but now plotted as a function of the halo orbits' Jacobi constant values (C-values) rather than their x_0-values. In this way, one can observe the minimum total ΔV required to reach any halo orbit of a particular Jacobi constant. Then, once in that orbit, one can theoretically transfer to a different desired orbit, provided the desired orbit has the same Jacobi constant. The left part of Fig. 3-28 shows transfers that may be used to reach only lunar L_1 halo orbits, since there are no lunar L_2 halo orbits with Jacobi constant values below 3.015. Figure 3-28 also shows that if a low-energy transfer can be found between halo orbits about L_1 and L_2 of a given Jacobi constant, it is almost always more efficient to transfer directly to the lunar L_1 halo orbit first, and then take the low-energy transfer over to the lunar L_2 halo orbit.

Halo orbits exist in two families: a northern and a southern family as illustrated in Fig. 2-25. Every lunar L_1 halo orbit explored in this work has been a member of the

Table 3-14 Characteristics of the numbered transfers identified in Fig. 3-23 [174] (Copyright © 2008 by American Astronautical Society Publications Office, all rights reserved, reprinted with permission of the AAS).

#	x_0 (km)	ΔV_{LEO} (m/s)	ΔV_{MI} (m/s)	Total ΔV (m/s)	Inc* (deg)	Transfer Δt (days)	Bridge Δt (days)	Manifold Δt (days)	τ
1	394370	3132.7	817.6	3950.3	31.9	38.6	4.7	34.0	0.951
2	394096	3149.0	824.6	3973.6	16.2	6.4	5.8	0.6	0.537
3	411239	3150.8	878.7	4029.4	12.4	12.4	5.9	6.5	0.948
4	429222	3152.0	953.8	4105.8	3.6	9.2	5.9	3.3	0.708
5	415075	3129.0	776.3	3905.3	19.9	25.4	4.5	20.9	0.288
6	425204	3126.7	705.0	3831.7	13.0	23.5	4.5	19.0	0.177
7	430641	3133.4	464.8	3598.2	5.6	18.3	5.2	13.1	0.034
8	420255	3112.2	605.2	3717.4	9.7	31.5	3.6	27.9	0.868
9	406534	3112.8	599.5	3712.3	20.9	49.2	3.8	45.4	0.112
10	403368	3113.6	676.7	3790.3	16.3	43.6	3.6	40.0	0.284
11	396769	3122.9	728.2	3851.1	17.9	31.3	4.1	27.2	0.356
12	393789	3135.7	792.6	3928.3	26.9	25.9	4.9	21.0	0.485

*The inclination of the LEO parking orbit in the CRTBP.

Table 3-15 Characteristics of example fast transfers observed in Figs. 3-22 and 3-23 [174] (Copyright © 2008 by American Astronautical Society Publications Office, all rights reserved, reprinted with permission of the AAS).

x_0 (km)	ΔV_{LEO} (m/s)	ΔV_{MI} (m/s)	Total ΔV (m/s)	Inc (deg)	Transfer Δt (days)	Bridge Δt (days)	Manifold Δt (days)	τ
383881	3147.0	802.9	3949.9	18.4	5.7	5.7	0.1	0.497
392708	3148.8	821.0	3969.8	16.5	5.9	5.8	0.1	0.495
401069	3149.9	844.3	3994.2	14.9	6.0	5.9	0.1	0.494
408865	3150.7	869.6	4020.3	13.1	6.0	5.9	0.1	0.494
416071	3151.2	896.0	4047.3	11.0	6.1	6.0	0.1	0.494
429548	3152.0	954.7	4106.7	3.2	6.1	6.1	0.0	0.493

northern family; every lunar L_2 halo orbit has been a member of the southern family. To access the symmetric family of halo orbits, in either case, the transfer must be reflected about the $z = 0$ plane. The only difference that would be noticeable in such a symmetric transfer would be that the LEO parking orbit's inclination relative to the Moon's orbital plane would have the opposite sign.

Table 3-16 Characteristics of example transfers within a collection of seven different sample families observed in Figs. 3-22 and 3-23. The families are identified by the number of the corresponding example plot shown around the perimeter of Fig. 3-23 [174] (Copyright © 2008 by American Astronautical Society Publications Office, all rights reserved, reprinted with permission of the AAS).

x_0 (km)	ΔV_{LEO} (m/s)	ΔV_{MI} (m/s)	Total ΔV (m/s)	Inc (deg)	Transfer Δt (days)	Bridge Δt (days)	Manifold Δt (days)	τ
Family (12)								
390299	3138.3	801.0	3939.3	26.9	27.6	5.0	22.5	0.607
395448	3134.5	787.5	3922.0	27.0	25.1	4.8	20.4	0.429
400189	3130.9	769.1	3900.0	25.6	23.7	4.6	19.1	0.302
405365	3128.0	744.0	3872.0	22.9	22.3	4.4	17.9	0.200
410586	3126.4	715.0	3841.4	19.9	21.3	4.4	16.9	0.126
415399	3125.9	685.8	3811.8	17.7	20.5	4.4	16.1	0.070
Family (1)								
394170	3132.7	818.8	3951.5	31.4	38.7	4.6	34.1	0.961
395343	3131.6	814.3	3945.9	35.1	38.2	4.6	33.6	0.873
395820	3130.3	812.9	3943.3	34.7	37.5	4.6	33.0	0.821
396907	3129.3	808.0	3937.4	31.4	35.7	4.5	31.2	0.737
Family (11)								
396738	3122.9	721.1	3844.0	14.9	30.2	4.1	26.1	0.240
402272	3119.7	717.4	3837.1	11.4	27.8	3.9	23.9	0.070
407548	3117.3	712.2	3829.5	8.4	25.8	3.8	22.0	0.973
413568	3116.3	695.0	3811.4	6.4	23.2	3.8	19.4	0.873
419902	3121.3	651.3	3772.6	10.0	21.0	4.2	16.7	0.808
425400	3126.1	600.1	3726.1	8.2	19.9	4.7	15.2	0.736
430618	3131.1	510.9	3642.0	0.2	19.2	5.2	14.0	0.648
Family (5, 6)								
401972	3121.6	782.7	3904.4	24.7	33.3	4.1	29.2	0.587
406699	3130.4	803.3	3933.7	22.0	28.3	4.6	23.7	0.440
411226	3130.0	793.3	3923.2	21.2	26.6	4.6	22.0	0.350
415688	3128.7	772.4	3901.0	19.3	25.3	4.5	20.8	0.281
420324	3127.5	742.3	3869.8	16.7	24.3	4.5	19.8	0.225
425204	3126.7	705.0	3831.7	13.0	23.5	4.5	19.0	0.177
429490	3126.4	668.3	3794.7	6.7	23.0	4.6	18.4	0.144
Family (8, 10)								
399413	3114.9	687.4	3802.3	15.7	41.4	3.7	37.7	0.145
403744	3112.9	679.1	3792.0	16.7	38.7	3.6	35.0	0.053
408970	3112.5	654.2	3766.7	15.7	35.1	3.5	31.5	0.999
413531	3112.7	626.8	3739.5	13.3	33.4	3.6	29.8	0.936
418227	3112.5	607.8	3720.3	10.7	31.9	3.5	28.4	0.887

Table 3-16 Continued.

x_0 (km)	ΔV_{LEO} (m/s)	ΔV_{MI} (m/s)	Total ΔV (m/s)	Inc (deg)	Transfer Δt (days)	Bridge Δt (days)	Manifold Δt (days)	τ
Family (8, 10) (cont'd)								
422292	3112.0	608.8	3720.7	8.5	31.3	3.5	27.8	0.853
425682	3111.4	640.5	3751.9	6.1	31.5	3.5	28.0	0.837
427668	3110.3	690.6	3801.0	4.2	32.4	3.5	28.9	0.830
Family (9)								
416120	3112.2	603.2	3715.5	10.4	43.7	3.5	40.2	0.950
418419	3111.7	600.1	3711.8	10.1	42.9	3.5	39.4	0.898
420564	3111.4	599.0	3710.4	9.2	42.4	3.5	38.9	0.873
423526	3110.8	605.7	3716.5	7.6	42.3	3.5	38.8	0.849
425847	3110.2	619.5	3729.7	6.0	43.0	3.5	39.5	0.836
Nearby (7)								
417319	3111.8	617.5	3729.3	13.0	43.4	3.5	39.9	0.194
420771	3110.7	586.4	3697.1	13.0	42.4	3.4	39.0	0.086
422285	3110.5	569.7	3680.1	12.4	42.2	3.4	38.8	0.055
425306	3110.8	536.4	3647.2	9.0	40.7	3.5	37.3	0.004
426565	3110.6	529.4	3640.0	7.5	40.5	3.5	37.0	0.981

3.3.6 Reducing the ΔV Cost

One notices that the transfers that require the least ΔV presented in the previous sections involve missions that perform the majority of the energy-changing maneuvers deep within either the Earth's or the Moon's gravity wells where the spacecraft is moving the fastest. The most convincing example of this is the trajectory labeled (7) in Fig. 3-23: the Earth-departure maneuver is large enough to send the spacecraft out to the radius of the Moon, and the manifold-insertion maneuver is performed quite close to the Moon.

The trajectories designed here do not purposefully place the manifold-insertion maneuver near the Moon, and in fact, may not converge well if the maneuver occurs nearby. However, the total transfer ΔV may be reduced if the manifold-insertion maneuver were indeed performed near the Moon, and recent research supports this [172].

Performing a maneuver near the Moon may have energy benefits, but it does increase the operational complexity of the mission. The manifold-insertion maneuver becomes very time-critical when performed close to the Moon, and any execution errors tend to exponentially increase afterward. Other operational considerations are discussed in Chapter 6.

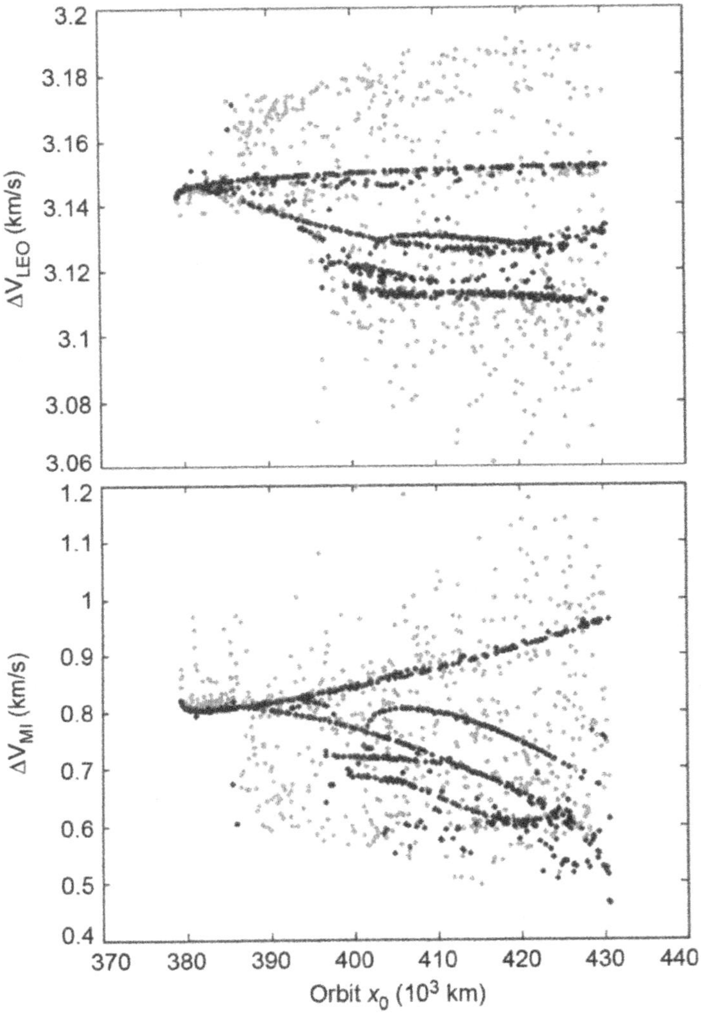

Figure 3-24 The two transfer maneuver magnitudes that combine to produce the total ΔV cost of the transfers shown in Figs. 3-22 and 3-23. Dark points correspond to locally optimal trajectories; faint points represent additional nonoptimal solutions. Top: The magnitudes of the trans-lunar injection maneuvers (ΔV_{LEO}) in each transfer; bottom: The magnitudes of the manifold-insertion maneuvers (ΔV_{MI}) in each transfer.

3.3.7 Conclusions

This section has explored direct transfers to lunar halo orbits. It has been found that short-duration transfers exist to both lunar L_1 and L_2 halo orbits, requiring approximately 5 days of transfer time. Such short-duration transfers require between

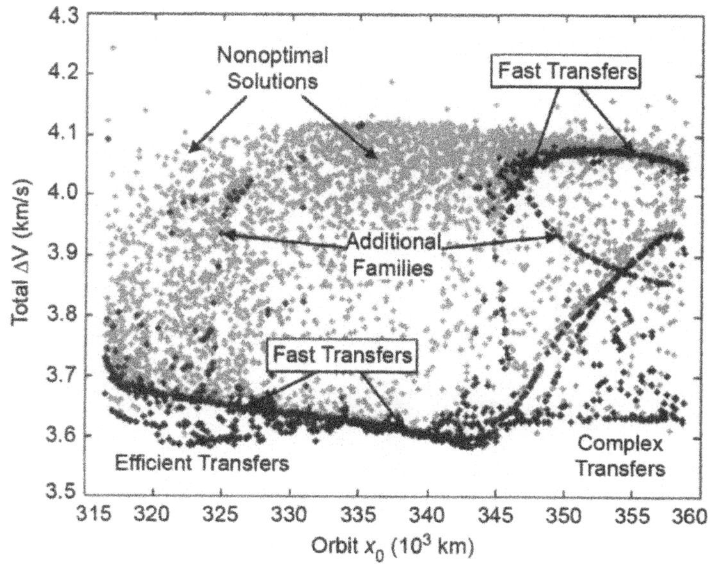

Figure 3-25 The total ΔV cost of many transfers to lunar L_1 halo orbits using either interior or exterior transfers.

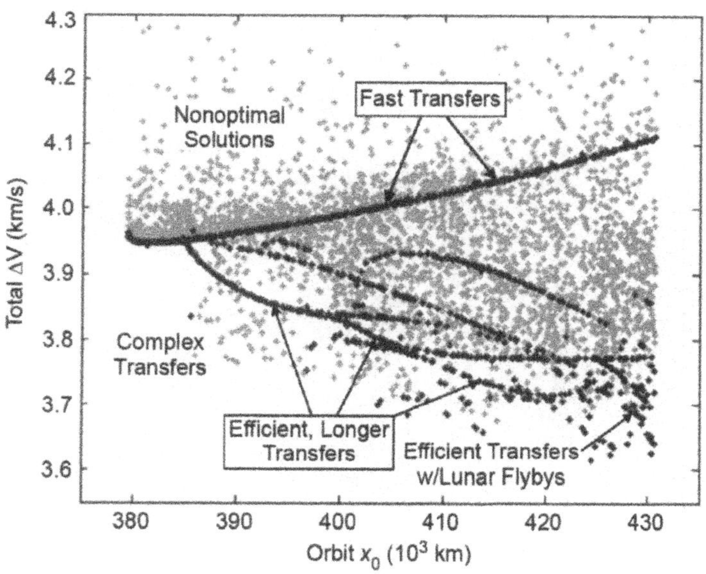

Figure 3-26 The total ΔV cost of many transfers to lunar L_2 halo orbits using either interior or exterior transfers.

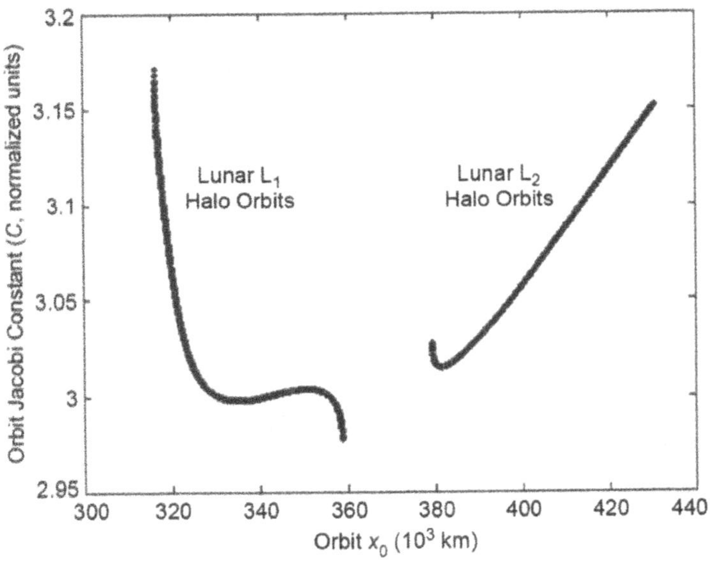

Figure 3-27 The Jacobi constant, C, of the lunar halo orbits surveyed in this work as a function of the halo orbits' x_0-values.

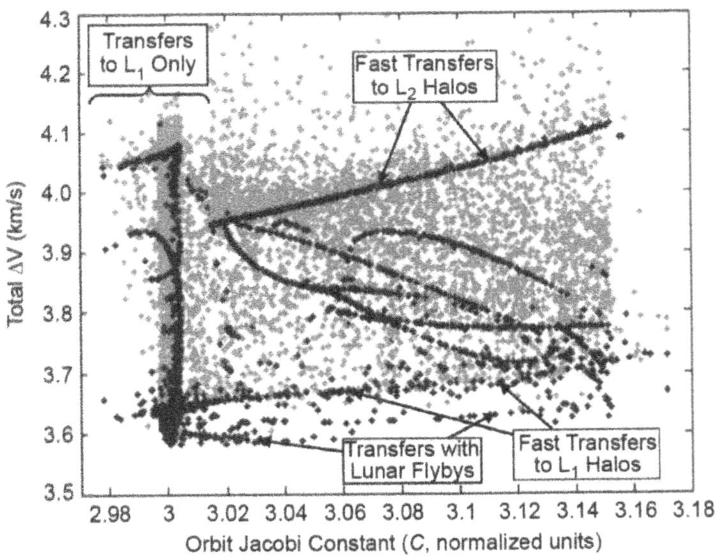

Figure 3-28 The total ΔV cost of direct lunar halo orbit transfers as a function of the halo orbits' Jacobi constant values.

3.6 and 4.1 km/s, depending on the halo orbit, when launched from a 185-km circular parking orbit. It has also been found that transfers exist between LEO and every halo orbit surveyed here that require as little as 3.59–3.65 km/s, although many of these transfers require 3 weeks or more of transfer time. Figure 3-29 summarizes the results, showing the least amount of total ΔV required to reach any halo orbit using the fastest optimized transfers, that is, transfers with a duration of approximately 5 days, as well as an envelope of longer low-ΔV transfers that require at most 2 months of transfer time. The curve representing the longer transfers is very approximate—it was produced by tracing out points that were produced successfully and interpolating between those points. Some of these transfers may be difficult to construct; other lower-cost transfers may also exist. Figure 3-30 summarizes the same results as a function of the halo orbits' C-values rather than their x_0-values.

3.4 LOW-ENERGY TRANSFERS BETWEEN EARTH AND LUNAR LIBRATION ORBITS

Transfers between the Earth and lunar libration orbits may be constructed that require less fuel than direct transfers by taking advantage of the gravity of the Sun. The scenario involves propelling a spacecraft beyond the orbit of the Moon, about 1–2 million kilometers away from the Earth, and letting the Sun's gravity raise the spacecraft's energy. When the spacecraft returns toward its perigee after 2–4 months, it encounters the Moon. The spacecraft encounters the Moon at a much lower relative velocity than that of a direct transfer. The trajectory is crafted such that the spacecraft approaches the Moon on the stable manifold of the target lunar libration orbit.

This section illustrates low-energy transfers that arrive at a variety of lunar libration orbits, such that they require no orbit insertion maneuver whatsoever. The performance of many low-energy transfers is surveyed. First, Section 3.4.1 demonstrates how to model a low-energy transfer using dynamical systems theory. Then Section 3.4.2 provides an energy analysis of an example transfer, which illuminates how energy shifts and how one may use both two-body and three-body tools to design and analyze a low-energy transfer. Sections 3.4.3 and 3.4.4 describe the process of constructing desirable low-energy transfers in the patched three-body and DE421 ephemeris models, respectively. The dynamical systems methods used to construct low-energy transfers may be extended to construct entire families of transfers. Section 3.4.5 surveys many families of transfers that have different geometries and performance characteristics. Section 3.4.6 discusses how these transfers vary from one month to the next. Finally, Section 3.4.7 presents several additional example analyses to design low-energy transfers to different three-body orbits, including an LL_1 halo orbit and a distant prograde orbit.

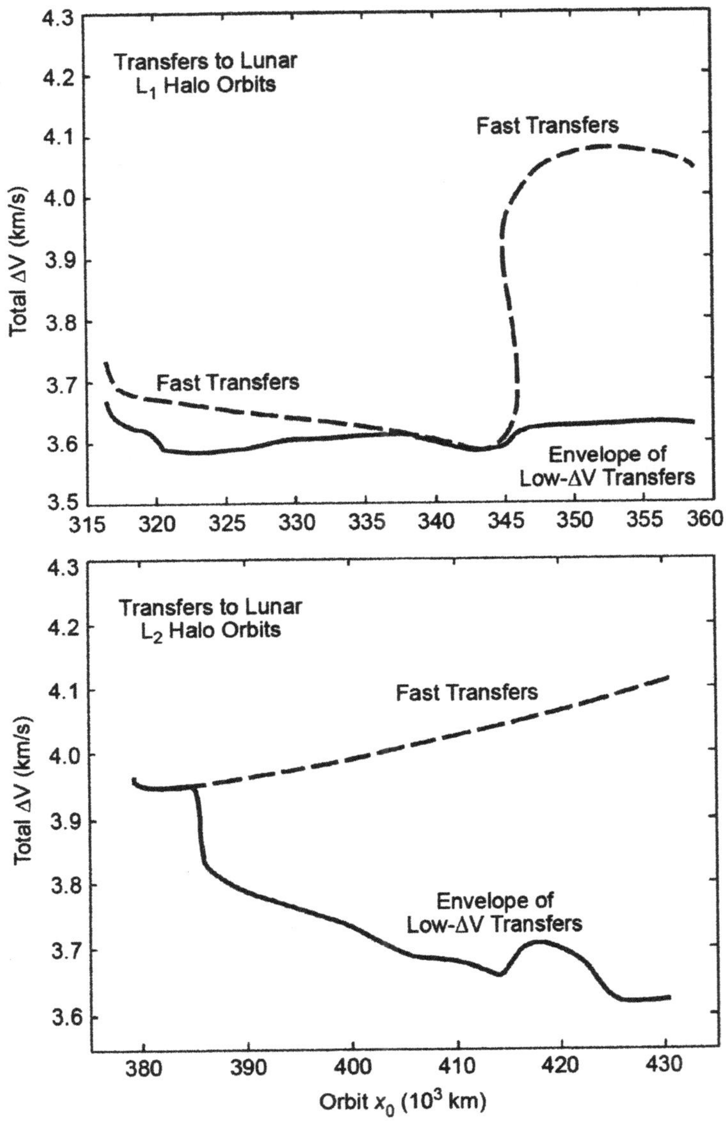

Figure 3-29 A summary of the minimum amount of total ΔV required to reach any lunar L_1 halo orbit (top) and any lunar L_2 halo orbit (bottom) surveyed here using the fastest optimized transfers (approximately 5 days) as well as an envelope of longer low-ΔV transfers (1–2 months).

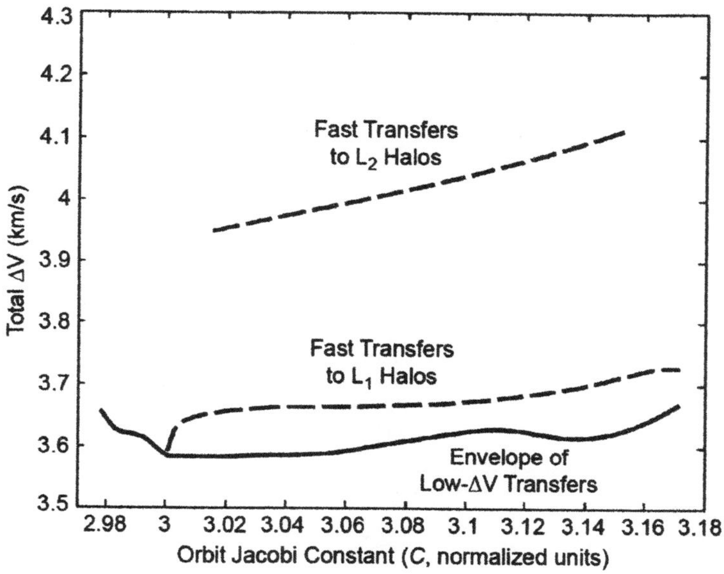

Figure 3-30 A summary of the minimum amount of total ΔV required to reach halo orbits with a given Jacobi constant.

3.4.1 Modeling a Low-Energy Transfer using Dynamical Systems Theory

Many types of low-energy transfers exist in any given month, and their characteristics tend to repeat from one month to the next. The most complex low-energy transfers typically do not appear in many consecutive months due to the asymmetries in the real Solar System; however, simple low-energy transfers reappear in a predictable fashion from one month to the next.

This section studies how to model low-energy transfers using dynamical systems theory and the Patched Three-Body Model (introduced in Section 2.5.2). It turns out that simple low-energy transfers are represented well in this simplified model of the Solar System, and that one may use the modeled trajectory as a guide to construct a realistic transfer in a more accurate model of the solar system. Because low-energy transfers may be represented in the Patched Three-Body Model, one may take advantage of tools within dynamical systems theory to analyze these transfers. The goal is to be able to build a useful low-energy transfer quickly to meet a mission's needs; dynamical systems tools provides an avenue to do this.

A low-energy ballistic transfer may be modeled as a series of heteroclinic transfers between unstable three-body orbits in the Sun–Earth system and the Earth–Moon system [39, 40, 45, 46]. Figure 3-31 illustrates these orbit transfers in the Patched Three-Body Model. One can see that a spacecraft departs the Earth on a trajectory that shadows the stable invariant manifold of an unstable three-body orbit in the

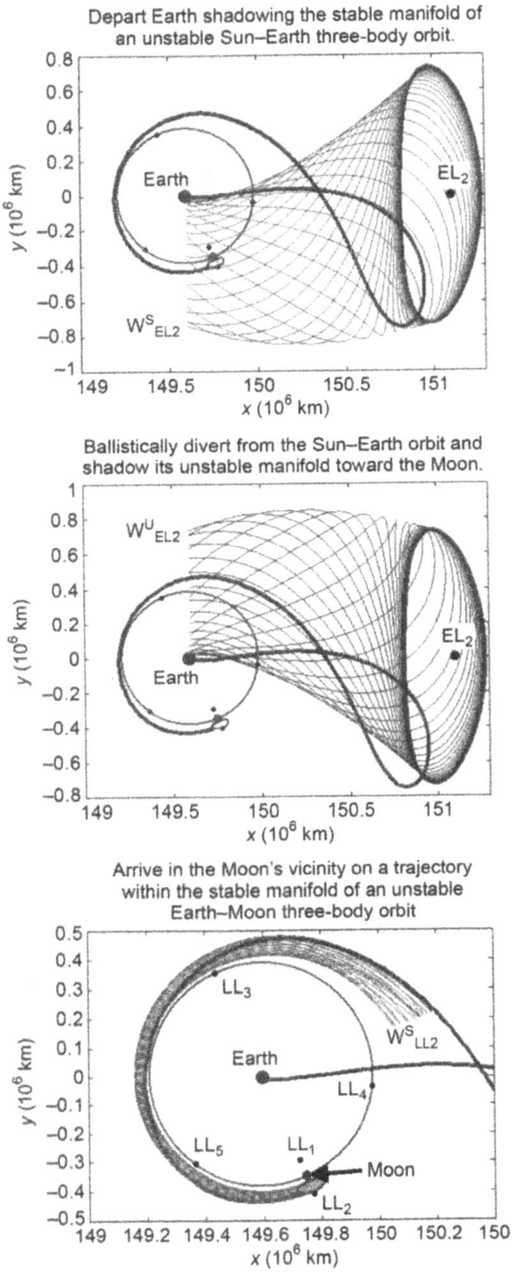

Figure 3-31 Modeling a ballistic lunar transfer as a series of heteroclinic transfers between unstable three-body orbits in the Patched Three-Body Model [97] (first published in Ref. [97]; reproduced with kind permission from Springer Science+Business Media B.V.). *(See insert for color representation of this figure.)*

Sun–Earth/Moon three-body system. The spacecraft does not arrive on that orbit, however, before it ballistically diverts and then shadows the unstable manifold of that orbit. The trajectory is designed to arrive in the stable manifold of a target three-body orbit in the Earth–Moon three-body system, for example, an LL_2 halo orbit. This process will be described in detail in this section.

A low-energy, ballistic lunar transfer may be modeled as a series of transfers from one three-body orbit to another. After the spacecraft launches from its LEO parking orbit, the spacecraft transfers to the vicinity of a three-body orbit in the Sun–Earth system, referred to in this section as the *Earth staging orbit.* The spacecraft's LEO departure trajectory follows the flow of the Earth staging orbit's stable manifold. Once in the vicinity of the Earth staging orbit, the spacecraft falls away from the staging orbit, following the flow of that orbit's unstable manifold. The trajectory is chosen so that it encounters the stable manifold of a three-body orbit in the Earth–Moon system, referred to in this section as the *lunar staging orbit.* The spacecraft may use the lunar staging orbit as a final destination or as a transitory orbit, as discussed later in Section 3.5. To generalize the modeling process even further, a ballistic lunar transfer may be modeled as a transfer from Earth to one or more Earth staging orbits to one or more lunar staging orbits and then to some final destination.

Earth Staging Orbits. Many types of three-body orbits may be used as Earth staging orbits in the process of modeling or constructing a low-energy transfer. A proper staging orbit must meet the following requirements:

1. The orbit must be unstable;
2. If the orbit is the first Earth staging orbit, then the orbit's stable manifold must intersect LEO or the launch asymptote; otherwise, the orbit's stable manifold must intersect the preceding staging orbit's unstable manifold;
3. The orbit's unstable manifold must intersect the following staging orbit's stable manifold, be it another Earth staging orbit or a lunar staging orbit.

A quasiperiodic Lissajous orbit has been selected to build the example transfer shown in this section, because it meets each of these requirements. Unfortunately, quasiperiodic orbits and their invariant manifolds are difficult to visualize since they never retrace their paths. This section illustrates the validity of a Lissajous orbit by showing that halo orbits are viable candidates to be used as Earth staging orbits.

Figure 3-32 shows four perspectives of the family of northern halo orbits centered about the Sun–Earth L_2 point. Lissajous orbits span a very similar region of space, but often do not extend as far in the z-axis.

Most libration orbits in the Sun–Earth system are unstable and hence meet Requirement 1 given above. This discussion will assume that a halo orbit from the family shown in Fig. 3-32 will be used as the only Earth staging orbit en route to a lunar staging orbit. Figure 3-33 shows two plots of an example halo orbit about the Sun–Earth L_2 point and the interior half of its stable manifold. One can see that this stable manifold intersects the Earth. Thus, a spacecraft may make a single maneuver to transfer from a LEO parking orbit to a trajectory on this halo orbit's stable

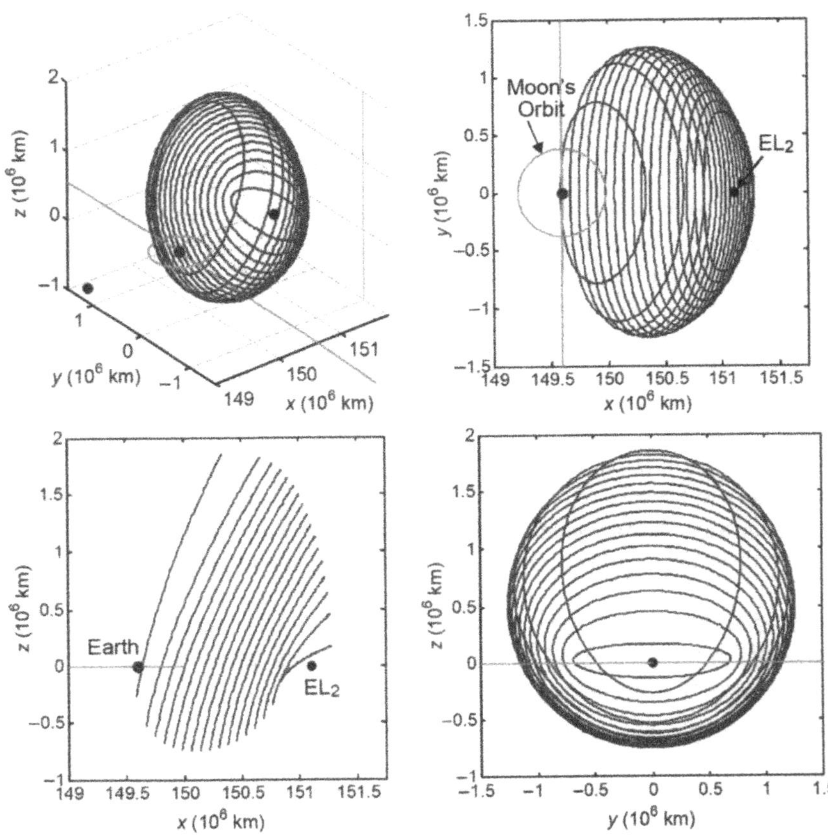

Figure 3-32 Four perspectives of the family of northern halo orbits about the Sun–Earth L_2 point.

manifold; this satisfies Requirement 2 for this itinerary. Similarly, Fig. 3-34 shows two plots of the same halo orbit's unstable manifold, showing that trajectories exist that intersect the Moon's orbit about the Earth. Thus, a spacecraft on, or sufficiently near, the halo orbit may use the orbit's unstable manifold to guide it to intersect the Moon (satisfying Requirement 3). The invariant manifolds of Lissajous orbits with similar Jacobi constants also demonstrate the same properties, making them viable candidates for low-energy staging orbits.

Lunar Staging Orbits. Many different Earth–Moon three-body orbits may be used as lunar staging orbits; the example low-energy transfer modeled in this section uses a halo orbit about the Earth–Moon L_2 point as its lunar staging orbit because it meets all of the requirements.

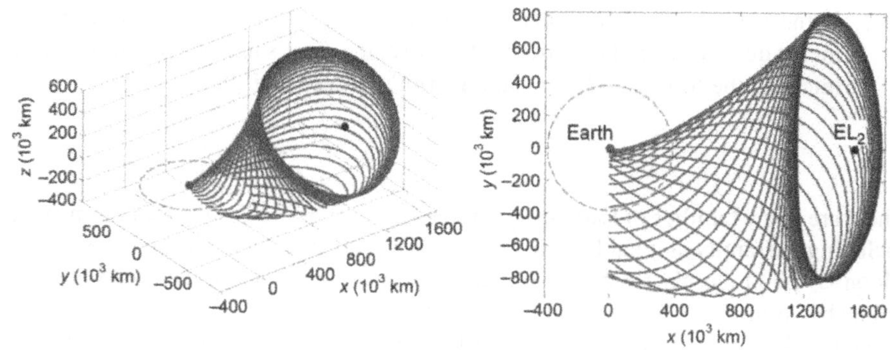

Figure 3-33 Two perspectives of an example Northern halo orbit about the Sun–Earth L_2 point, shown with the interior half of its stable manifold. One can see that the stable manifold intersects the Earth.

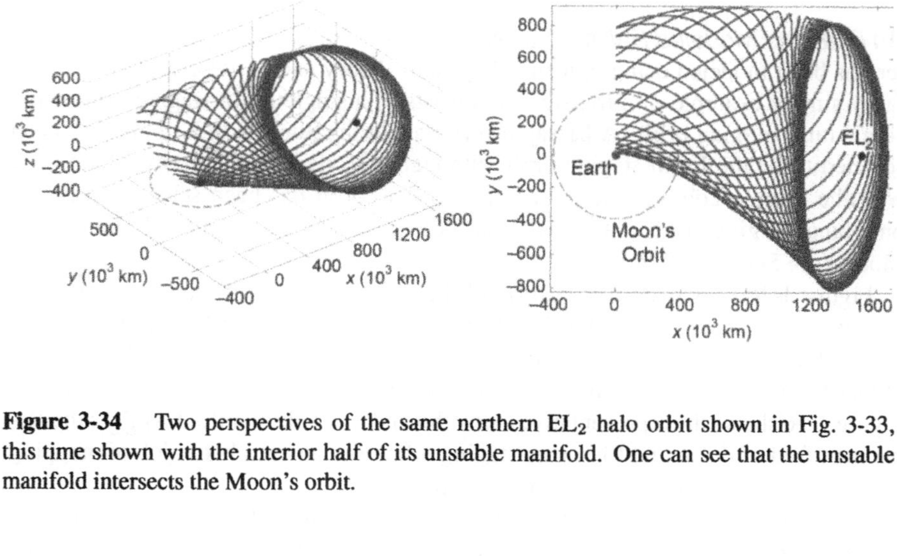

Figure 3-34 Two perspectives of the same northern EL_2 halo orbit shown in Fig. 3-33, this time shown with the interior half of its unstable manifold. One can see that the unstable manifold intersects the Moon's orbit.

The requirements for a lunar staging orbit typically come from the requirements of the mission itself. The following list summarizes the additional requirements imposed on the lunar staging orbit:

1. The orbit must be unstable;

2. The orbit's stable manifold must intersect the unstable manifold of the preceding staging orbit, be it the previous lunar or the previous Earth staging orbit;

3. If the orbit is the final lunar staging orbit, then it must meet any requirements derived from the mission; otherwise, the orbit's unstable manifold must intersect the following lunar staging orbit's stable manifold.

There are many families of Earth–Moon three-body orbits that satisfy Requirement 1, including the family of lunar L_2 halo orbits. The family of halo orbits about the Earth–Moon L_2 point closely resembles the family of halo orbits about the Sun–Earth L_2 point shown in Fig. 3-32 and won't be shown here for brevity.

Figure 3-35 shows two perspectives of an example LL_2 halo orbit along with its exterior stable manifold, propagated in the Patched Three-Body Model. If a spacecraft were to target a trajectory on this manifold it would asymptotically approach and eventually arrive onto the staging orbit. Thus, if a spacecraft were able to transfer from the Earth staging orbit's unstable manifold onto this LL_2 halo orbit's stable manifold, then the spacecraft would have achieved a ballistic transfer to this lunar orbit from LEO.

An Example Modeled Ballistic Lunar Transfer. An example ballistic lunar transfer has been modeled using dynamical systems theory and is presented here. It is a fairly simple example of a transfer: it consists of a single Earth staging orbit and a single lunar staging orbit. A Lissajous orbit about the Sun–Earth L_2 point has been selected to be the Earth staging orbit, although it is visualized here by a halo orbit with the same Jacobi constant. A lunar L_2 halo orbit has been selected to be the only lunar staging orbit. The transfer has been produced in the Patched Three-Body Model (see Section 2.5.2).

Figure 3-36 shows the first portion of the three-dimensional transfer in two perspectives. The spacecraft is launched from a 185-km low Earth orbit, travels outward toward the Sun–Earth L_2 point along a trajectory that shadows the stable manifold of an EL_2 libration orbit, skims the periodic orbit, and then travels toward the Moon. Figure 3-36 shows the representative halo orbit and its stable manifold, $W^S_{EL_2}$; the stable manifold of the actual Lissajous staging orbit does an even better job of mapping out the flow of the spacecraft's motion in space.

Figure 3-37 shows two perspectives of the same transfer trajectory, but this time plotted with the Earth staging orbit's unstable manifold, $W^U_{EL_2}$. One can see that as the spacecraft departs the vicinity of the Earth staging orbit and approaches the Moon, its trajectory shadows the unstable manifold of the Earth staging orbit.

Figure 3-38 shows the same two perspectives of the three-dimensional low-energy transfer plotted alongside the lunar staging orbit's stable manifold, $W^S_{LL_2}$. One can see that the low-energy transfer intersects the manifold in full phase space, indicating that the spacecraft has injected into the LL_2 halo orbit. Once in the final Earth–Moon halo orbit, the spacecraft has all of the options presented in Section 3.5 available to it.

Figure 3-39 shows a top-down perspective of the entire three-dimensional low-energy transfer with all three manifolds displayed.

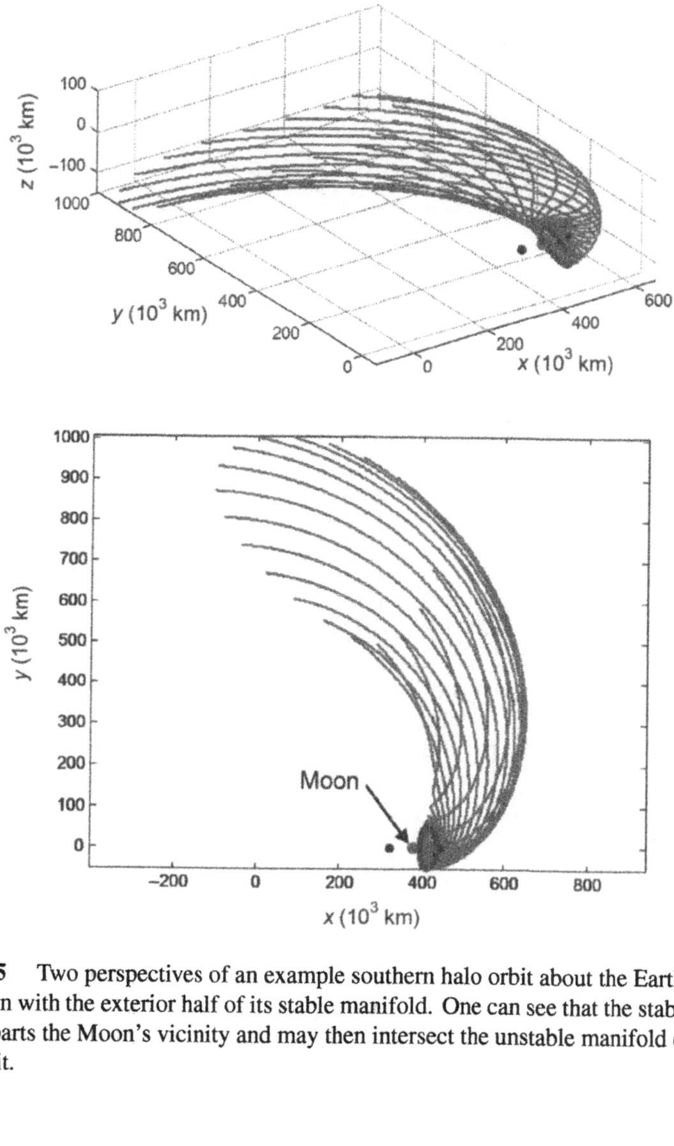

Figure 3-35 Two perspectives of an example southern halo orbit about the Earth–Moon L_2 point, shown with the exterior half of its stable manifold. One can see that the stable manifold quickly departs the Moon's vicinity and may then intersect the unstable manifold of the Earth staging orbit.

3.4.2 Energy Analysis of a Low-Energy Transfer

Low-energy lunar transfers harness the Sun's gravity to reduce the ΔV requirements of a lunar transfer. It is useful to observe how the two-body energy of the spacecraft with respect to each of the massive bodies changes throughout the transfer. It is also useful to observe how the Moon affects the spacecraft's Sun–Earth Jacobi constant and especially how the Sun affects the spacecraft's Earth–Moon Jacobi constant. These energy changes are explored in this section, applied to the example transfer

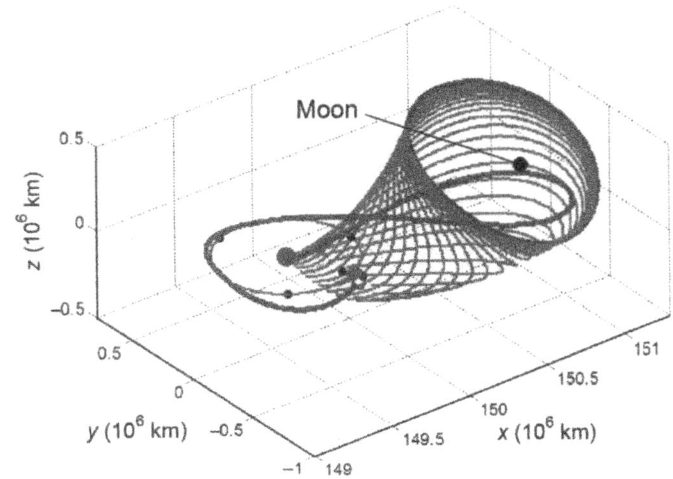

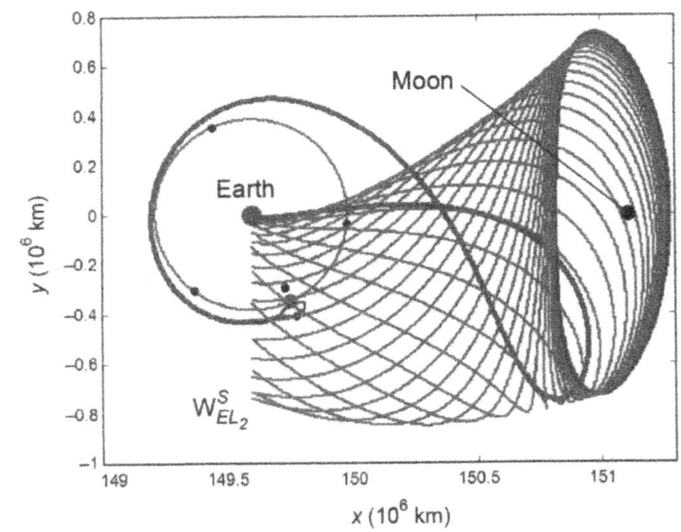

Figure 3-36 Two perspectives of the first portion of the example low-energy transfer, modeled using the stable manifold of a halo orbit about the Sun–Earth L_2 point. One can see that the spacecraft's outbound motion shadows the halo orbit's stable manifold.

produced in the previous section. Other low-energy transfers have been found to behave in a very similar fashion.

To begin this analysis, Fig. 3-40 shows plots of the distance between the spacecraft and both the Earth and Moon as the spacecraft traverses the low-energy ballistic lunar transfer. This is a useful illustration since both the spacecraft's two-body energy and its Jacobi constant vary as functions of distance to these bodies. By observing

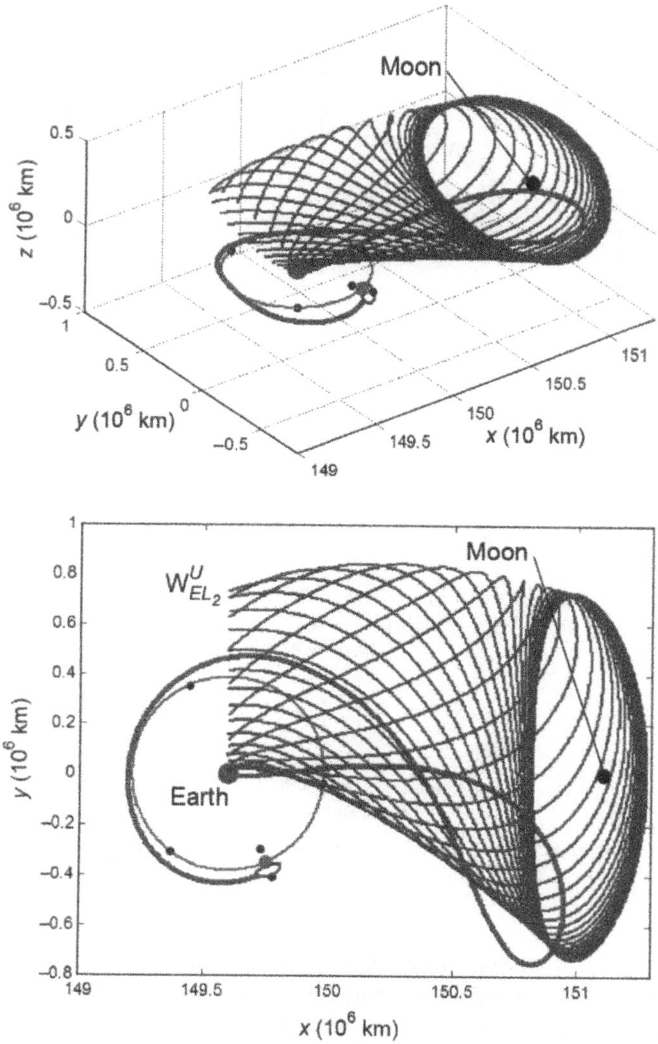

Figure 3-37 Two perspectives of the second portion of the example low-energy transfer, modeled using the unstable manifold of a halo orbit about the Sun–Earth L_2 point. One can see that as the spacecraft departs the vicinity of the Earth staging orbit and approaches the Moon, its trajectory shadows the unstable manifold of the Earth staging orbit.

Fig. 3-40, one can determine the time at which the spacecraft arrives at its lunar halo orbit destination.

It is expected that the two-body energy of a spacecraft with respect to the Earth increases over time due to the Sun's gravity, since the spacecraft's perigee radius gradually rises throughout the transfer. Figure 3-41 shows the two-body specific

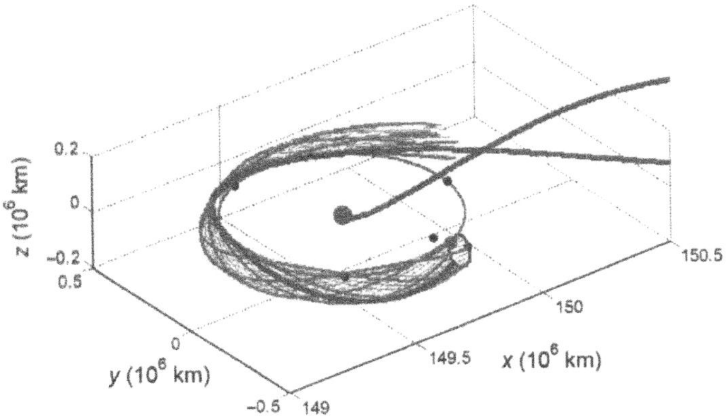

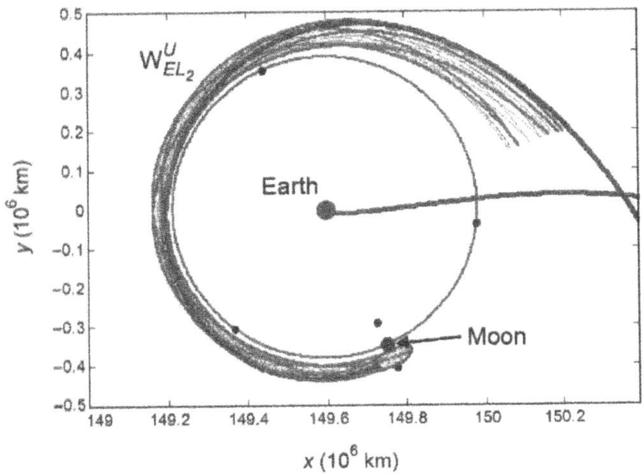

Figure 3-38 Two perspectives of the third portion of the example low-energy transfer, modeled using the stable manifold of a halo orbit about the Earth–Moon L_2 point. Every fourth trajectory has been darkened for visualization purposes. One can see that the transfer intersects the manifold in full phase space, indicating that the spacecraft has injected into the LL_2 halo orbit.

energy of the spacecraft with respect to the Earth throughout the transfer. One can see that the spacecraft's energy does indeed rise while it is in the vicinity of the Earth staging orbit. The energy then begins to vary wildly once it enters the lunar halo orbit, which makes sense because the halo orbit only exists in the presence of both the Earth and the Moon, balancing the gravity of both bodies. Figure 3-42 shows four other two-body orbital elements of the spacecraft with respect to the Earth as the spacecraft traverses the ballistic transfer, including the spacecraft's semi-major axis, perigee radius, eccentricity, and ecliptic inclination. One can see that

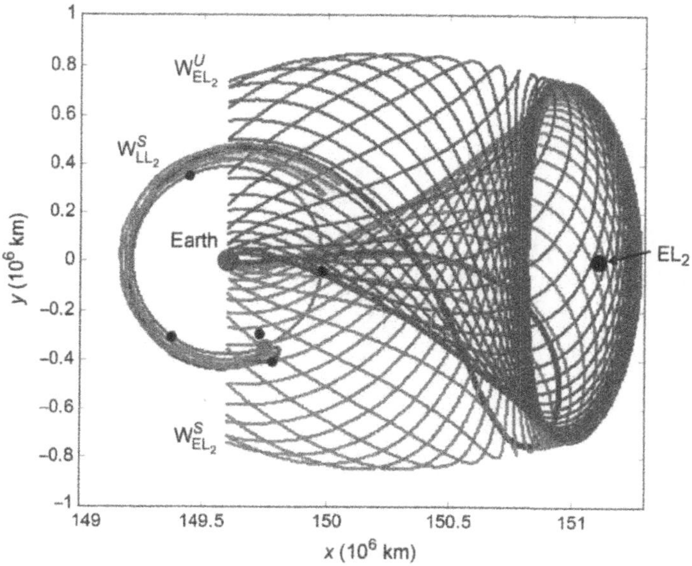

Figure 3-39 A top-down perspective of the example low-energy transfer, shown with all three manifolds that were used to model it (blue = the low-energy transfer, green = stable manifold of a three-body orbit, brown = unstable manifold of a three-body orbit). (See color insert.)

the Sun's gravity increases the spacecraft's semi-major axis and perigee radius as the spacecraft traverses the Earth staging orbit. The Sun's gravity reduces the spacecraft's eccentricity and inclination with respect to the Earth. The spacecraft enters the lunar halo orbit at approximately 110 days after launch, beyond which the Moon's gravity is the dominant source causing each of the spacecraft's orbital elements to vary over time.

It is interesting to notice that the spacecraft's inclination changes dramatically during the first half of the transfer, while the perigee radius remains near zero; then during the second half of the transfer the perigee radius rises dramatically while the spacecraft's inclination settles down. These effects may be correlated with the location of the spacecraft relative to the four quadrants of the Sun–Earth state space. In this particular transfer, the spacecraft spends several weeks near the boundary of the first and fourth quadrants before moving definitively into the fourth quadrant, where the spacecraft's perigee radius rises rapidly. Other low-energy transfers have varying geometries and their two-body orbital elements change in correspondingly different fashions.

It is also expected that the spacecraft's two-body energy with respect to the Moon decreases as the spacecraft approaches and ballistically inserts into the lunar halo orbit. Figure 3-43 shows the two-body specific energy of the spacecraft with respect to the Moon throughout the low-energy lunar transfer. One can clearly see that the

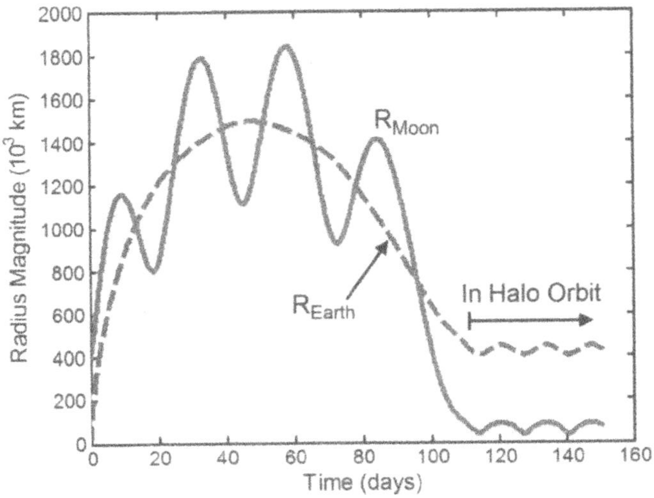

Figure 3-40 The magnitude of the radius vector of the spacecraft with respect to the Earth and the Moon as the spacecraft traverses the example low-energy lunar transfer.

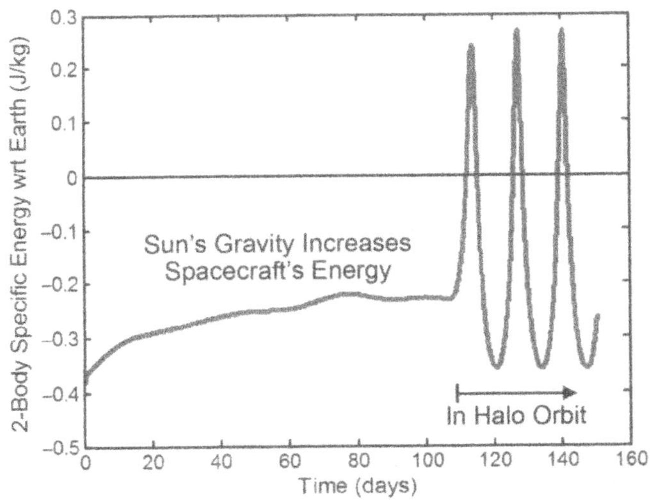

Figure 3-41 The two-body specific energy of a spacecraft with respect to the Earth over time as it traverses an example low-energy lunar transfer.

spacecraft's specific energy drops as it approaches the lunar halo orbit. Furthermore, its energy drops below zero, satisfying some authors' requirements to be temporarily captured by the Moon [29, 46, 182].

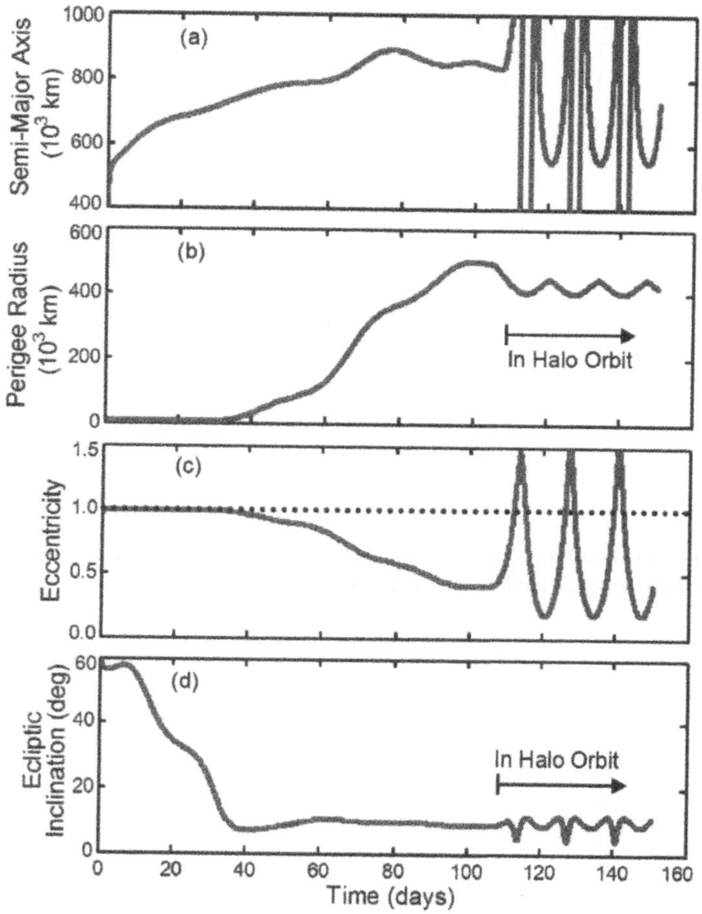

Figure 3-42 Four two-body orbital elements of the spacecraft with respect to the Earth as the spacecraft traverses the example lunar transfer: (a) the spacecraft's semi-major axis, (b) perigee radius, (c) eccentricity, and (d) ecliptic inclination.

Figures 3-44 and 3-45 show the evolution of the spacecraft's Jacobi constant with respect to the Sun–Earth and Earth–Moon three-body systems, respectively, as the spacecraft traverses the example lunar transfer. The spacecraft's trajectory has been constructed in the Patched Three-Body Model; hence, the spacecraft's Jacobi constant will be constant in one or the other three-body system at any given time, depending on which three-body system is responsible for the given segment of the spacecraft's trajectory. The spacecraft's motion has been modeled by the Sun–Earth three-body system during the first 105 days of the transfer. After the spacecraft has crossed the Earth–Moon three-body sphere of influence (3BSOI), its motion is then modeled by the Earth–Moon three-body system.

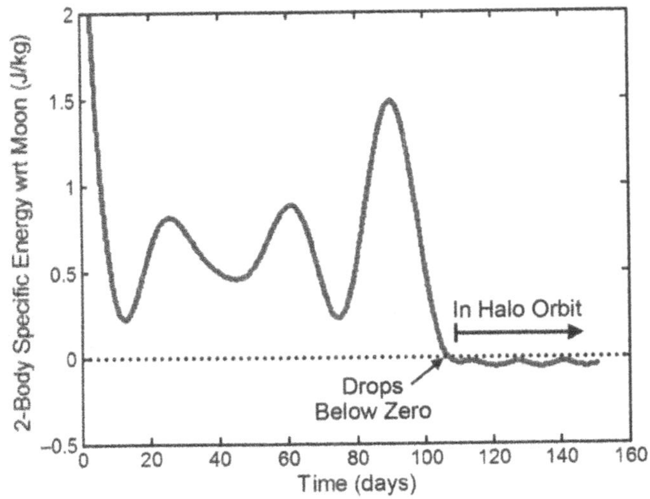

Figure 3-43 The two-body specific energy of a spacecraft with respect to the Moon over time as it traverses an example low-energy lunar transfer.

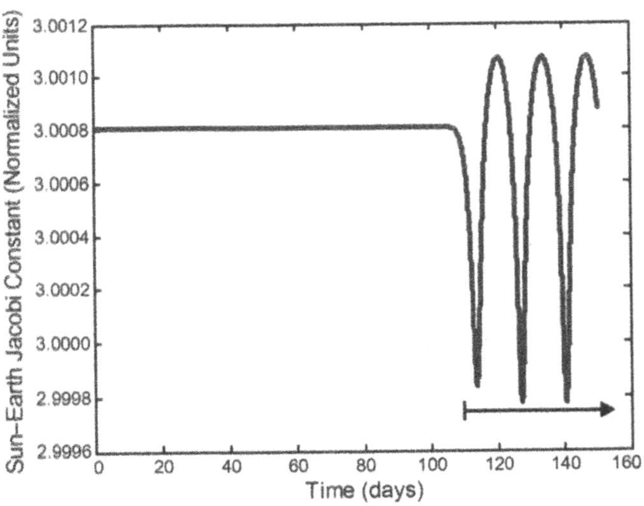

Figure 3-44 The evolution of the spacecraft's Jacobi constant with respect to the Sun–Earth three-body system as the spacecraft traverses the example lunar transfer.

Figure 3-45 presents a compelling case that it is possible to build low-energy transfer to lunar halo orbits, or other unstable Earth–Moon three-body orbits, with a wide variety of different Jacobi constants. If the spacecraft traversing the example transfer had arrived at the Moon slightly earlier or slightly later, it could have transferred to a

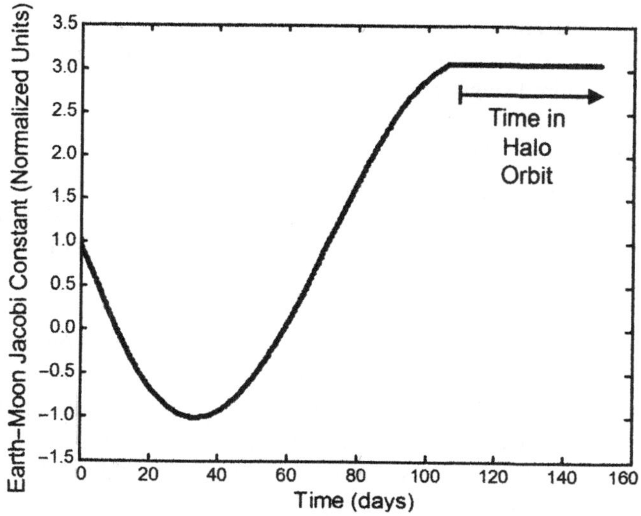

Figure 3-45 The evolution of the spacecraft's Jacobi constant with respect to the Earth–Moon three-body system as the spacecraft traverses the example lunar transfer.

lunar halo orbit with a different Jacobi constant. Furthermore, it may be possible for a spacecraft to depart one lunar halo orbit, traverse through the Sun–Earth environment for some time, and return to the Moon on the stable manifold of a different lunar halo orbit. Section 3.4.5 demonstrates that it is indeed possible to build low-energy transfers to lunar halo orbits within a wide range of Jacobi constants [46], but more work needs to be accomplished to determine how to take advantage of the time series shown in Fig. 3-45 to target a lunar halo orbit with a specified Jacobi constant.

3.4.3 Constructing a Low-Energy Transfer in the Patched Three-Body Model

Modeling a low-energy transfer using dynamical systems theory involves the use of several staging orbits and their corresponding invariant manifolds in the Earth–Moon and Sun–Earth systems. If a mission designer wishes to construct a transfer that intentionally visits certain staging orbits, then the transfer may be constructed in the same manner that it is modeled. More often, a mission designer only wishes for the spacecraft to reach the final lunar orbit, no matter its route through the Sun–Earth system. In that case, the methods used to construct a low-energy transfer may be simplified.

Ballistic lunar transfers are constructed here by propagating the stable manifold of the final lunar halo orbit backward in time for a set amount of time. After each trajectory has been propagated, the perigee point of the trajectory is identified. A proper transfer may be identified as one whose perigee point corresponds to some desired value, for example, an altitude of 185 km. In this manner, a practical transfer may be constructed between the Earth and the lunar three-body orbit without identifying any required staging orbit.

3.4.3.1 Parameters The dynamical systems method of constructing ballistic lunar transfers provides a natural set of six parameters that may be used to define each transfer. In the Patched Three-Body Model, this set may be described by the parameters: $[F, C, \theta, \tau, p, \Delta t_m]$. Each of these parameters is described in this section.

Orbit Family Parameter: F. Depending on the mission requirements, one may wish to target any type of Earth–Moon three-body orbit. The parameter F is a discrete variable that describes the orbit family that contains the desired target orbit. The example transfer presented previously has had the parameter F set to describe the family of southern LL_2 halo orbits. There are certainly symbolic ways to represent each family of three-body orbits, but using text to do so provides a clear description of which family is being used.

Orbit Parameter: C. The Jacobi constant, C, of the targeted orbit is used in this work to specify which orbit is being targeted within the family. There are numerous ways to identify a particular three-body orbit within its family [108, 113]. The Jacobi constant is used here because it also provides information about the corresponding forbidden regions and allowable motion of spacecraft with that Jacobi constant [46].

Sun–Earth–Moon Angle: θ. The parameter θ is defined to be the angle between the Sun–Earth line and the Earth–Moon line. It is a required parameter needed to convert between the two three-body systems in the Patched Three-Body Model. Figure 3-46 shows an example of the geometry and the definition of θ.

Arrival Location: τ. Each point on a periodic orbit may be uniquely described by the parameter τ, a parameter analogous to a conic orbit's true anomaly. This parameter was introduced in Section 2.6.2.3, but is described again here. The parameter τ may range from 0 to 1, representing a revolution number, or from 0 deg to 360 deg,

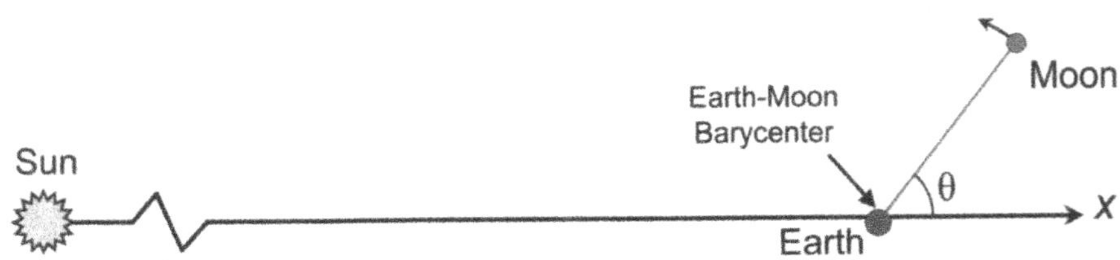

Figure 3-46 An illustration of θ, the Sun–Earth–Moon angle.

representing an angle like the true anomaly [46]. Figure 3-47 shows a plot of the definition of τ when applied to two halo orbits. For halo orbits, it is intuitive to use an angle and model τ off of a conic orbit's true anomaly; for other three-body orbits it is confusing using an angle. In any case, the only use of τ here is to identify each point about a three-body orbit, and either representation may be used.

Perturbation Direction: p. To construct a trajectory in the stable invariant manifold of a given unstable orbit, one takes the state of the orbit at a given τ-value and perturbs that state along the direction of the stable eigenvector [46, 147]. The perturbation may occur in two directions: an interior or an exterior direction, as illustrated in Fig. 3-4. The parameter p is a discrete variable that may be set to *interior* or *exterior*, indicating the direction of the perturbation.

Manifold Propagation Duration: Δt_m. The trajectory in the given three-body orbit's stable manifold is propagated backward in time for an amount of time equal to Δt_m. Typically when propagated backward in time, the trajectories that lead to desirable low-energy transfers depart the vicinity of the Moon, traverse their apogee, fall toward the Earth, and then intersect a desirable altitude above the surface of the Earth. However, transfers may also be constructed that pass near the Earth once or several times before intersecting the desirable altitude above the surface of the Earth. Such trajectories must be propagated long enough to allow the desirable perigee passage to occur. Thus, the parameter Δt_m is important in order to ensure that the proper perigee passage is being implemented by the low-energy transfer.

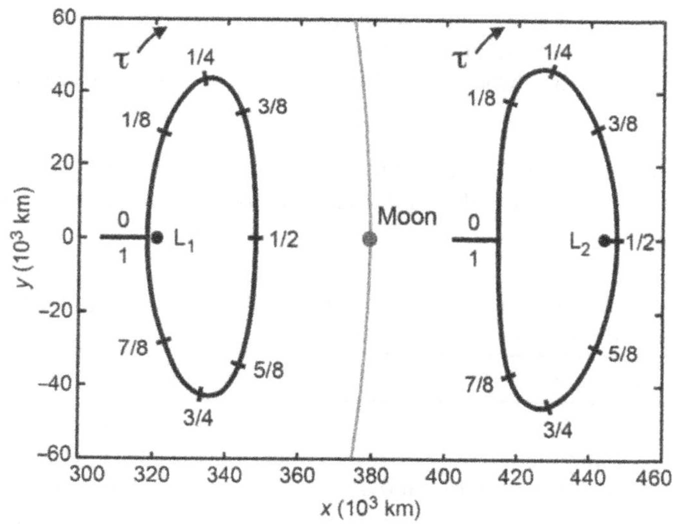

Figure 3-47 The two halo orbits shown demonstrate how the parameter τ moves from 0 to 1 about an orbit [174] (Copyright © 2008 by American Astronautical Society Publications Office, all rights reserved, reprinted with permission of the AAS).

Discussion Regarding Parameters. The set of parameters used here does not contain all continuous variables as other sets of orbital elements do, such as the Keplerian orbital element set of a two-body orbit. The present parameter set also requires knowledge about how to use it, for example, how to build the target lunar orbit given the parameters F and C. Nonetheless, this set may be used to uniquely describe any low-energy ballistic transfer between the Earth and an unstable lunar three-body orbit. Table 3-17 summarizes the parameter set.

3.4.3.2 Producing the low-energy transfer The process of producing a low-energy transfer given the parameter set $[F, C, \theta, \tau, p, \Delta t_m]$ is very simple and is described henceforth.

Step 1. First, one must build the target Earth–Moon orbit. The desired orbit must be unstable and may be identified using the parameters F and C, as defined above. The example low-energy transfer presented in this section has been produced using an orbit in the family, F, of southern halo orbits about the Earth–Moon L_2 point. The specific orbit has been identified in its family by the value of C, equal to 3.05.

Step 2. The parameter θ specifies the location of the Moon, and hence the target orbit, with respect to the Earth and Sun in the Patched Three-Body Model. The example transfer has used an initial θ-value of approximately 293.75 deg. This may be verified by inspecting the final location of the Moon in Figs. 3-36–3-39. Since the transfer is generated backward in time, the value of θ specifies the *final* position of the Moon.

Step 3. The parameter τ specifies a particular state in the unstable three-body orbit. The example transfer has implemented a τ-value of approximately 0.74, corresponding to a point roughly three quarters around the orbit from the orbit's reference point (the point where the orbit crosses the $y = 0$ plane with positive $\dot{y}$) [46, 108].

Step 4. The particular state in the target orbit is then perturbed in order to construct a single trajectory in the stable manifold of the orbit. The magnitude of this

Table 3-17 A summary of the six parameters used to produce low-energy transfers in the Patched Three-Body Model.

Parameter	Domain	Description
F	Discrete	Target three-body orbit family
C	Continuous	Jacobi constant of target orbit
θ	Continuous [0 deg, 360 deg]	Sun–Earth–Moon angle
τ	Continuous [0,1]	Arrival location on the target orbit
p	Discrete	Perturbation direction
Δt_m	Continuous	Propagation duration

perturbation is given by ϵ; the direction is given by the orbit's monodromy matrix [131] and the parameter p. The orbit's monodromy matrix is used to compute the orbit's stable and unstable eigenvectors; the stable eigenvector is then mapped to the given τ-value using the orbit's state transition matrix [46, 147]. The example lunar transfer has implemented a trajectory in the halo orbit's *exterior* manifold with the value of ϵ set proportional to a 100-km perturbation.

Step 5. The resulting state is then used as the initial condition to construct a trajectory in the stable manifold of the three-body orbit. This trajectory is propagated backward in time for a duration of time equal to Δt_m. The trajectory that has produced the example transfer has been propagated for approximately 28.53 non-dimensional Earth–Moon time units (approximately 123.9 days) before encountering the desired perigee point, that is, the desired LEO injection point.

Step 6. The final step in the construction of a low-energy transfer is to connect this trajectory with a prescribed LEO parking orbit or with the surface of the Earth. It is unlikely that an arbitrary set of parameters will yield a lunar transfer that connects with its prescribed LEO starting conditions. In such a case, either the parameters should be adjusted [46], or a bridge must be constructed to connect the spacecraft's origin with the lunar transfer, as discussed in Section 3.3 [174].

3.4.3.3 Discussion The parameter set derived here is very useful if a mission designer needs to build a transfer to a specific lunar orbit that cannot exceed some maximum transfer time. In that case, the parameters F, C, and Δt_m are fixed. By setting Δt_m to the maximum transfer duration, one ensures that no transfers are constructed that require excessive transfer time, but one still permits transfers that require less transfer time. The three remaining parameters are conveniently well defined. The parameter p is binary and the parameters θ and τ are cyclic. Thus, mission designers can explore all possible low-energy transfers to a target orbit by producing two maps: one map of θ vs. τ with p set to "Exterior," and another identical map with p set to "Interior." Examples of these two maps that survey all possible low-energy transfers to an example halo orbit about the LL_2 point, along with several representative transfers, are illustrated in Figs. 3-48 and 3-49. The exploration of these maps will be the purpose of Section 3.4.5, and further description of these figures will appear there.

Other methods have been described in the literature that also describe parameter sets to target low-energy lunar transfers. The majority of these methods start with a spacecraft in orbit about the Earth and target a maneuver for that spacecraft to perform in order to reach the Moon's vicinity via a low-energy transfer. For instance, Belbruno and Carrico have developed a set of parameters that describe the six-dimensional state that a spacecraft would need to obtain to reach the Moon's vicinity via a low-energy transfer [27]. Five parameters are specified, including an epoch (t), the spacecraft's radial distance from Earth (r_E), its longitude (α_E), its latitude (δ_E), and its flight path azimuth (σ_E). Then, the spacecraft's speed (V_E) and flight path angle (γ_E) are varied to target a prescribed radial distance from the Earth (r_M) and a prescribed inclination (i_M), which would ultimately send the spacecraft in the general direction

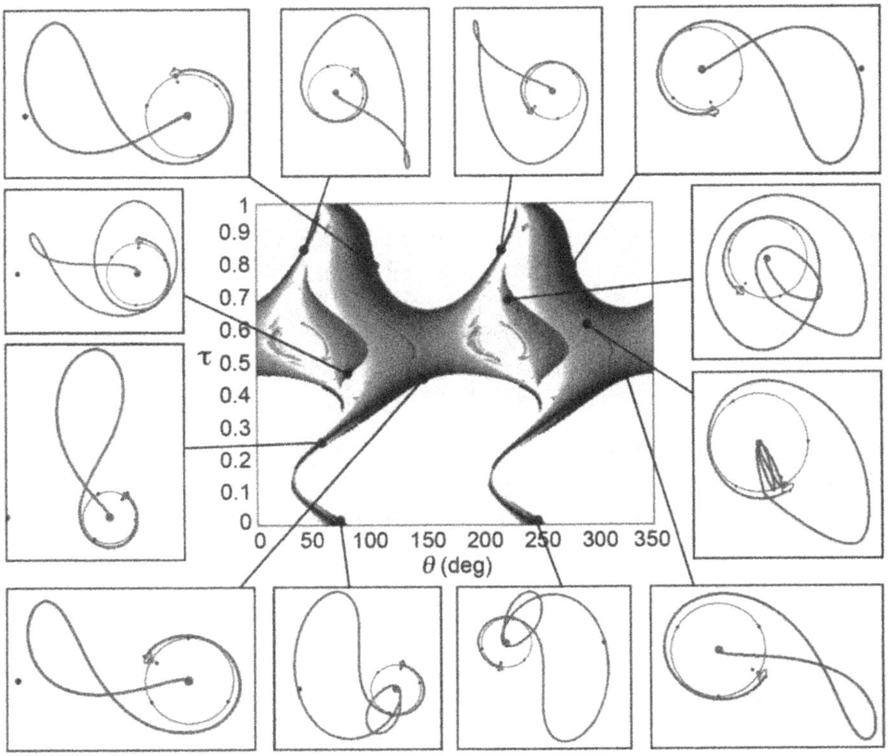

Figure 3-48 An example state space map, capturing a wide variety of low-energy transfers that exist between the Earth and an example LL_2 halo orbit. Each trajectory arrives at the halo orbit from the exterior direction, and arrives at the orbit in a geometry according to the given (θ, τ) combination. The color of the map indicates how close to the Earth the trajectory gets when propagated from the LL_2 halo orbit backward in time. All black points represent viable low-energy transfers. *(See insert for color representation of this figure.)*

of a low-energy transfer. The advantage of this method is that the spacecraft's initial orbit at the Earth is well-defined, which is useful when a transfer must be designed for a spacecraft that is already in orbit about the Earth. However, the technique requires a great deal of predetermined knowledge of the problem, including *a priori* estimates for the values of r_M, i_M, V_E, γ_E, and t (t is specified to obtain a proper Sun–Earth–Moon angle). The procedure is therefore constrained to build a transfer with a predefined geometry that may not be ideal.

Operationally, it is likely that a combination of these two approaches will work the best to produce practical low-energy transfers. A transfer may then be constructed that starts from a prescribed orbit, ends at a specified lunar orbit, and probably includes one or two small trajectory correction maneuvers to connect the segments.

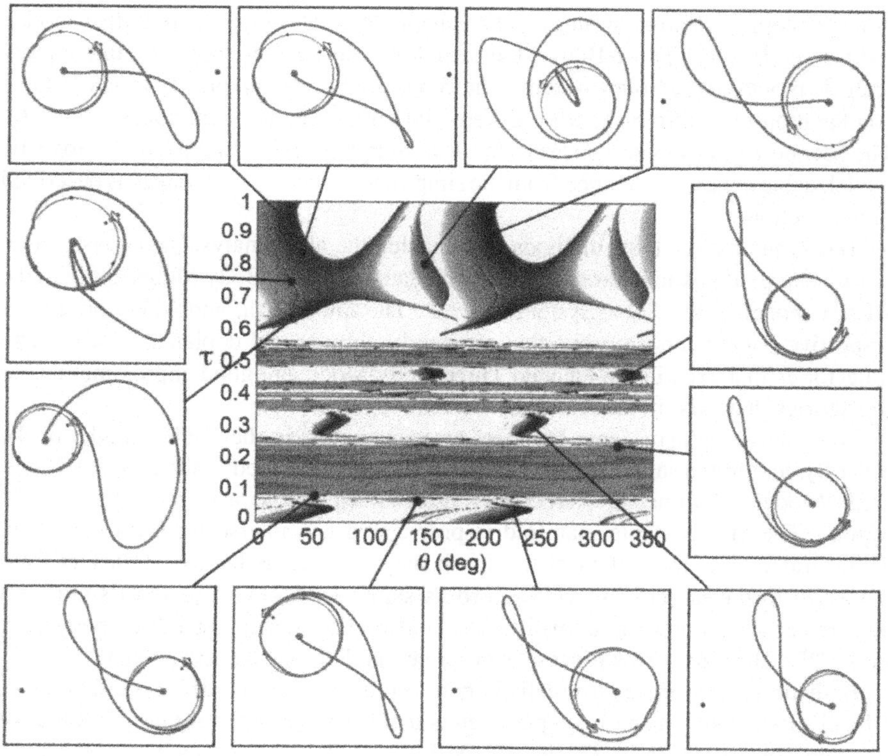

Figure 3-49 An example state space map that is identical to the map illustrated in Fig. 3-48, except that the trajectories arrive at the LL_2 halo orbit from the interior direction. *(See insert for color representation of this figure.)*

Chapter 6 presents such an algorithm, and the results demonstrate that it generates very successful trajectories.

3.4.4 Constructing a Low-Energy Transfer in the Ephemeris Model of the Solar System

The previous sections demonstrated how to analyze and construct a low-energy lunar transfer to a libration orbit using the Patched Three-Body Model; this section describes how to do so in the more accurate DE421 ephemeris model of the Solar System.

There are two main strategies that have been shown to work to generate a low-energy transfer in a realistic model of the Solar System, such as a model that uses the JPL Ephemerides to approximate the motion of the planets and the Moon in the

Solar System. The first strategy is to generate the transfer in a simplified model, such as the Patched Three-Body Model, and then convert the transfer into the more realistic model of the Solar System. The conversion process typically involves some combination of multiple shooting differential correction and continuation [40, 46]. The second strategy is to construct the low-energy transfer directly in the realistic model, using experience gained from the simplified models. This strategy is described in this section.

The dynamical systems methods that enabled the clear analysis and construction of low-energy ballistic lunar transfers in the Patched Three-Body Model apply to the DE421 model of the Solar System as well. The Sun, Earth, and Moon orbit their respective barycenters in orbits that are nearly circular and coplanar. Thus, many trajectories that exist in the Patched Three-Body Model are good approximations of trajectories that exist in the real Solar System.

Low-energy ballistic lunar transfers are constructed in the DE421 model of the Solar System in the same way that they have been constructed in the Patched Three-Body Model. An unstable three-body orbit is selected as a target orbit near the Moon. The orbit's stable manifold is propagated and intersected with the Earth. Those trajectories that intersect the Earth may be used as ballistic transfers from the Earth to the target orbit via the orbit's stable manifold. The most significant adjustment to this procedure involves the construction of the target three-body orbit in the DE421 model. This process is described in detail in Section 2.6.6.3.

Ballistic lunar transfers to realistic halo orbits may be uniquely specified in the DE421 model using a set of six parameters that is similar to the set used to describe transfers constructed in the Patched Three-Body Model. This set includes the parameters: $\{F, A_z, T_{\text{ref}}, p, \tau, \Delta t_m\}$, where A_z replaces the Jacobi constant and T_{ref} replaces the parameter θ from the previous set of parameters. It is very straightforward to generate a halo orbit in the DE421 model using an analytical approximation as an initial guess to the multiple shooting differential corrector (Section 2.6.5.2). The parameter A_z specifies the z-axis amplitude of the halo orbit in the analytical approximation specified by Richardson [123]. The parameter T_{ref} specifies the reference epoch that ties the initial guess of the states of the halo orbit to the DE421 model.

Table 3-18 summarizes the set of parameters that generates an example transfer in the DE421 model, shown in Fig. 3-50. The parameters F, A_z, and T_{ref} define the southern LL_2 halo orbit that is shown in Fig. 3-51. One can see that the multiple shooting differential corrector adjusted the state of the analytical approximation of the halo orbit such that the reference epoch is no longer at the $\tau = 0$ deg point, but at the $\tau \approx 3.84$ deg point. A particular trajectory in the halo orbit's stable manifold is then generated that corresponds to the parameters τ and p in Table 3-18, which propagates backward in time to a perigee with an altitude of 185 km. The distance between this trajectory and the Moon is shown in Fig. 3-52. One can see that this trajectory asymptotically arrives at the orbit from the exterior direction.

Table 3-18 The parameters used to produce the low-energy transfer shown in Fig. 3-50.

Parameter	Value
F	The family of southern Earth–Moon L_2 halo orbits
A_z	30,752 km (0.08 normalized distance units)
T_{ref}	15 January 2017 12:57:36 Ephemeris Time
τ	280.2 deg
p	Exterior
Δt_m	115.9 days

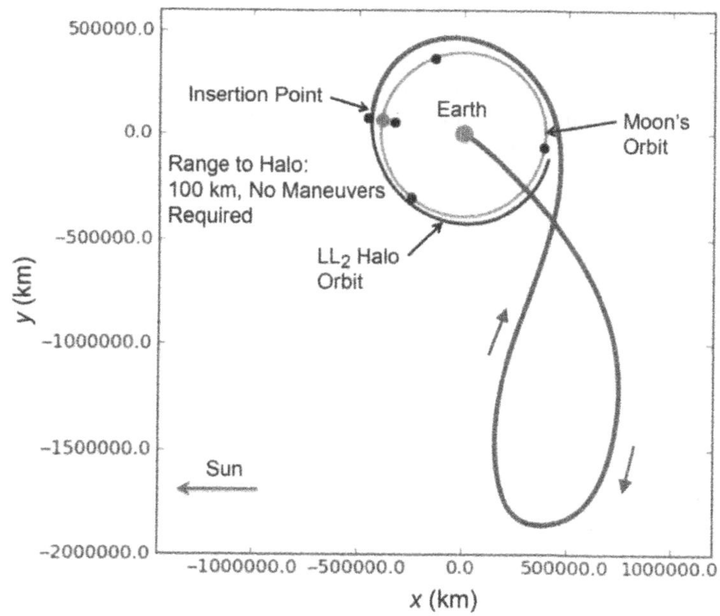

Figure 3-50 An example low-energy transfer produced in the DE421 model using the parameters specified in Table 3-18 [44] (Copyright © 2009 by American Astronautical Society Publications Office, San Diego, California (Web Site: http://www.univelt.com), all rights reserved; reprinted with permission of the AAS).

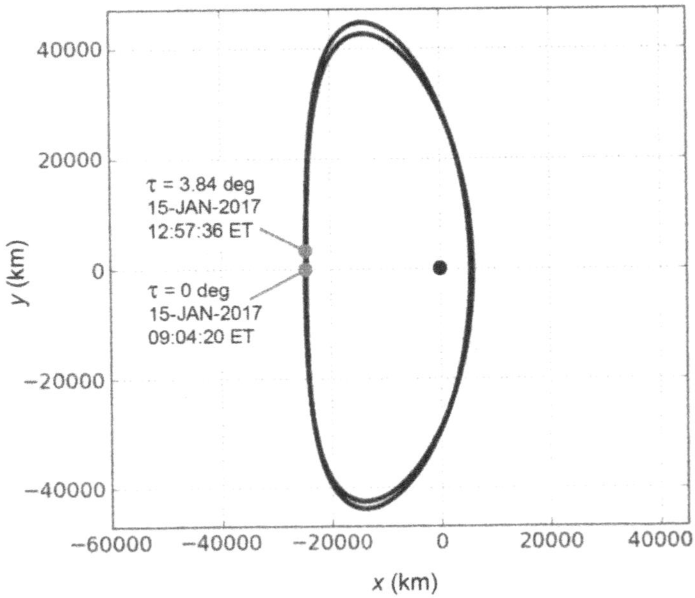

Figure 3-51 The halo orbit specified by F, A_z, and T_{ref} in Table 3-18 [44] (Copyright © 2009 by American Astronautical Society Publications Office, San Diego, California (Web Site: http://www.univelt.com), all rights reserved; reprinted with permission of the AAS).

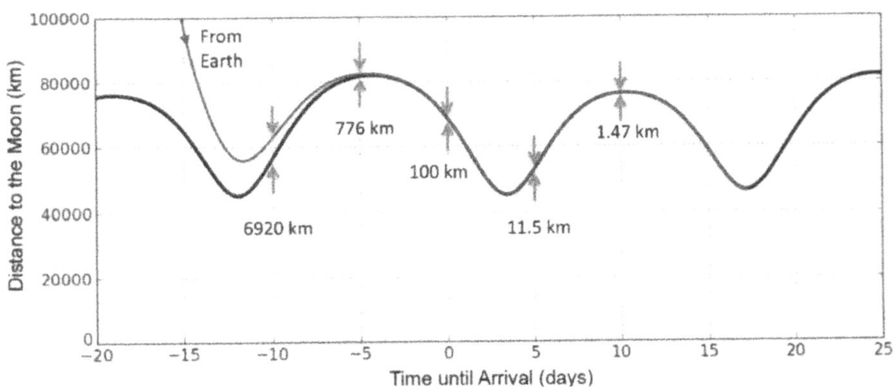

Figure 3-52 The distance between the transfer and the Moon as the trajectory approaches and arrives at the LL_2 halo orbit [44] (Copyright © 2009 by American Astronautical Society Publications Office, San Diego, California (Web Site: http://www.univelt.com), all rights reserved; reprinted with permission of the AAS).

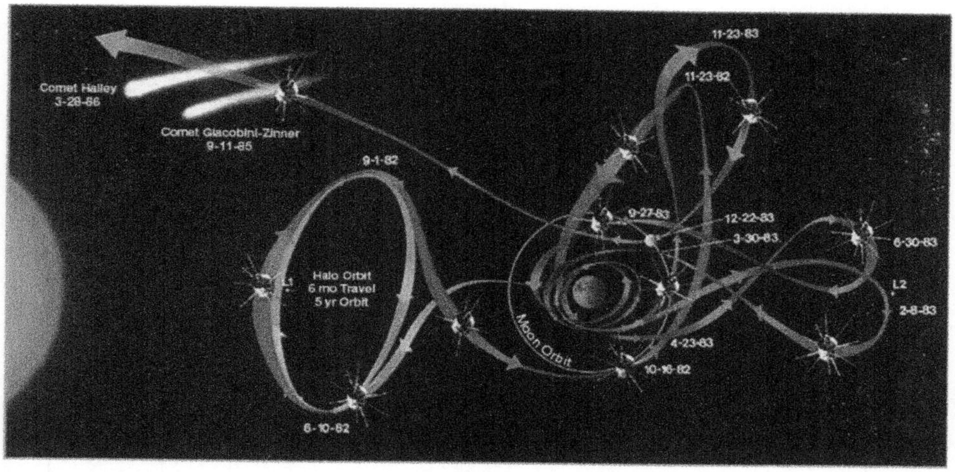

Figure 1-8 The trajectory of *ISEE-3/ICE* [62].

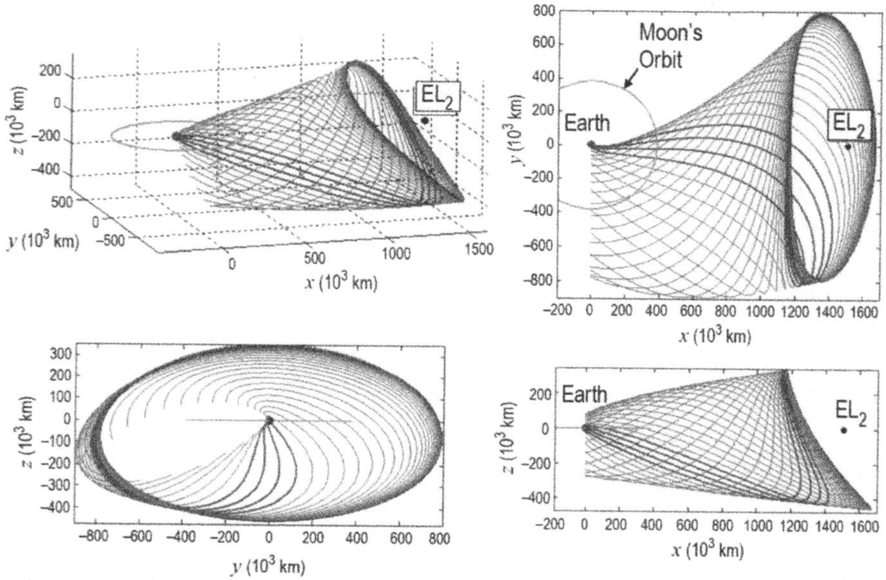

Figure 2-37 An example unstable halo orbit (red) about the Sun–Earth L_2 point and its stable invariant manifold (green/blue). Trajectories shown in blue intersect the Earth.

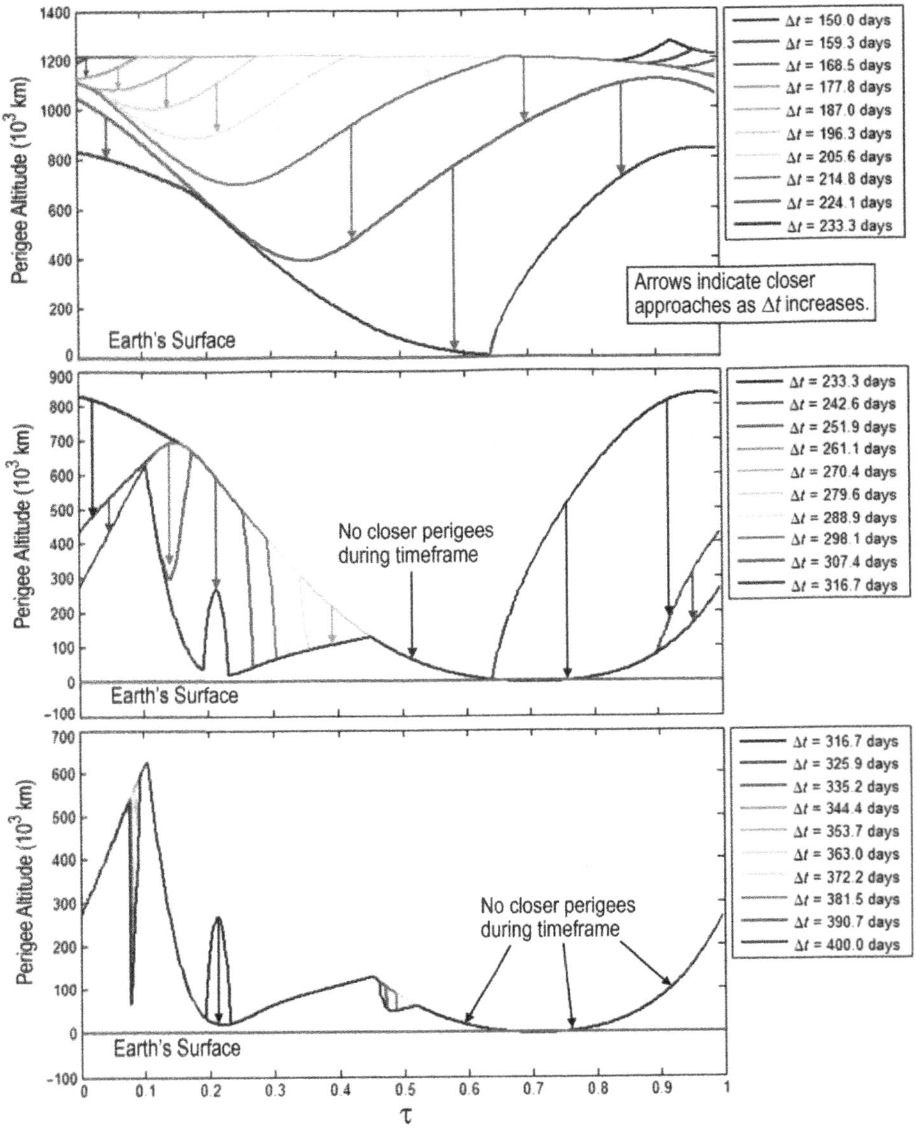

Figure 2-40 The altitude of closest approach of each trajectory in the stable manifolds shown in Fig. 2-39 with respect to the Earth. The propagation times of each manifold are shown in the legend. Longer propagation times yield closer perigee passages.

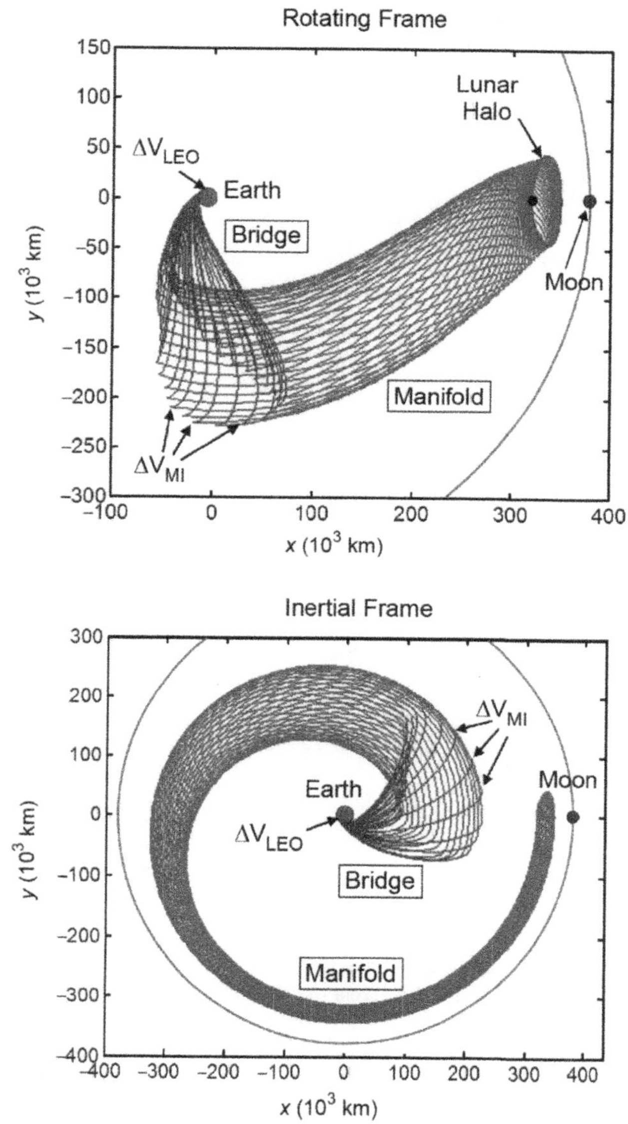

Figure 3-6 Example trajectories that implement the *perigee-point* scheme to directly transfer from LEO to a lunar L_1 halo orbit. The transfers are shown in the Earth–Moon rotating frame (top) and the corresponding inertial frame of reference (bottom).

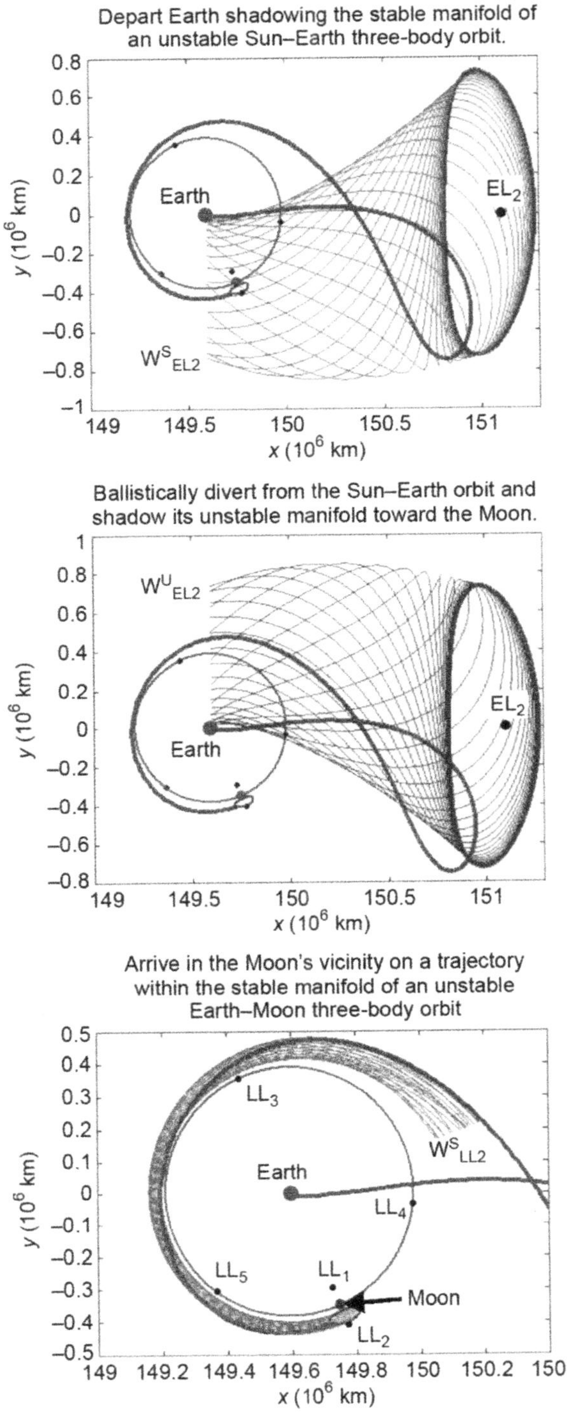

Figure 3.31 Modeling a ballistic lunar transfer as a series of heteroclinic transfers between unstable three-body orbits in the Patched Three-Body Model [97] (first published in Ref. [97]; reproduced with kind permission from Springer Science + Business Media B.V.).

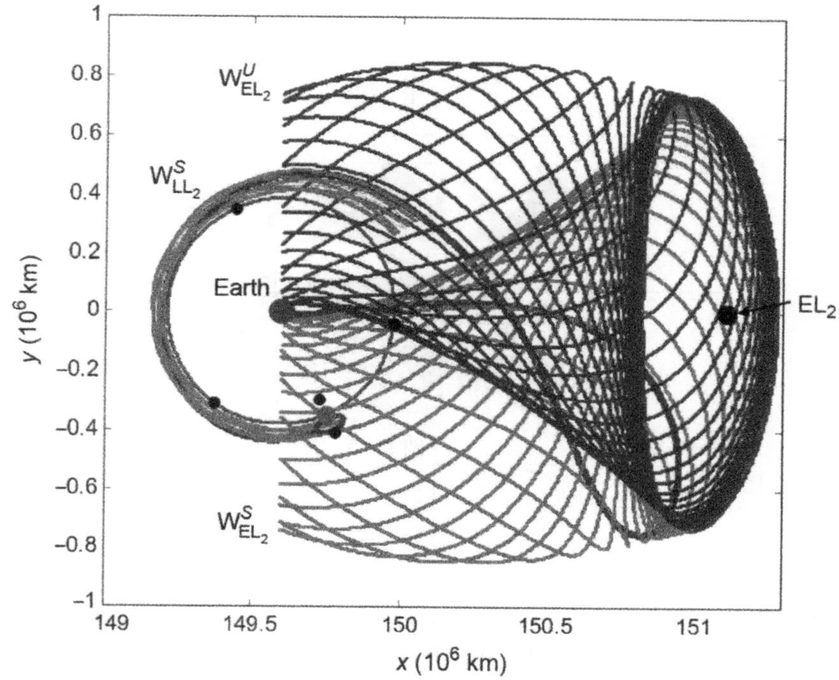

Figure 3-39 A top-down perspective of the example low-energy transfer, shown with all three manifolds that were used to model it (blue = the low-energy transfer, green = stable manifold of a three-body orbit, brown = unstable manifold of a three-body orbit).

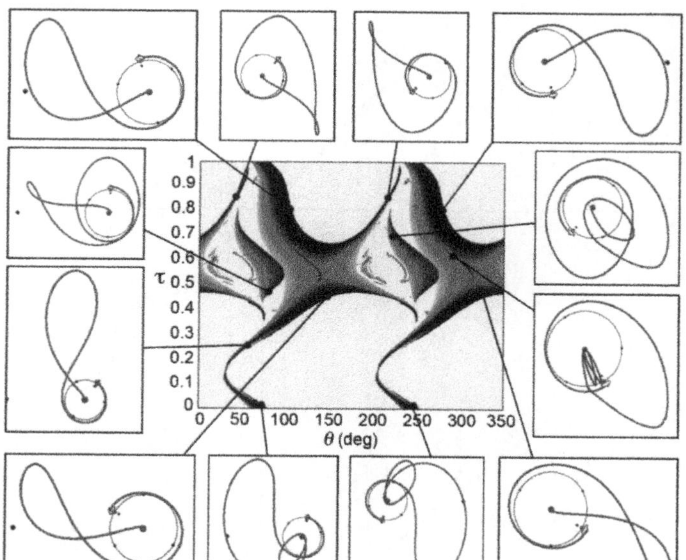

Figure 3-48 An example state space map, capturing a wide variety of low-energy transfers that exist between the Earth and an example LL_2 halo orbit. Each trajectory arrives at the halo orbit from the exterior direction, and arrives at the orbit in a geometry according to the given (θ, τ) combination. The color of the map indicates how close to the Earth the trajectory gets when propagated from the LL_2 halo orbit backward in time. All black points represent viable low-energy transfers.

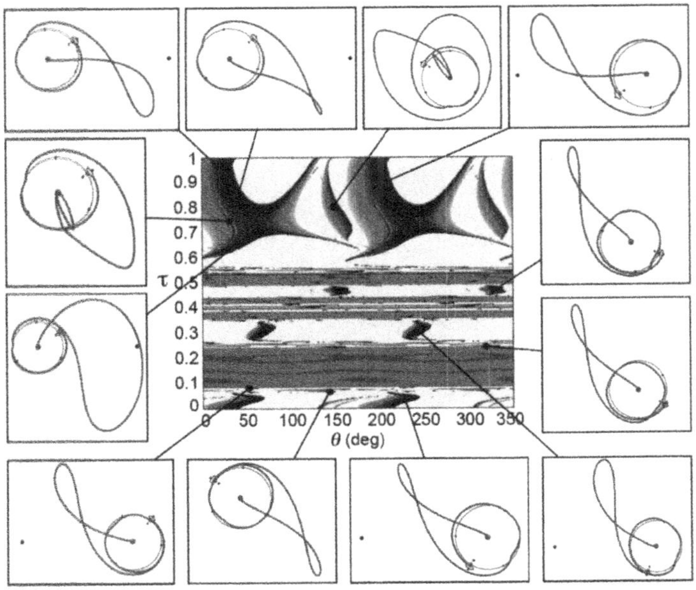

Figure 3-49 An example state space map that is identical to the map illustrated in Fig. 3-48, except that the trajectories arrive at the LL_2 halo orbit from the interior direction.

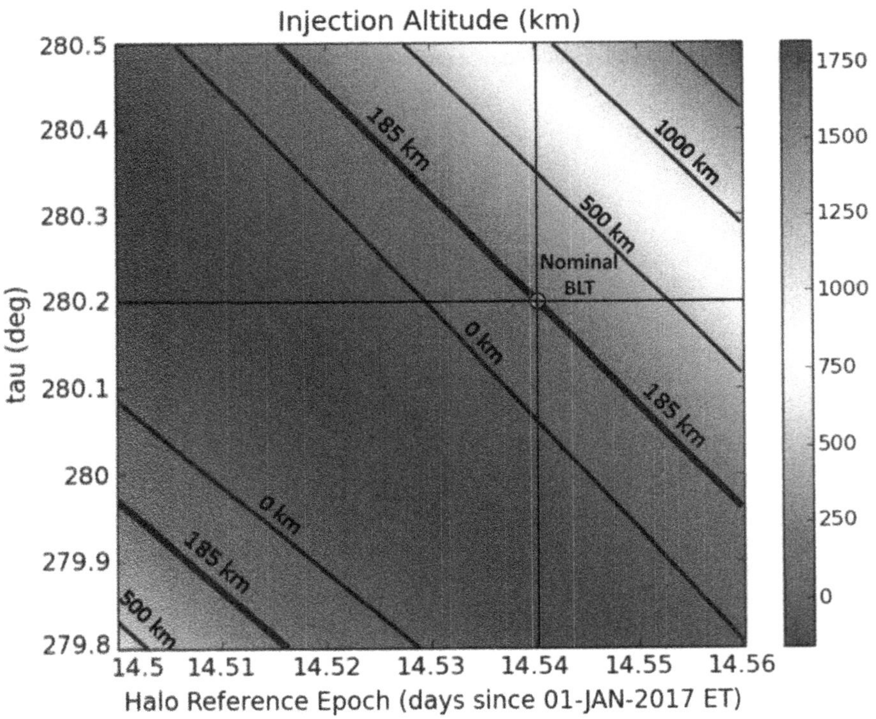

Figure 3-53 A map of the perigee altitude that each low-energy trajectory encounters as a function of T_{ref} and τ. The 185-km contour is highlighted, which includes the nominal ballistic lunar transfer presented in Table 3-18 [44] (Copyright © 2009 by American Astronautical Society Publications Office, San Diego, California (Web Site: http://www.univelt.com), all rights reserved; reprinted with permission of the AAS).

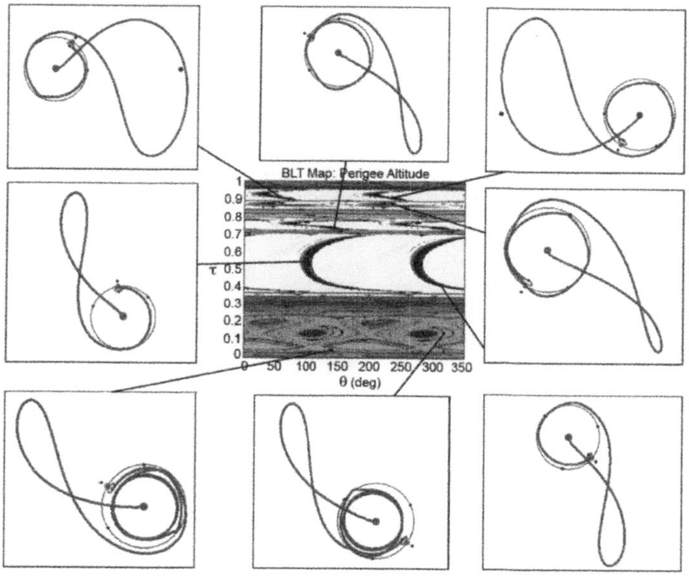

Figure 3-85 The interior BLT map for low-energy transfers to the example lunar L_1 halo orbit. Eight example low-energy transfers are shown around the BLT map to demonstrate some of the types of transfers that may be constructed between 185-km LEO orbits and this halo orbit.

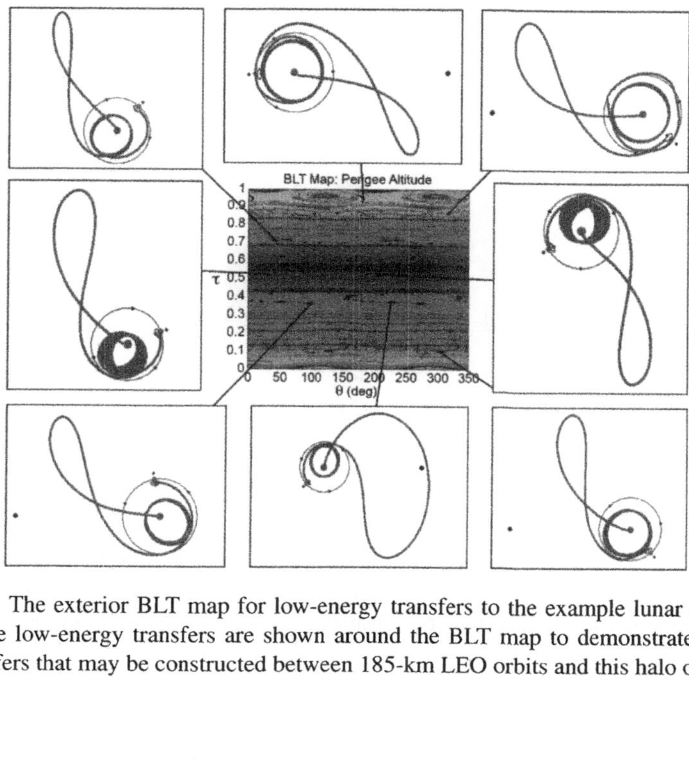

Figure 3-86 The exterior BLT map for low-energy transfers to the example lunar L_1 halo orbit. Eight example low-energy transfers are shown around the BLT map to demonstrate some of the types of transfers that may be constructed between 185-km LEO orbits and this halo orbit.

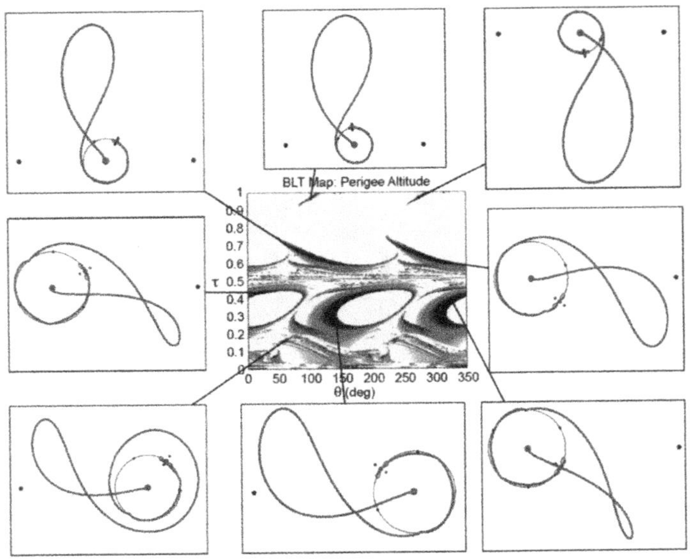

Figure 3-89 The exterior BLT map for low-energy transfers to the example distant prograde orbit about the Moon. Eight example low-energy transfers are shown around the BLT map to demonstrate some of the types of transfers that may be constructed between 185-km LEO orbits and this lunar orbit.

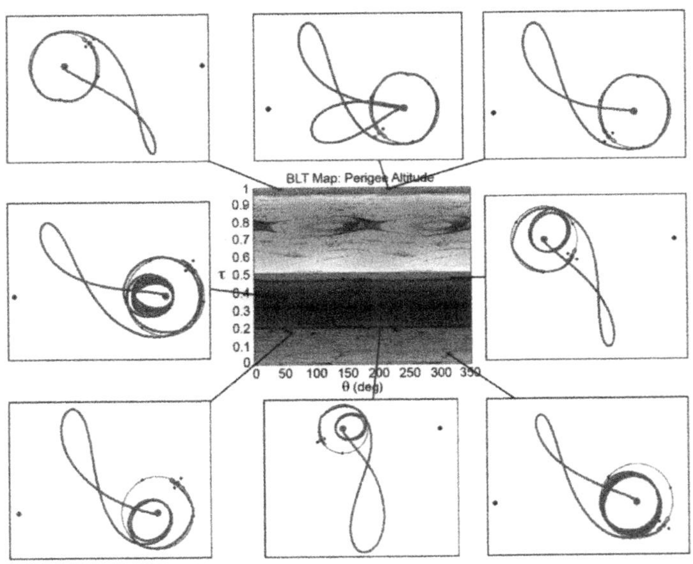

Figure 3-90 The interior BLT map for low-energy transfers to the example distant prograde orbit about the Moon. Eight example low-energy transfers are shown around the BLT map to demonstrate some of the types of transfers that may be constructed between 185-km LEO orbits and this lunar orbit.

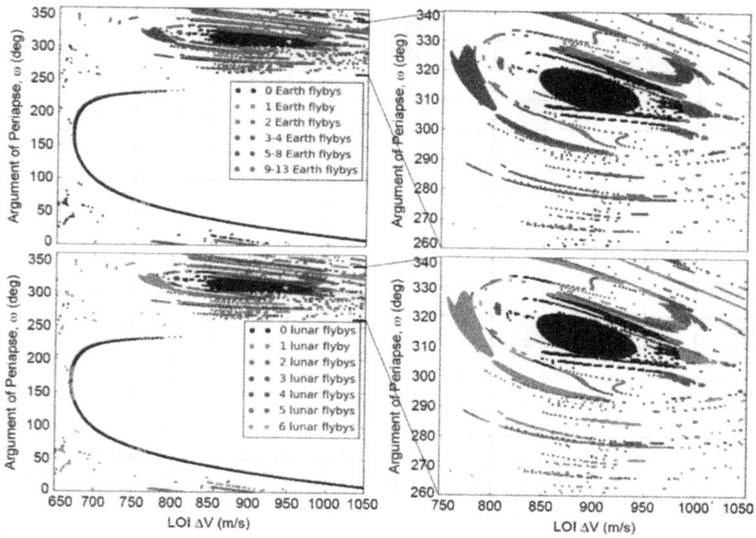

Figure 4-7 The transfers shown in Fig. 4-6 that approach within 1000 km of the Earth, shaded according to the number of Earth phasing orbits (top) and lunar flybys (bottom) that they make [2]. The upper portion of each map, corresponding with direct lunar transfers, is magnified in the plots on the right.

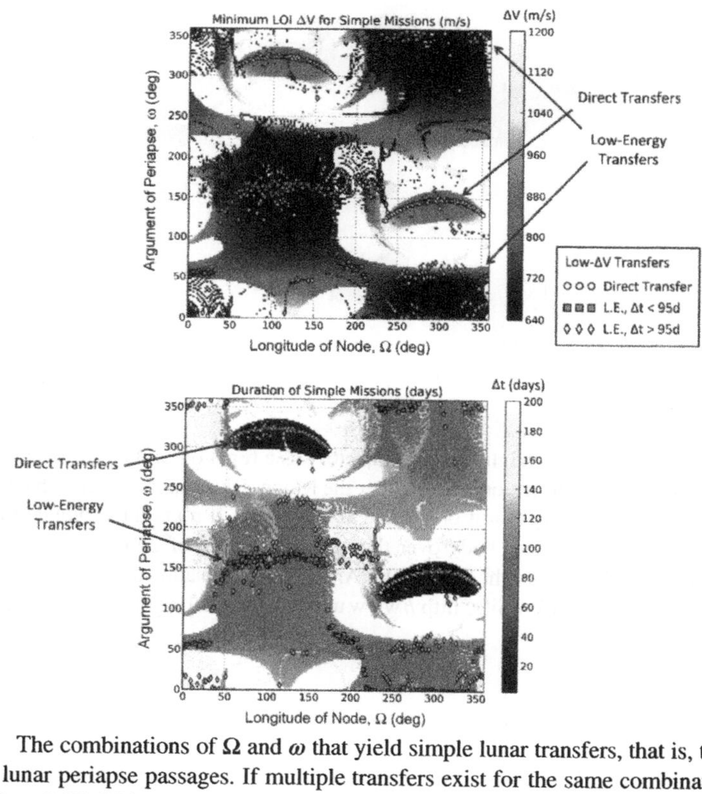

Figure 4-13 The combinations of Ω and ω that yield simple lunar transfers, that is, those without low Earth or lunar periapse passages. If multiple transfers exist for the same combination, then the one with the least LOI ΔV is shown. All of these transfers arrive at a first-quarter Moon. The low-ΔV transfers shown in Fig. 4-12 are indicated by dots in each map [2]

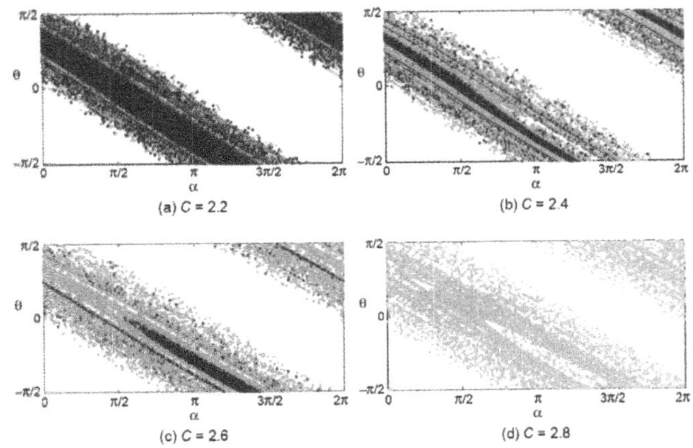

Figure 5-4 Plots showing the origin of each trajectory as a function of the position and orientation of the velocity vectors as the trajectories encounter the Moon's surface. Blue points indicate that the trajectory originated at the Earth and gray that it originated at the Moon. If no point is plotted the integrated trajectory did not encounter the surface of either body over the given time span of 200 days (Earth–Moon CRTBP) [192] (Copyright © 2011 by American Astronautical Society Publications Office, San Diego, California (Web Site: http://www.univelt.com), all rights reserved; reprinted with permission of the AAS).

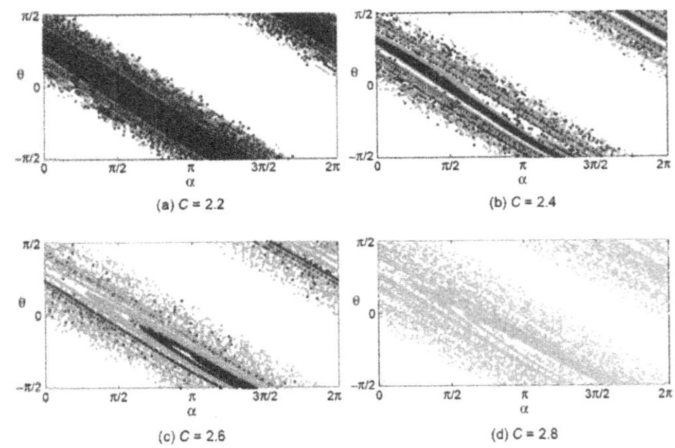

Figure 5-6 Plots showing the origin of each trajectory as a function of the position and orientation of the velocity vectors as the trajectories encounter the Moon. Blue points indicate that the trajectory originated at the Earth and gray that it originated at the Moon. If no point is plotted the integrated trajectory did not encounter either body over the given time span of 200 days (Earth–Moon only Ephemeris system) [192] (Copyright © 2011 by American Astronautical Society Publications Office, San Diego, California (Web Site: http://www.univelt.com), all rights reserved; reprinted with permission of the AAS).

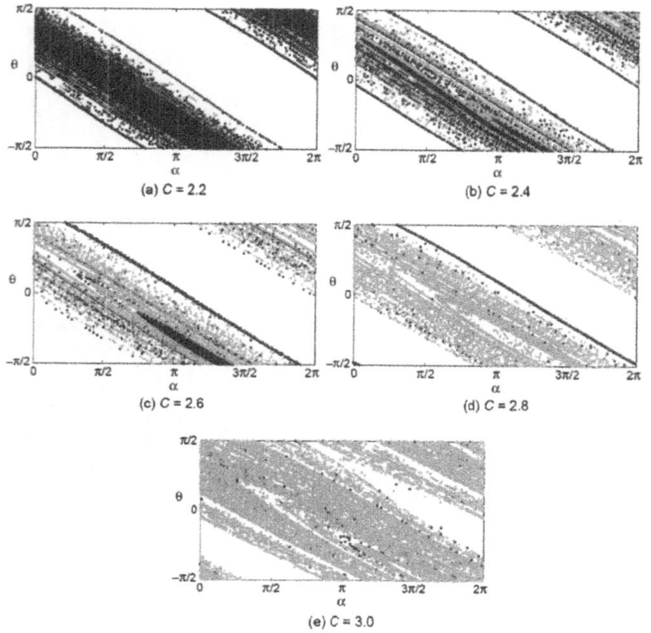

Figure 5-8 Plots showing the origin of each trajectory as a function of the position and orientation of the velocity vectors as the trajectories encounter the Moon. Blue points indicate that the trajectory originated at the Earth and gray that it originated at the Moon. If no point is plotted the integrated trajectory did not encounter either body over the given time span of 200 days (Sun–Earth–Moon Ephemeris system) [192] (Copyright © 2011 by American Astronautical Society Publications Office, San Diego, California (Web Site: http://www.univelt.com), all rights reserved; reprinted with permission of the AAS).

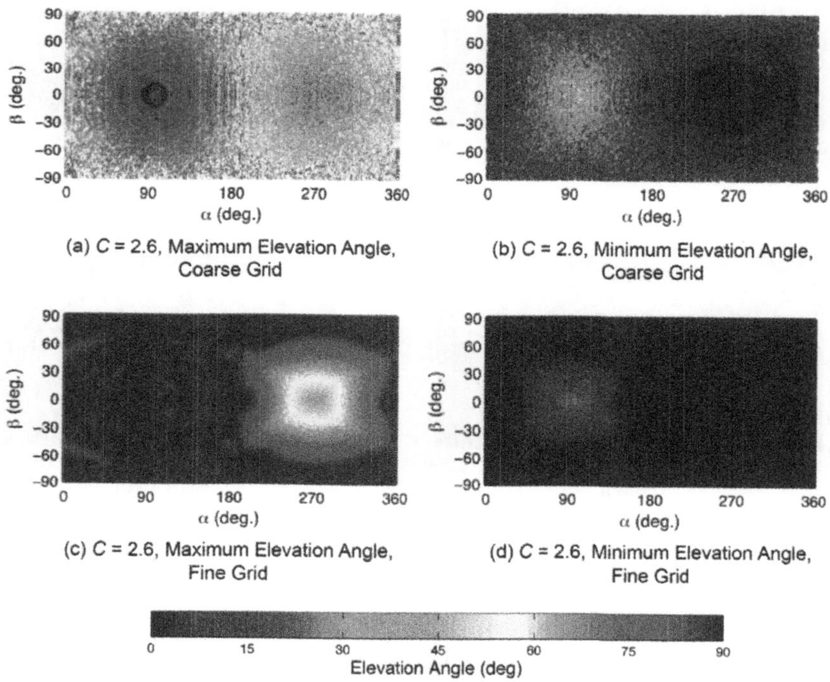

Figure 5-17 Maximum and minimum elevation angles for trajectories originating at the Earth and encountering the Moon at each point on the surface. These cases are computed in the CRTBP for $C = 2.6$. Results from two different grids (in elevation and azimuth angle) are shown [193] (Copyright © 2011 by American Astronautical Society Publications Office, San Diego, California (Web Site: http://www.univelt.com), all rights reserved; reprinted with permission of the AAS).

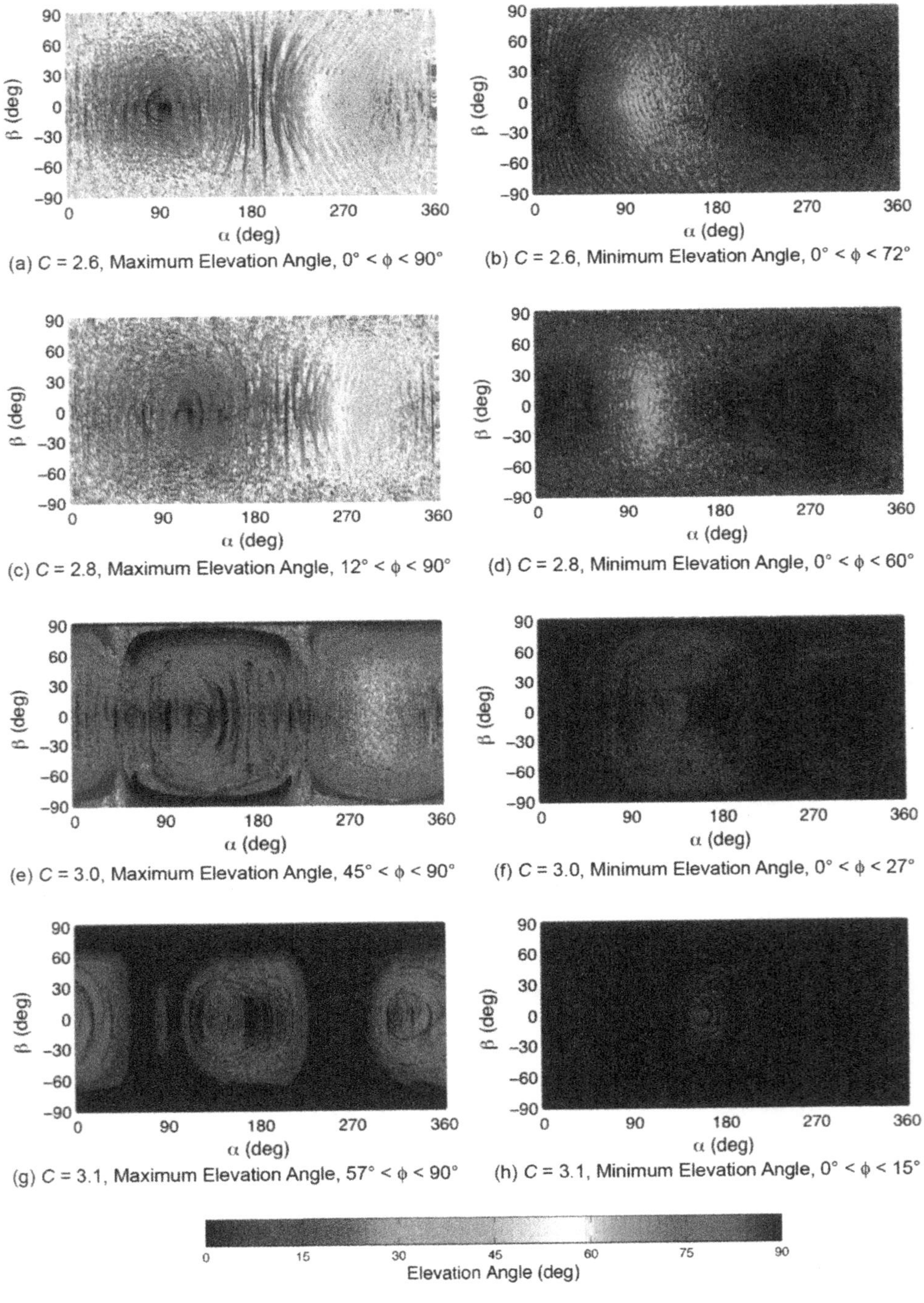

Figure 5-18 The minimum and maximum elevation angles of trajectories originating at the Earth for each point on the lunar surface. These trajectories are computed in the Earth–Moon ephemeris system including the Sun's perturbations [193] (Copyright © 2011 by American Astronautical Society Publications Office, San Diego, California (Web Site: http://www.univelt.com), all rights reserved; reprinted with permission of the AAS).

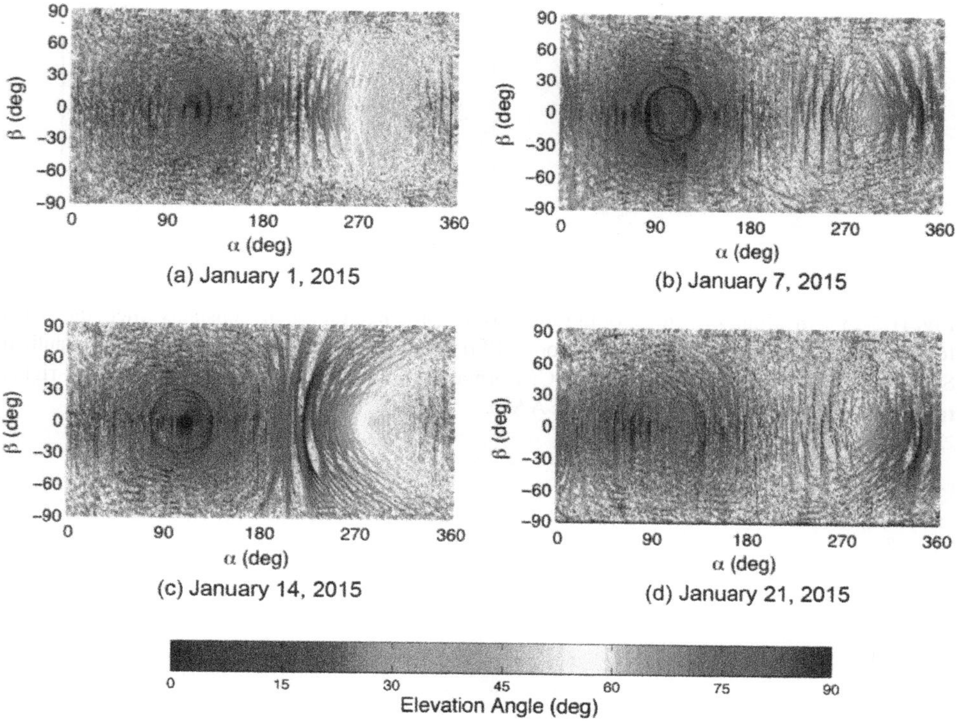

Figure 5-19 Comparison of maximum elevation angle results around the lunar orbit at seven day intervals for $C = 2.8$ [193] (Copyright © 2011 by American Astronautical Society Publications Office, San Diego, California (Web Site: http://www.univelt.com), all rights reserved; reprinted with permission of the AAS).

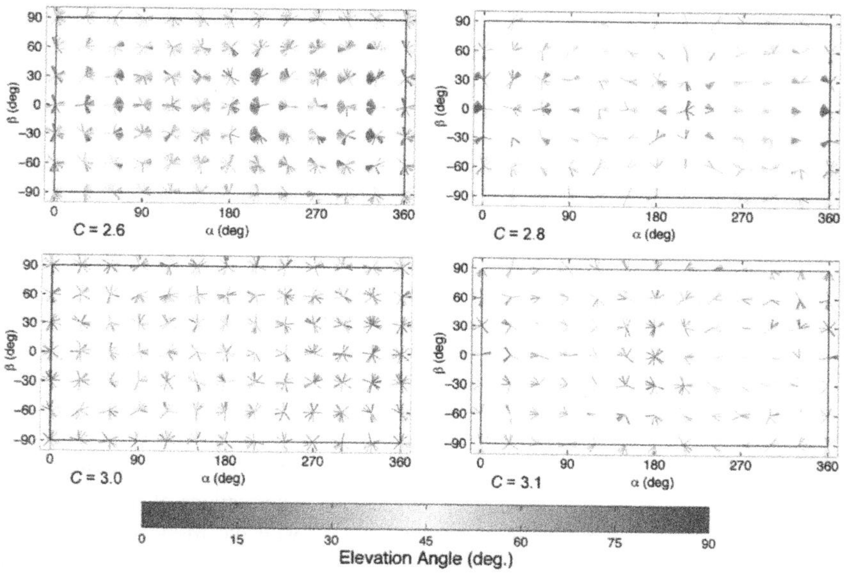

Figure 5-20 Azimuth angles at points on a 30-deg grid on the lunar surface. The plotted lines at each gridpoint are oriented in the proper azimuth direction for each individual trajectory. The color corresponds to the elevation angle of that trajectory. The trajectories shown here are limited to those with $C_3 < 0.0$ km^2/s^2 at the Earth [194] (first published in Ref. [194]; reproduced with kind permission from Springer Science + Business Media B.V.).

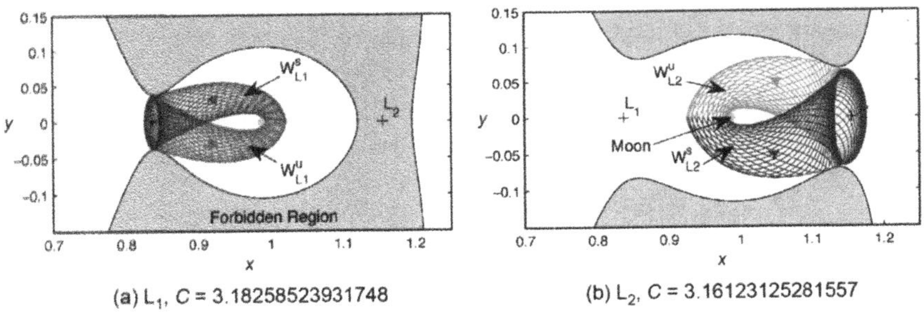

Figure 5-21 Invariant manifolds of libration orbits computed for Jacobi constants where the manifolds are tangent to the surface of the Moon [193] (Copyright © 2011 by American Astronautical Society Publications Office, San Diego, California (Web Site: http://www.univelt.com), all rights reserved; reprinted with permission of the AAS).

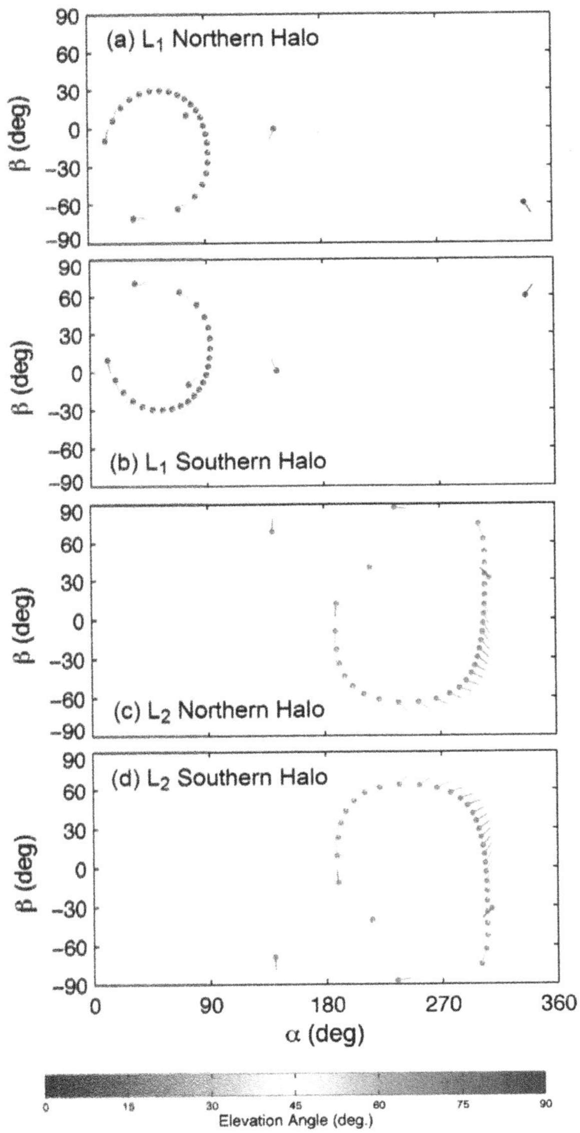

Figure 5-22 Unstable manifold intersections of the specified orbits with the Moon for $C = 3.1$ [193] (Copyright © 2011 by American Astronautical Society Publications Office, San Diego, California (Web Site: http://www.univelt.com), all rights reserved; reprinted with permission of the AAS).

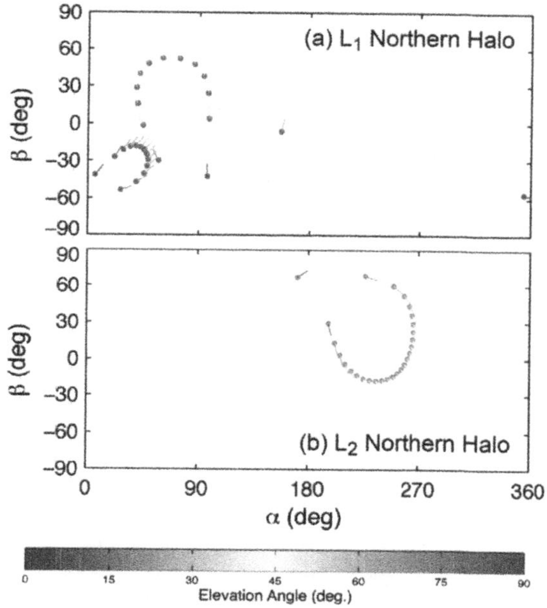

Figure 5-23 Unstable manifold intersections of the specified orbits with the Moon for $C = 3.08$ [193] (Copyright © 2011 by American Astronautical Society Publications Office, San Diego, California (Web Site: http://www.univelt.com), all rights reserved; reprinted with permission of the AAS).

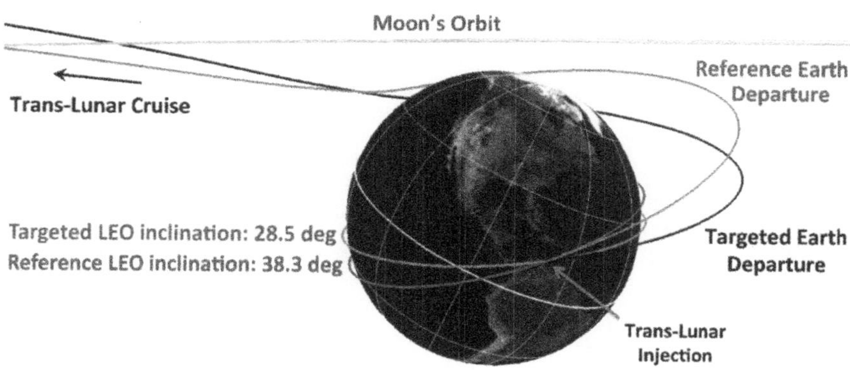

Figure 6-2 The targeted Earth departure compared with the reference Earth departure [191] (*Acta Astronautica* by International Academy of Astronautics, reproduced with permission of Pergamon in the format reuse in a book/textbook via Copyright Clearance Center).

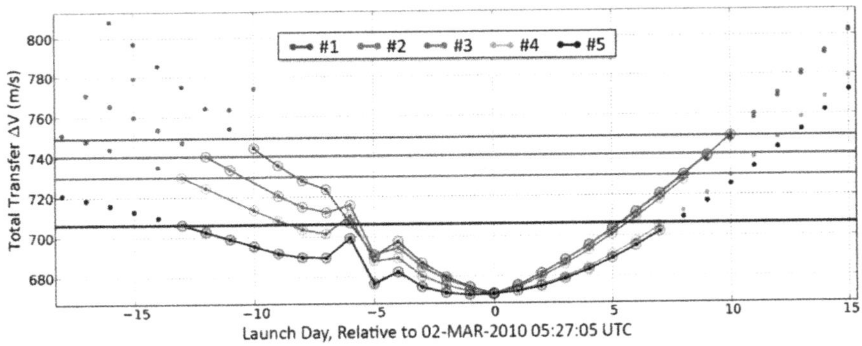

Control	#1	#2	#3	#4	#5	Comments
Launch Time	Vary	Vary	Vary	Vary	Vary	This permits the node of the parking orbit to vary.
Parking Orbit Inc.	Fixed	Fixed	Fixed	Fixed	Fixed	All held the parking orbit inclination fixed at 28.5 deg.
TLI mag, location	Vary	Vary	Vary	Vary	Vary	The magnitude and location of the TLI maneuver.
TCM 1 Epoch	L+21d	L+21d	L+21d	L+21d	Vary	The epoch of TCM1 (most 21 days after launch).
TCM 2 Epoch	L+60d	LOI−40d	LOI−30d	LOI−40d	Vary	The epoch of TCM2, either relative to launch or LOI.
LOI Epoch	Fixed	Fixed	Fixed	Fixed	Fixed	The epoch of the LOI.
LOI ΔV	Fixed	Fixed	Fixed	Vary	Vary	The magnitude and direction of the LOI.
Total ΔV (m/s)	749.2	740.4	729.7	705.9	706.2	Including two TCMs and the full LOI (impulsive).

Figure 6-5 Several example launch periods for the example lunar mission, depending on which controls are fixed, their fixed values, and which controls are permitted to vary.

3.4.5 Families of Low-Energy Transfers

A set of low-energy parameters may be used to generate the initial conditions of a trajectory that is propagated backward in time to construct a ballistic lunar transfer. If one set of parameters $\{F', A'_z, T'_{\text{ref}}, p', \tau', \Delta t'_m\}$ generates a trajectory that originates from a LEO with an altitude of 185 km, then it is typically the case that a small deviation in either T'_{ref} or τ' will generate a trajectory that originates from a LEO with a slightly different altitude. However, small deviations in both of those parameters may often be designed to generate a new trajectory that originates from a LEO with the same 185 km altitude. In that case, the two sets of parameters define two different ballistic lunar transfers that are in the same family of transfers.

Figure 3-53 illustrates how transfers may be organized into families. In this example, the lunar transfer shown in Fig. 3-50 with the parameters given in Table 3-18 is used as a reference trajectory. The transfer's parameters are all held constant, except for the parameters T_{ref} and τ, which are systematically varied through all combinations of values shown in Fig. 3-53. At each combination, a new trajectory is propagated and analyzed to determine its new perigee altitude. One can see that by reducing both T_{ref} and τ, one builds trajectories that come closer to the Earth at

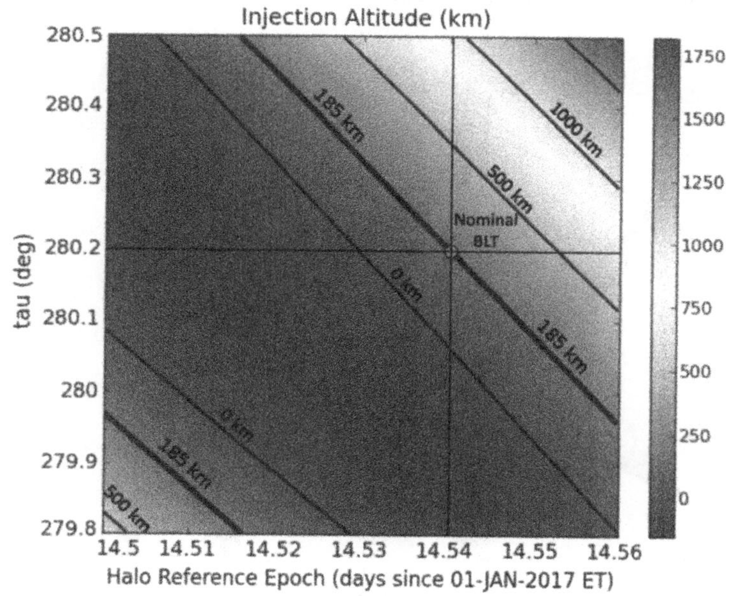

Figure 3-53 A map of the perigee altitude that each low-energy trajectory encounters as a function of T_{ref} and τ. The 185-km contour is highlighted, which includes the nominal ballistic lunar transfer presented in Table 3-18 [44] (Copyright © 2009 by American Astronautical Society Publications Office, San Diego, California (Web Site: http://www.univelt.com), all rights reserved; reprinted with permission of the AAS). *(See insert for color representation of this figure.)*

their perigee point, and vice versa. By reducing T_{ref} and increasing τ an appropriate amount, one can produce new trajectories that also have a perigee altitude of 185 km.

The exercise given above may be extended to allow T_{ref} to vary across an entire month and τ to vary across 360 deg to observe full families of low-energy lunar transfers. Figure 3-54 shows such a Ballistic Lunar Transfer (BLT) state space map given the parameter set summarized in Table 3-19. The figure shows a plot that maps the perigee altitude of each trajectory generated using each combination of T_{ref} and τ. The darkest regions contain the parameters that produce useful transfers; the white fields contain parameters that generate trajectories that do not approach the Earth. Figure 3-55 shows the same map with several trajectories plotted to illustrate the trajectories that may be generated using these parameters.

Families of transfers may be identified in the BLT state space map shown in Fig. 3-54 by tracing those combinations of T_{ref} and τ that have a perigee altitude of some desirable value, for example, 185 km. Figure 3-56 shows samples of the combinations of T_{ref} and τ that generate ballistic transfers with injection altitudes of 185 km. The points displayed in black correspond to trajectories that traverse closer to EL_2 than EL_1 and vice versa. Table 3-20 presents a summary of the characteristics of a sample of the transfers identified in Fig. 3-56. Each of these transfers is a member of a family of similar trajectories for which the characteristics vary smoothly away from those presented in the table. There are certainly many families of ballistic

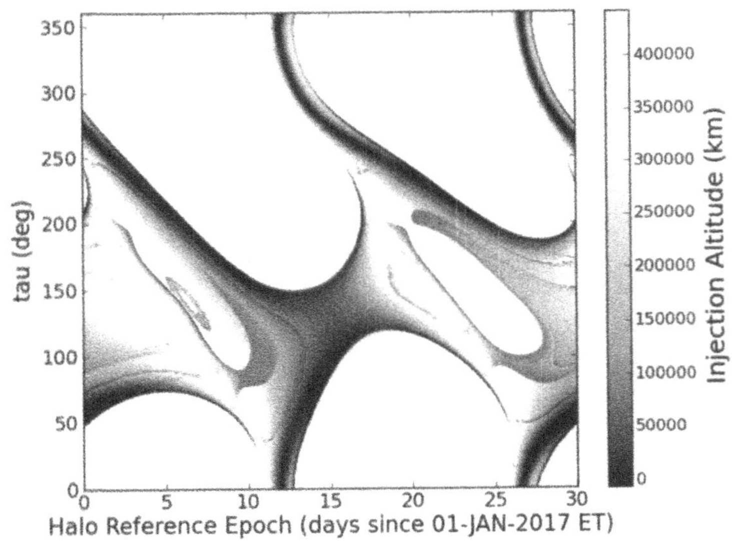

Figure 3-54 A BLT state space map that shows the perigee altitude of each generated trajectory as a function of T_{ref} and τ. The darkest regions include the combinations of T_{ref} and τ that yield transfers that begin from low Earth orbits [44] (Copyright © 2009 by American Astronautical Society Publications Office, San Diego, California (Web Site: http://www.univelt.com), all rights reserved; reprinted with permission of the AAS).

Table 3-19 The parameters used to produce the results shown in Figs. 3-54–3-56.

Parameter	Value
F	The family of southern Earth–Moon L_2 halo orbits
A_z	30,752 km (0.08 normalized distance units)
T_{ref}	1 Jan 2017 00:00:00 ET $\leq T_{ref} \leq$ 31 Jan 2017 00:00:00 ET
τ	0 deg $\leq \tau \leq$ 360 deg
p	Exterior
Δt_m	180 days

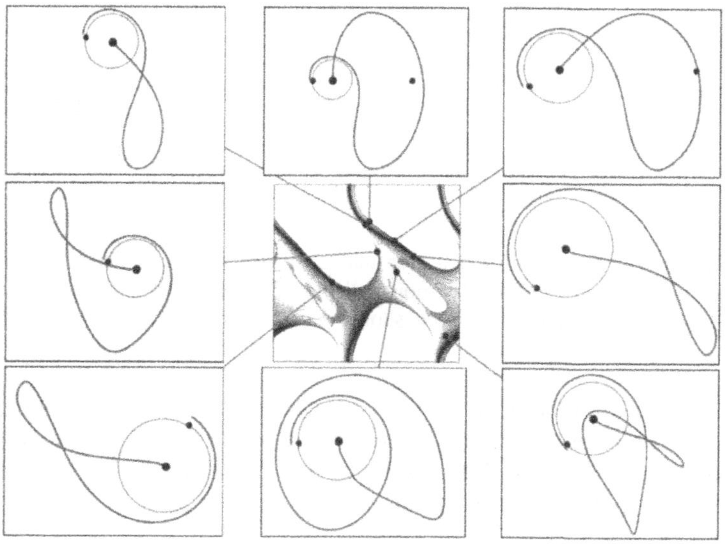

Figure 3-55 The same BLT state space map shown in Fig. 3-54 with example transfers shown around the perimeter [44] (Copyright © 2009 by American Astronautical Society Publications Office, San Diego, California (Web Site: http://www.univelt.com), all rights reserved; reprinted with permission of the AAS).

transfers unrepresented in the table. Figure 3-57 illustrates six example families of low-energy transfers. One can see that the general characteristics of each family varies in a smooth fashion from one transfer to the next in the family.

The quickest transfer identified in Fig. 3-56 requires fewer than 83 days between the injection and the point when the trajectory has arrived within 100 km of the lunar halo orbit. The vast majority of the transfers shown require a launch energy in the range of -0.75 km^2/s^2 $\leq C_3 \leq -0.35$ km^2/s^2. The transfers that include a lunar flyby often require less launch energy, particularly those that involve a lunar flyby on the outbound trajectory soon after injection. Several transfers have been identified that require a C_3 as low as -2.1 km^2/s^2, implementing a lunar flyby at an altitude of approximately 2000 km. Figure 3-58 shows the relationship between the required injection C_3 and the transfer duration; Fig. 3-59 compares the required injection C_3

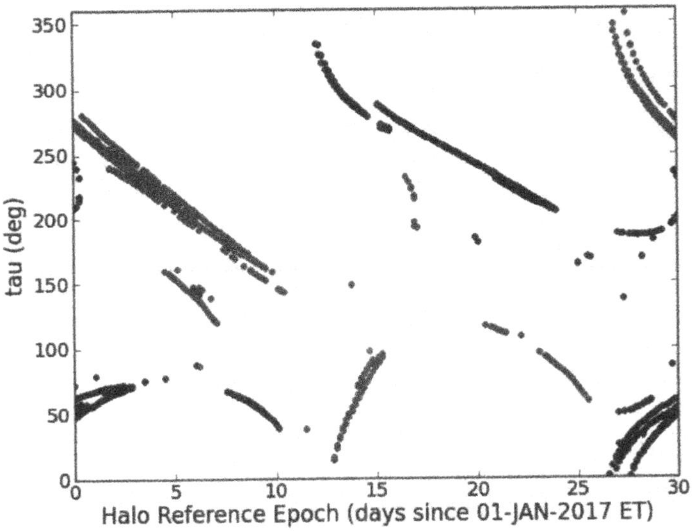

Figure 3-56 Combinations of T_{ref} and τ that generate ballistic transfers with perigee injections at an altitude of 185 km. The points displayed in black correspond to trajectories that traverse closer to EL_2 than EL_1; points shown in gray travel closer to EL_1 than EL_2.

with the lowest lunar periapse altitude. One can see a clear correlation in Fig. 3-59 that the closer a trajectory gets to the Moon during the transfer, the lower the injection C_3 may be. Additional lunar flybys or Earth phasing orbits may help provide the geometry needed for a particular mission.

3.4.6 Monthly Variations in Low-Energy Transfers

The BLT state space map shown in Figs. 3-54–3-56 will repeat perfectly from one synodic month to the next in the Patched Three-Body Model, since the model is symmetric. The characteristics of the BLT state space map generated in the DE421 model of the solar system will not repeat perfectly each month, although similar features will be present in each month. Figure 3-60 shows a map of the perigee altitude of trajectories generated from the same set of parameters presented in Table 3-18. But for a wider range of T_{ref} and τ, T_{ref} is varied over 3 months, and τ is varied over two halo orbit revolutions. One can see the same features from cycle to cycle, but the details of the state space map vary. Significant variations are observed between the first halo orbit revolution (0 deg $\leq \tau \leq$ 360 deg) and the second halo orbit (360 deg $\leq \tau \leq$ 720 deg), mostly as a consequence of the nonzero eccentricity of the Moon's orbit about the Earth–Moon barycenter.

3.4.6.1 12-Month Survey The state space map has been further extended to 12 months to study the variations that exist throughout an entire year. It has been observed that the most prominent features continue to persist, and repeat regularly,

Table 3-20 Summary characteristics for a sample of the ballistic transfers identified in Fig. 3-56 [44] (Copyright © 2009 by American Astronautical Society Publications Office, San Diego, California (Web Site: http://www.univelt.com), all rights reserved; reprinted with permission of the AAS).

#	Δ Reference Epoch* (days)	τ (deg)	EL_1 / EL_2	C_3 (km^2/s^2)	Transfer Δt (days)	# Earth Flybys	# Lunar Flybys	Injection Inclination (deg) Equatorial	Ecliptic
1	12.060302	334.519	2	−0.2902	133.76	0	0	23.225	28.192
2	12.211055	333.736	2	−0.3457	132.91	0	0	131.701	151.274
3	14.170854	283.655	2	−0.3444	118.36	0	0	51.319	69.045
4	15.226131	271.069	2	−0.4944	108.76	0	1	32.329	51.431
5	15.829146	279.419	2	−0.4296	171.62	0	0	85.326	103.860
6	20.351759	238.347	2	−0.6556	130.11	0	0	115.737	137.694
7	20.351759	239.232	2	−0.5856	145.20	0	0	21.877	22.738
8	22.311558	221.171	2	−0.6904	137.51	0	1	35.973	13.527
9	23.819095	206.901	2	−0.7153	129.17	0	0	22.180	10.275
10	20.050251	180.970	2	−1.8533	171.79	0	1	97.684	92.972
11	25.025126	164.113	2	−1.9222	146.35	0	1	20.490	4.271
12	27.286432	137.373	2	−2.0307	176.72	0	2	38.302	36.809
13	28.190955	168.405	2	−2.0880	122.46	2	2	19.325	30.359
14	28.040201	185.608	2	−1.0318	145.08	0	1	34.251	11.315
15	28.040201	185.630	2	−1.6144	145.75	0	2	103.995	126.244
16	0.000000	55.325	2	−0.9032	179.35	2	1	143.590	121.792
17	0.150754	63.382	2	−0.6429	97.90	0	0	23.372	0.836
18	0.452261	54.781	2	−0.6608	132.55	0	0	145.538	168.969
19	1.507538	66.990	2	−1.1266	113.39	0	1	166.454	144.152
20	8.592965	59.539	2	−0.8393	178.32	0	1	99.214	87.676
21	8.592965	59.962	2	−0.6791	165.37	0	0	14.732	20.434
22	6.030151	144.580	2	−0.6940	170.11	0	3	23.140	17.669
23	27.889447	53.118	2	−0.9637	140.22	1	2	11.452	28.632
24	28.040201	15.470	2	−0.4261	172.37	0	1	27.743	40.712
25	28.190955	34.787	2	−0.5891	105.30	0	0	148.336	171.495
26	28.341709	43.756	2	−0.5740	96.55	0	0	20.962	3.797
27	2.110553	245.420	1	−0.5465	91.66	0	0	20.003	4.747
28	2.412060	247.372	1	−0.6290	172.42	1	0	54.249	30.825
29	2.110553	251.704	1	−0.6311	178.46	1	2	59.547	36.213
30	2.261307	255.586	1	−0.5150	154.75	0	0	65.164	44.035
31	6.934673	122.568	1	−0.7340	165.38	0	0	20.624	28.138
32	6.783920	138.709	1	−0.5098	164.58	0	2	124.809	129.384
33	11.457286	38.141	1	−1.1299	167.55	0	2	39.917	26.275
34	14.170854	65.695	1	−0.5599	143.25	0	0	19.771	14.374
35	14.170854	70.107	1	−0.6869	123.22	0	0	106.493	129.791
36	14.170854	73.417	1	−0.6246	115.20	0	0	87.048	110.261
37	16.733668	222.850	1	−0.7658	179.64	0	1	137.534	126.323
38	16.733668	223.945	1	−0.6178	171.17	0	0	11.994	14.627
39	17.035176	192.365	1	−1.5154	156.53	1	1	28.596	51.902
40	22.160804	108.406	1	−2.0107	129.17	0	1	18.754	5.377
41	23.819095	87.587	1	−0.6915	167.13	0	0	50.748	32.372
42	28.190955	313.713	1	−0.4043	177.60	0	0	140.309	130.765
43	28.492462	285.732	1	−0.4568	109.17	0	1	10.097	14.214
44	3.165829	227.614	1	−1.9572	169.47	7	2	153.358	172.197

*The reference epoch is given as a duration of time, in days, away from 1 Jan 2017 00:00:00 Ephemeris Time.

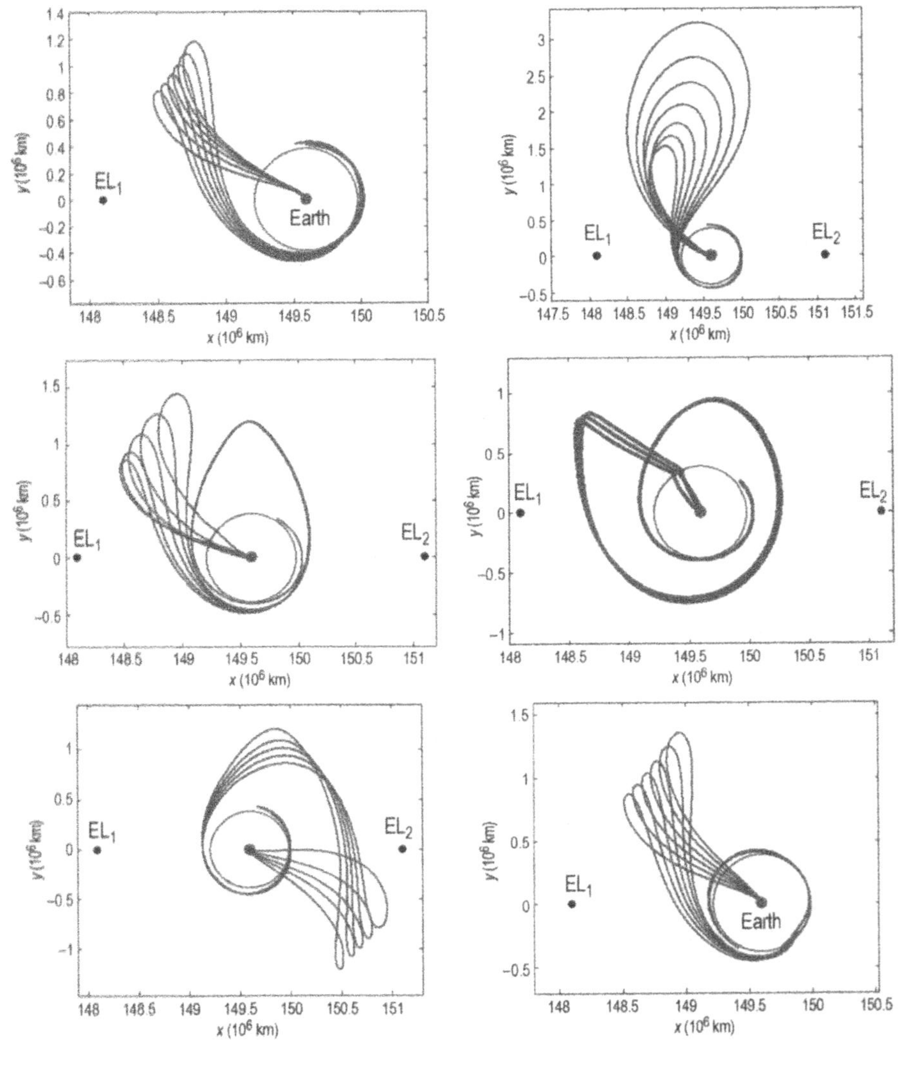

Figure 3-57 Example trajectories within six families of low-energy transfers that each may be used to transfer a spacecraft from a 185-km altitude state above the Earth to the same LL_2 halo orbit, though at different arrival times.

while subtle features appear and disappear from month to month. Figure 3-61 shows a plot of samples of the combinations of T_{ref} and τ that yield low-energy transfers between 185-km LEO parking orbits and the lunar halo orbit.

The reference epoch of each transfer shown in Fig. 3-61 may be wrapped into one synodic month to observe the changes that occur in the state space map from one synodic month to the next. Figure 3-62 shows the resulting plot, revealing the

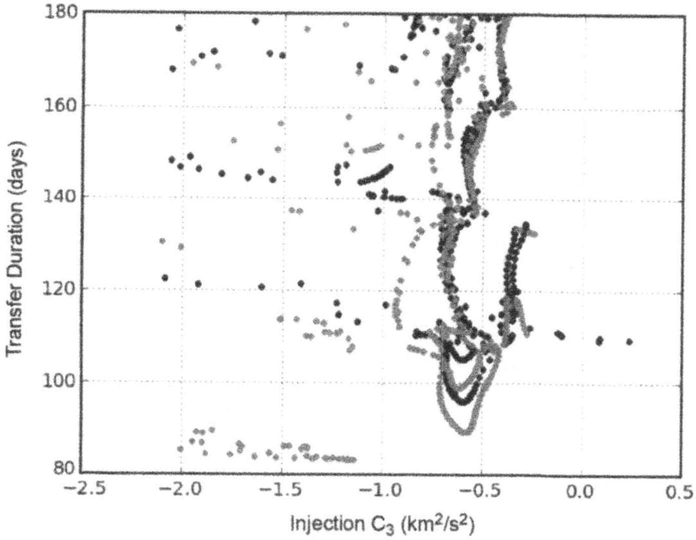

Figure 3-58 The relationship between the injection C_3 value and the duration of the transfer for each transfer identified in Fig. 3-56. The points displayed in black correspond to trajectories that traverse closer to EL_2 than EL_1.

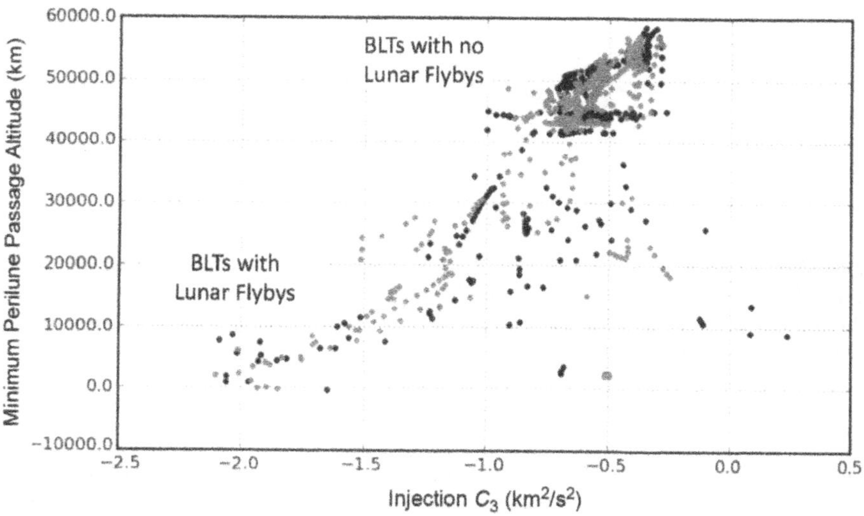

Figure 3-59 The relationship between the injection C_3 value and the lowest lunar periapse altitude during each lunar transfer identified in Fig. 3-56. The points displayed in black correspond to trajectories that travel closer to EL_2 than EL_1.

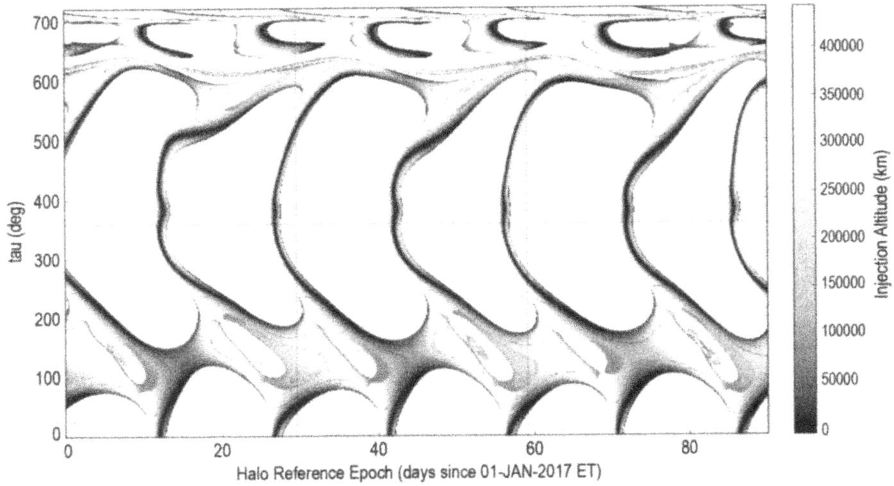

Figure 3-60 The same state space map shown in Fig. 3-54 extended to cover 90 days of reference epochs and two revolutions of the halo orbit.

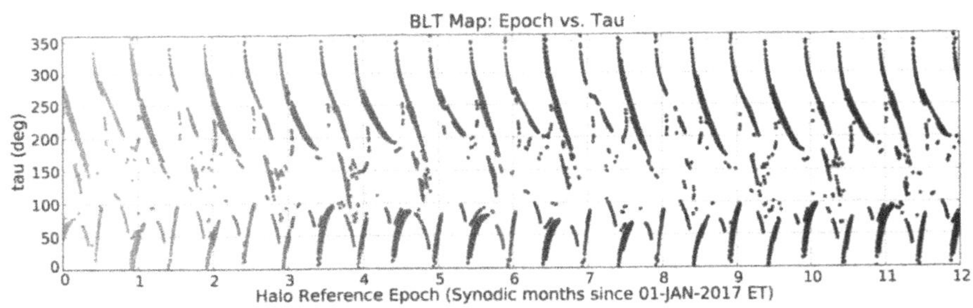

Figure 3-61 Sample combinations of T_{ref} and τ that yield low-energy transfers between 185-km LEO parking orbits and the lunar halo orbit for reference dates that span the year 2017. From lightest to darkest, the shading corresponds to reference dates from 1/1/2017 to 1/1/2018 [47] (first published by the American Astronautical Society).

variations in the locations of the curves as they shift throughout the 12 months. The transfers are shaded in Fig. 3-62 in the same manner as they are in Fig. 3-61, that is, the transfers that exist in the first month, which starts at a reference epoch of January 1, 2017, are shown in the lightest shade and the transfers in each consecutive synodic month thereafter are plotted in a darker shade. One can see that certain features repeat very closely from one synodic month to the next. Other features only appear in a subset of synodic months.

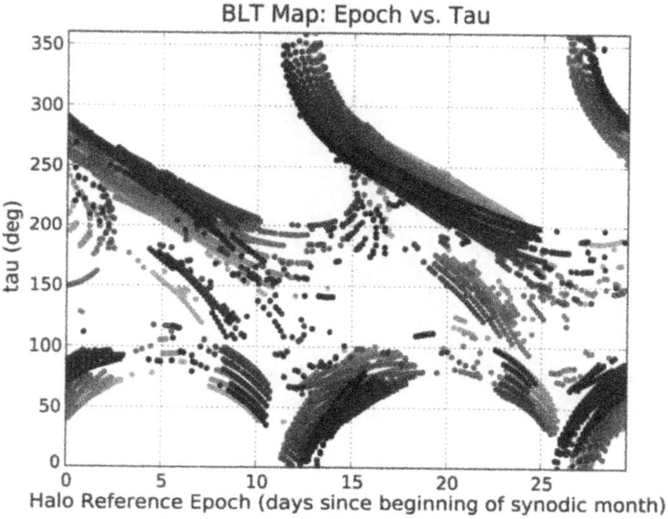

Figure 3-62 The combinations of T_{ref} and τ that yield transfers during 12 synodic months, relative to the beginning of each synodic month. The first month, which starts at a reference epoch of 1 Jan 2017 00:00:00 Ephemeris Time, is shown in the lightest shade and each consecutive synodic month thereafter is plotted in a darker shade. From lightest to darkest, the shading corresponds to reference dates from 1/1/2017 to 1/1/2018 [47] (first published by the American Astronautical Society).

Quite a few patterns exist in the families of transfers that are observed. First of all, the most pronounced curves observed in Figs. 3-61 and 3-62 correspond to transfers that do not include any lunar flybys or Earth phasing orbits. Most of them require between 90 and 110 days to transfer between the Earth and 100 km from their target orbit. Examples of these sorts of transfers may be seen in Fig. 3-55.

Several relationships exist between the launch energy of a low-energy lunar transfer and how close it gets to the Moon on its Earth-departure leg. If the transfer does not encounter the Moon, it typically requires a launch energy in the range of $-0.75\ \mathrm{km}^2/\mathrm{s}^2 \leq C_3 \leq -0.35\ \mathrm{km}^2/\mathrm{s}^2$. If a spacecraft traversing a low-energy transfer does encounter the Moon as it departs the Earth's vicinity, one finds that the Moon may either boost or reduce the spacecraft's energy, depending on how the spacecraft passes by the Moon. If it boosts the spacecraft's energy, then the lunar transfer's required launch energy drops to as low as $-2.1\ \mathrm{km}^2/\mathrm{s}^2$. Figure 3-63 shows a plot of the relationship between the launch energy of each low-energy transfer observed in Fig. 3-62 and how close the transfer passes by the Moon.

One can also glean a great deal of understanding about the characteristics of these transfers by observing the relationship between each transfer's injection energy and the transfer's duration. Figure 3-64 shows this relationship for each transfer in the 12-month survey. One can see that the trends in this relationship are nearly

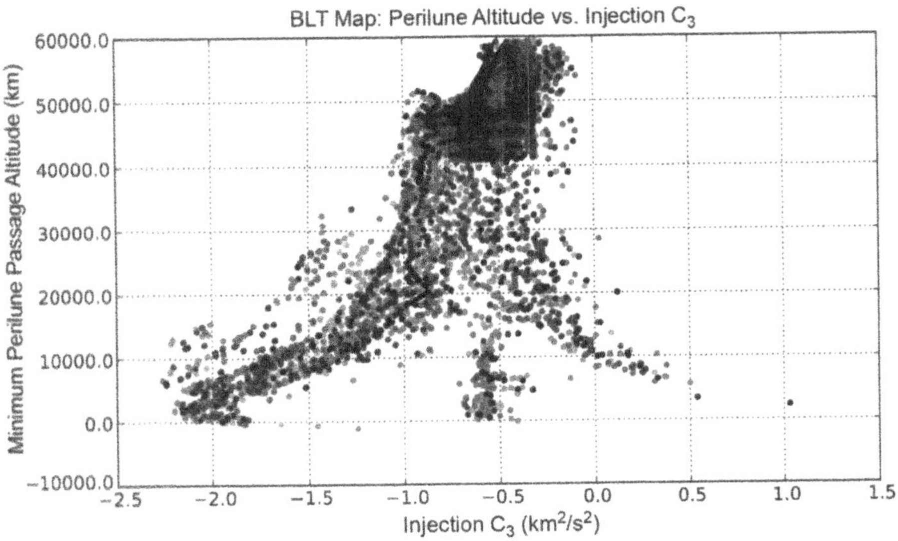

Figure 3-63 The relationship between injection C_3 and the lowest perilune altitude for each transfer in the 12-month survey. The trajectories near the top of the plot do not include any lunar flyby; trajectories toward the bottom do, where those toward the bottom-left receive an energy boost from the Moon and those toward the bottom right have energy removed by the Moon. From lightest to darkest, the shading corresponds to reference dates from 1/1/2017 to 1/1/2018 [44] (Copyright © 2009 by American Astronautical Society Publications Office, San Diego, California (Web Site: http://www.univelt.com), all rights reserved; reprinted with permission of the AAS).

independent of the month of the transfer. Typical mission designs prefer the transfer duration to be as short as possible. One can see that there are two types of transfers that require fewer than 100 days to perform: those that require an injection C_3 on the order of -2.1 to -1.5 km^2/s^2 and those that require an injection C_3 on the order of -0.7 to -0.5 km^2/s^2. Clearly, those that require less injection C_3 pass near the Moon on the way out of the Earth's vicinity.

The *inertial* orientation of each low-energy transfer observed in this 12-month survey clearly depends on which month the transfer departs the Earth. However, the orientation of each similar low-energy transfer is fairly constant throughout the year when observed in the Sun–Earth rotating frame. One way to observe that is to track each transfer's departure from Earth in the Sun–Earth rotating frame. Figure 3-65 shows a plot that compares the departure state of each transfer in the 12-month survey by plotting the relationship of each transfer's right ascension of apogee vector (RAV) and declination of apogee vector (DAV) parameters of the transfer's initial apogee vector. The RAV and DAV values have been computed at the instant of the trans-lunar injection, before any perturbations change the orbit. Each transfer departs the Earth

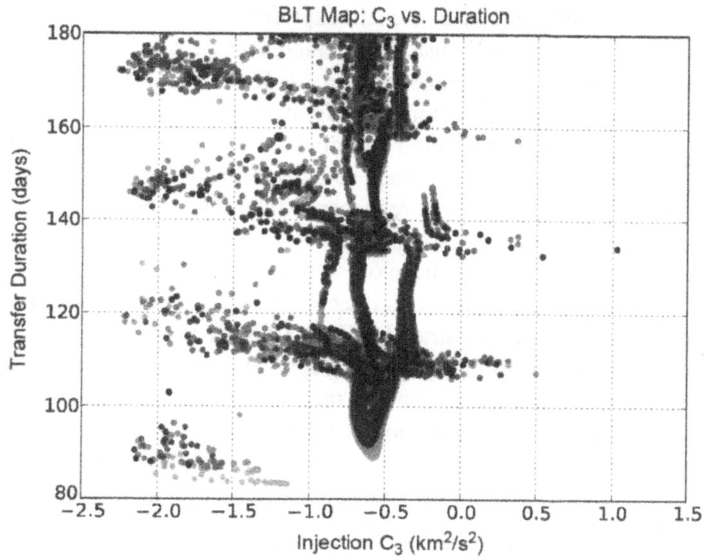

Figure 3-64 The relationship between injection C_3 and duration for each transfer in the 12-month survey. From lightest to darkest, the shading corresponds to reference dates from 1/1/2017 to 1/1/2018 [47] (first published by the American Astronautical Society).

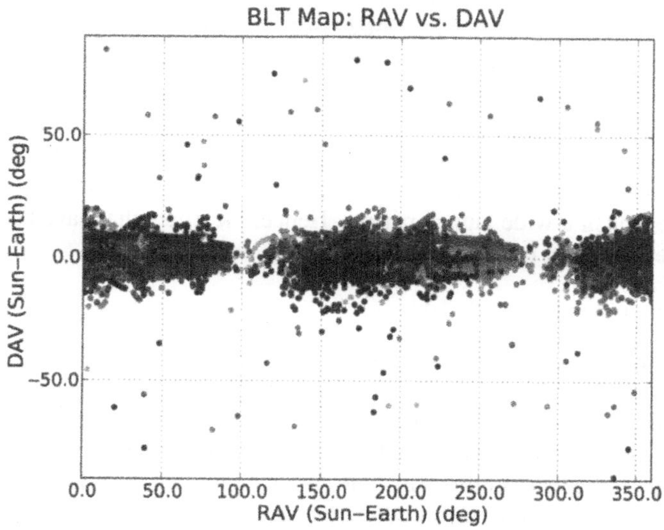

Figure 3-65 The relationship between the right ascension and declination of the apogee vector, RAV and DAV, respectively, for each transfer in the 12-month survey. From lightest to darkest, the shading corresponds to reference dates from 1/1/2017 to 1/1/2018 [47] (first published by the American Astronautical Society).

on an orbit that is highly eccentric, but still captured by the Earth. From Fig. 3-65, one can see that this initial orbit is usually oriented near the ecliptic plane and usually oriented either toward or away from the Sun. A RAV value of 0 deg corresponds to an orbit that has its apogee vector pointing away from the Sun, in the direction of positive x in the Sun–Earth rotating coordinate frame. The outlying points in the figure correspond to transfers that include some combination of Earth phasing loops and lunar flybys and typically do not reappear in the same region of this figure from one month to the next.

The largest variations observed from one synodic month to the next correspond to differences in the low-energy transfer's injection inclination, in both equatorial and ecliptic reference frames, as illustrated in Fig. 3-66. It is apparent when studying the plots shown in Fig. 3-66 that transfers depart the Earth from orbital planes at nearly any inclination during each synodic month. It is expected that the equatorial inclination of the transfers' injection points will vary from one synodic month to the next due to the Earth's obliquity angle; however, significant variations also exist from month to month when observing the transfers' injection points' ecliptic inclination values. The variations in the geometry during the year have a more pronounced effect when the trajectories fly near the Earth or Moon.

3.4.6.2 Tracking One Family Through 12 Months The figures shown in the previous sections, as well as analyses in the literature [46] show that one can trace hundreds of different families of low-energy lunar transfers in any given reference month. The characteristics of these families often stack on top of each other in each relationship presented in Figs. 3-62–3-65, making it difficult to discern how the characteristics of one family evolve from month to month. This section studies a subset of transfers of the 12-month survey, filtered to isolate a particular set of practical low-energy transfers. It is often the case that a practical spacecraft mission benefits by shorter transfer durations; it is also usually beneficial to avoid outbound lunar flybys because they add geometrical constraints to the system that make it more difficult to establish a wide launch period. Hence, the filters that have been applied to the transfer selection include:

- Maximum duration: 105 days
- Minimum perilune altitude: 20,000 km

In addition, the set of all transfers that meets these criteria has been divided into two subsets, split such that one subset includes those transfers that travel closer to EL_1 than EL_2 and vice versa. In this way, one can compare practical EL_1 transfers and practical EL_2 transfers from one month to the next.

Figure 3-67 identifies the transfers that meet the filter criteria in the state space map. A visual comparison will confirm that these transfers exist in the most prominent features of the state space maps shown in Figs. 3-54, 3-55, 3-61, and 3-62. One can see that the location of the curves of each family on these plots varies from month to month; the variations appear to be contained within approximately 50 deg in τ and at most 5 days in the orbit's reference date, T_{ref}.

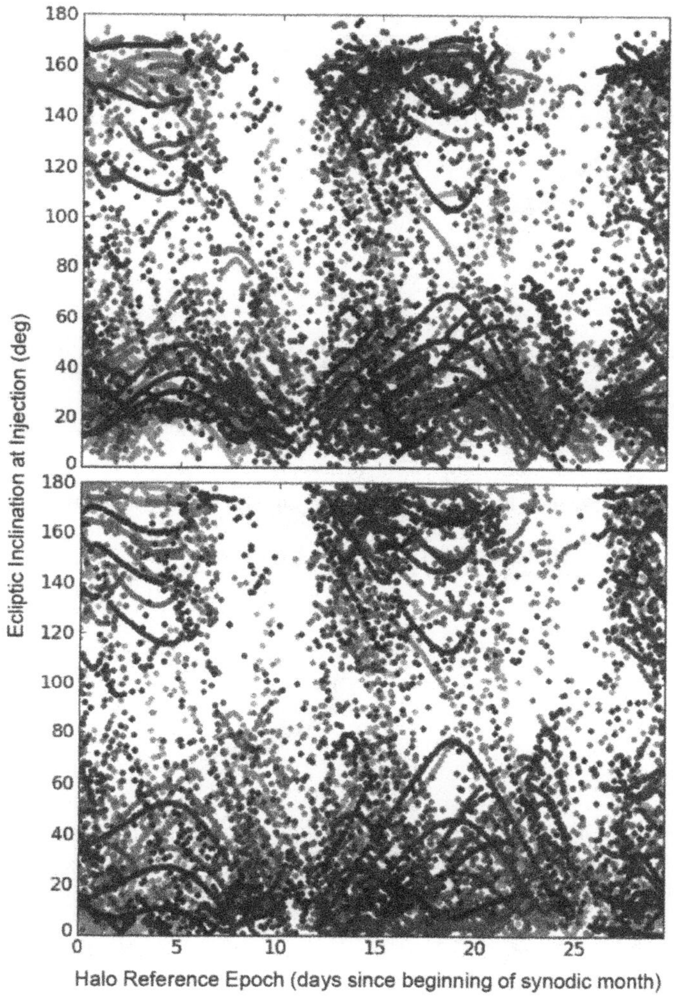

Figure 3-66 The equatorial (top) and ecliptic (bottom) inclination of the transfers' injection point for each low-energy lunar transfer identified in Fig. 3-62. The first month, which starts at a reference epoch of 1 Jan 2017 00:00:00 Ephemeris Time, is shown in the lightest shade and each consecutive synodic month thereafter is plotted in a darker shade. From lightest to darkest, the shading corresponds to reference dates from 1/1/2017 to 1/1/2018 [44] (Copyright © 2009 by American Astronautical Society Publications Office, San Diego, California (Web Site: http://www.univelt.com), all rights reserved; reprinted with permission of the AAS).

Figure 3-68 shows the relationship of each transfer's injection C_3 and its duration for every transfer that satisfies the filter criteria. One can clearly see that the transfers' performance parameters vary along a curve for each month, and the performance curve does not vary significantly from one month to the next. The transfer duration

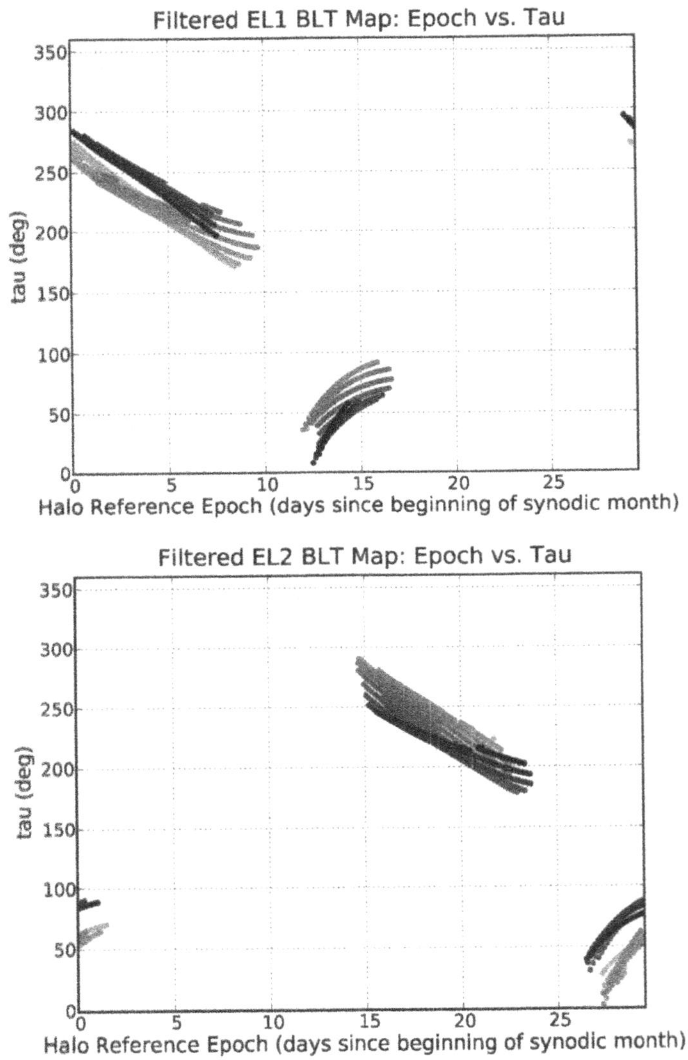

Figure 3-67 The relationship between the reference epoch and τ for each EL_1 (top) and EL_2 (bottom) transfer in the 12-month survey that satisfies the filter criteria. From lightest to darkest, the shading corresponds to reference dates from 1/1/2017 to 1/1/2018 [47] (first published by the American Astronautical Society).

may vary by several days between months, but the curves span very similar ranges of injection C_3.

It is very interesting to plot the relationship between each transfer's injection date and its injection energy, C_3. Figure 3-69 shows this comparison for the EL_1 and

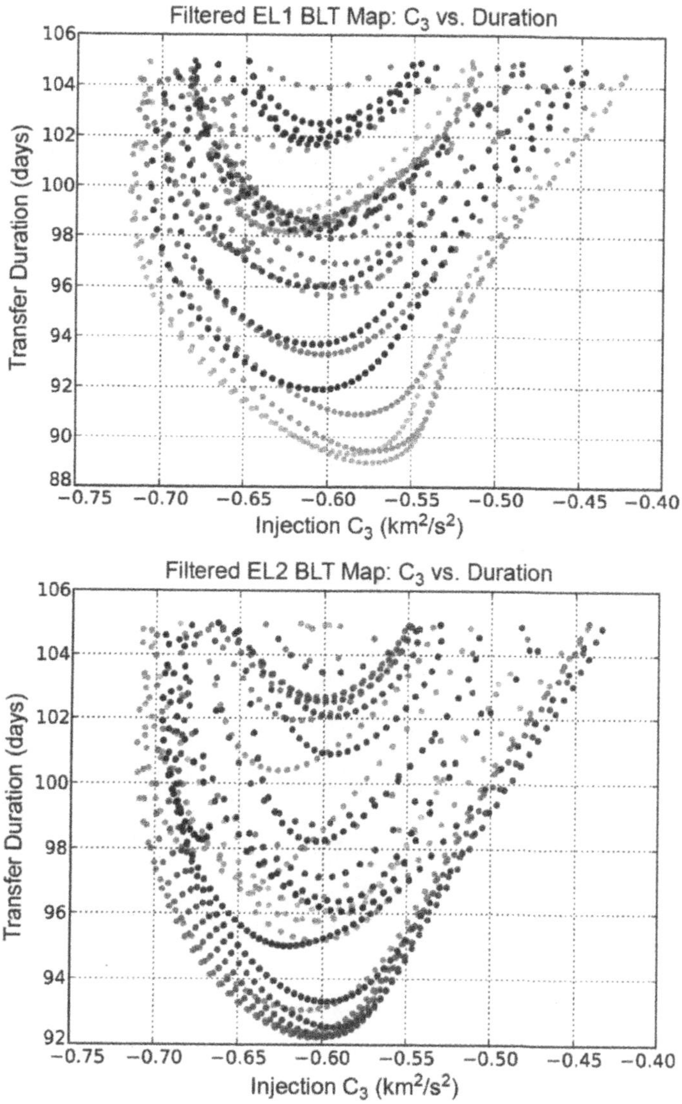

Figure 3-68 The relationship between the injection C_3 and transfer duration for each EL_1 (top) and EL_2 (bottom) transfer in the 12-month survey that satisfies the filter criteria. From lightest to darkest, the shading corresponds to reference dates from 1/1/2017 to 1/1/2018 [47] (first published by the American Astronautical Society).

EL_2 transfers. One can see that the families of transfers shift on this plot from month to month. The comparison also shows that most families of transfers span an injection date of 10 to 15 days. This suggests that there are 10 to 15 days in a launch

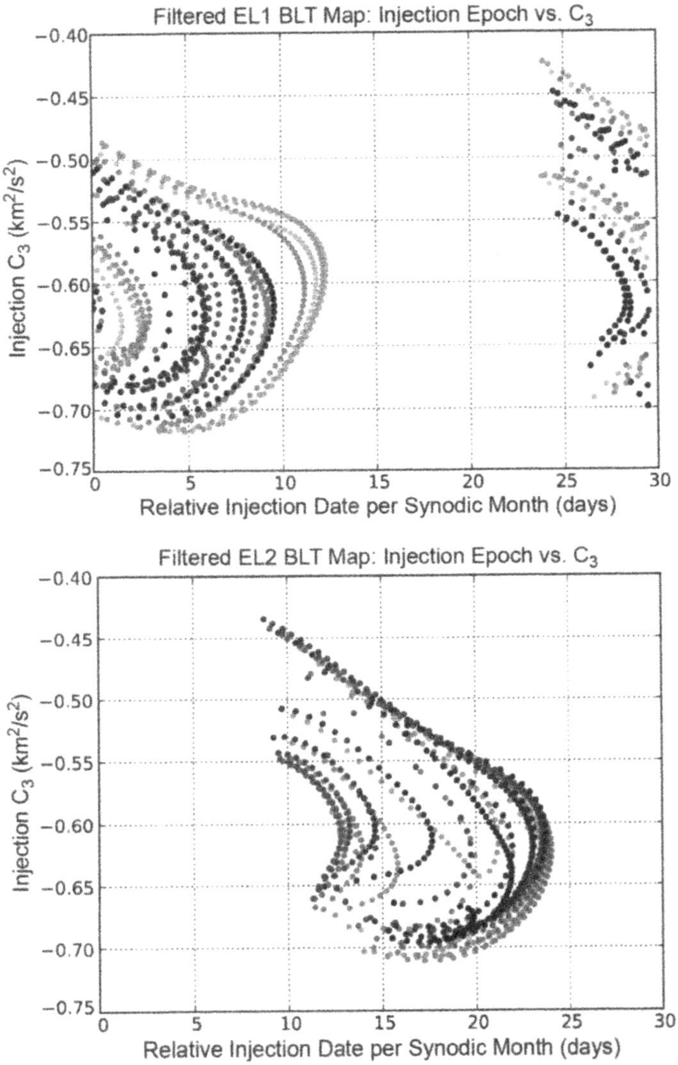

Figure 3-69 The relationship between the injection date and the injection C_3 for each EL_1 (top) and EL_2 (bottom) transfer in the 12-month survey that satisfies the filter criteria. From lightest to darkest, the shading corresponds to reference dates from 1/1/2017 to 1/1/2018 [47] (first published by the American Astronautical Society).

period to this lunar libration orbit via this type of transfer before the deep space ΔV cost increases. This relationship, however, does not take into account differences in the injection inclination throughout the family. Figure 3-69 also verifies that EL_1 transfers and EL_2 transfers depart approximately two weeks apart from each other.

The departure geometry of the filtered transfers is very consistent and predictable from month to month, given the proper analysis. Figures 3-70 and 3-71 show the RAV and DAV parameters for the EL_1 and EL_2 transfers, respectively, computed in the Sun–Earth rotating coordinate frame at the instant of trans-lunar injection. One can immediately observe that the ranges of RAV and DAV values are very limited for each set of transfers: the EL_1 transfers are confined to the approximate range of $\sim$140 deg $\leq$ RAV $\leq$ $\sim$170 deg, the EL_2 transfers are confined to the range of $\sim$320 deg $\leq$ RAV $\leq$ $\sim$355 deg, and both sets are confined in DAV to the approximate range $\sim$$-10$ deg $\leq$ DAV $\leq$ $\sim$10 deg. The RAV values appear to cover a very similar span of values for each month, but there appears to be an annual signal in the DAV values. This systematic variation may be isolated by observing the relationship between a transfer's DAV value and the orientation of the Moon's orbital pole vector at the arrival time. The Moon's orbit has an inclination of approximately 5.1 deg relative to the ecliptic. The Moon's orbital plane is approximately fixed in inertial space, but rotates in the Sun–Earth rotating frame. Figure 3-72 shows the relationship between the transfer's injection DAV value and the right ascension of the lunar orbit pole vector in the Sun–Earth rotating coordinate frame at the time of arrival. One sees a clear annual signal in the data. A mission designer may be able to use this information to improve an initial estimate of the trans-lunar injection geometry. The injection DAV value still varies by approximately 10 deg throughout a family after accounting for the annual variation. This remaining variation may be explained by the z-axis motion of the target orbit at the time of arrival, though that relationship has not been studied sufficiently yet.

A relationship has also been observed between the injection RAV value and the injection C_3. Figure 3-73 shows this relationship for both the EL_1 and EL_2 transfers. One can see that higher RAV values require less injection energy and there is very little monthly variation in the observed data.

Another parameter that depends closely on the relative orientation of the Moon's orbit about the Earth at the time of the transfer is the inclination of the LEO parking orbit that is used to transfer onto these low-energy transfers. The transfers are constructed by building an initial state at the Moon and propagating backward in time until they intersect a 185-km parking orbit above the Earth's surface. The inclination of that parking orbit is driven by the geometry of the transfer. A real mission launched from Cape Canaveral, Florida, would likely launch from an orbit with an equatorial inclination near 28.5 deg and perform maneuvers to target the desirable low-energy transfer [183, 184]. This is the subject of Section 6.5. That section shows that the closer the natural transfer is to having a parking orbit with a particular, desired inclination, the less ΔV is required to target that transfer from the desired parking orbit, though extended launch periods reduce the ΔV significance.

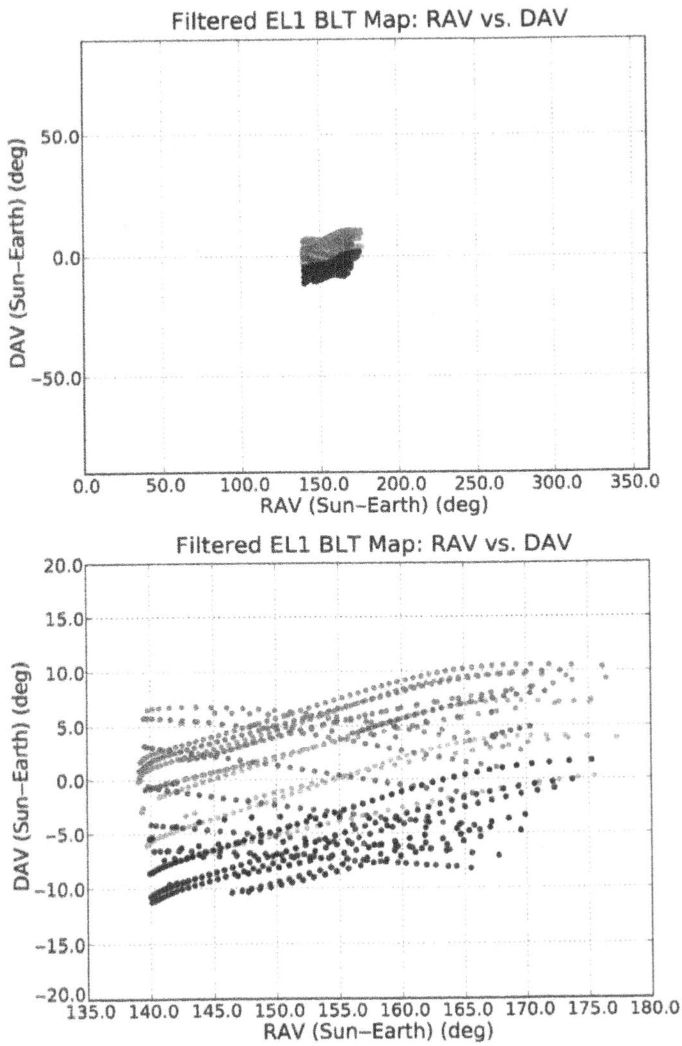

Figure 3-70 The relationship between RAV and DAV (the right ascension and declination of the apogee vector) at the time of trans-lunar injection for the filtered EL_1 transfers. From lightest to darkest, the shading corresponds to reference dates from 1/1/2017 to 1/1/2018. Top: one can see that RAV and DAV are confined in a narrow box for these transfers; bottom: a closer look at the parameter space [47] (first published by the American Astronautical Society).

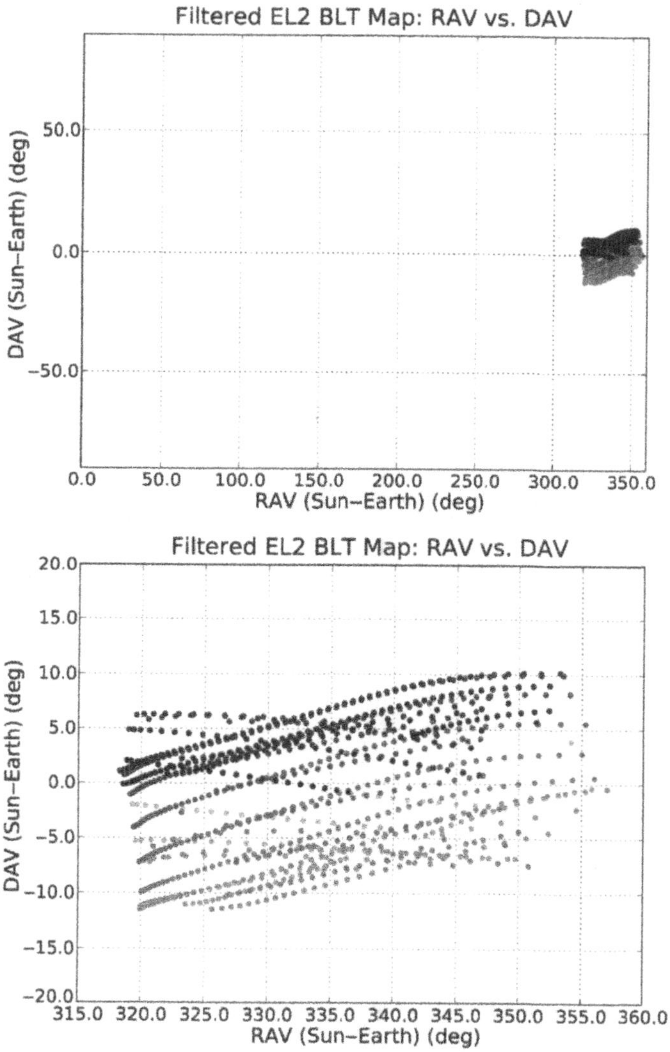

Figure 3-71 The same relationship between RAV and DAV as Fig. 3-70, but for the filtered EL_2 transfers. From lightest to darkest, the shading corresponds to reference dates from 1/1/2017 to 1/1/2018 [47] (first published by the American Astronautical Society).

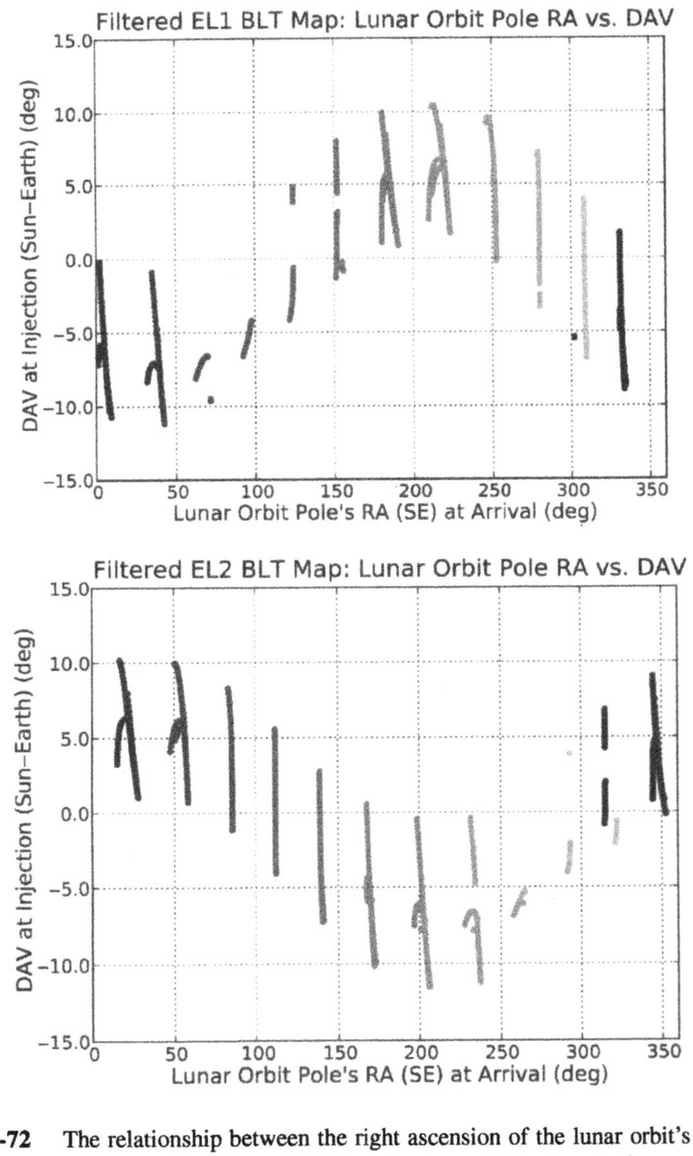

Figure 3-72 The relationship between the right ascension of the lunar orbit's pole vector at the time of arrival and the value of DAV at the time of injection, both computed in the Sun–Earth rotating coordinate frame. From lightest to darkest, the shading corresponds to reference dates from 1/1/2017 to 1/1/2018.This relationship is shown for each EL_1 (top) and EL_2 (bottom) transfer in the 12-month survey that satisfies the filter criteria [47] (first published by the American Astronautical Society).

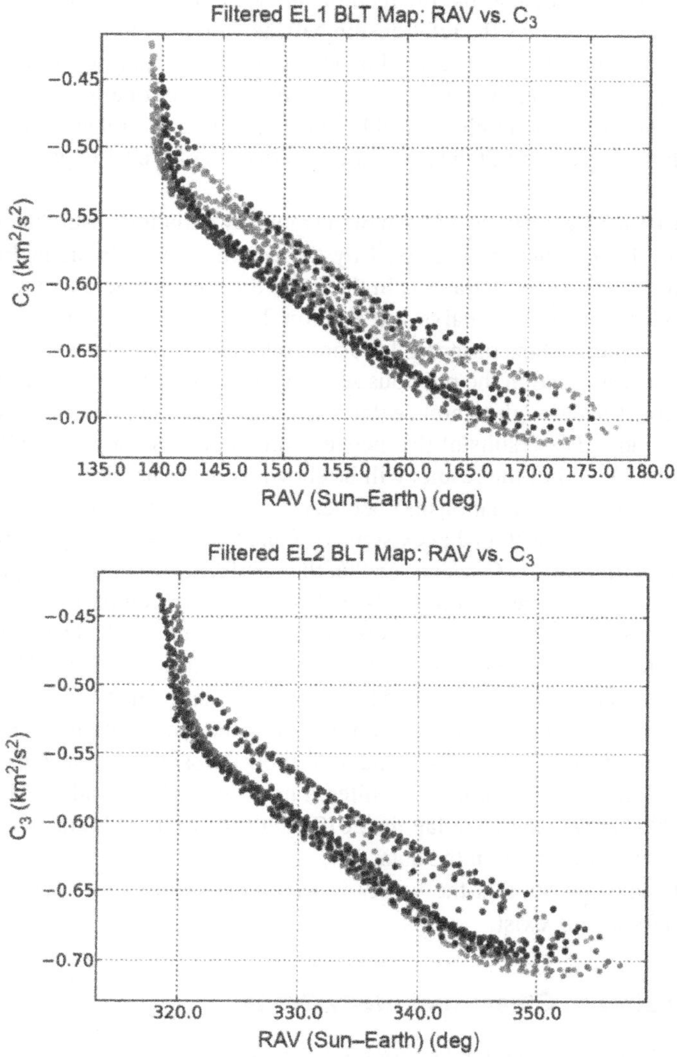

Figure 3-73 The relationship between the right ascension of the apogee vector, RAV, at the time of trans-lunar injection and the injection energy, C_3, for each EL_1 (top) and EL_2 (bottom) transfer in the 12-month survey that satisfies the filter criteria. From lightest to darkest, the shading corresponds to reference dates from 1/1/2017 to 1/1/2018 [47] (first published by the American Astronautical Society).

Figure 3-74 shows the relationship between the reference date of the lunar halo orbit and the equatorial inclination of the natural LEO parking orbit needed to perform the transfer. One can see that the inclination varies significantly from one month to the next. Figure 3-75 shows the relationship between the right ascension of the lunar orbit pole vector and the ecliptic inclination of the LEO parking orbit. One can clearly see that there is an evolution of the inclination from one month to the next. Figure 3-76 shows the same plot, but this time presenting the relationship between the lunar orbit pole vector and the equatorial inclination of the LEO parking orbit.

3.4.6.3 Annual Variations Much of the monthly variation observed in families of low-energy lunar transfers is caused by the Moon's noncircular, inclined orbit relative to the Earth. Other variations in the Solar System change over the course of several years, evident in the analysis in Section 2.5.3. It is therefore of interest to ensure that the relationships observed here hold over the course of several years. The same analyses performed in the previous section have been performed again on a set of transfers constructed with reference dates spanning the year 2021, four years after the previous study. The results of this new examination coincide very well with the previous study. Not all of the results will be shown here for brevity.

Figure 3-77 shows the relationship between T_{ref} and τ, where the lighter shaded points are low-energy transfers that exist in 2017 and the darker points are low-energy transfers that exist in 2021. One can see that the combinations of the two parameters are very similar for both years. Figure 3-78 shows a similar comparison between the injection C_3 and duration of the transfers in both 2017 and 2021. One can see that there is very little noticeable difference between the points in 2017 and 2021.

The transfers that exist in 2021 have been filtered in the same way as the transfers presented in Section 3.4.6.2 in order to observe how the family might change during the course of four years. Figures 3-79 and 3-80 show the same relationships as shown in Figs. 3-72 and 3-75, except now for filtered transfers in 2017 and 2021. One can see that the 2021 parameters overlap the 2017 data very well, including the dramatic monthly variations observed in the data.

The evidence suggests that the yearly variations are much more subtle than the monthly variations that exist.

3.4.7 Transfers to Other Three-Body Orbits

All of the analyses performed in Sections 3.4.3 through 3.4.6 have used the family of halo orbits about the LL_2 point as the example destination, but these analyses work for any unstable three-body orbit in the Earth–Moon system.

Section 3.4.7.1 explores low-energy lunar transfers that target an example lunar L_1 halo orbit. Since this orbit is on the interior side of the Moon, the trajectories that target it must transfer from the lunar L_2 region past the Moon before encountering the target orbit.

Section 3.4.7.2 explores low-energy lunar transfers that target an example distant prograde orbit about the Moon. Orbits in this family traverse both the near and far sides of the Moon. Hence, transfers that target these orbits may demonstrate

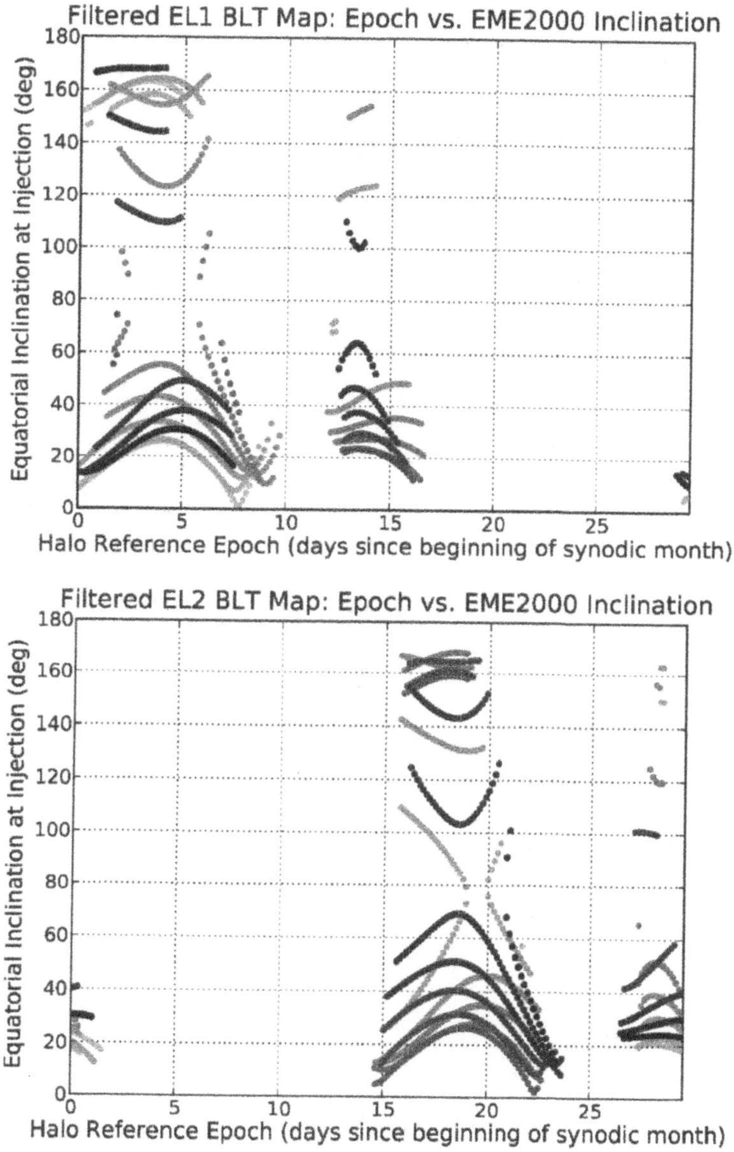

Figure 3-74 The relationship between the reference date of the lunar halo orbit and the equatorial inclination of the LEO parking orbit needed to perform the transfer. From lightest to darkest, the shading corresponds to reference dates from 1/1/2017 to 1/1/2018. This relationship is shown for each EL_1 (top) and EL_2 (bottom) transfer in the 12-month survey that satisfies the filter criteria [47] (first published by the American Astronautical Society).

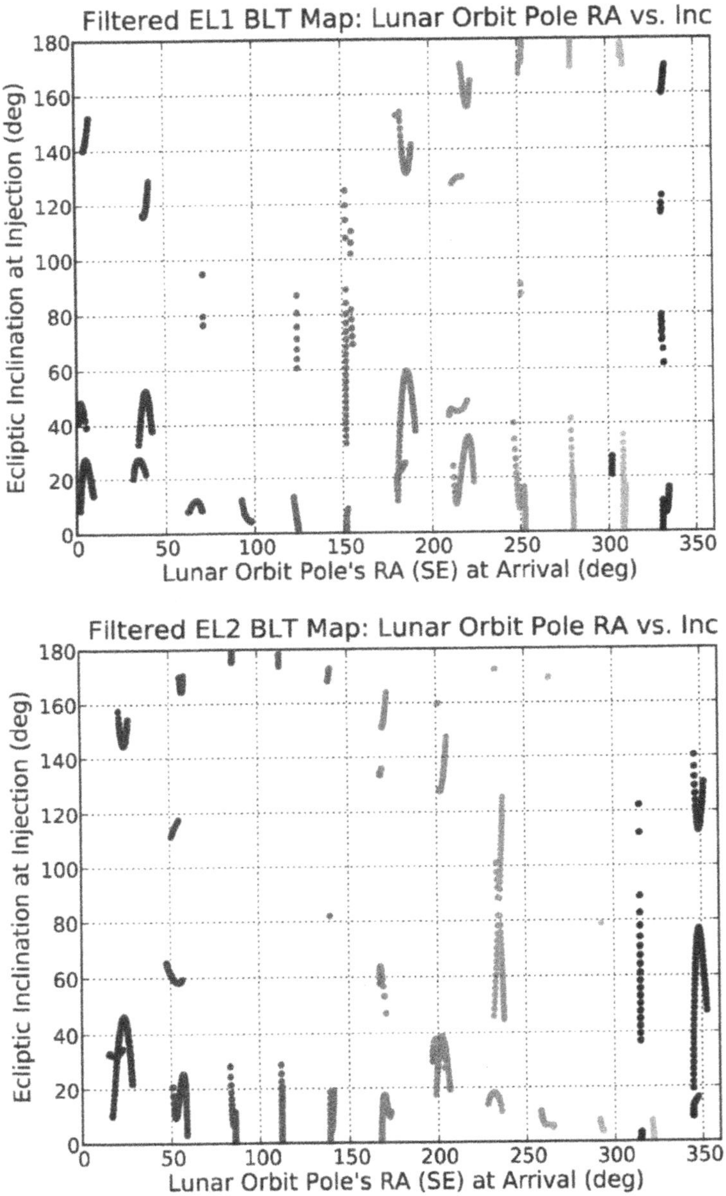

Figure 3-75 The relationship between the right ascension of the lunar orbit pole vector and the ecliptic inclination of the LEO parking orbit. From lightest to darkest, the shading corresponds to reference dates from 1/1/2017 to 1/1/2018. This relationship is shown for each EL_1 (top) and EL_2 (bottom) transfer in the 12-month survey that satisfies the filter criteria [47] (first published by the American Astronautical Society).

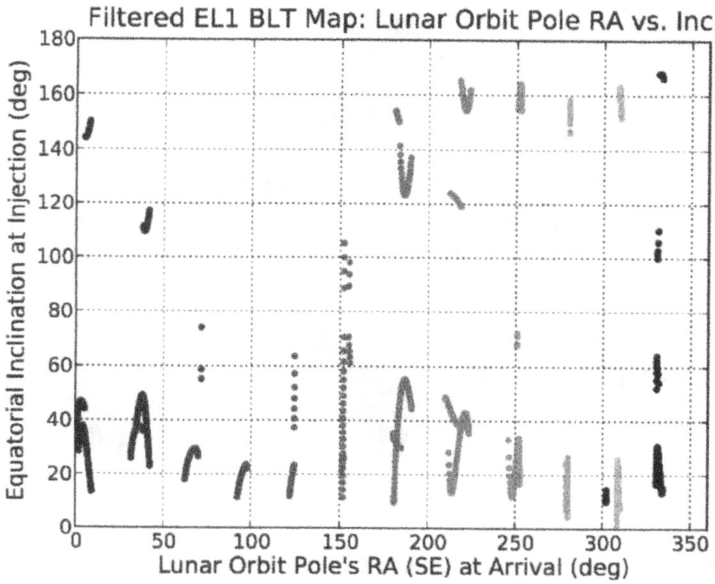

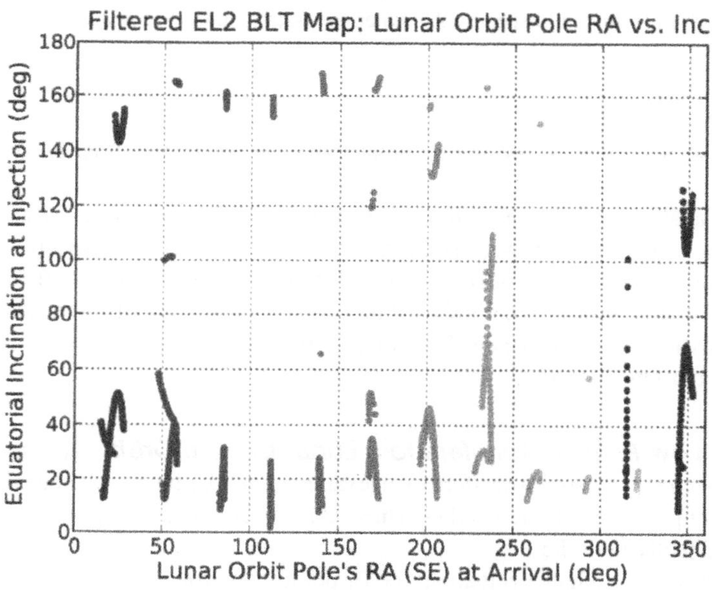

Figure 3-76 The relationship between the right ascension of the lunar orbit pole vector and the equatorial inclination of the LEO parking orbit. From lightest to darkest, the shading corresponds to reference dates from 1/1/2017 to 1/1/2018. This relationship is shown for each EL_1 (top) and EL_2 (bottom) transfer in the 12-month survey that satisfies the filter criteria [47] (first published by the American Astronautical Society).

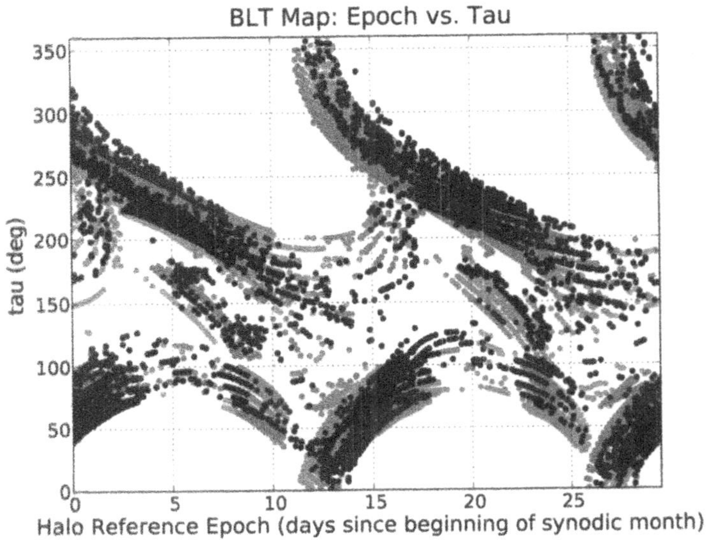

Figure 3-77 The combinations of T_{ref} and τ that yield low-energy transfers between 185-km LEO parking orbits and the target lunar libration orbit during 2017 (gray points) and 2021 (black points) [47] (first published by the American Astronautical Society).

characteristics similar to low-energy lunar transfers that target either L_1 or L_2 halo orbits.

These analyses are merely additional examples to demonstrate these analysis techniques. All analyses will likely need to be repeated given specific mission design requirements. That is, a given mission may require a spacecraft to transfer to a particular unstable three-body orbit, perhaps for communication, staging, or rendezvous reasons, and a new BLT map will need to be generated to study the trajectory options that exist.

3.4.7.1 Low-Energy Transfers to a Lunar L_1 Halo Orbit This section explores low-energy ballistic transfers to an example lunar L_1 halo orbit. For simplicity in this example analysis, the Patched Three-Body Model is used; hence, the L_1 halo orbit is perfectly periodic.

In order to reach a halo orbit about the L_1 point via a typical low-energy transfer, a spacecraft must depart the Earth and arrive in the lunar L_2 vicinity in much the same way as a spacecraft following a low-energy transfer to a lunar L_2 halo orbit. Then from the vicinity of L_2, the spacecraft must transfer past the Moon before arriving at its target L_1 halo orbit. As usual, there are two types of transfers: transfers that implement either the exterior or the interior stable manifold of the L_1 halo orbit. Interior transfers may arrive on the L_1 halo orbit immediately after passing by the Moon since the interior stable manifold is propagated in that direction. Exterior

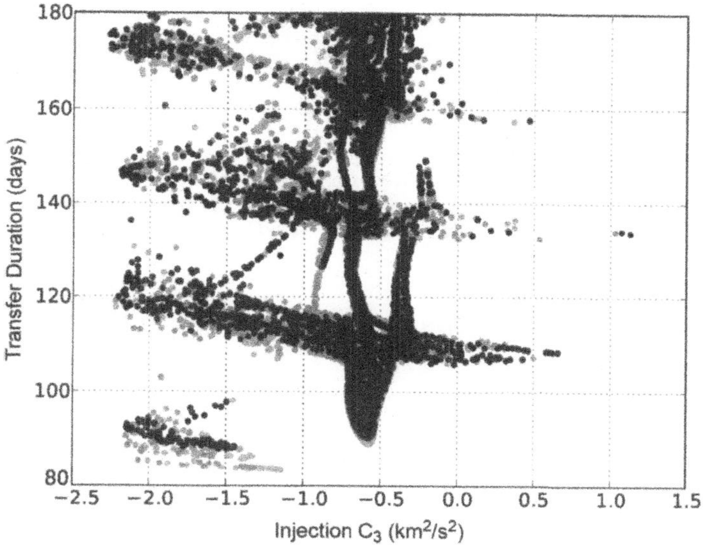

Figure 3-78 The combinations of injection C_3 and transfer duration that yield viable low-energy lunar transfers in 2017 (gray points) and 2021 (black points) [44] (Copyright © 2009 by American Astronautical Society Publications Office, San Diego, California (Web Site: http://www.univelt.com), all rights reserved; reprinted with permission of the AAS).

transfers to most L_1 halo orbits must first traverse some sort of Earth staging orbit prior to arriving on the L_1 halo orbit.

Figure 3-81 shows an example interior low-energy transfer to a lunar L_1 halo orbit in the Sun–Earth synodic reference frame. Figure 3-82 shows the same transfer in the Earth–Moon synodic reference frame. The characteristics of this example transfer are very similar to many of the low-energy transfers previously studied in this work that have transferred to L_2 halo orbits. The only major difference is that this example low-energy transfer passes through the L_2 region en route to the L_1 region, where it encounters its target L_1 halo orbit.

Figures 3-83 and 3-84 show an example exterior low-energy transfer to a lunar L_1 halo orbit in the Sun–Earth and Earth–Moon synodic reference frames, respectively. One can see that the transfer involves an Earth staging orbit, which permits it to encounter the L_1 halo orbit along the orbit's exterior stable manifold. Every exterior low-energy transfer that has been constructed in this work between the Earth and this L_1 halo orbit requires the use of at least one Earth staging orbit. When propagated backward in time, the exterior lunar transfers depart the L_1 halo orbit away from the Moon; hence, they must return to the Moon via an Earth staging orbit in order to transfer out of the Earth–Moon system and into the Sun–Earth system.

Figures 3-85 and 3-86 show the interior and exterior BLT maps, respectively, for low-energy transfers to this halo orbit, making it possible to characterize many

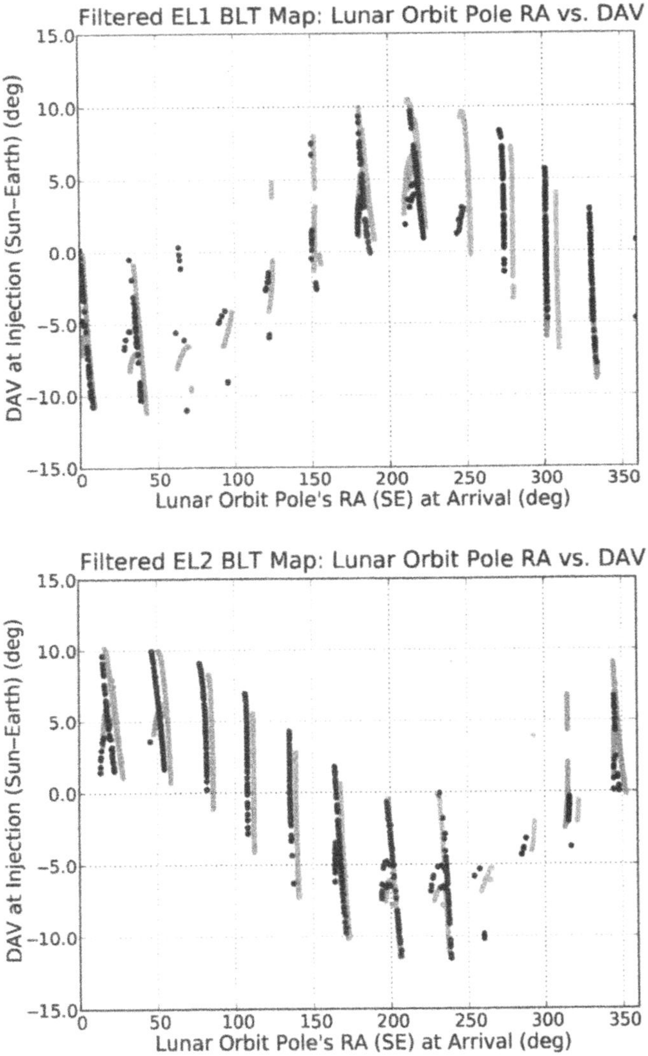

Figure 3-79 The relationship between the right ascension of the lunar orbit pole vector and the declination of the apogee vector at the time of injection. This relationship is shown for each EL_1 (top) and EL_2 (bottom) transfer in both the 2017 (light) and 2021 (dark) surveys that satisfies the filter criteria [47] (first published by the American Astronautical Society).

transfers to this orbit simultaneously. Each figure also shows eight example transfers to display some of the available transfer options that exist to this halo orbit. The BLT maps are colored according to the altitude of closest approach that each trajectory makes, given the values of θ and τ, when propagated backward in time at most

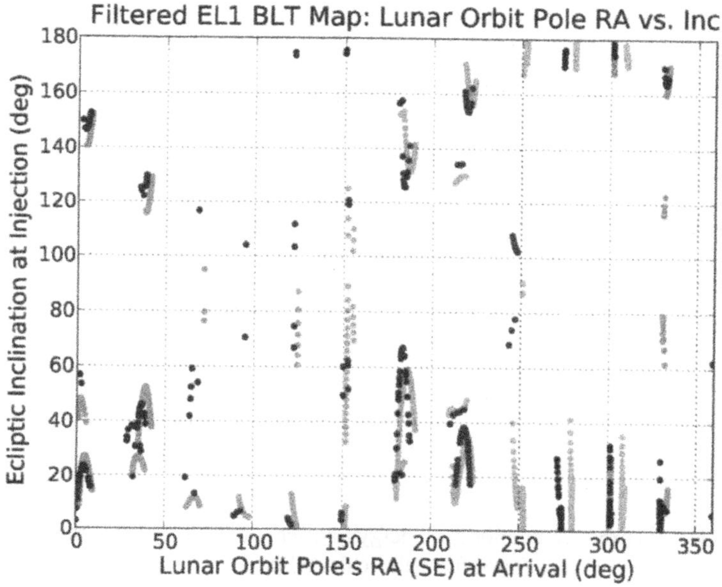

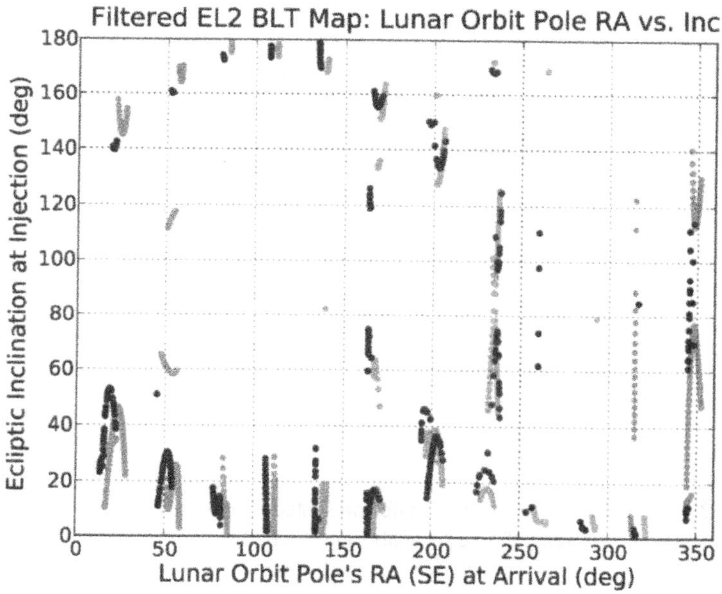

Figure 3-80 The relationship between the right ascension of the lunar orbit pole vector and the ecliptic inclination of the LEO parking orbit. This relationship is shown for each EL_1 (top) and EL_2 (bottom) transfer in both the 2017 (light) and 2021 (dark) surveys that satisfies the filter criteria [47] (first published by the American Astronautical Society).

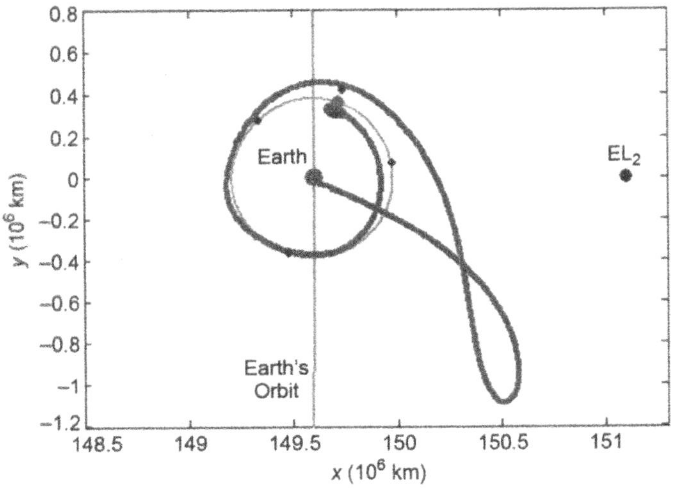

Figure 3-81 An example interior low-energy transfer to a lunar L_1 halo orbit, shown in the Sun–Earth synodic reference frame from above the ecliptic.

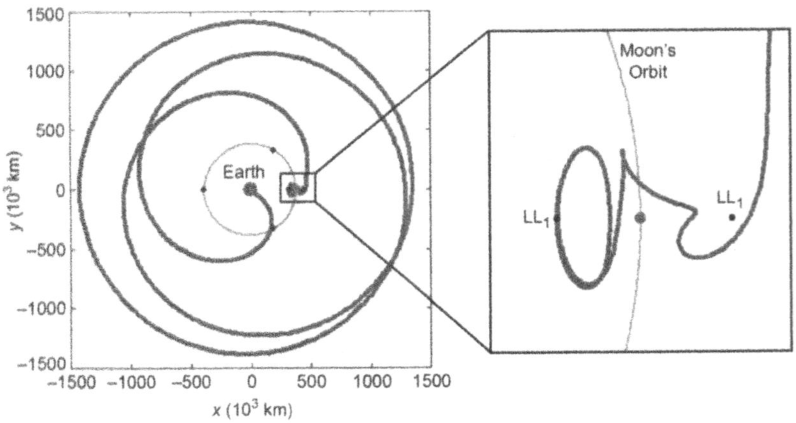

Figure 3-82 The same low-energy transfer presented in Fig. 3-81, but now shown in the Earth–Moon synodic reference frame from above the ecliptic.

195 days. Points colored black in each BLT map correspond to transfers that may be used to depart the Earth from a low-altitude orbit, or from the surface directly. The lightest colors correspond to transfers that do not approach any closer to the Earth than the L_1 orbit itself when propagated backward in time. As usual, we are only interested in the darkest regions of the BLT maps because those regions correspond with trajectories that depart from practical low Earth orbits.

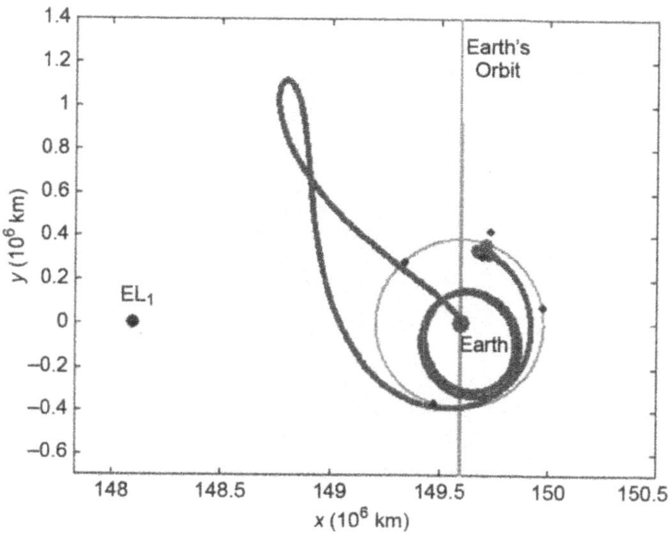

Figure 3-83 An example exterior low-energy transfer to a lunar L_1 halo orbit, shown in the Sun–Earth synodic reference frame from above the ecliptic.

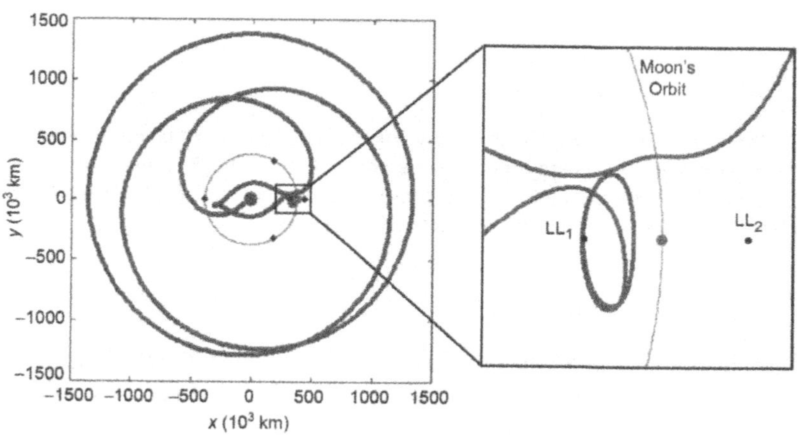

Figure 3-84 The same low-energy transfer presented in Fig. 3-83, but now shown in the Earth–Moon synodic reference frame from above the ecliptic.

One can see that the two BLT maps shown in Figs. 3-85 and 3-86 are very complex. This makes sense because the only ways to construct ballistic transfers between the Earth and this lunar L_1 halo orbit require some combination of lunar passages and Earth staging orbits.

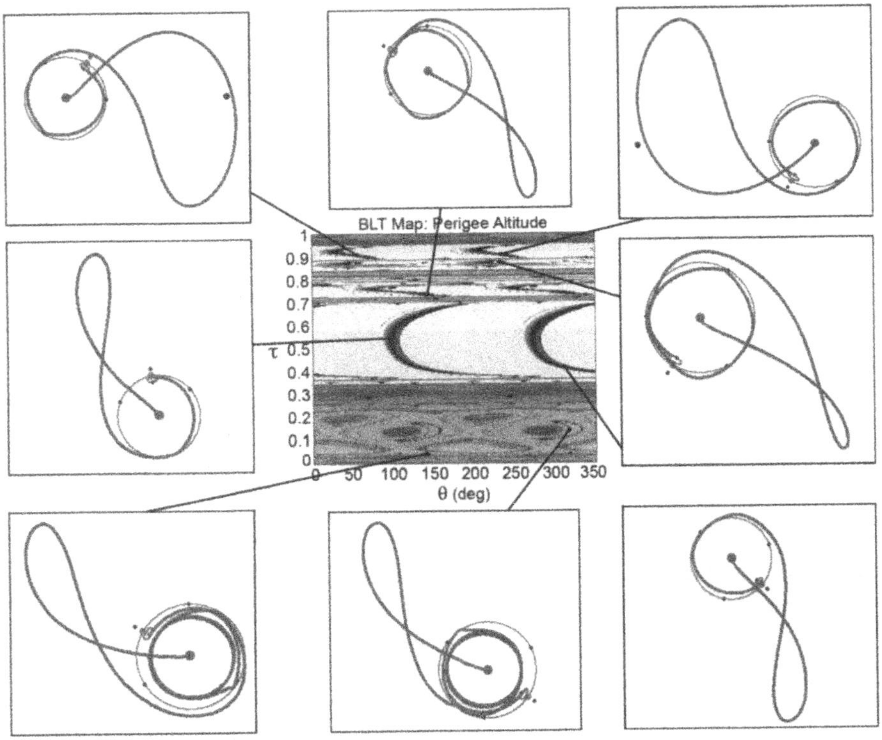

Figure 3-85 The interior BLT map for low-energy transfers to the example lunar L_1 halo orbit. Eight example low-energy transfers are shown around the BLT map to demonstrate some of the types of transfers that may be constructed between 185-km LEO orbits and this halo orbit. *(See insert for color representation of this figure.)*

When studying Fig. 3-85, one notices many things. First, the BLT map is rather simple in the range of τ-values between 0.4 and 0.7. This region of τ-values includes ballistic lunar transfers that make only a single lunar passage en route to the L_1 halo orbit. These transfers resemble the simplest low-energy transfers to lunar L_2 halo orbits and have very similar performance parameters. Somewhat more complex transfers are shown in the BLT map for τ-values between 0.7 and 0.96: most of these involve several close lunar passages en route to the L_1 halo orbit. Every transfer constructed with a τ-value between 0 and 0.35 involves at least one Earth staging orbit, as may be seen in the two example transfers shown on the lower-left edge of the figure.

The exterior BLT map shown in Fig. 3-86 is more complex than the interior BLT map. This is because each transfer must implement at least one Earth staging orbit in addition to whatever lunar passages are required to complete the low-energy transfer.

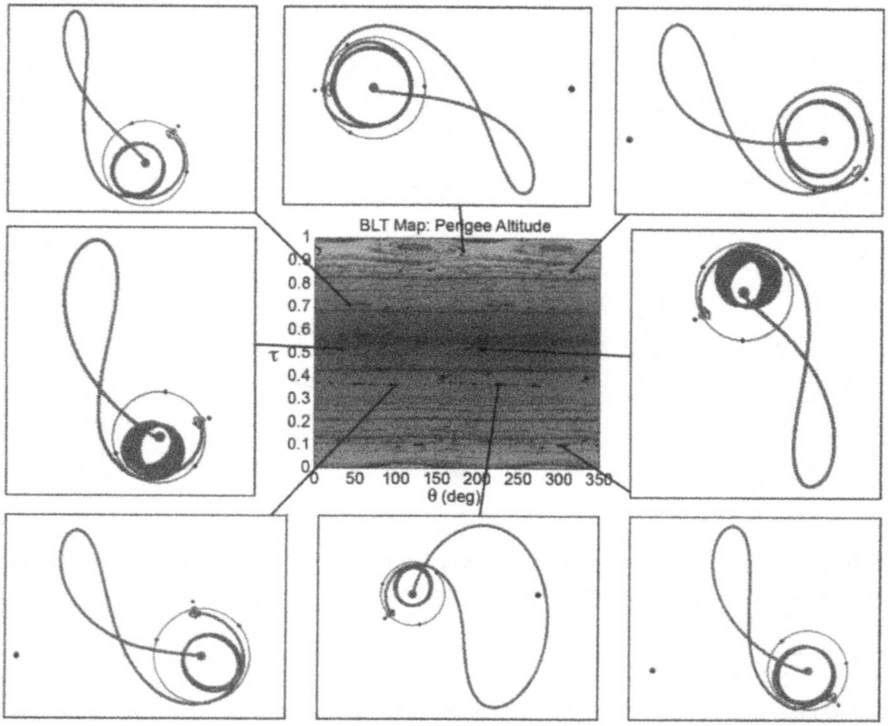

Figure 3-86 The exterior BLT map for low-energy transfers to the example lunar L_1 halo orbit. Eight example low-energy transfers are shown around the BLT map to demonstrate some of the types of transfers that may be constructed between 185-km LEO orbits and this halo orbit. *(See insert for color representation of this figure.)*

One may verify this by observing that every example trajectory shown around the edge of Fig. 3-86 includes at least one Earth staging orbit. Otherwise, these transfers are very similar to other lunar transfers previously studied.

3.4.7.2 Low-Energy Transfers to a Distant Prograde Orbit This section explores low-energy ballistic transfers to an example distant prograde orbit (DPO) about the Moon. Like the previous section, this analysis is performed using the Patched Three-Body Model, making the DPO perfectly periodic. Distant prograde orbits are interesting because they traverse both the near and far sides of the Moon. One might suspect that the qualitative nature of a low-energy transfer to such an orbit might take on characteristics of transfers to either L_1 or L_2 halo orbits, depending on how the specific transfer arrives at the orbit.

An example DPO has been generated here that has fairly large lobes and is easy to view in the example transfers presented here. Figure 3-87 shows an example

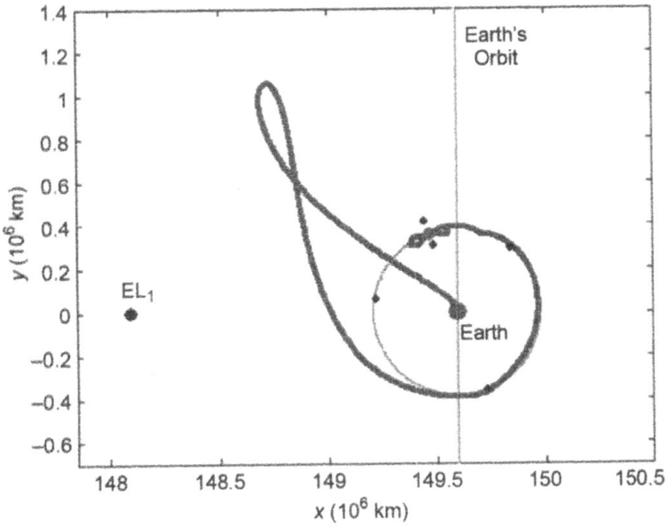

Figure 3-87 An example low-energy transfer to a distant prograde orbit, shown in the Sun–Earth synodic reference frame from above the ecliptic.

low-energy transfer to this distant prograde orbit in the Sun–Earth synodic reference frame. Figure 3-88 shows the same example transfer in the Earth–Moon synodic reference frame. One can see that this transfer does not enter any staging orbits, nor make any lunar flybys, but rather injects immediately into the distant prograde orbit. Other ballistic transfers may be produced that do use staging orbits or other complex lunar flybys en route to the orbit.

Because of the symmetry in the distant prograde orbit's shape, the two halves of the orbit's stable manifold are not clearly identifiable based on their immediate motion. That is, both halves of the stable manifold include both interior and exterior trajectories. However, the majority of one half of the distant prograde orbit's stable manifold propagates toward the Earth, and the majority of the other half propagates away from the Earth. This discussion refers to the half that propagates toward the Earth as the interior stable manifold and the other half as the exterior manifold. Using this nomenclature, Figs. 3-89 and 3-90 show the exterior and interior BLT maps, respectively, for low-energy transfers to this distant prograde orbit.

Along with the exterior BLT map, Fig. 3-89 also shows eight example exterior transfers that exist to this distant prograde orbit. One can see that these transfers are very simple—they don't require any lunar flybys or staging orbits to reach the target orbit. Because such simple transfers are prevalent in this exterior BLT map, the map is consequently not nearly as chaotic as some of the previous BLT maps studied in this chapter. The interior BLT map shown in Fig. 3-90, however, presents more complex transfers to this distant prograde orbit, including several examples of low-energy transfers that require Earth staging orbits.

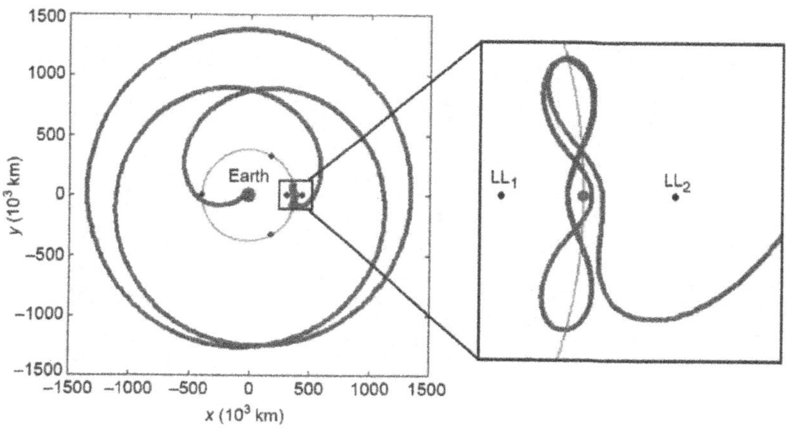

Figure 3-88 The same transfer presented in Fig. 3-87, but now shown in the Earth–Moon synodic reference frame from above the ecliptic.

The characteristics of the exterior transfers shown in Fig. 3-89 resemble the characteristics of the exterior transfers to the lunar L_2 halo orbit. The only real complexity that may be introduced into the majority of such transfers is the addition of a lunar flyby en route to the transfers' apogee passages. Conversely, the characteristics of many of the interior transfers shown in Fig. 3-90 resemble the characteristics of the exterior transfers to the lunar L_1 halo orbit shown in Fig. 3-86. This makes sense because the majority of both types of transfers involve Earth staging orbits, among other features.

3.4.7.3 Discussion This section has demonstrated that the methodology presented in this examination may be applied to many different families of unstable three-body orbits. The same techniques may be applied to quasiperiodic and aperiodic orbits as well, such as Lissajous orbits, though the parameters that generate the BLT maps will not be perfectly cyclical. The low-energy transfers and BLT maps constructed using different target orbits may appear very different. Nonetheless, families of low-energy transfers may still be identified and systematically evaluated in order to identify good candidates for practical lunar missions.

3.5 THREE-BODY ORBIT TRANSFERS

Once a spacecraft has arrived at a lunar three-body orbit, the spacecraft has several options. First, it may remain there for as long as desired, or at least until its station-keeping fuel budget is exhausted (which may be years). Lunar halo orbits may be a desirable location for communication and/or navigation satellites; they may also be a desirable location for space stations or servicing satellites.

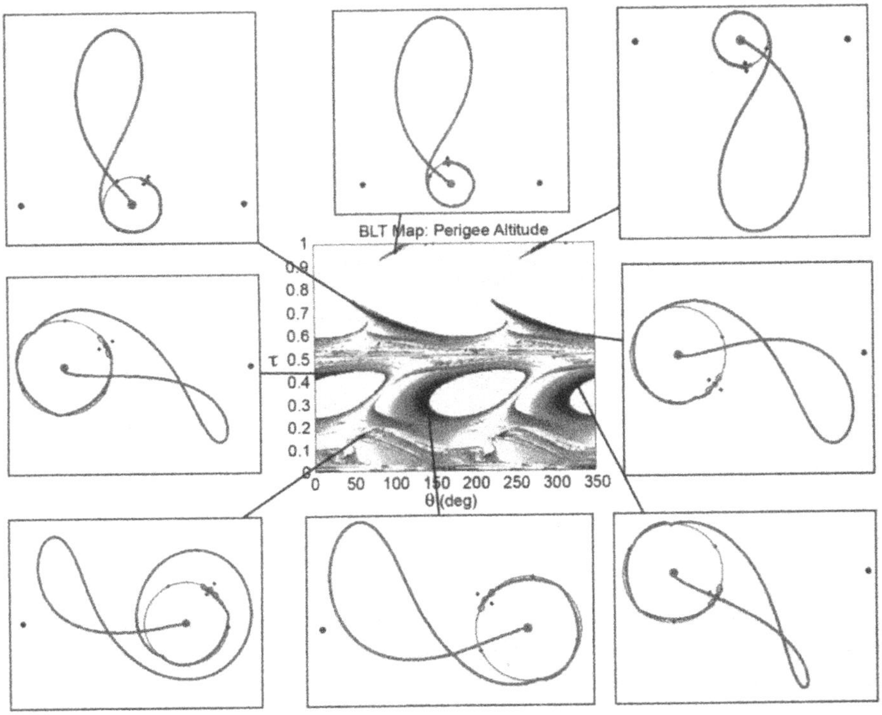

Figure 3-89 The exterior BLT map for low-energy transfers to the example distant prograde orbit about the Moon. Eight example low-energy transfers are shown around the BLT map to demonstrate some of the types of transfers that may be constructed between 185-km LEO orbits and this lunar orbit. *(See insert for color representation of this figure.)*

The spacecraft may transfer from the three-body orbit to a different three-body orbit in the Earth–Moon system for very little energy, provided that both orbits are unstable and have the same Jacobi constant [162, 185, 186]. For instance, the spacecraft might arrive at a lunar L_2 halo orbit and then later transfer to a lunar L_1 halo orbit. Section 2.6.11 presents several methods that one may use to identify and construct such transfers.

The spacecraft may also transfer from the nominal three-body orbit onto its unstable manifold and follow that trajectory to a desirable stable lunar orbit. It has been found that nearly any low lunar orbit is accessible in this way, and every transfer studied has required a smaller orbit-insertion maneuver than any conventional, direct transfer to the same low lunar orbit [46]. An example of such a transfer will be described in more detail below.

Similarly, the spacecraft may follow the unstable manifold of the three-body orbit down to the surface of the Moon. It has been found that any point on the surface

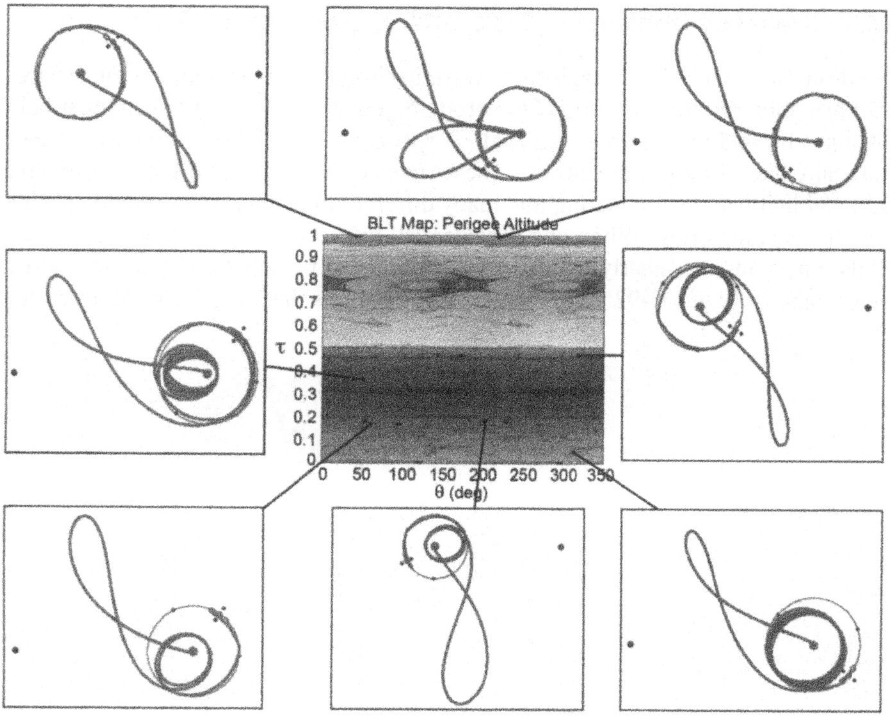

Figure 3-90 The interior BLT map for low-energy transfers to the example distant prograde orbit about the Moon. Eight example low-energy transfers are shown around the BLT map to demonstrate some of the types of transfers that may be constructed between 185-km LEO orbits and this lunar orbit. *(See insert for color representation of this figure.)*

of the Moon may be reached, although some points require several orbits about the Moon prior to touch-down [11, 46]. Again, the required ΔV to land from the lunar three-body orbit is smaller than the required ΔV to land following a conventional, direct transfer from the Earth.

Finally, the spacecraft has the option to return to the Earth following a low-energy Earth-return trajectory. Every low-energy lunar transfer has a symmetric Earth-return counterpart; the Earth-return trajectory does not need to be a mirror image of the trajectory used to arrive at the lunar orbit.

If the spacecraft's final destination is not the lunar three-body orbit, then the spacecraft does not need to inject into that orbit. Instead, the orbit's stable manifold may be used to guide the spacecraft to its final destination rather than to inject the spacecraft onto the three-body orbit. The stable manifold may be used as an initial guess into a trajectory optimization routine, such as a multiple-shooting differential corrector (Section 2.6.5.2).

3.5.1 Transfers from an LL_2 Halo Orbit to a Low Lunar Orbit

The discussion henceforth graphically illustrates some example options that a spacecraft has upon arriving at a lunar halo orbit. Figure 3-91 shows one such lunar halo staging orbit and its unstable manifold. A spacecraft on this halo orbit may depart along any one of these trajectories. These trajectories fly by the Moon at different radii and inclinations, indicating that many different final lunar orbits are accessible from this staging orbit. When one considers all halo orbits in the family of L_2 halo orbits, one finds that nearly any low lunar orbit may be accessed by a low-energy lunar transfer. Figure 3-92 shows the available options that have been identified for

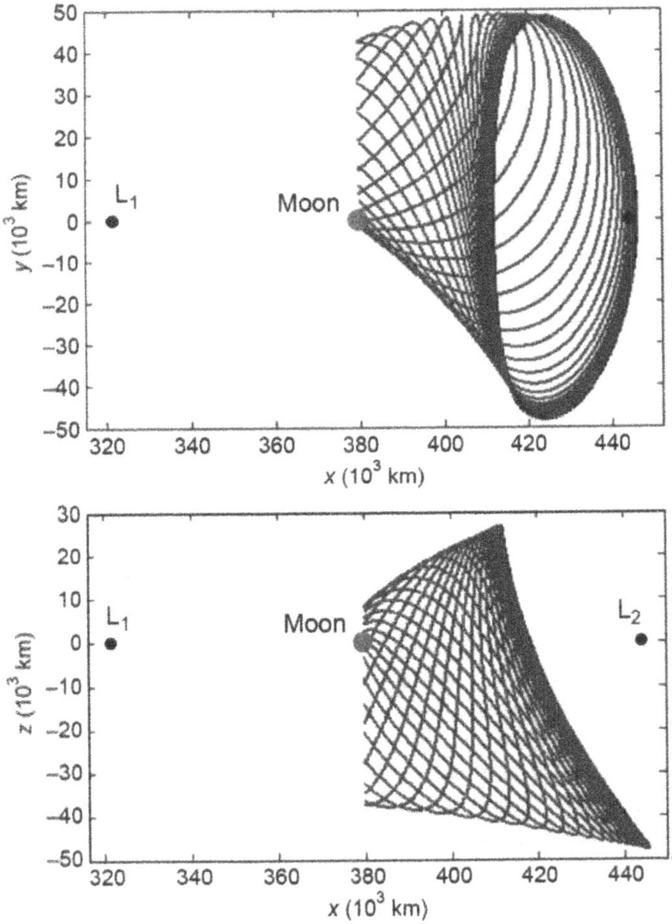

Figure 3-91 An example lunar halo staging orbit and its unstable manifold, viewed in the Earth–Moon rotating frame from above (top) and from the side (bottom). A spacecraft on this halo orbit may depart along any one of the trajectories shown.

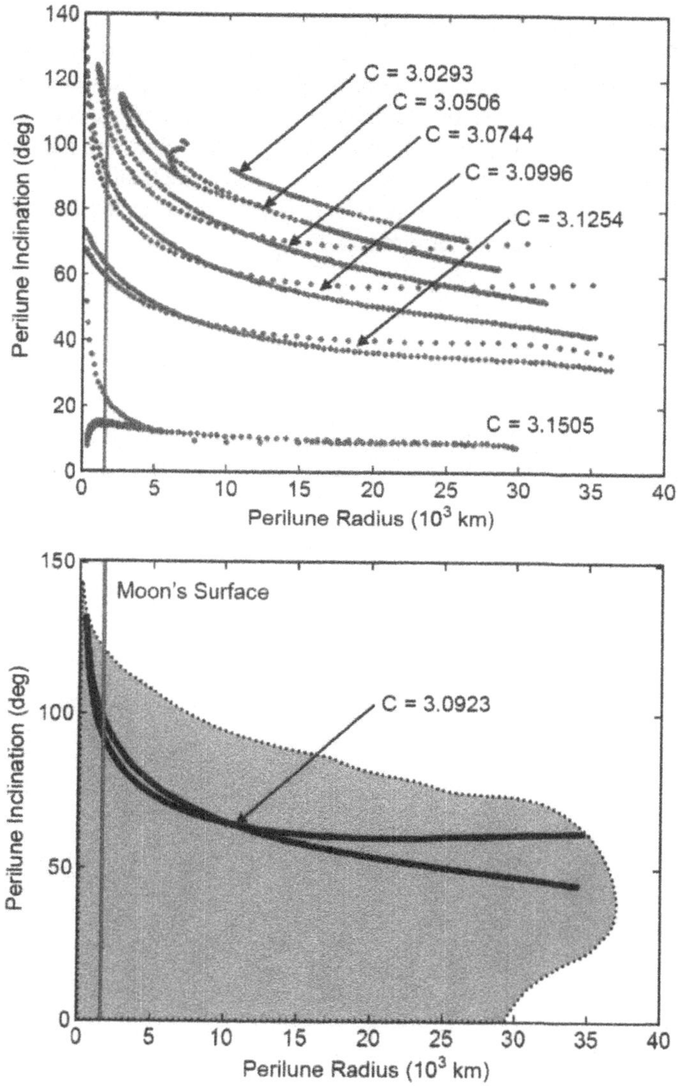

Figure 3-92 Available options identified for the radius and inclination of lunar orbits accessed by southern lunar L_2 halo orbits. Top: The radii and inclination combinations that may be obtained at perilune of the unstable manifolds of six different lunar L_2 halo orbits, where each orbit's available options are labeled with that orbit's Jacobi constant. Bottom: The radii and inclination combinations that may be obtained at perilune of the unstable manifolds of many orbits in the family of southern halo orbits. The highlighted options in the plot at right correspond to the available options for the halo orbit shown in Fig. 3-91.

the radius and inclination of lunar orbits that may be accessed by southern lunar L_2 halo orbits. The shaded field in the right plot has been constructed by sampling the unstable manifolds of hundreds of halo orbits and interpolating between the results. The highlighted points in the plot on the right are those points that are accessible from the example southern halo staging orbit shown in Fig. 3-91. Northern halo orbits can access the same set of lunar orbits except with a negative inclination. In each case, it is assumed that the orbit-insertion maneuver is performed at the perilune of the unstable manifold, but this is not required.

CHAPTER 4

TRANSFERS TO LOW LUNAR ORBITS

4.1 EXECUTIVE SUMMARY

This chapter examines low-energy transfers that target low, 100-kilometer (km), polar lunar orbits. The analyses presented here may be applied to any lunar orbit insertion; polar orbits are used as examples since mapping missions have historically been frequently sent to near-polar orbits about the Moon. This chapter presents surveys of direct transfers as well as low-energy transfers to low lunar orbits, and provides details about how to construct a desirable transfer, be it a short-duration direct transfer or a longer duration low-energy transfer.

Figure 4-1 shows an example direct transfer, compared with an example low-energy transfer to low lunar orbits. Much like the transfers presented in Chapter 3, these trajectories are ballistic in nature; they require a standard trans-lunar injection (TLI) maneuver, a few trajectory correction maneuvers, and an orbit insertion maneuver. One may again add Earth phasing orbits and/or lunar flybys to the trajectories, if needed, which change their performance characteristics.

Many thousands of direct and low-energy trajectories are surveyed in this chapter. Table 4-1 provides a quick guide for several types of transfers that are presented here, much like Table 3-1 from Chapter 3, comparing their launch energy costs, the breadth

Low-Energy Lunar Trajectory Design. By Jeffrey S. Parker and Rodney L. Anderson

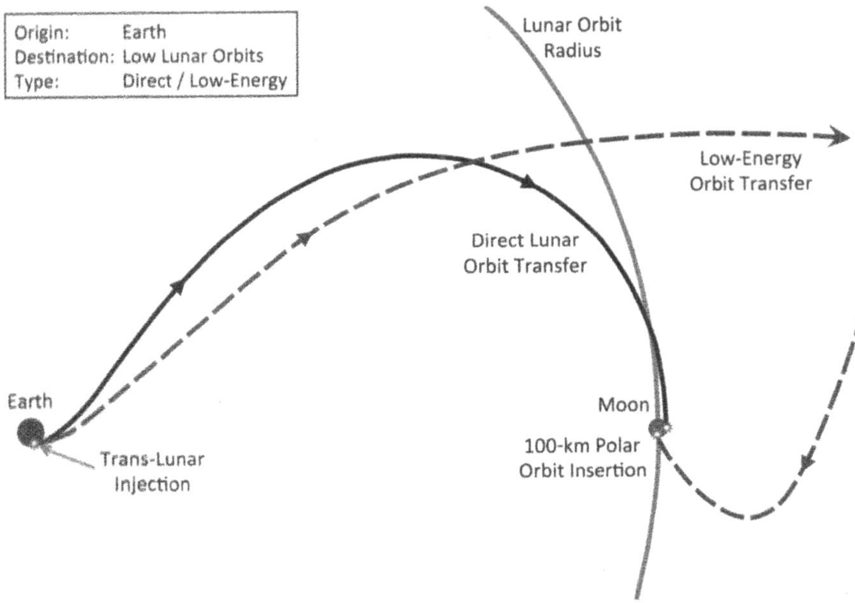

Figure 4-1 The profiles for both a direct and a low-energy transfer from the Earth to a low lunar orbit.

Table 4-1 A summary of several parameters that are typical for different mission scenarios to low lunar orbits. EPOs = Earth Phasing Orbits, BLT = Low-Energy Ballistic Lunar Transfer.

Mission Element	Direct Transfer	Direct w/EPOs	Simple BLT	BLT w/Outbound Lunar Flyby	BLT w/EPOs
Launch C_3 (km^2/s^2)	−2.2 to −1.5	< −1.5	−0.7 to −0.4	−2.1 to −0.7	< −1.5
Launch Period	Short	Extended	Extended	Short	Extended
Transfer Duration (days)	2–6	13+	70–120+	70–120+	80–130+
Outbound Lunar Flyby	No	No	No	Yes	Yes
Lunar Orbit Insertion ΔV (m/s)	~820+	~820+	~640+	~640+	~640+

of their launch period, that is, the number of consecutive days they may be launched, their transfer duration, and the relative magnitude of the orbit insertion change in velocity (ΔV) upon arriving at the lunar orbit. The performance parameters are very similar to low-energy transfers to lunar libration orbits, except for the orbit insertion

ΔV. These parameters are representative and may be used for high-level mission design judgements, though the details will likely vary from mission to mission.

Conventional lunar mission design is presented in Section 4.3 as a reference for the analyses of low-energy lunar transfers. The trajectories shown in that section require trans-lunar injection parameter (C_3) values of at least -2.06 kilometers squared per second squared (km^2/s^2), realistic transfer durations between 2 and 6 days, and lunar orbit insertion ΔV values of at least 813 m/s. One can certainly construct quicker or longer transfers, but the injection C_3 and lunar orbit insertion ΔV values increase rapidly.

Direct transfers and low-energy transfers to low lunar orbits are directly compared and analyzed in Section 4.4. The surveys include many thousands of lunar transfers, arriving at the Moon in any orientation and arriving at different times. The surveys demonstrate that direct transfers must arrive at the Moon in a geometry such that the orbital plane is roughly normal to the Earth–Moon line at the time of arrival. Whereas low-energy transfers may be constructed that arrive at any orbital plane. If a mission must enter a lunar orbit with a particular node, then only certain values of the orbit's argument of periapse may be targeted, depending on the lunar arrival date; further, those values are different for low-energy transfers than they are for direct lunar transfers. It has been found that low-energy transfers require trans-lunar injection C_3 values of about -0.6 km^2/s^2, compared with typical direct transfers that require C_3 values of about -2.0 km^2/s^2. Low-energy transfers require about 70–120 days of transfer duration, compared with direct transfers that require 2–6 days, though either type of transfer may be designed to take more time. The lunar orbit insertion ΔV is at least 640 m/s for low-energy transfers, assuming an impulsive maneuver to immediately target a 100-km circular lunar orbit. Direct lunar transfers require at least 120 m/s more ΔV, and often significantly more ΔV than that to target the same arrival conditions. Finally, low-energy lunar transfers exist in families, such that very similar transfers exist to neighboring libration orbits. Very similar transfers also exist to the same orbit when the arrival time or arrival geometry is adjusted.

4.2 INTRODUCTION

This chapter is devoted to the analysis and construction of low-energy transfers to low lunar orbit. This is a rich problem; it is far too complex to present all possible examinations of such transfers in a concise form. To simplify the problem, while retaining a connection to practical spacecraft mission design, this book limits the scope of this study and only examines low-energy transfers to low-altitude, 100-km circular, polar orbits about the Moon. These orbits are very similar to many mapping orbits flown by historical lunar missions, including *Lunar Prospector* [56], *Kaguya/SELENE* [187], *Chang'e 1* [58], *Chandrayaan-1* (*CH-1*) [3], the *Lunar Reconnaissance Orbiter* (*LRO*) [188], and *Gravity Recovery and Interior Laboratory* (*GRAIL*) [83]. The procedures presented in this chapter may easily be applied to transfers that implement an eccentric capture orbit about the Moon: in that case the argument of periapse of the target orbit becomes a design constraint and the orbit insertion ΔV is reduced appropriately. This chapter contains all of the information to design such orbit insertions, assuming that the mission performs lunar orbit insertion (LOI) at an altitude of 100 km and an inclination of 90 degrees (deg). Even so, the procedures presented here may be applied to orbit insertions at other altitudes and in other inclinations, though in those cases the design space will have to be reconstructed by the mission designer. The surveys presented here provide a good representation of the trade space of any direct and low-energy transfer to any low orbit about the Moon.

Although the general characteristics of low-energy transfers to low lunar orbits are similar to the characteristics of low-energy transfers to lunar libration orbits, such as those presented in Chapter 3, the geometry of transfers that arrive at polar orbits is still significantly different. Therefore, the analysis in this chapter is independent of Chapter 3 and specifically tailored to study missions to low lunar orbit.

The *GRAIL* mission is the only mission in history, prior to 2012, to implement a low-energy transfer to a low-altitude orbit about the Moon as part of its primary mission. Its design features will be used as a reference in many of the discussions in this chapter [83–85]. *GRAIL*'s trajectory design is illustrated in Fig. 4-2, including the first and last launch opportunity in a 21-day launch period. This is the launch period published in Ref. 83; however, it was actually extended by many days as the mission developed. The *GRAIL* mission launched on September 10, 2011, on the third day of its launch period. *GRAIL*'s mission design includes two significant

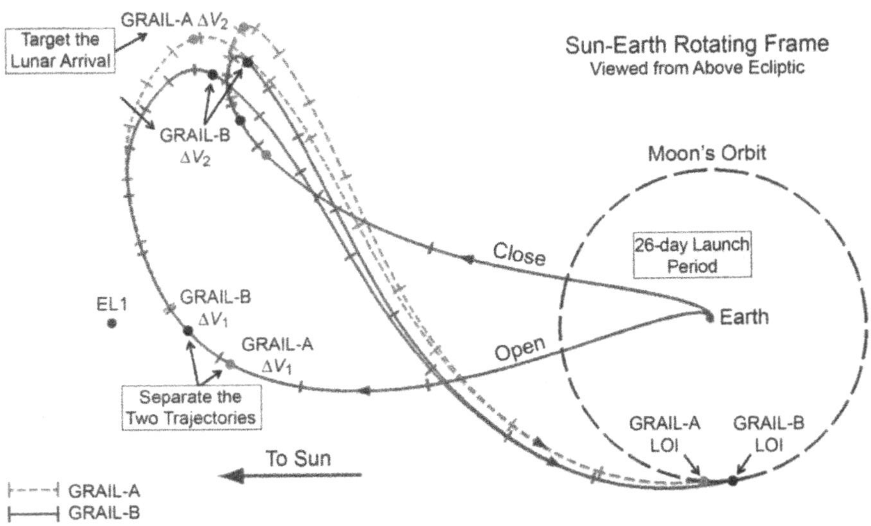

Figure 4-2 An illustration of *GRAIL*'s mission design, including a 26-day launch period and two deterministic maneuvers for both *GRAIL*-A and *GRAIL*-B, designed to separate their lunar orbit insertion times by 25 hours [83] (Originally published by the American Astronautical Society).

deterministic maneuvers performed per spacecraft during the cruise, performed primarily to separate their lunar orbit insertion dates. The trajectories generated in this chapter do not include these sorts of maneuvers. Chapter 6 explores the addition of maneuvers like those in *GRAIL*'s design.

4.3 DIRECT TRANSFERS BETWEEN EARTH AND LOW LUNAR ORBIT

The purpose of this book is to illustrate the costs, benefits, and characteristics of low-energy lunar transfers; the primary referent is the *direct* lunar transfer, which has been used so frequently in lunar missions that it is known as the conventional method. The first spacecraft launched toward the Moon, *Luna 1*, followed a direct transfer: a trajectory that required only 34 hours to reach the Moon, passing by within 6000 km of the surface. Since then, dozens of missions have implemented direct lunar transfers with durations ranging from 1.4 to 5.5 days, not including any staging orbits. Table 1-2 on page 16 summarize many example missions that implemented such direct transfers. Many resources exist that describe these direct lunar transfers in great detail [189]. This section only considers the ΔV of basic transfers as a function of the transfer duration to be used as a reference when describing low-energy lunar transfers.

Direct lunar transfers are trajectories that depend only on the gravity of the Earth and Moon. The Sun's gravity is accounted for, but only as a perturbation to the transfer. A very short-duration direct transfer departs the Earth on a hyperbola that encounters the Moon. The most efficient direct transfers typically require 4–5 days, depending on the location of the Moon in its elliptical orbit, and resemble Hohmann transfers. Figure 4-3 illustrates several direct lunar transfers that have varying transfer durations.

Figure 4-3 illustrates how the ΔV cost of a direct transfer increases away from the optimal transfer duration. But the cost doesn't rise very rapidly until the transfer duration has changed by several days. Recent spacecraft have taken advantage of the optimal transfer durations to maximize the amount of payload sent to the Moon. Conversely, it is apparent why the Apollo mission planners opted for a shorter transfer: the ΔV cost does not rise very much by decreasing the transfer duration from 4.5 days to 3.0 days, but the other consumables (including items such as food, water, and electrical power) required 1.5 days less support time on both the outbound and return transfer segments.

Since a spacecraft following a direct transfer only requires a few days to reach the Moon, it must be prepared to perform a maneuver within hours, or perhaps at most a day, to perform a trajectory correction maneuver. If this is an undesirably short amount of time, the mission may implement an Earth phasing orbit to extend the transfer duration. The spacecraft would be launched into an orbit that does not encounter the Moon, and only after one or more perigee passages would the trajectory finally arrive at the Moon.

The launch periods for many historical direct transfers were very short: only a handful of opportunities to launch per month, when the geometry was aligned

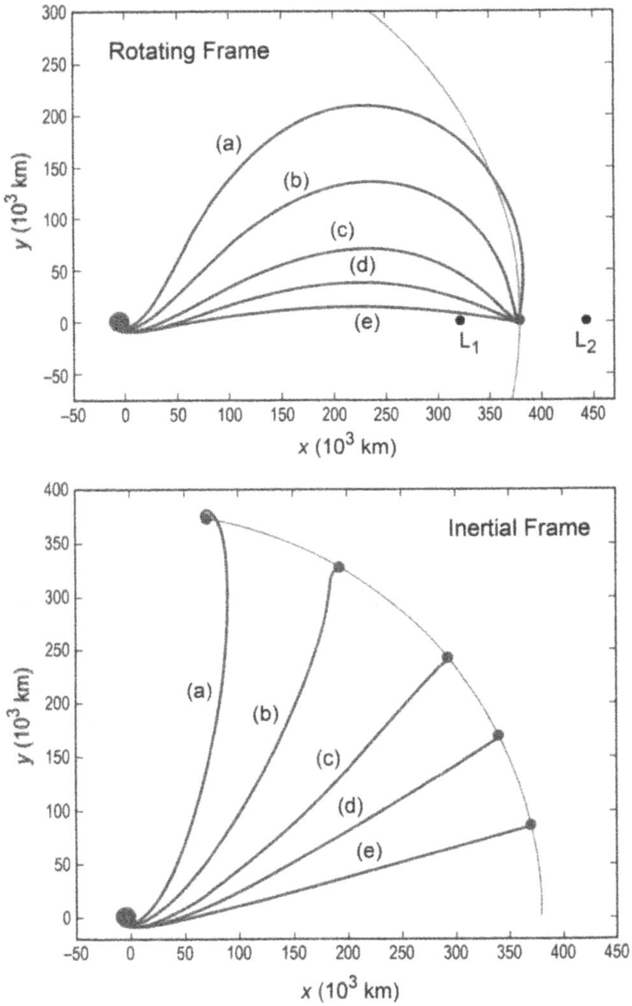

Figure 4-3 Five example direct transfers from 185-km circular Earth orbits to 100-km prograde lunar orbits, shown in the rotating frame (top) and inertial frame (bottom). These trajectories have been generated in the planar circular restricted three-body system. The following information applies to the labeled trajectories:

Traj.	Duration (days)	C_3 (km^2/s^2)	ΔV_{TLI} (km/s)	ΔV_{LOI} (km/s)	Total ΔV (km/s)
(a)	6.0	−1.976	3.138	0.829	3.966
(b)	4.5	−2.064	3.134	0.813	3.948
(c)	3.0	−1.670	3.152	0.893	4.045
(d)	2.0	0.264	3.240	1.248	4.488
(e)	1.0	13.654	3.831	3.024	6.854

properly. The *Clementine* and *Chandrayaan-1* missions implemented Earth phasing orbits, which extended the launch periods. *Chandrayaan-1*'s nominal mission profile included half a dozen Earth orbits prior to the lunar encounter. If the mission launched a day late, then the orbital period of one or more of these orbits would be adjusted to compensate for the change in transfer duration. The drawbacks of Earth phasing orbits include an extended operational timeline, which may add to the costs of the mission, and an increased dose of radiation as the spacecraft passes through the Van Allen Belts multiple times.

4.4 LOW-ENERGY TRANSFERS BETWEEN EARTH AND LOW LUNAR ORBIT

This section discusses how to build a low-energy ballistic transfer between the Earth and a low lunar orbit. The algorithms and methodology used to build a low-energy transfer are first described. Then, several example surveys are conducted, examining low-energy transfers that arrive at the Moon in some particular geometry at some given arrival time. The surveys become more general as this analysis continues. It then shows how to construct a map that tracks the minimum transfer ΔV cost required for a spacecraft to target any lunar orbit at a particular arrival time. Finally, the arrival time is opened up and transfers are examined that arrive at the Moon at many different times. The goal is to capture the transfer ΔV cost for transfers to any polar orbit about the Moon at any given arrival time in order to guide mission planners as they define the orbits and timeline for a given mission.

4.4.1 Methodology

Each transfer in the surveys presented here departs the Earth, coasts to the Moon, and injects directly into a low lunar orbit. To reduce the scope of the problem while still yielding practical data, the surveys presented here assume that the mission targets a circular 100-km polar orbit about the Moon. This lunar orbit is akin to the mapping orbits of several spacecraft, including *Lunar Prospector* [56], *Kaguya/SELENE* [187], *Chang'e 1* [58], *Chandrayaan-1* [3], the *LRO* [188], and *GRAIL* [83].

The LOI is modeled as a single impulsive maneuver that is performed at the periapse point and places the vehicle directly into a circular orbit. This is not a realistic maneuver, but it is useful to directly compare the total insertion cost of one transfer to another. The orbit insertion cost needed to place a satellite into an elliptical orbit, rather than a circular orbit, may be determined via the Vis-Viva equation [97].

The surveys presented here have been generated using a method that does not make many assumptions about what the lunar transfers look like. This permits each survey to reveal trajectories that may not have been expected. Each trajectory in each survey is constructed using the following procedure:

1. Construct the target lunar orbit. The following parameters are used in this study, specified in the International Astronomical Union (IAU) Moon Pole coordinate frame (see Section 2.4.4).

Periapse radius, r_p :	1837.4 km (~100-km altitude)
Eccentricity, e :	ϵ
Equatorial inclination, i :	90 deg
Argument of periapse, ω :	Specified value
Longitude of the ascending node, Ω :	Specified value
True anomaly, ν :	0 deg

The argument of periapse is undefined for a circular orbit. However, since all practical missions to date have inserted into elliptical orbits, and some missions remain in a highly elliptical orbit, the target orbit's argument of periapse, ω, is presented here rather than the true anomaly, which is kept at 0 deg to indicate that LOI is performed at periapse. The orbit's eccentricity is given as ϵ: it is approximately zero (1×10^{-9}) while permitting ω to be defined. One may also use the argument of latitude, which is defined for a circular orbit.

2. Construct the LOI state.
   (a) Specify the date of the LOI, t_{LOI}. Dates are given here in Ephemeris Time (ET).
   (b) Specify the magnitude of the impulsive orbit insertion maneuver, ΔV_{LOI}. Apply the ΔV in a tangential fashion to the LOI state.
3. Propagate the state backward in time for 160 days.
4. Identify the perigee and perilune passages that exist in the trajectory.
   (a) If the trajectory flies by the Moon within 500 km, label the trajectory as undesirable.
   (b) The latest perigee passage that approaches within 500 km of the Earth is considered the earliest opportunity to inject into that trajectory.
   (c) If no low perigees are observed, then the lowest perigee is identified as the trans-lunar injection (TLI) location.
5. Characterize the performance of the trajectory, making note of the following values:
   - TLI altitude, inclination, and C_3;
   - Duration of the transfer;
   - Periapse altitude of any/all Earth and Moon flybys; and
   - LOI ΔV magnitude.

This procedure requires four inputs: the longitude of the ascending node of the target orbit (Ω), the argument of periapse of the target orbit (ω), the ΔV of the impulsive LOI (ΔV_{LOI}), and the date of the LOI (t_{LOI}). Figures 4-4 and 4-5 show two examples of lunar transfers generated with this procedure using the inputs summarized in Table 4-2. Figure 4-4 illustrates a direct 4-day transfer and Fig. 4-5 illustrates an 84-day low-energy transfer.

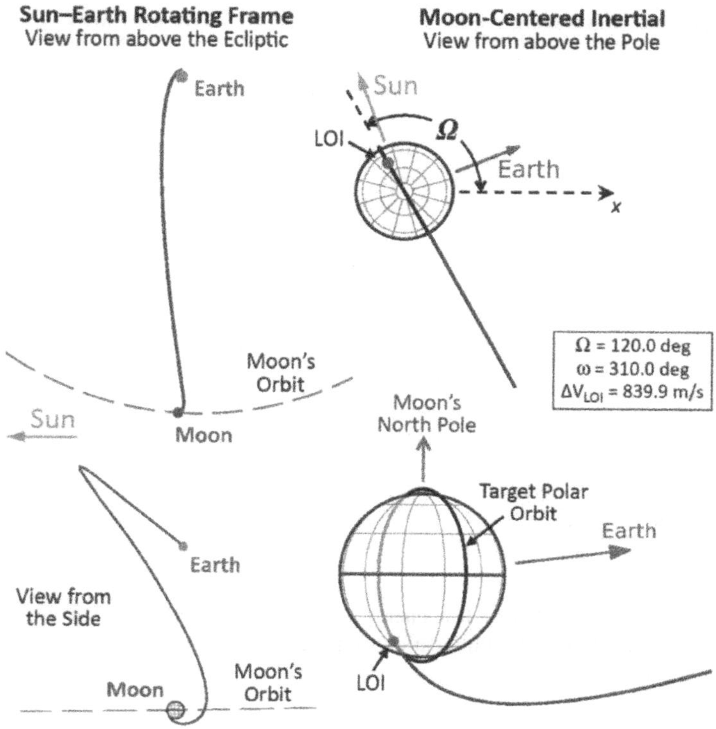

Figure 4-4 An example 4-day direct lunar transfer [2] (Copyright © 2011 by American Astronautical Society Publications Office, all rights reserved, reprinted with permission of the AAS).

All integrations performed here have been performed using a DIVA integrator (Section 2.7.1) with tolerance set to 1×10^{-10}; the force model includes the Sun, Earth, Moon, and each of the planets, all configured as point-mass gravitating bodies whose positions are estimated from JPL's DE421 Planetary and Lunar Ephemeris (Section 2.5.3).

Many surveys have been conducted, searching for practical lunar transfers. In general, a survey fixes the parameters Ω and t_{LOI} and systematically varies the other two parameters. This process generates a two-dimensional map displaying a parameter—typically the TLI altitude—which changes smoothly as either Ω or t_{LOI} shift. These surveys are described in more detail in the next sections.

4.4.2 Example Survey

Figure 4-6 shows the results of an example survey of lunar transfers. In this example, Ω is set to 120 deg, the LOI date is set to 18 July 2010 09:50:08 ET, the value

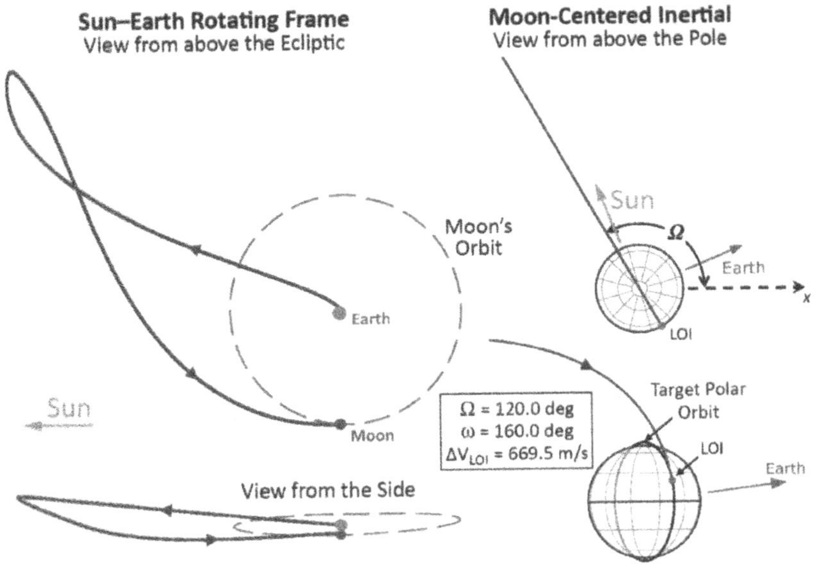

Figure 4-5 An example 84-day low-energy lunar transfer [2] (Copyright © 2011 by American Astronautical Society Publications Office, all rights reserved, reprinted with permission of the AAS).

Table 4-2 The inputs and performance parameters of the two example lunar transfers shown in Figs. 4-4 and 4-5. Both transfers begin in a 185-km circular low Earth orbit (LEO) parking orbit before their injections, and both transfers arrive at the Moon at a time t_{LOI} of 18 July 2010 9:50:08 ET.

Figure	Ω	ω	ΔV_{LOI}	Duration	LEO Inclination (deg)		C_3
#	(deg)	(deg)	(m/s)	(days)	Equatorial	Ecliptic	(km^2/s^2)
4-4	120.0	310.0	839.878	4.036	62.114	39.761	−2.064
4-5	120.0	160.0	669.543	83.706	28.093	5.921	−0.725

of ω is systematically varied from 0–360 deg, and ΔV_{LOI} is systematically varied from 650–1050 meters per second (m/s), a range empirically determined to generate practical transfers. Figure 4-6 shows the altitude of the trans-lunar injection point for each combination of ω and ΔV_{LOI}, assuming a spherical Earth with radius of 6378.136 km. The points shaded white correspond to trajectories that arrive at the Moon such that when propagated backward in time they never come any closer to the Earth than the orbit of the Moon itself. The points shaded black correspond to trajectories that arrive at the Moon such that when propagated backward in time they approach within 10,000 km of the Earth: trajectories that may be used to generate real missions [183, 184, 190, 191], assuming the departure time and geometry are

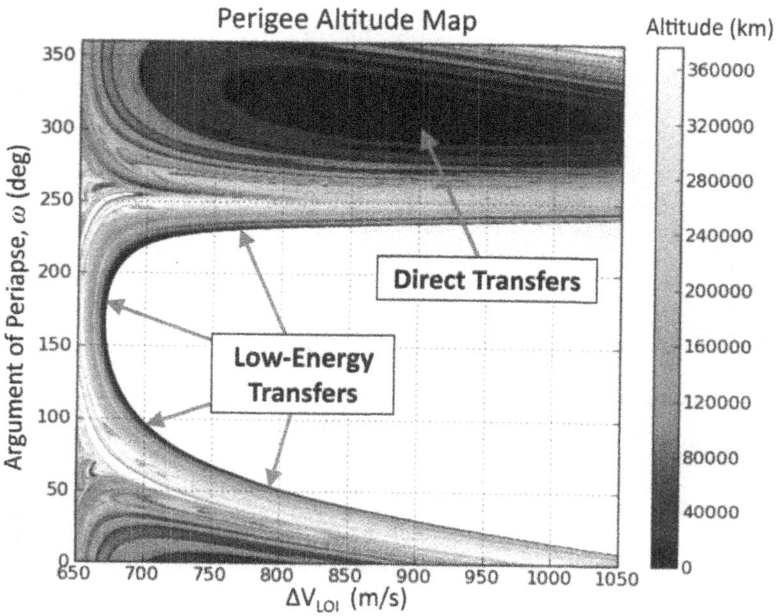

Figure 4-6 The altitude of the TLI location for each combination of ν and ΔV_{LOI}, given a lunar orbit insertion on July 18, 2010 into a lunar orbit with Ω equal to 120 deg [2] (Copyright © 2011 by American Astronautical Society Publications Office, all rights reserved, reprinted with permission of the AAS).

acceptable (see Section 6.5 for more information about generating a real mission using a ballistic guess).

The plot shown in Fig. 4-6 contains many interesting features. First, roughly half of the state space is white, corresponding to trajectories that arrive at the Moon from heliocentric orbits. With a quick investigation, one finds that the large black field toward the top of the plot corresponds to direct transfers to the Moon, that is, trajectories that take 2–12 days to reach the Moon, in family with the transfers that were implemented by the Apollo program and *LRO*; though most of the trajectories include Earth phasing orbits that extend the transfer's duration. The black curve that outlines the large white field corresponds to low-energy lunar transfers that require 80–120 days. There are many other curves throughout the plot that correspond to trajectories that enter some sort of large Earth orbit, or perform a combination of one or more flybys.

The direct transfers that are observed in the upper part of the plot shown in Fig. 4-6 require ΔV_{LOI} values from 760 m/s to 1000 m/s or more. The direct transfers that don't involve any Earth phasing orbits or any sort of lunar flyby require at least 818 m/s, though nearly all require 845 m/s or more. Figure 4-7 explores the structure of the direct transfer state space, presenting two additional maps that only show those

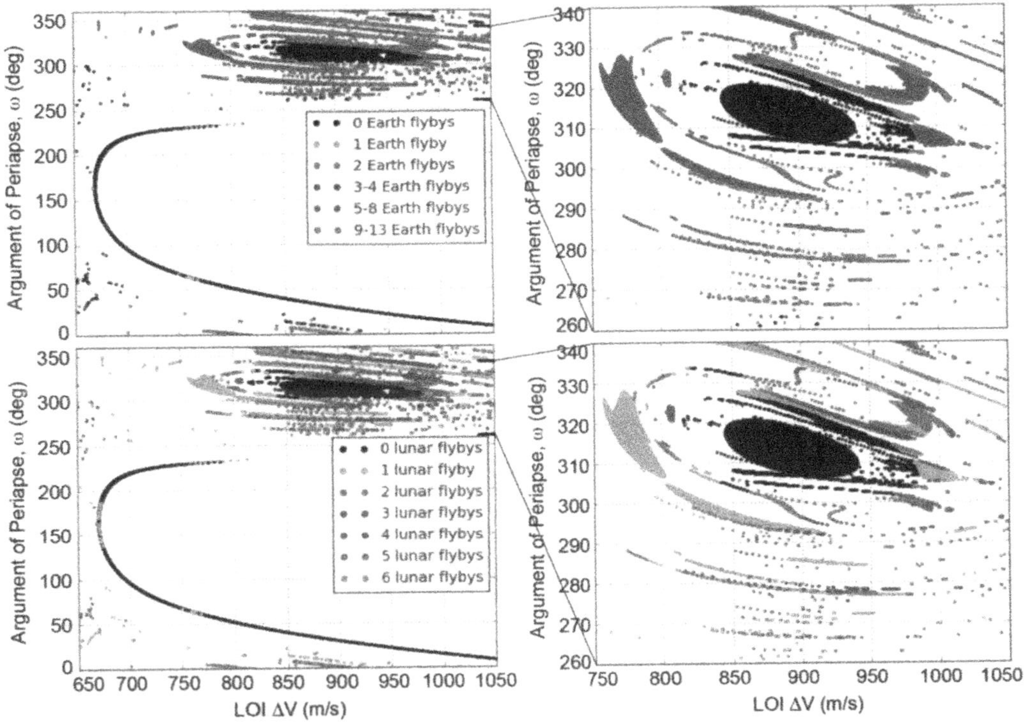

Figure 4-7 The transfers shown in Fig. 4-6 that approach within 1000 km of the Earth, shaded according to the number of Earth phasing orbits (top) and lunar flybys (bottom) that they make [2]. The upper portion of each map, corresponding with direct lunar transfers, is magnified in the plots on the right. (Copyright © 2011 by American Astronautical Society Publications Office, all rights reserved, reprinted with permission of the AAS). *(See insert for color representation of this figure.)*

trajectories that approach within 1000 km of the Earth; the two maps are shaded according to the number of Earth perigee passages (top) and lunar flybys (bottom) that they make before arriving at their target orbit. One notices that direct transfers with more phasing orbits and/or lunar flybys may require less orbit insertion ΔV than the most basic lunar transfers. In any case, simple low-energy trajectories exist that require as little as 669 m/s, ~100 m/s less than most multi-rev direct transfers observed and ~170 m/s less than most simple direct transfers.

Tables 4-3 and 4-4 summarize the performance parameters of several example direct lunar transfers and low-energy lunar transfers, respectively. Several examples of these trajectories are shown in Figs. 4-8 and 4-9, respectively. One can see that the value of ΔV_{LOI} is generally over 100 m/s lower for low-energy transfers in nearly all examples, though the TLI injection energy, C_3, is higher. The injection energy of direct lunar transfers is very close to -2.0 km^2/s^2, compared to a value of approximately -0.7 km^2/s^2 for low-energy transfers. Both types of transfers include missions with a wide range of TLI inclinations, both relative to the Earth's Equator and to the ecliptic. This suggests that transfers can begin from any inclination about the Earth. Section 6.5 demonstrates that one can add one to three maneuvers and adjust a trajectory to depart from a specified TLI inclination rather than the ballistic inclination value shown in the tables for a very modest ΔV cost. The total ΔV required to make this adjustment is on the order of 1 m/s per degree of inclination change.

4.4.3 Arriving at a First-Quarter Moon

All of the transfers presented in the previous section arrive at the Moon at a particular time into a particular orbit, namely, a circular, polar orbit with a longitude of the ascending node, Ω, of 120 deg and a time of arrival, t_{LOI}, of 18 July 2010 at 9:50:08 ET. This time of arrival corresponds to a moment in time when the Sun–

Table 4-3 A summary of the performance parameters of several direct lunar transfers shown in Fig. 4-6 and illustrated in Fig. 4-8 [2] (Copyright © 2011 by American Astronautical Society Publications Office, all rights reserved, reprinted with permission of the AAS).

Traj #	Ω (deg)	ω (deg)	ΔV_{LOI} (m/s)	Duration (days)	LEO Inclination (deg) Equatorial	LEO Inclination (deg) Ecliptic	C_3 (km^2/s^2)	# Earth Flybys	# Moon Flybys
D1	120.0	321.3	818.0	4.111	22.147	8.551	−2.078	0	0
D2	120.0	326.4	860.4	4.155	43.459	62.667	−2.058	0	0
D3	120.0	304.8	867.5	4.004	85.516	63.963	−2.045	0	0
D4	120.0	301.5	947.7	3.942	142.173	123.280	−2.006	0	0
D5	120.0	311.7	971.8	4.009	131.320	154.340	−2.002	0	0
D6	120.0	321.0	813.3	13.941	24.717	6.435	−2.095	1	0
D7	120.0	326.4	868.0	14.005	52.683	72.504	−2.071	1	0
D8	120.0	279.0	870.0	32.759	19.407	31.944	−2.046	2	1
D9	120.0	325.5	758.0	67.175	37.135	13.784	−2.292	6	1
D10	120.0	327.0	810.1	84.747	62.694	39.723	−2.055	7	3
D11	120.0	354.9	828.8	85.441	75.489	54.465	−2.061	7	1
D12	120.0	268.2	861.4	141.341	46.894	63.333	−2.054	8	1

Table 4-4 A summary of the performance parameters of several low-energy lunar transfers shown in Fig. 4-6 and illustrated in Fig. 4-9 [2] (Copyright © 2011 by American Astronautical Society Publications Office, all rights reserved, reprinted with permission of the AAS).

Traj #	Ω (deg)	ω (deg)	ΔV_{LOI} (m/s)	Duration (days)	LEO Inclination (deg) Equatorial	LEO Inclination (deg) Ecliptic	C_3 (km^2/s^2)	# Earth Flybys	# Moon Flybys
L1	120.0	169.2	669.3	83.483	29.441	6.129	−0.723	0	0
L2	120.0	103.8	692.1	85.287	25.688	34.778	−0.723	0	0
L3	120.0	70.2	743.9	93.598	57.654	74.955	−0.667	0	0
L4	120.0	225.3	716.0	93.621	134.322	112.840	−0.657	0	0
L5	120.0	99.9	697.5	110.060	83.127	61.624	−0.697	0	0
L6	120.0	186.9	673.2	122.715	23.941	3.088	−0.712	0	0
L7	120.0	61.5	660.4	143.360	18.624	35.412	−0.572	0	1
L8	120.0	59.7	651.3	129.422	73.143	96.544	−0.612	0	3
L9	120.0	36.3	661.5	144.417	146.592	138.491	−0.658	0	1
L10	120.0	348.6	675.1	155.107	36.598	16.583	−0.645	5	1
L11	120.0	262.2	656.1	141.982	153.641	176.867	−0.608	0	3
L12	120.0	244.2	657.8	136.687	179.084	156.890	−0.640	0	6

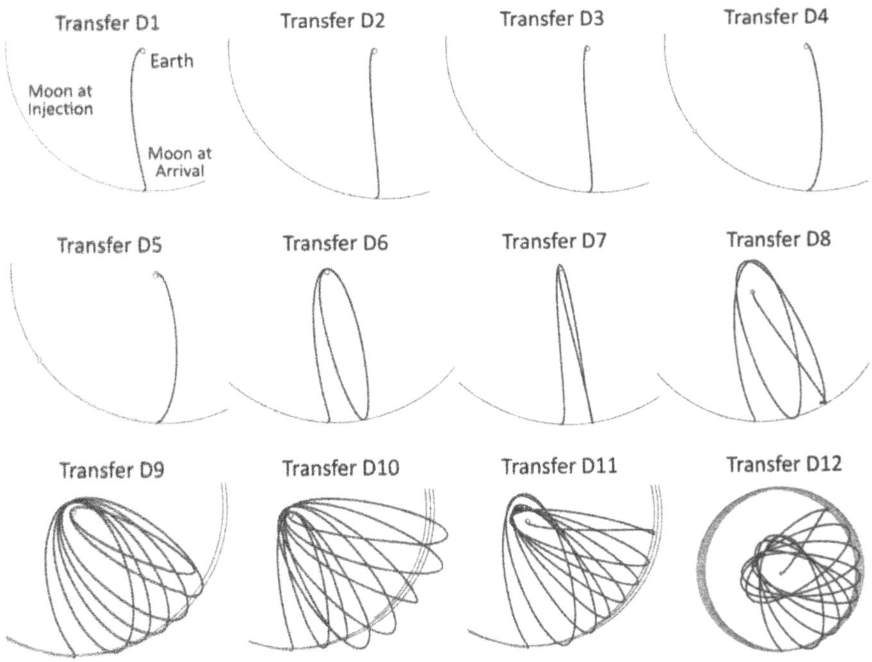

Figure 4-8 Example plots of several of the transfers summarized in Table 4-3. The trajectories are shown in the Sun–Earth rotating frame, such that the Sun is fixed on the x-axis toward the left [2] (Copyright © 2011 by American Astronautical Society Publications Office, all rights reserved, reprinted with permission of the AAS).

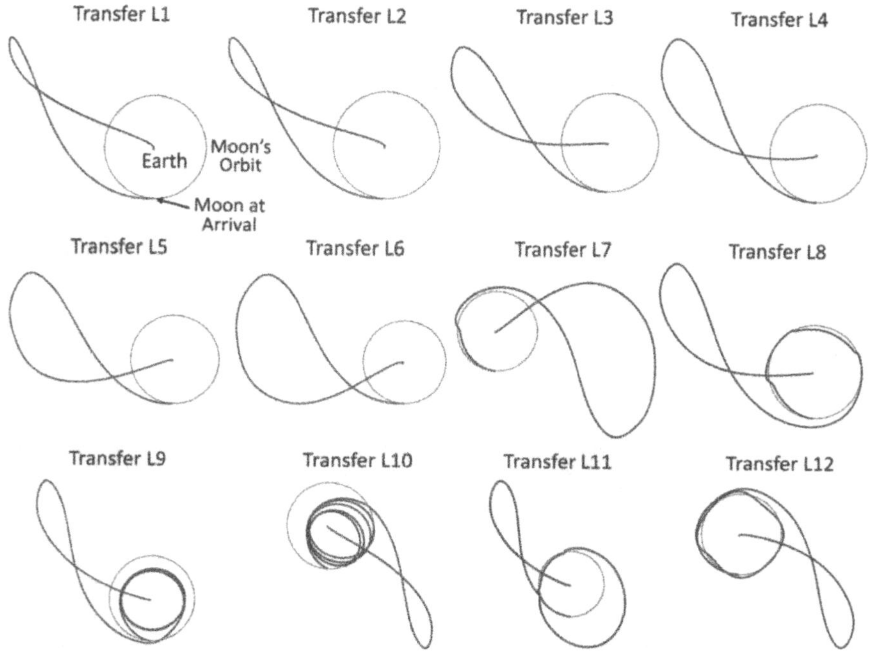

Figure 4-9 Example plots of several of the transfers summarized in Table 4-4. The trajectories are shown in the Sun–Earth rotating frame, such that the Sun is fixed on the x-axis toward the left [2] (Copyright © 2011 by American Astronautical Society Publications Office, all rights reserved, reprinted with permission of the AAS).

Earth–Moon angle is approximately equal to 90 deg at the Moon's first quarter. This is very similar to the arrival geometry of the two *GRAIL* spacecraft, though in a different month. In addition, the plane of the target orbit is nearly orthogonal to the Earth–Moon line. A polar orbit with an Ω-value of 111.9 deg (also 291.9 deg) is in a plane that is as close to orthogonal to the Earth–Moon line as a polar orbit can get on this date. The surveys presented in this section keep the time of arrival the same and explore the changes to the lunar transfers that occur as the target orbit's Ω-value is varied.

Figures 4-10 and 4-11 show surveys of the lunar transfer state space as Ω varies from 0–80 deg and 160–270 deg, respectively. There is a clear progression of the state space as Ω varies. Locations where direct and low-energy transfers exist are indicated. The state space varies much less discernibly when Ω is within $\sim$30 deg of 111.9 deg or 291.9 deg, namely, when the orbit is close to being orthogonal to the Earth–Moon line.

Many features are quickly apparent when studying the maps shown in Figs. 4-10 and 4-11. First, a large portion of each map is white, corresponding to combinations of ΔV_{LOI} and ω that result in trajectories that depart the Moon backward in time and

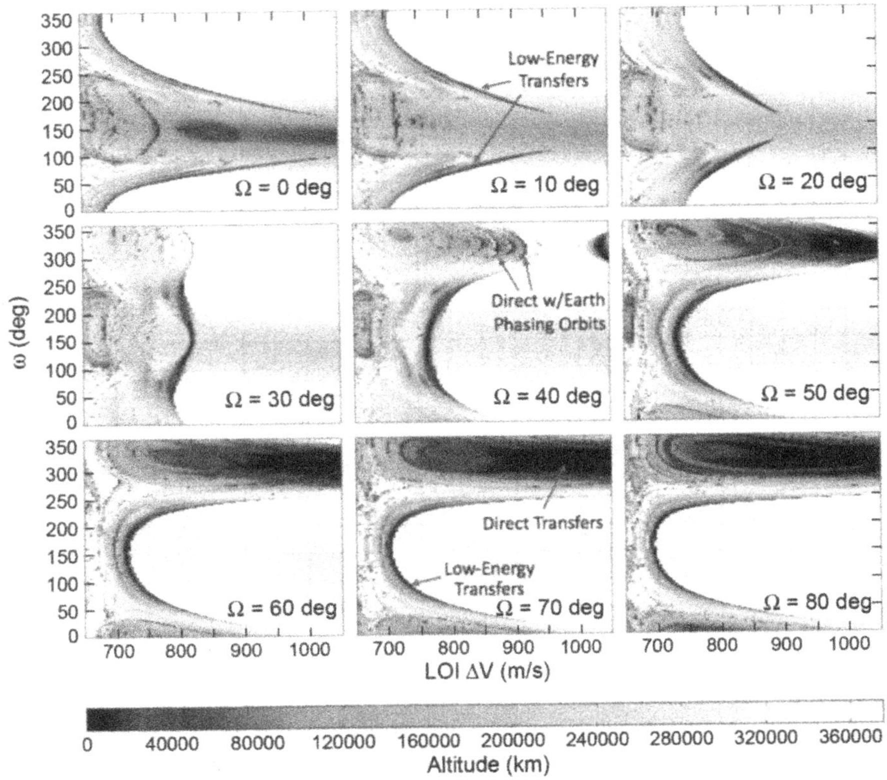

Figure 4-10 Nine surveys of trajectories that arrive at the first-quarter Moon, where the target orbit's Ω varies from 0–80 deg. Points in black originate from the Earth; other points are shaded according to how close they come to the Earth when propagated backward, using the light–dark shading scheme presented in Fig. 4-6 [2] (Copyright © 2011 by American Astronautical Society Publications Office, all rights reserved, reprinted with permission of the AAS).

traverse away from the Earth–Moon system. At lower ΔV_{LOI}-values, the trajectories depart the Moon backward in time and later impact the Moon or remain very near the Moon. One can see curves of black in each map, corresponding to trajectories that depart the Moon backward in time and eventually come very near the Earth; hence, making viable Earth–Moon transfers. The features are observed to shift in a continuous fashion across the range of Ω-values.

If one surveys these maps, one finds that low-energy transfers exist to any lunar orbit plane, but simple direct transfers only exist for certain ranges of Ω-values. Direct transfers can only reach orbits with Ω-values between approximately 50 deg and 170 deg and between approximately 230 deg and 350 deg for this particular

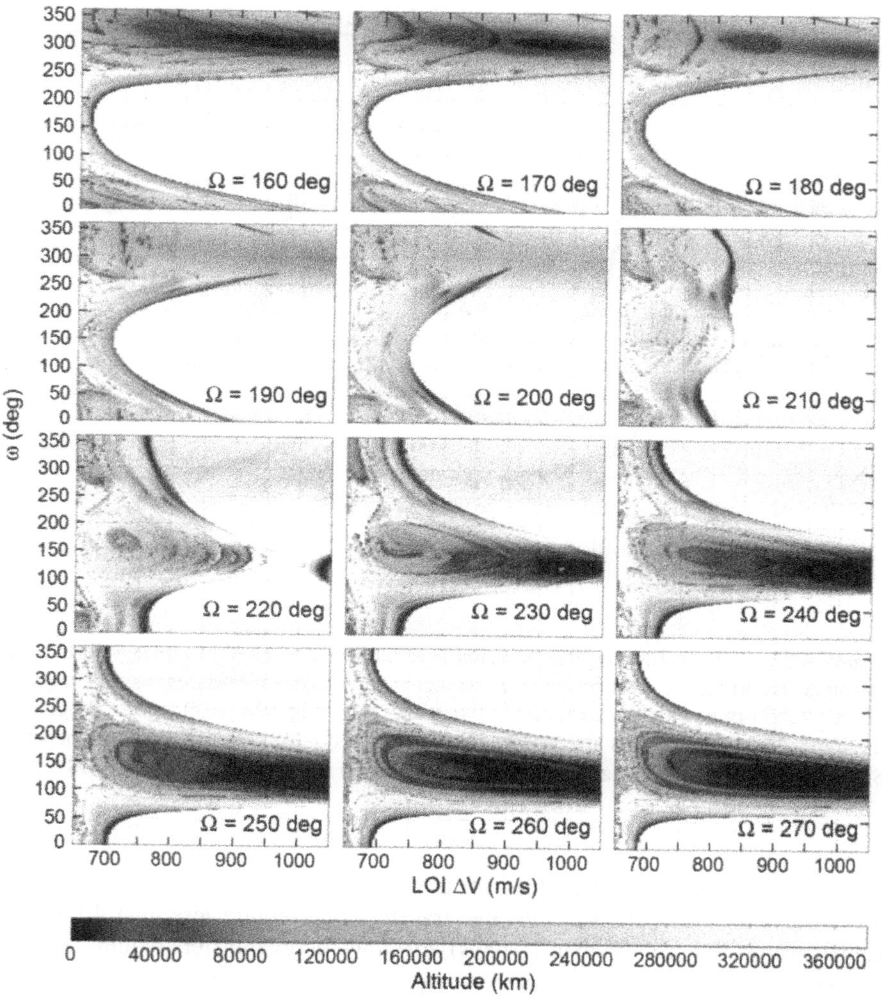

Figure 4-11 Twelve surveys of missions that arrive at the first-quarter Moon, where the target orbit's Ω varies from 160–270 deg. The maps are shaded according to the closest approach distance that the trajectories make with the Earth, as illustrated in Figs. 4-6 and 4-10 [2] (Copyright © 2011 by American Astronautical Society Publications Office, all rights reserved, reprinted with permission of the AAS).

arrival date. These orbit planes are within about 60 deg of being orthogonal to the Earth–Moon line; furthermore, direct lunar transfers require less ΔV for their orbit insertions the closer they are to being orthogonal to the Earth–Moon line.

Figure 4-12 captures the least-expensive ΔV_{LOI} for simple direct lunar transfers, as well as simple low-energy lunar transfers (that is, transfers that do not involve

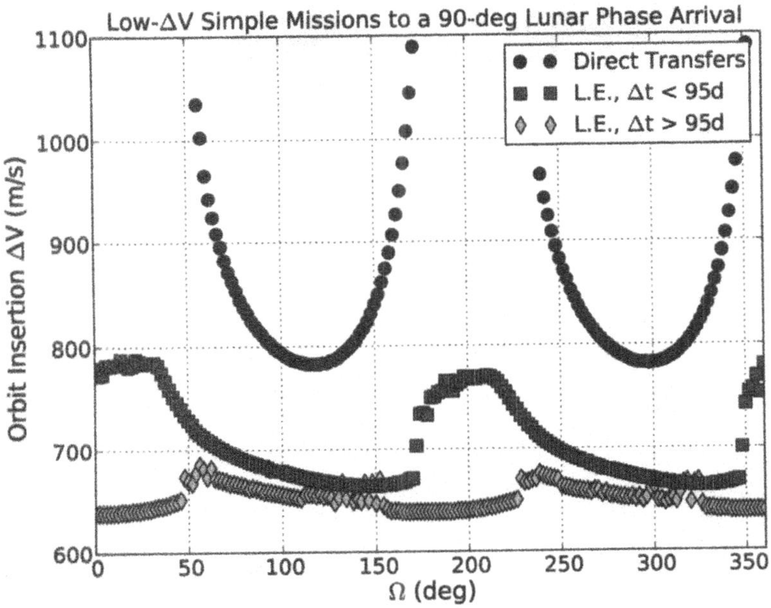

Figure 4-12 The minimum lunar orbit insertion ΔV for direct and low-energy (L.E.) lunar transfers, requiring no Earth phasing orbits nor lunar flybys for transfers to a first-quarter Moon. Polar orbits with Ω-values of 111.9 deg and 291.9 deg are very close to orthogonal to the Earth–Moon line on this arrival date [2] (Copyright © 2011 by American Astronautical Society Publications Office, all rights reserved, reprinted with permission of the AAS).

lunar flybys or Earth phasing orbits) for any target orbit plane studied. Three curves are presented: *direct* transfers involve transfers that require less than 40 days to achieve (most require less than 10 days), *fast low-energy* transfers require less than 95 days, and *long low-energy* transfers require more than 95 days to achieve. The transfer durations are not permitted to exceed 160 days in this study. There are many trajectories that require more ΔV than what is shown in Fig. 4-12; the illustration tracks the least expensive transfer in each case. Trajectories with Earth phasing orbits and/or lunar flybys may require even less ΔV, but those are not tracked here since there are so many paths that a spacecraft can take through the system. One observes that low-energy transfers do indeed reach any target orbit, though the insertion ΔV costs vary as the orbit plane changes. Direct lunar transfers are indeed limited to certain orbital planes, and they require at least 120 m/s more LOI ΔV than a low-energy transfer to the same orbit. Further, the cost of longer low-energy transfers remains very constant—within 50 m/s of ΔV—for any target lunar orbit plane.

The lunar transfers with the least LOI ΔV and no low Earth or lunar periapse passages have been identified for each combination of Ω and ω; their performance parameters are plotted in Fig. 4-13. The left plot shows a map of the LOI ΔV cost of these transfers; the plot on the right shows the corresponding transfer duration for

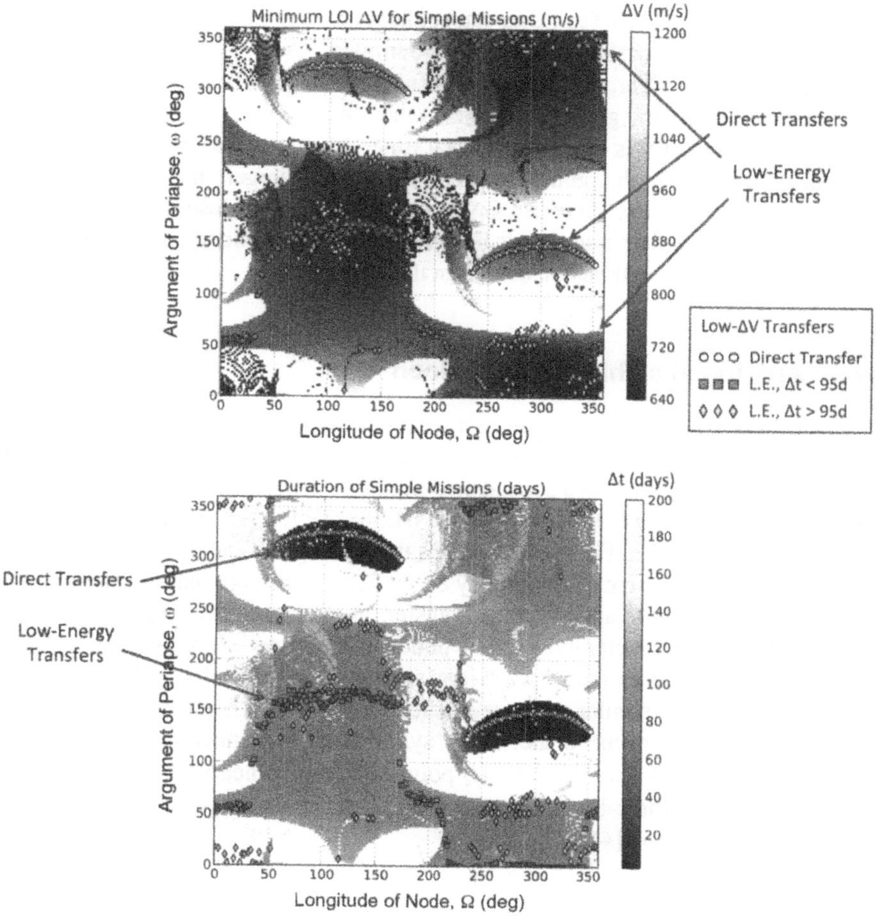

Figure 4-13 The combinations of Ω and ω that yield simple lunar transfers, that is, those without low Earth or lunar periapse passages. If multiple transfers exist for the same combination, then the one with the least LOI ΔV is shown. All of these transfers arrive at a first-quarter Moon. The low-ΔV transfers shown in Fig. 4-12 are indicated by dots in each map [2] (Copyright © 2011 by American Astronautical Society Publications Office, all rights reserved, reprinted with permission of the AAS). *(See insert for color representation of this figure.)*

each trajectory. The low-ΔV solutions identified in Fig. 4-12 are plotted in these maps for reference, and to identify their ω-values and durations. Direct transfers are easily discerned by observing the dark fields in the plot on the right, corresponding to short-duration transfers. One can see that there are large fields of combinations of Ω and ω that yield low-energy transfers, though the costs increase as one moves away from the low-ΔV curves. One can see that the combinations of Ω and ω that yield practical direct transfers are much more limited.

The maps shown in Fig. 4-13 are very useful: they illustrate what sorts of transfers may be used to reach any given polar orbit at the Moon, given that the transfers must arrive at the Moon at this particular arrival time. Missions that target an elliptical orbit must consider which argument of periapsis value to target; missions that aim to enter a circular orbit may likely use any ω for the initial orbit insertion, simplifying the trade space. Similar maps may be generated for any lunar arrival time: two different arrival times will be considered in the next sections.

4.4.4 Arriving at a Third-Quarter Moon

All of the transfers studied so far have arrived at the Moon at the same time, when the Moon is at its first quarter. Yet spacecraft missions may need to arrive at the Moon at any time of the month. As a second step in this survey, lunar transfers are studied that arrive at the Moon on 3 August 2010 at 04:38:29 ET: a time when the Moon has reached its third quarter. Figure 4-14 shows two example transfers that arrive at the third-quarter Moon, where the trajectory on the left is a direct lunar transfer and the trajectory on the right is a low-energy transfer. Neither transfer requires any extra Earth phasing orbits or lunar flybys. One notices that the low-energy transfer extends away from the Sun rather than toward it as seen in Figs. 4-5 and 4-9. Otherwise the transfers appear very similar to those studied previously. The symmetry observed here is expected according to the nearly symmetrical dynamics in the Sun–Earth system [86]. The Sun–Earth L_1 and L_2 points are located nearly the same distance from the Earth, and three-body libration orbits about those Lagrange points behave in a very similar fashion [46].

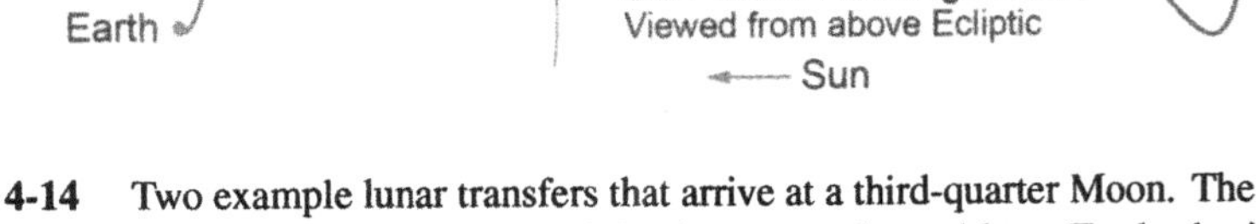

Figure 4-14 Two example lunar transfers that arrive at a third-quarter Moon. The transfers are simple, direct (left) and low-energy (right) lunar transfers with no Earth phasing orbits nor lunar flybys. The transfers are viewed from above in the Sun–Earth rotating frame of reference [2] (Copyright © 2011 by American Astronautical Society Publications Office, all rights reserved, reprinted with permission of the AAS).

One may construct state space maps for transfers to a third-quarter Moon in the same way that maps have been constructed previously to a first-quarter Moon. Figures 4-15 and 4-16 plot state space maps for transfers to target orbits with Ω-values of 0–80 deg and 180–260 deg, respectively. These ranges of Ω-values track the interesting features as the orbit plane changes; the maps of the Ω-values between those plotted in the figures vary little across the range. An orbit with an Ω-value of 126.9 deg (also 306.9 deg) is as close to orthogonal to the Earth–Moon axis as a polar orbit can be at this time. Transfers within about 60 deg of this angle are all very similar, though the cost of those transfers rises as the orbital plane moves away from this optimal Ω-value. When one compares the maps shown in Figs. 4-15 and 4-16 to those constructed earlier in Figs. 4-10 and 4-11, one sees that the maps are very

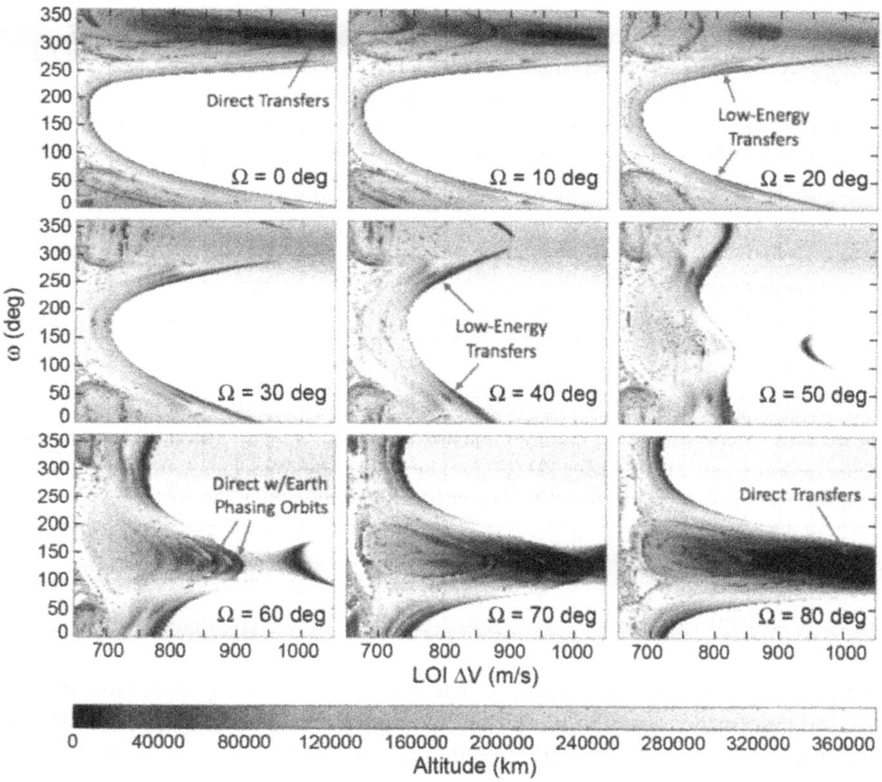

Figure 4-15 Nine surveys of missions that arrive at the third-quarter Moon, where the target orbit's Ω varies from 0–80 deg. The points are again shaded according to how close they approach to the Earth when propagated backward in time, using the same light–dark shading scheme applied in previous maps [2] (Copyright © 2011 by American Astronautical Society Publications Office, all rights reserved, reprinted with permission of the AAS).

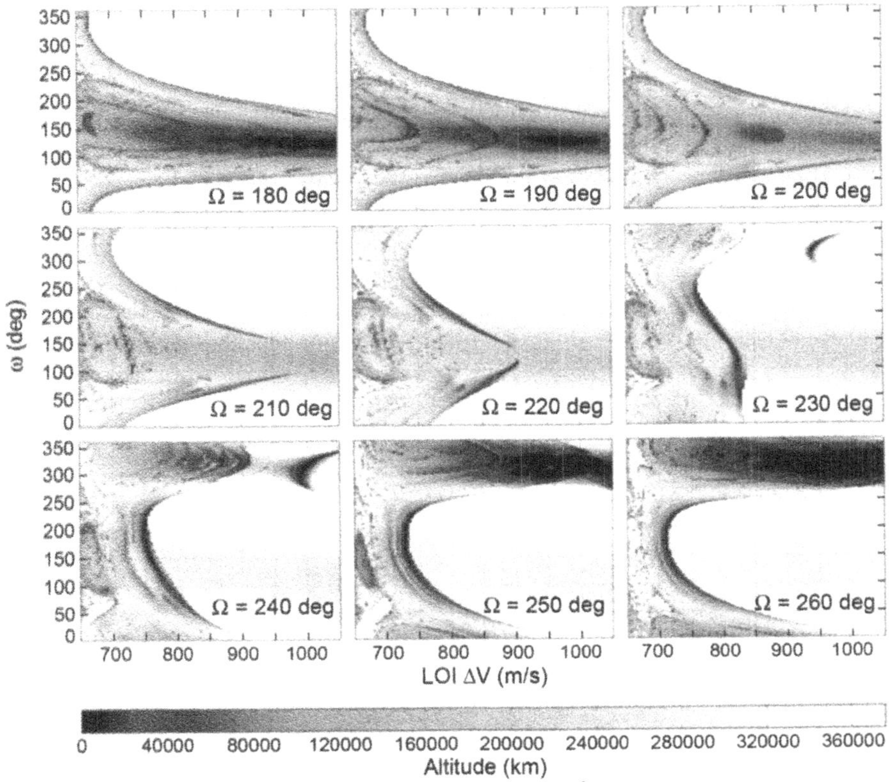

Figure 4-16 Nine surveys of missions that arrive at the third-quarter Moon, where the target orbit's Ω varies from 180–260 deg [2] (Copyright © 2011 by American Astronautical Society Publications Office, all rights reserved, reprinted with permission of the AAS).

similar with a 195 deg plane change. The transfers are arriving at the Moon when it is 180 deg further along in its orbit in the Sun–Earth synodic frame and 195 deg further in its orbit inertially, while the inertial coordinate axes that define Ω and ω have not changed.

Figure 4-17 shows the same two plots as shown in Fig. 4-13 for these third-quarter lunar arrival transfers. The maps show the LOI ΔV cost and transfer duration for simple lunar transfers that target different lunar orbits. As before, if there are multiple lunar transfers that may be used to arrive at the same lunar orbit, then the maps present the parameters for the transfer with the least LOI ΔV. The maps illustrate that the same trends exist to third-quarter lunar arrivals as do to first-quarter lunar arrivals, but with a 195-deg shift in Ω.

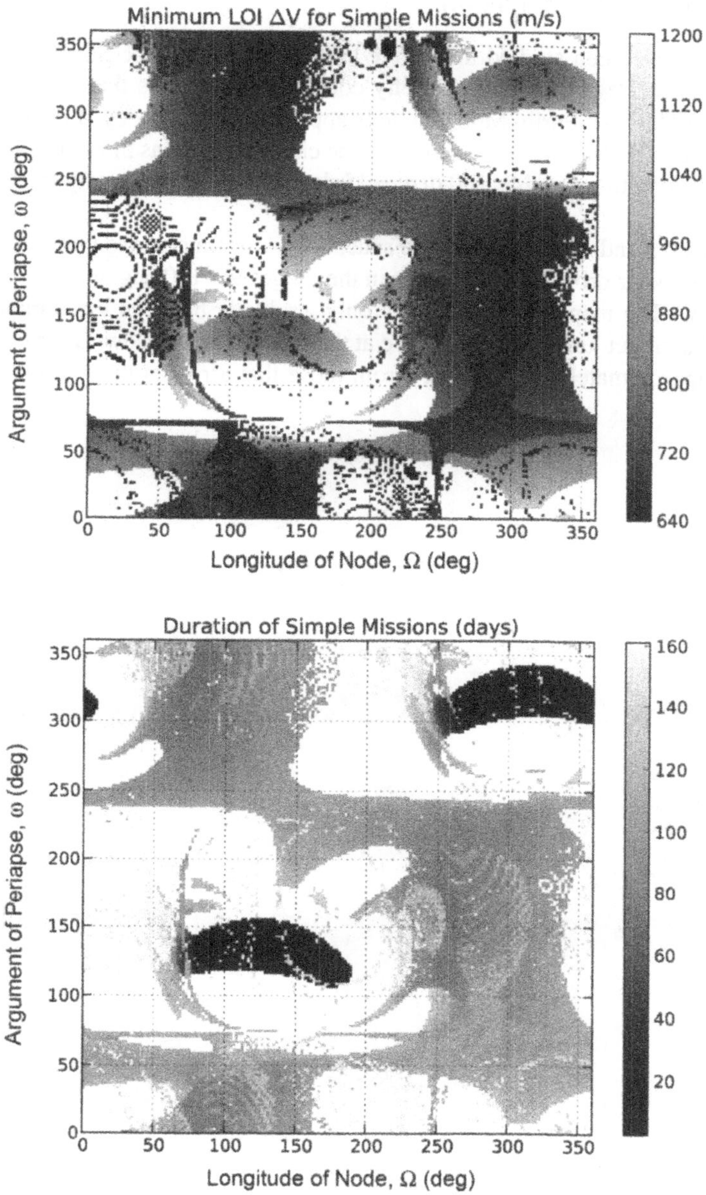

Figure 4-17 The combinations of Ω and ω that yield simple lunar transfers, that is, those without low Earth or lunar periapse passages. If multiple transfers exist for the same combination, then the one with the least LOI ΔV is shown. All of these transfers arrive at a third-quarter Moon [2] (Copyright © 2011 by American Astronautical Society Publications Office, all rights reserved, reprinted with permission of the AAS).

4.4.5 Arriving at a Full Moon

Trajectories have been studied that arrive at the Moon when the Sun–Earth–Moon angle is near 90 deg; this section briefly considers trajectories that arrive at a full Moon, when the Sun–Earth–Moon angle is approximately 180 deg. Lunar transfers that arrive at a new Moon have much the same characteristics as those that arrive at a full Moon, but with a familiar 180 deg ± 15 deg shift in Ω; for brevity they will not be shown here.

Figures 4-18 and 4-19 present state space maps for trajectories that arrive at the full Moon in polar orbits with Ω-values in the ranges 90–170 deg and 270–350 deg, respectively. The maps not shown vary only gradually between these maps. One observes that direct lunar transfers arrive at the full Moon with low-ΔV insertions at Ω-values approximately 90 deg apart from those that arrive at the first-quarter and

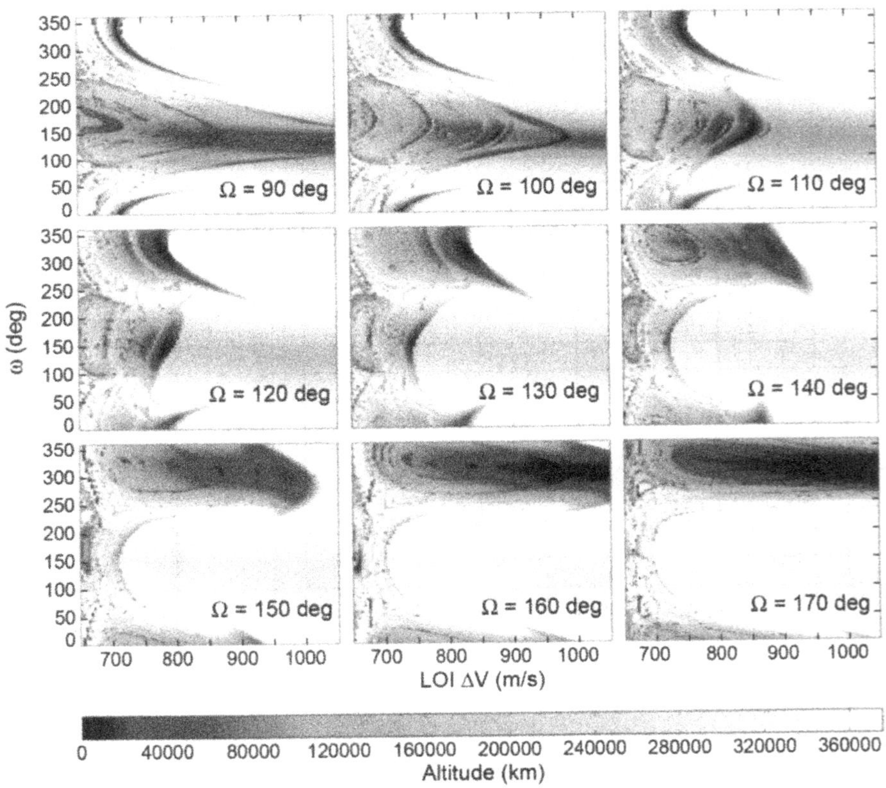

Figure 4-18 Nine surveys of missions that arrive at a full Moon, where the target orbit's Ω varies from 90–170 deg [2] (Copyright © 2011 by American Astronautical Society Publications Office, all rights reserved, reprinted with permission of the AAS).

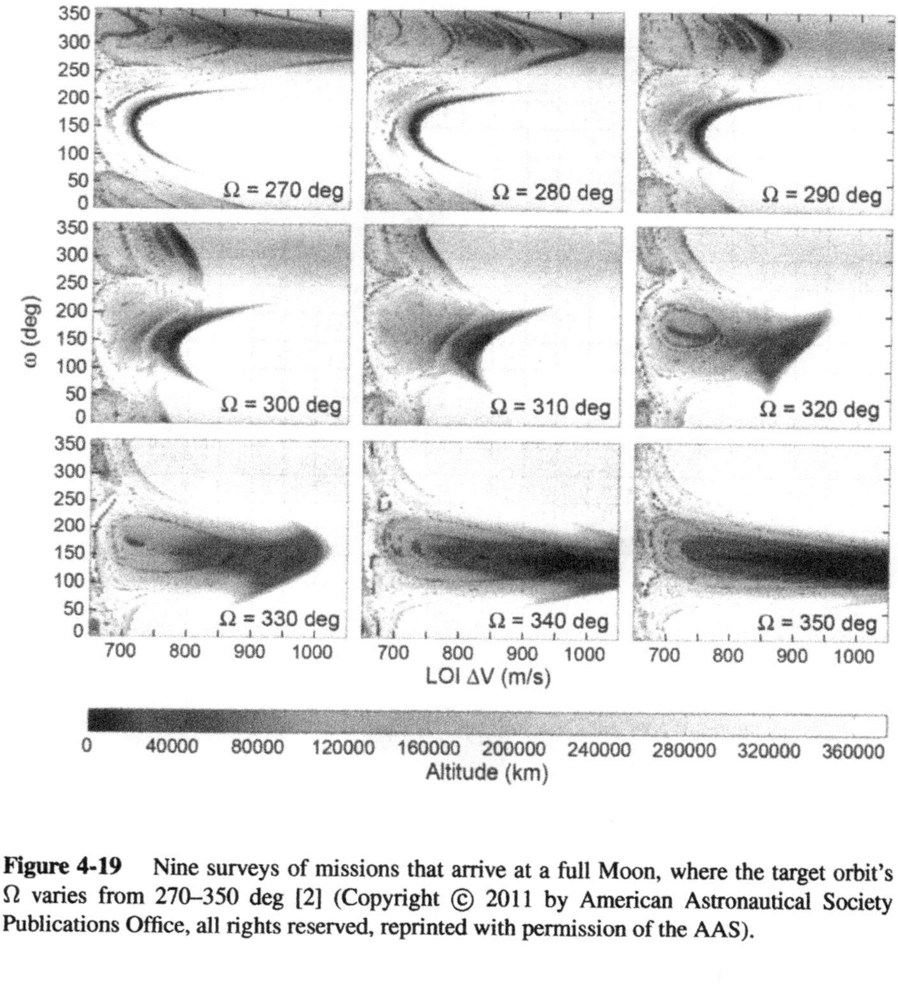

Figure 4-19 Nine surveys of missions that arrive at a full Moon, where the target orbit's Ω varies from 270–350 deg [2] (Copyright © 2011 by American Astronautical Society Publications Office, all rights reserved, reprinted with permission of the AAS).

third-quarter Moons. This demonstrates additional evidence that the minimum orbit insertion ΔV requirements for direct lunar transfers occurs when the orbit's plane is nearly orthogonal to the Earth–Moon line.

The low-energy lunar transfers' locations in the full-Moon state space maps evolve somewhat differently as Ω varies compared with their evolutions in the state space maps for first- and third-quarter Moons. Low-energy transfers still arrive at the Moon for any Ω-value, but the range of ω-values that may be used are bifurcated along the range of Ω-values. Many of the low-energy transfers that require the least LOI ΔV arrive at the full Moon at ω-values near 75 deg and 255 deg. These transfers fly further out of the plane of the Moon's orbit than others; those transfers that remain closer to the Moon's orbital plane require more ΔV and target ω-values near 165 deg and 345 deg. These characteristics are also apparent in Fig. 4-20, which shows the LOI ΔV and transfer duration state space maps for lunar transfers to this arrival time.

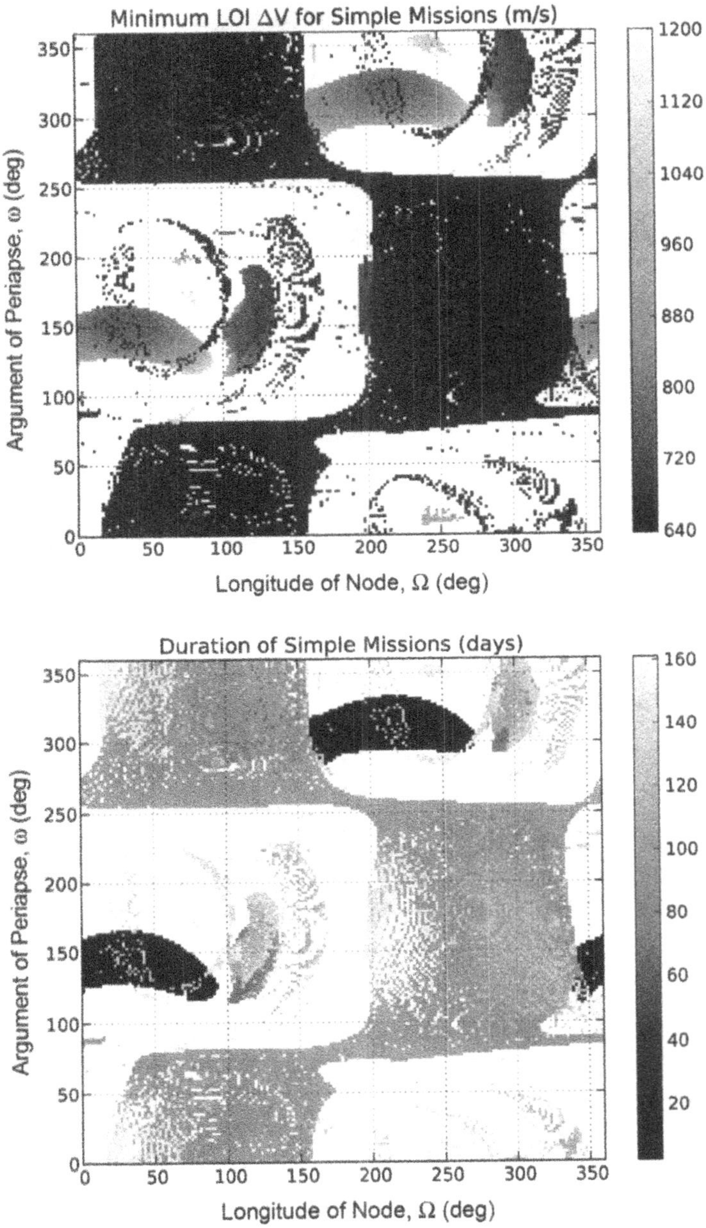

Figure 4-20 The combinations of Ω and ω that yield simple lunar transfers, that is, those without low Earth or lunar periapse passages. If multiple transfers exist for the same combination, then the one with the least LOI ΔV is shown. All of these transfers arrive at a full Moon [2] (Copyright © 2011 by American Astronautical Society Publications Office, all rights reserved, reprinted with permission of the AAS).

4.4.6 Monthly Trends

The work presented here describing transfers between the Earth and low lunar polar orbits has been extended, surveying transfers that arrive at the Moon at eight points in its orbit for several consecutive months. This section first presents results from surveys throughout one month and then considers similarities and variations that exist in lunar transfers across multiple months. The goal is to be able to predict the performance of lunar transfers for any given month.

Figure 4-21 shows eight state space maps, including those previously studied in Figs. 4-13, 4-17, and 4-20. These maps include simple transfers that arrive at the Moon at eight different points in a synodic month. Each map only tracks lunar transfers with no close lunar flybys or Earth phasing orbits, though each map does include both direct and low-energy transfers.

These maps are very useful to identify the combinations of Ω and ω that may be accessed via direct or low-energy transfers for a particular lunar arrival time. Similarly, the collection of these maps may be used to identify when to perform the lunar orbit insertion for a transfer to a particular combination of Ω and ω. One can see that low-energy transfers with LOI ΔV values below 700 m/s may be constructed that arrive at the Moon at any time. One also observes strong symmetry in the state space maps. First, each map shows a strong symmetrical mapping by shifting both Ω and ω by ± 180 deg. This shift corresponds to the difference between arriving at the Moon over the North Pole and arriving at the Moon over the South Pole. Second, a strong symmetry appears between two maps that correspond to arrivals ± 180 deg apart in the Moon's orbit: the maps show very similar characteristics when their arrival position and their Ω-values are both shifted by ± 180 deg. This shift corresponds to the symmetry that exists in the Sun–Earth three-body system: the dynamics are very similar, with a 180 deg rotation about the Earth, for the case where a spacecraft traverses from the Earth toward the Sun and for the case where a spacecraft traverses away from the Sun.

Figure 4-22 shows eight scatter plots, corresponding to the same arrival times presented in Fig. 4-21. The plots illustrate the relationships between each transfer's duration and its lunar orbit insertion ΔV. One can clearly see that direct transfers and low-energy transfers exist at every arrival time: direct transfers are shown on the far left of each plot, corresponding with short transfer durations and raised LOI ΔV requirements; low-energy transfers are similarly shown toward the bottom-right of each plot, corresponding with longer transfers and lower LOI ΔV requirements. Intermediate transfers exist for some arrival times, with transfer durations on the order of 60 days. One can see the same symmetry described above, between a given plot and the one that corresponds to a lunar arrival ± 180 deg apart. These plots are useful to quickly identify the limits of transfer duration and LOI ΔV for each type of transfer at any given lunar arrival time.

Most characteristics of ballistic two-burn lunar transfers repeat from one month to the next. The Moon's orbital plane is nearly coplanar to the Earth's, and the orbits of the bodies involved are nearly circular. However, since these conditions are not perfectly met, the characteristics of these lunar transfers do vary from one month

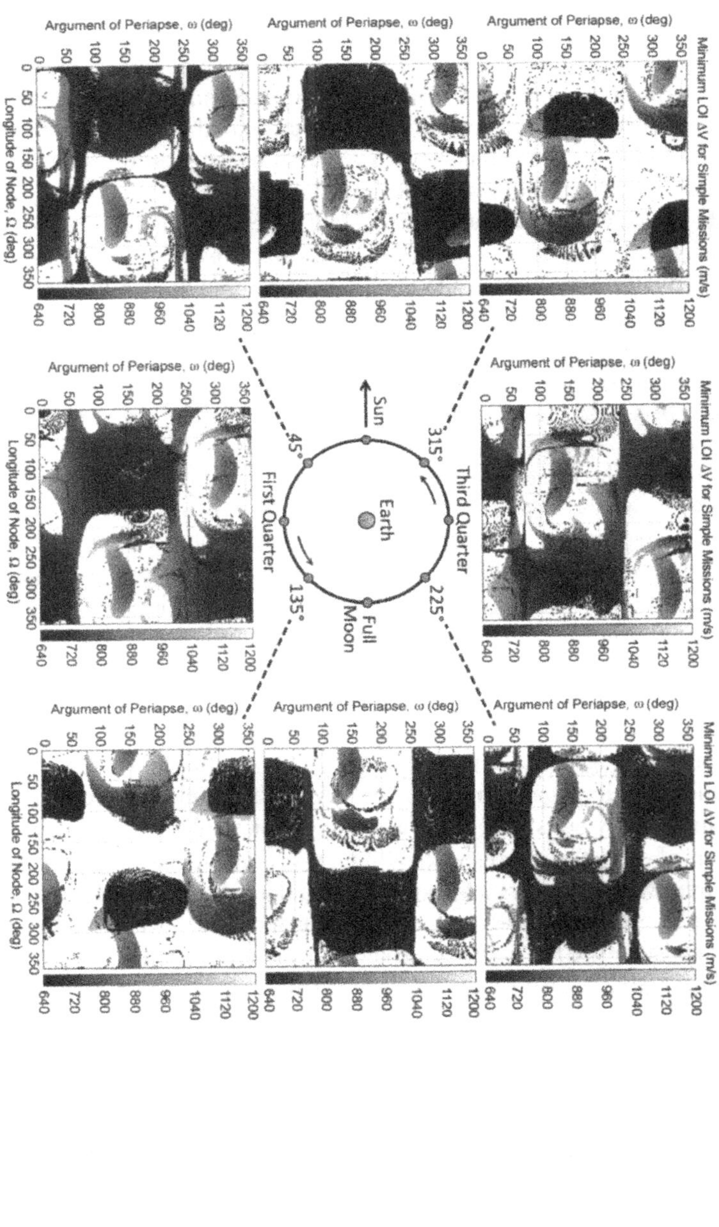

Figure 4-21 State space maps, illustrating the LOI ΔV required to transfer from the Earth to low polar lunar orbit at different arrival times in a month [2] (Copyright © 2011 by American Astronautical Society Publications Office, all rights reserved, reprinted with permission of the AAS).

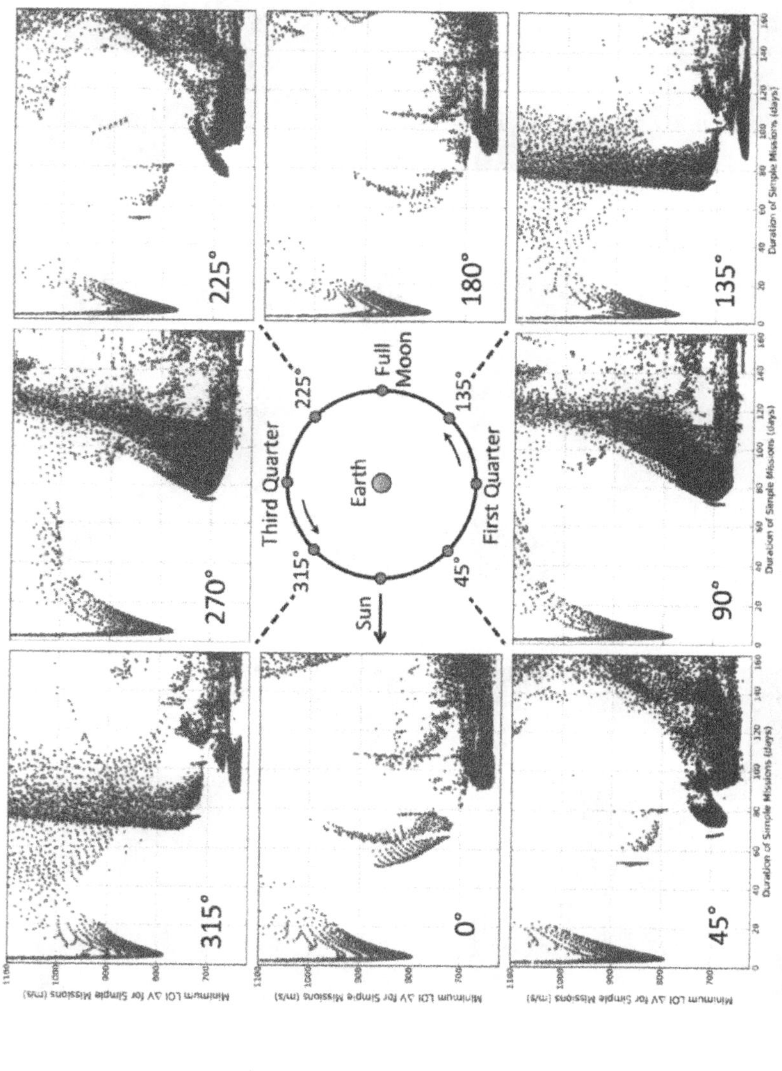

Figure 4-22 Scatter plots showing the relationship between transfer duration and a transfer's LOI ΔV cost for simple lunar transfers arriving at the Moon at different points about its orbit.

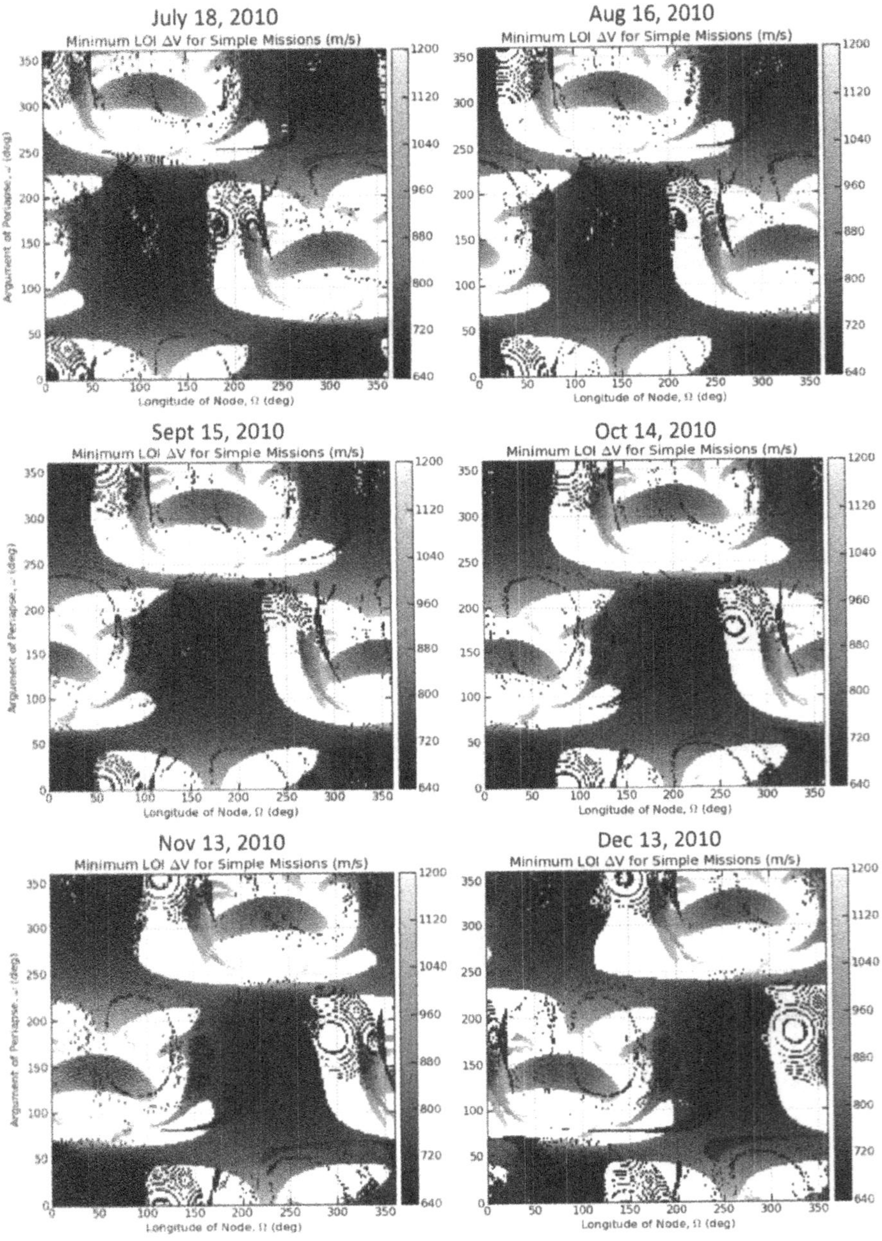

Figure 4-23 State space maps that illustrate the LOI ΔV for transfers to each combination of Ω and ω that arrive at the Moon at its first quarter in each of six consecutive months [2] (Copyright © 2011 by American Astronautical Society Publications Office, all rights reserved, reprinted with permission of the AAS).

to the next. The Moon's orbital plane and equatorial plane are tilted approximately 5.1 deg and 1.5 deg, respectively, relative to the ecliptic. One may therefore assume that the characteristics observed in the state space maps presented here vary by several degrees in their ω-values in a given month. The inclination of the trans-lunar departure state for a given type of lunar transfer may vary by many degrees from one month to the next, particularly on account of the obliquity of the Earth's spin axis.

It has been found that most types of *simple* lunar transfers appear in any given month and their characteristics remain relatively constant relative to the ecliptic. Figure 4-23 illustrates how little the state space maps vary from one month to the next, when evaluating simple lunar transfers. The six state space maps shown capture the LOI ΔV for transfers to each combination of Ω and ω that arrive at the Moon at its first quarter in each of six consecutive months. The only major apparent variation is that the features in each map shift approximately 30 deg in Ω from one month to the next. This is because Ω is defined inertially and the Earth moves approximately 30 deg in its orbit from one month to the next, rotating the Sun–Earth geometry. The more complex lunar transfers, such as those with multiple lunar flybys, vary much more on a monthly basis and may not even appear at all in a given month.

4.4.7 Practical Considerations

The surveys presented here study trajectories that are entirely ballistic—they do not contain any correction maneuvers or targeting maneuvers of any sort. When propagated backward in time from the Moon, if a trajectory arrives at the Earth without impacting the Moon, then it is considered a viable Earth–Moon transfer. However, the trajectory may have arrived at the Earth with an inclination that is unsuitable for a mission that launches from a particular launch site. Ideally, a mission would start in a low-Earth parking orbit with an inclination very close to that of the latitude of the launch site, for example, near 28.5 deg for missions that launch from Cape Canaveral, Florida. It is undesirable to perform a large plane change during launch and trans-lunar injection. Section 6.5 shows that one can add 1–3 small trajectory correction maneuvers to depart the Earth from a particular LEO parking orbit and transfer onto a desirable *low-energy* transfer to the Moon; and doing so requires only about 1 m/s per degree of inclination change. This works for low-energy transfers particularly well since low-energy transfers travel far from the Earth and spend many weeks doing so. This method does not work well for direct lunar transfers, which require far more ΔV to change planes.

Mid-course maneuvers may also be implemented to establish a launch period for a low-energy transfer to the Moon, extending or shrinking its transfer duration. Missions that implement direct lunar transfers may establish a launch period using Earth phasing orbits, making those sorts of transfers more desirable in the surveys presented here.

4.4.8 Conclusions for Low-Energy Transfers Between Earth and Low Lunar Orbit

The surveys presented in this section characterize two-burn lunar transfers that arrive at the Moon, targeting 100-km polar orbits with any orientation. Transfers are studied that arrive at an example first-quarter Moon, an example full Moon, and an example third-quarter Moon. Additional results are also presented for transfers that arrive at eight different times during a month and for several consecutive months. Many types of transfers are observed, including low-energy transfers, short-duration direct transfers, and variations that involve any number of lunar flybys and Earth phasing orbits, provided that they do not involve any deterministic maneuvers. The only two burns considered are the trans-lunar injection maneuver and orbit insertion maneuver.

It has been found that lunar transfers consistently require trans-lunar injection C_3 values on the order of -2.0 km^2/s^2 for direct transfers and -0.6 km^2/s^2 for low-energy transfers. Simple transfers typically require 2–12 days for direct transfers and 70–120 days for low-energy transfers, though both types can require more time. The low-energy transfers that require the least LOI ΔV require 640 m/s, or more depending on the target orbit and the arrival time; direct lunar transfers require at least 120 m/s more ΔV than low-energy transfers to the same arrival conditions. Further, low-energy transfers can reach many arrival conditions that direct transfers cannot reach without additional maneuvers. Practical simple direct transfers only exist that target a lunar orbit that is within 60 deg of being orthogonal to the Earth–Moon line, though the ΔV cost rises significantly when the orbit is beyond 30 deg of orthogonal. Low-energy transfers can target polar orbits with any argument of periapse, ω, or with any longitude of ascending node, Ω; targeting one such parameter restricts the other for a particular arrival date as illustrated in the state space maps presented here.

4.5 TRANSFERS BETWEEN LUNAR LIBRATION ORBITS AND LOW LUNAR ORBITS

Many mission designs may benefit by transferring a spacecraft from the Earth to a lunar libration orbit prior to descending to a low lunar orbit. For instance, Hill et al. [11], designed a mission where two satellites transferred to a halo orbit about the lunar L_2 point. One satellite remained there as a navigation and communication relay and the other satellite transferred to a low lunar orbit. Information about such transfers is summarized in Section 3.5.1 on page 224.

4.6 TRANSFERS BETWEEN LOW LUNAR ORBITS AND THE LUNAR SURFACE

Many historical missions have performed maneuvers to transfer a spacecraft from a low lunar orbit to the lunar surface, for example, the Apollo missions [1]. A few spacecraft, including Apollo missions, have then risen from the lunar surface and returned to lunar orbit. These maneuvers are very straightforward and may even be well approximated by conic sections; nevertheless, it is useful to briefly describe their designs here.

Let's assume we have a spacecraft in a 100-km circular lunar orbit. That spacecraft is traveling approximately 1633.5 m/s in its orbit and revolves about the Moon once every 117.8 minutes. The minimum ΔV required to place the spacecraft on a collision course with the Moon would reduce the spacecraft's orbital periapse to an altitude of 0 km, at which point it would just graze the surface, that is, a Hohmann transfer. This transfer requires a ΔV of approximately 23 m/s, sending the spacecraft on a 180-deg transfer in about 56.5 minutes. The spacecraft's grazing velocity upon arriving at its orbital periapse is approximately 1703.2 m/s. If the spacecraft performs a larger braking burn from its 100 km orbit, then its transfer orbit will strike the surface of the Moon at a steeper flight path angle in less time.

Figure 4-24 illustrates the flight path angles that may be achieved at the mean radius of the Moon as a function of the de-orbit burn ΔV for trajectories starting from an altitude of 100 km. One can see that a ΔV of 23 m/s is indeed required to obtain a flight path angle of 0 deg, which is the limit of trajectories that have a passive abort option, not including local geometry variations. Of course, by performing a braking burn ΔV of 1633.5 m/s, the spacecraft completely removes its orbital velocity and falls straight down to the surface, achieving a vertical impact.

Figure 4-25 illustrates the velocities that the spacecraft will have at the impact point, assuming the impact point occurs at a radius of 1737.4 km, for example, the mean radius of the Moon. Figure 4-26 shows the duration of time required to reach the impact point.

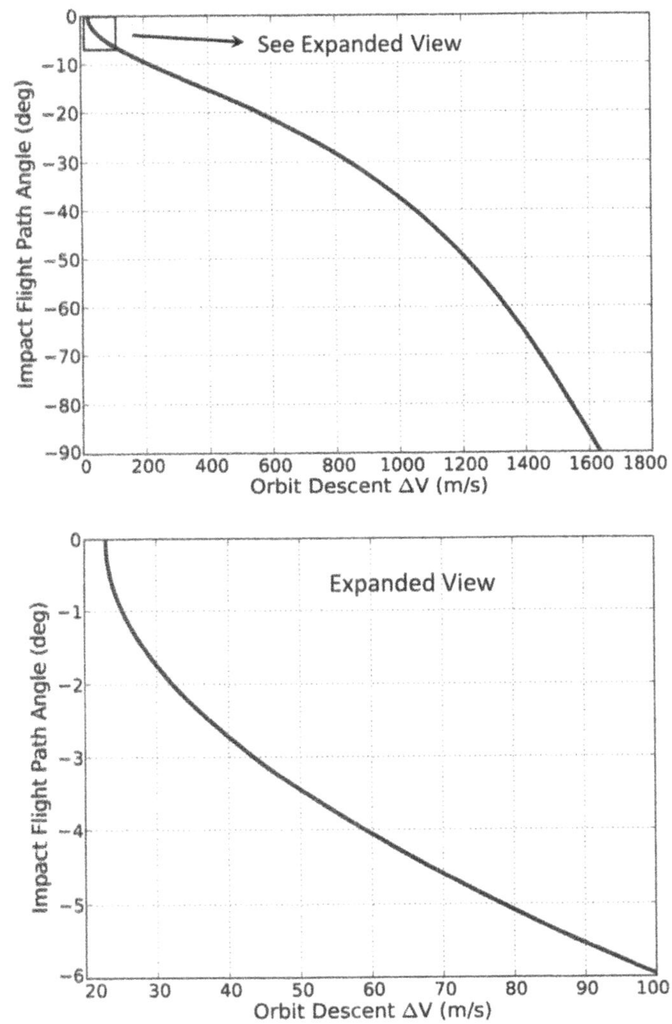

Figure 4-24 The flight path angles that may be achieved at the mean surface of the Moon as a function of the de-orbit burn ΔV for trajectories starting from a circular orbit at an altitude of 100 km.

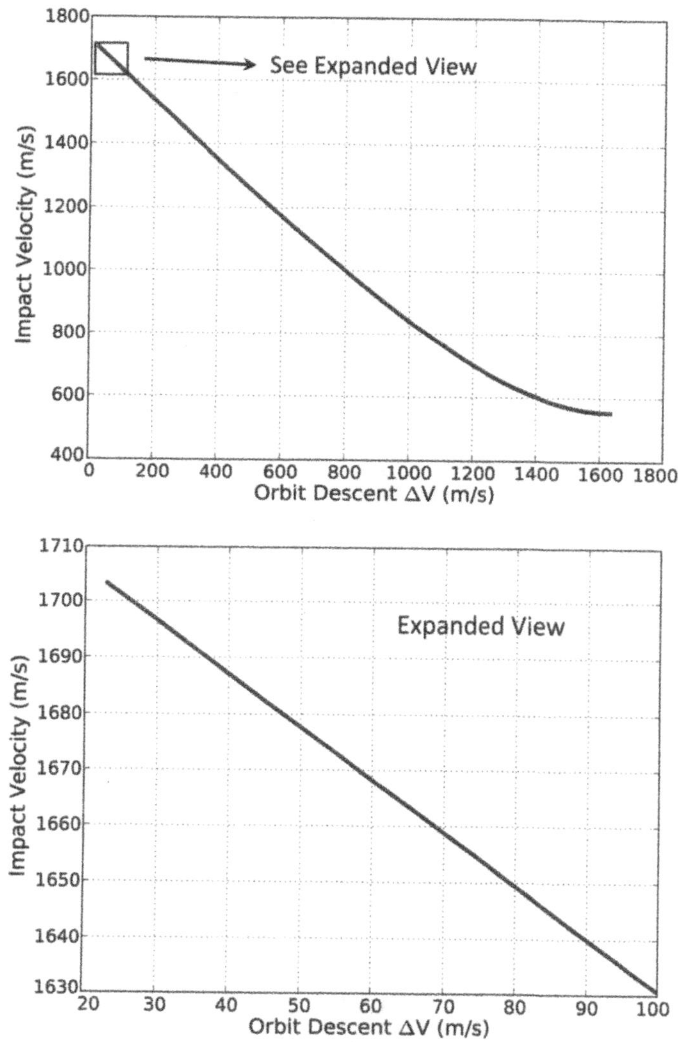

Figure 4-25 The impact velocity values that may be achieved at the mean surface of the Moon as a function of the de-orbit burn ΔV for trajectories starting from a circular orbit at an altitude of 100 km.

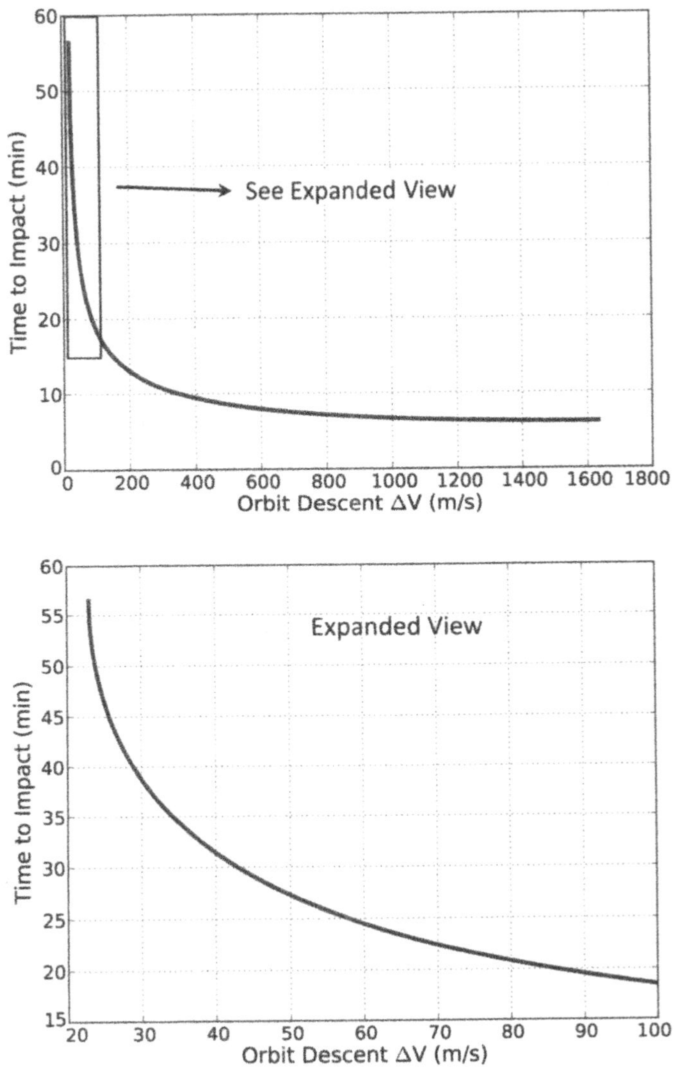

Figure 4-26 The duration of time required to reach the mean surface of the Moon as a function of the de-orbit burn ΔV for trajectories starting from a circular orbit at an altitude of 100 km.

CHAPTER 5

TRANSFERS TO THE LUNAR SURFACE

5.1 EXECUTIVE SUMMARY

In this chapter techniques are developed that allow an analysis of a range of different types of transfer trajectories from the Earth to the lunar surface. Trajectories ranging from those obtained using the invariant manifolds of unstable orbits to those derived from collision orbits are analyzed. These techniques allow the computation of trajectories encompassing low-energy trajectories as well as more direct transfers. A conceptual illustration of the types of trajectories discussed in this chapter is given in Fig. 5-1. The range of possible trajectory options is summarized, and a broad range of trajectories that exist as a result of the Sun's influence are computed and analyzed. The results are classified by type, and trades between different measures of cost are discussed. The information in this chapter is largely derived from papers presented by Anderson and Parker [192–195], and the results as presented here are oriented as a guide for mission design.

The problem of designing transfers to the lunar surface is approached here as follows. First, an analysis is given showing the types of trajectories that exist as a result of the Sun's influence in both the planar and spatial problems. A significant set of trajectories at high Jacobi constants, or low energies, is found to exist when the Sun's

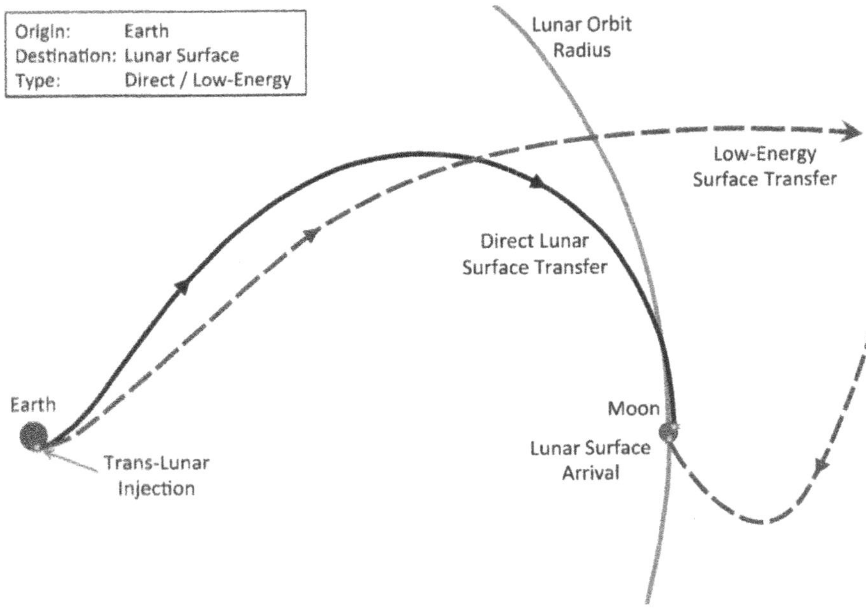

Figure 5-1 The profiles for both a direct and a low-energy transfer from the Earth to the lunar surface.

influence is taken into account. This result indicates that for trajectory design in this energy regime, trajectories traveling to the Sun–Earth Lagrange points and following the invariant manifolds of orbits around these points deserve careful consideration. Monthly variations are examined, and it is determined that the monthly variations capture the majority of the variations seen in the studied transfer trajectories. The greatest variation over a month occurs between the cases traveling to the Moon when it is at its apoapse and periapse. These trajectories are described for the spatial case initially with trajectories normal to the surface, which illustrates in a succinct manner the types of options available. A more detailed analysis of trajectories arriving at various angles to the surface is also presented, and these trajectory options are summarized using several different plots of various parameters. These trajectories may serve as initial guesses for future mission design, and they provide a general overview of the range of trajectory options. Invariant manifold trajectories traveling to the lunar surface are also described, and some sample trajectory options traveling from libration orbits to the lunar surface are given.

The numerical results presented in this chapter are given primarily relative to the Jacobi constant (C) of trajectories encountering the lunar surface. The velocity of the trajectories varies little over the surface of the Moon for each Jacobi constant, and in each case an approximate value of the velocity corresponding to each Jacobi constant may be obtained by referring to Fig. 5-2. From this plot, a Jacobi constant of 2.5 gives

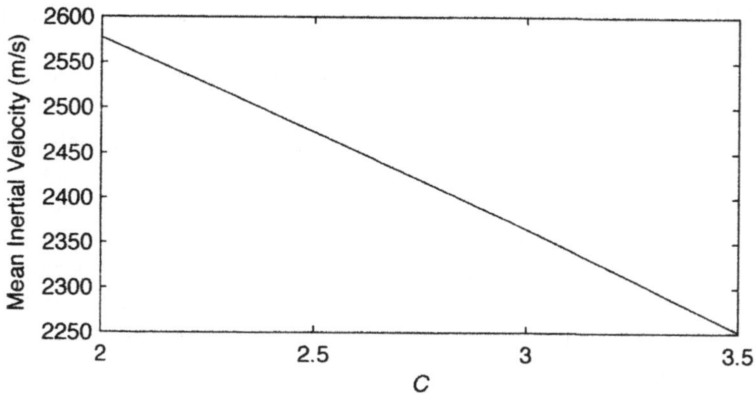

Figure 5-2 Mean inertial velocities relative to the Moon at each Jacobi constant for the cases with velocities normal to the surface [192] (Copyright © 2011 by American Astronautical Society Publications Office, San Diego, California (Web Site: http://www.univelt.com), all rights reserved; reprinted with permission of the AAS).

a mean velocity at the Moon of approximately 2473.3 meters per second (m/s), while a Jacobi constant of 3.0 gives a mean velocity of 2365.0 m/s. In the planar analysis it is shown that trajectories from the Earth to the Moon exist without the Sun's influence up to a Jacobi constant of 2.78, while they exist up to a Jacobi constant of 3.16 when the Sun's influence is included in the computations. As expected, the time of flight of the trajectories generally increased and the launch injection energy parameter (C_3) generally decreased as the Jacobi constant increased. For the spatial case with trajectories encountering the Moon perpendicular to the surface, the minimum time of flight varied from approximately 3.4 days at C = 2.2 to around 101.0 days and 78.7 days at C = 3.0 and 3.1, respectively. The range of possible elevation angles that generate missions to the surface is very dependent on the Jacobi constant and the target location on the surface. For the selected grid size, the maximum elevation angle range for points on the surface changed from 0 deg to 90 deg at C = 2.6 to between 57 deg and 90 deg at C = 3.1. Likewise the minimum elevation angle range decreased from 0 deg to 72 deg at C = 2.6 to between 0 deg and 15 deg at C = 3.1. These numbers are given to present a rough idea of the kinds of results that are discussed in this chapter. The details of these cases along with a wide variety of launch and approach parameters are contained in the plots presented throughout the chapter.

5.2 INTRODUCTION FOR TRANSFERS TO THE LUNAR SURFACE

A wide variety of Earth–Moon trajectories have been employed for past missions, ranging from the more direct transfers used for the Apollo missions [196] to more recent missions such as *ARTEMIS* [4] that make use of the multi-body dynamics of the Earth–Moon and Sun–Earth–Moon systems. The design of trajectories in multi-body systems is a particularly rich problem because the two-body model is often insufficient to compute accurate trajectories, and the gravity of the Sun, Earth, and Moon combine to form a highly nonlinear dynamical environment. These facts limit the applicability of traditional patched-conic techniques commonly used for interplanetary missions, and the three-dimensional aspects of the problem further complicate real-world missions. Mission designers must take into account the orientation of each body in addition to the relative orientations of the orbits of the Earth and the Moon over time. Parker et al. [44, 47, 183] have studied trajectories that include many of these complicating factors for insertion into a variety of orbit types near the Moon. This chapter focuses on an analysis including these types of effects with a focus on trajectories traveling to the lunar surface.

Lunar landing trajectories often have a different set of constraints from those of orbiters, and the nature of this problem makes it possible to approach it with a different set of techniques. Indeed, a theoretical basis for analyzing lunar landing trajectories may be found in the computation of collision orbits. Collision orbits have been studied extensively in the mathematical community by Easton [197] and McGehee [198]. Anderson and Lo [199], Villac and Scheeres [200], and Von Kirchbach et al. [201] have previously analyzed collision orbit trajectories for the Jupiter–Europa system and categorized the different regions and trajectories that exist for orbits that terminate or originate at Europa's surface. While the theoretical basis for collision orbits is focused on trajectories that intersect the surface of the selected body normal to the surface, this type of analysis can be extended to trajectories coming in at the various flight path angles and declinations of interest to mission designers. A study of these trajectories is almost directly applicable to impactor missions such as the *Lunar Crater Observation and Sensing Satellite* (*LCROSS*) [202]. This mission used an 86 deg impact angle relative to the lunar surface for the impact trajectory. The techniques developed here are also easily applied to systems including the full ephemeris and multiple bodies. Much of the work to design low-energy trajectories from the Earth to the Moon has focused on the use of libration point orbits along with their stable and unstable manifolds [39, 45, 51, 203]. These techniques have proven to be quite successful, and they are increasingly used for the design of Earth–Moon trajectories. The invariant manifolds of libration orbit trajectories are also studied here with an emphasis on their applicability to landing trajectories.

A wide range of trajectory types for lunar landing trajectories were computed for the results given here, and presenting a complete picture of the possible trajectory categories while remaining easily accessible was a goal of this research. In keeping with this goal, the problem is approached and presented using several different levels of analysis with increasing complexity. Presenting all different combinations of velocities encountering the surface of the Moon with all different magnitudes and orientations makes it difficult to see the relevant structures in the solution space, so two divisions were made in the approach to the analysis. The problem is first approached by analyzing planar cases covering selected velocities or energies with the trajectories encountering the Moon at various angles relative to the surface. The characteristics of these trajectories are observed to lay the groundwork for understanding the spatial trajectory cases. The spatial case is then attacked using the results from the planar analysis to understand the dynamics in this more complex model. These trajectories are categorized with the goal of providing a broad survey of the trajectory types that may be available for transfers to the lunar surface from the Earth. The specific trades between launch costs and time of flight (TOF) are quantified and summarized in addition to the topological characteristics of the trajectories. Other parameters relevant to mission design such as the launch orientation are computed. The regions of the Moon attainable using different types of trajectories are also characterized. These results are summarized with the goal of providing a tool for mission designers to quickly understand the trades between various measures of cost and time when a particular mission is being designed to land on the Moon.

The results in this chapter are also presented using two key concepts. The first is to view the problem in terms of the limiting bounds that a mission designer could use to refine the search space. This practically takes the form of computing parameters such as the velocities for which trajectories exist that travel from the Earth to the Moon or the launch energies required to reach such trajectories. The second, which is an overall theme of this work and one of the primary results, is related to computationally examining in a more comprehensive sense the trajectories available when the Sun's perturbations are taken into account. To achieve this objective, comparisons involving several different models are made. Many traditional trajectories were computed using the Earth–Moon model or the circular restricted three-body problem (CRTBP), and in general, similar types of trajectories exist in the full ephemeris model. If simpler models are used, however, some solutions in the full problem may be ignored. Some particular solutions employing the effects of the Sun for transfers in the Earth–Moon system have been examined more recently. The 1991 Japanese mission *MUSES-A* (*Hiten*) used the effects of the Earth, Moon, and Sun for its transfer to the Moon [30]. Koon et al. provided techniques for systematically reproducing missions similar to Hiten using the invariant manifolds of libration orbits [38]. In each of these techniques the Sun's effects were included in the mission design. Parker and Lo examined trajectories within the Sun–Earth–Moon spatial problem and looked at multiple trajectories for transfer to lunar halo orbits [39, 46]. The work here seeks to broaden the search space for landing trajectories traveling to the Moon and characterize the effects of the fourth-body perturbations of the Sun on the potential trajectories that may be used. A direct approach isolating the effects of the Sun is taken here by comparing trajectories in the CRTBP, the Earth–Moon system, and the Sun–Earth–Moon system.

5.3 METHODOLOGY

Two primary models are used for the analyses contained in this chapter. The first model, the CRTBP, closely approximates real world systems, and a significant set of tools exists within this model to bring to bear on the problem. The qualitative insights gained in this model are very helpful in providing an overview of the categories of trajectories that are available. The trajectories developed within the CRTBP are also generally transferable to the full ephemeris although trajectories developed with the effects of other bodies may not be transferable to the CRTBP. Refer back to Section 2.5.1 for a more complete description of the CRTBP. The ephemeris model, implemented using point masses, is used to capture additional types of trajectories that are not found using the CRTBP model. Although the variations in the orbits of the Earth and Moon are important, this model is primarily used to search for members of the broad category of trajectories utilizing the Sun's perturbation for Earth–Moon transfers. See Section 2.5.3 for more details on the use of the ephemeris.

5.4 ANALYSIS OF PLANAR TRANSFERS BETWEEN THE EARTH AND THE LUNAR SURFACE

The procedure described next involves varying the location of the landing site on the Moon, the orientation of the incoming trajectory, and the energy/velocity of the trajectory. Each trajectory must also be characterized or evaluated using some figure of merit. While this can provide a relatively complete picture of the potential trajectory options, it is helpful to first gain insight into the dynamics by limiting the scope of the problem to allow the results to be easily visualized.

Several different techniques have been used to achieve this goal in the Jupiter–Europa system, and it is useful to consider their application here. One technique used by Anderson and Lo [199] varied the Jacobi constant for trajectories intersecting Europa on a sphere for several different trajectory orientations and characterized the origin of the trajectories. Von Kirchbach et al. [201] examined the planar case for the Jupiter–Europa system for additional velocity orientations leaving the surface. Both of these techniques are applied here to the Earth–Moon system, and it is interesting to start with the planar problem in order to gain some initial insight. First, the planar results are computed in the Earth–Moon CRTBP system to allow for a comparison with the results from Von Kirchbach et al. [201] in the Jupiter–Europa system. This technique is then used to extend the analysis to the ephemeris case with the Earth and Moon and then to the case where the Sun is included. The effect of adding the Sun is examined in detail over a range of Jacobi constants.

For this planar analysis, a set of trajectories was integrated backward in time from the surface of the Moon. Specifying the Jacobi constant gives the velocity magnitude for each trajectory, while the location of the trajectory and the orientation of the velocity are specified using α and θ as shown in Fig. 5-3. Multiple simulations have been performed using these techniques, and the results for several selected Jacobi constant values are given in Fig. 5-4. The resulting points are colored according to the original location of each trajectory. Note that if a trajectory integrated backward in time were to intersect the Moon and then encounter the Earth at an earlier time, the trajectory would be gray. The (α, θ) point corresponding to the intermediate intersection of the Moon would then be blue. The fact that points with only slightly different initial conditions in the plot can travel to either the Earth or the Moon confirms the known existence of chaos in this problem. Comparison with the results from the Jupiter–Europa system in Von Kirchbach et al. [201] reveals that the divid-

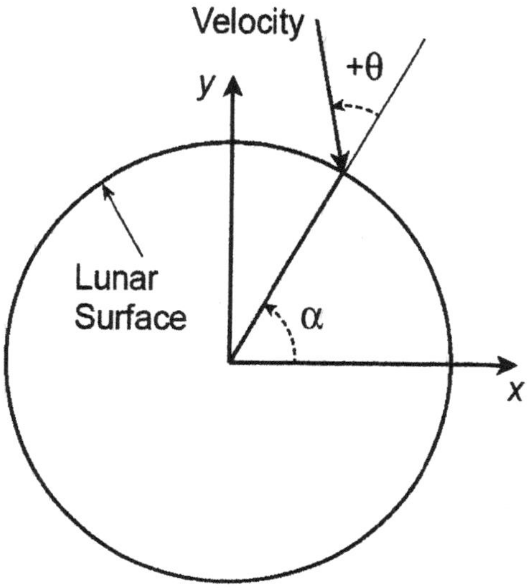

Figure 5-3 Diagram showing location and orientation of the velocity vector as it intersects the lunar surface. The xy axes shown here are centered on the Moon in the same orientation as the axes in the rotating frame [192] (Copyright © 2011 by American Astronautical Society Publications Office, San Diego, California (Web Site: http://www.univelt.com), all rights reserved; reprinted with permission of the AAS).

ing lines between different regimes of motion are less distinct at equivalent Jacobi constants for the Earth–Moon system. This existence of chaos indicates that it may be possible to design trajectories that cover a relatively wide range of the surface by carefully selecting landing sites. It is useful to note here again that Moon–Earth transfers may be derived from Earth–Moon transfers, and the plot corresponding to these trajectories may be obtained from Fig. 5-4 for the planar problem using $x \to x$, $y \to -y$, $\dot{x} \to -\dot{x}$, and $\dot{y} \to \dot{y}$. The transformation in position gives $\alpha \to 2\pi - \alpha$ and then from examination of the transformed velocity vector, $\theta \to -\theta$.

As can also be seen from the results, a significant percentage of the trajectories do not encounter either the primary or the secondary over the given time span. However, it is useful to note that for low Jacobi constants (higher energies), a significant percentage of the trajectories do originate at the Earth. Determining the Jacobi constant where Earth–Moon transfer trajectories no longer exist in the planar problem can help provide a rough limit on energies or velocities for these trajectories and provide a method for determining the potential benefits of perturbations from other bodies in trajectory design. To determine the approximate Jacobi constant above which Earth–Moon trajectories computed in this simulation no longer exist, a series of runs were made in parallel to step through the Jacobi constant. The grid resolution used for this step was one degree in both α and θ. The percent of the total number of trajectories that encountered the Earth for each Jacobi constant was computed and

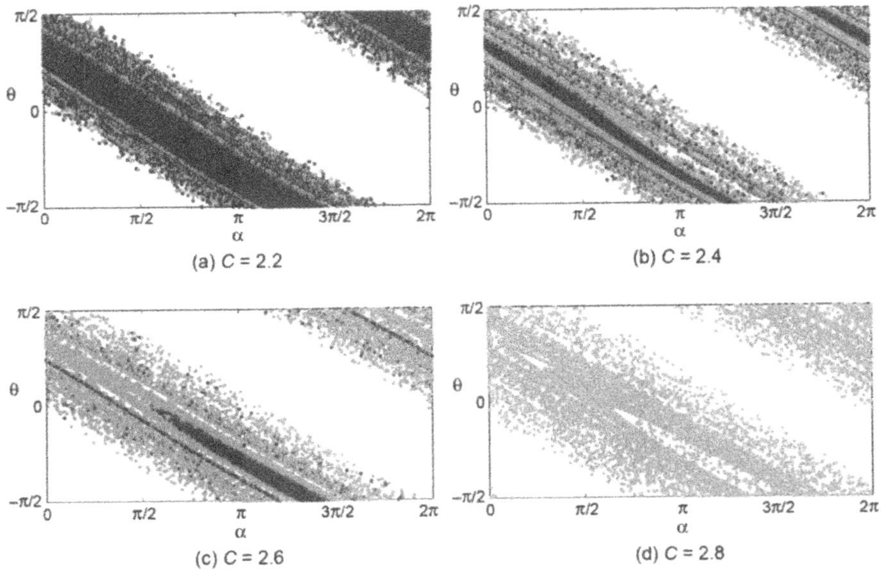

Figure 5-4 Plots showing the origin of each trajectory as a function of the position and orientation of the velocity vectors as the trajectories encounter the Moon's surface. Blue points indicate that the trajectory originated at the Earth and gray that it originated at the Moon. If no point is plotted the integrated trajectory did not encounter the surface of either body over the given time span of 200 days (Earth–Moon CRTBP) [192] (Copyright © 2011 by American Astronautical Society Publications Office, San Diego, California (Web Site: http://www.univelt.com), all rights reserved; reprinted with permission of the AAS). *(See insert for color representation of this figure.)*

plotted in Fig. 5-5. As expected from the previous plots, the number of trajectories originating at the Earth generally decreases with increasing Jacobi constant, but it is interesting that the slope of the curve varies significantly over the plotted range. It is also interesting that although the curve approaches zero percent near a Jacobi constant of 2.7, for Jacobi constants as high as 2.78, the percent of trajectories originating at the Earth remains at approximately 0.03 percent or approximately 19 out of 64800 trajectories. So even for this relatively low energy, some trajectories manage to travel from the Earth to the Moon.

For mission design, it is helpful to be aware of the velocities of the trajectories as they intersect the surface. They will actually vary somewhat as the constant for the computations so far has been the Jacobi constant rather than velocity. In general the inertial velocities relative to the Moon only vary at the m/s level. Figure 5-2 shows the average velocities for the case with velocities normal to the surface as a reference for each Jacobi constant.

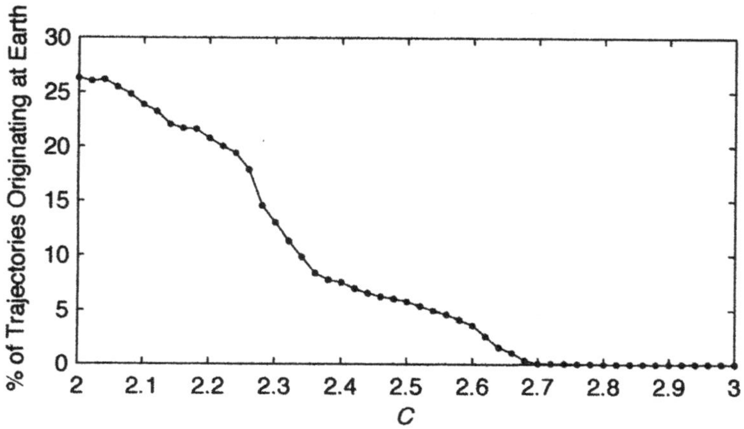

Figure 5-5 Percent of trajectories originating at the Earth for each Jacobi constant (CRTBP) [192] (Copyright © 2011 by American Astronautical Society Publications Office, San Diego, California (Web Site: http://www.univelt.com), all rights reserved; reprinted with permission of the AAS).

Although the CRTBP is known to provide an accurate approximation to real-world trajectories, an obvious question for mission designers is related to how much the inclusion of real-world effects would affect selected trajectories. This question can be addressed by using planetary ephemerides and replicating the analysis for the CRTBP in this model. This analysis was first performed in the ephemeris model initially including only the gravity of the Earth and Moon. The initial velocities were computed for a given Jacobi constant in the CRTBP in the rotating frame, and then the states were initialized in the integrator relative to the Moon in an instantaneous rotating frame aligned with the Earth–Moon frame on an epoch of January 1, 2015. As the distance between the Earth and the Moon varies over the course of the orbit, it is difficult to obtain a direct comparison to the results from the CRTBP, but this method was selected because it was found to provide a good approximate comparison using the important mission design parameter of velocity at the lunar surface. Although the Jacobi constant will vary along the trajectory in this model, the final impact velocities at the Moon will be the same in each system. So the Jacobi constant labels in the ephemeris model plots in this study serve to indicate the velocities that were used at the lunar surface as they were computed in the CRTBP. The initial conditions were originally planar for this case, but the trajectory was free to vary in three dimensions for this problem. Using this method for a system including the Earth and Moon ephemerides, the trajectories were integrated, and the results are plotted in Fig. 5-6. Comparing the results for this system with the results in Fig. 5-4 reveals few obvious differences. The Earth impacting cases for $C = 2.6$ have some slight differences, but in general the trajectories match the expectation that the CRTBP is a good approximation to the three-body problem including the ephemerides. If the

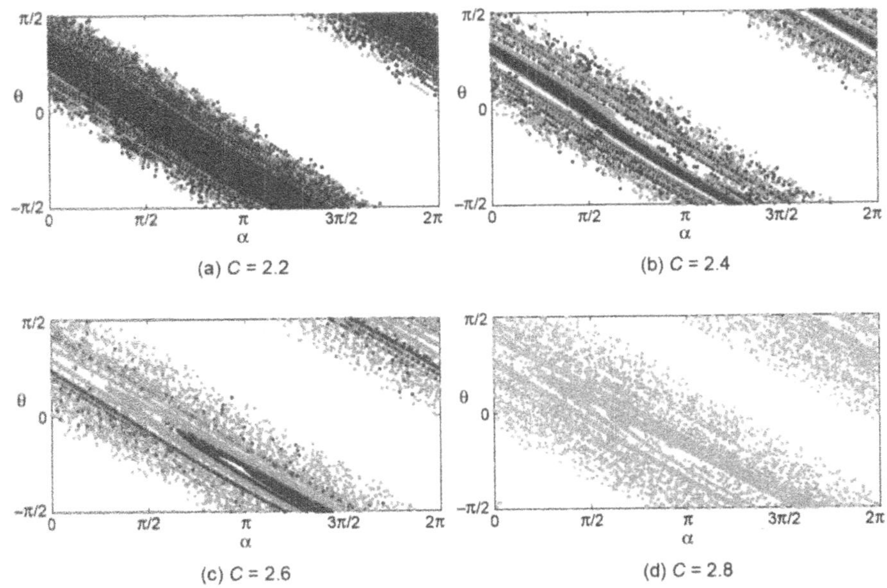

Figure 5-6 Plots showing the origin of each trajectory as a function of the position and orientation of the velocity vectors as the trajectories encounter the Moon. Blue points indicate that the trajectory originated at the Earth and gray that it originated at the Moon. If no point is plotted the integrated trajectory did not encounter either body over the given time span of 200 days (Earth–Moon only Ephemeris system) [192] (Copyright © 2011 by American Astronautical Society Publications Office, San Diego, California (Web Site: http://www.univelt.com), all rights reserved; reprinted with permission of the AAS). *(See insert for color representation of this figure.)*

percent of trajectories originating at the Earth are compared, some differences in the shape of the curve can be found, but the overall trends are very similar. In this case, the percent of trajectories in Fig. 5-7 originating at the Earth decreases down to 0.006 percent at $C = 2.76$, approximately the same Jacobi constant cutoff as the CRTBP.

Next, the same procedure was performed including the Sun in the integration, and the results are plotted in Fig. 5-8. Now, comparison with the results in both the CRTBP and the Earth–Moon systems reveal some obvious differences. Several new bands of trajectories originating at the Earth spring into existence. The overall structure remains generally similar, but the points appear chaotic. A new band of solutions remains for $C = 2.8$ and a significant number of Earth origin trajectories still exist at $C = 3.0$. Remember that the final velocities at the Moon are the same as the other models, but the Jacobi constant will vary as a result of the Sun's influence. In this sense, the Sun may be thought of as changing the trajectory's energy or Jacobi constant to provide the transfer. If the percent of trajectories originating at the Earth

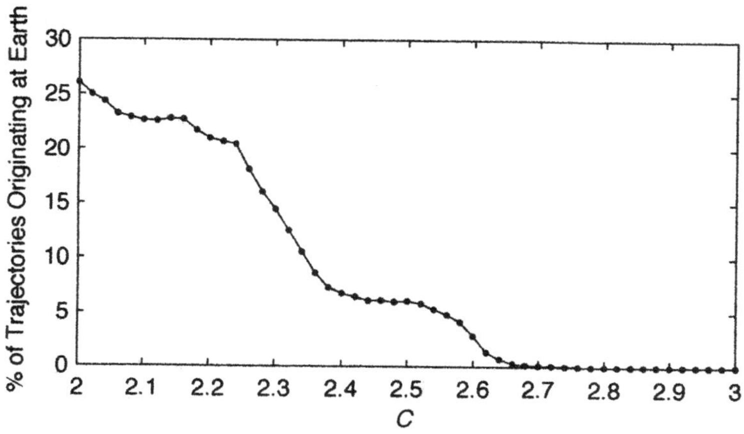

Figure 5-7 Percent of trajectories originating at the Earth for each Jacobi constant (Earth–Moon only Ephemeris system) [192] (Copyright © 2011 by American Astronautical Society Publications Office, San Diego, California (Web Site: http://www.univelt.com), all rights reserved; reprinted with permission of the AAS).

is examined in Fig. 5-9(a), it can be seen that at $C = 3.0$, 0.3 percent of the trajectories still originate at the Earth. Indeed, as high as $C = 3.16$, 0.15 percent of the trajectories still originate at Earth.

One immediate question that arises is whether the selected epoch for lunar arrival would significantly affect these results, so three additional cases seven days apart were computed and plotted in Fig. 5-9(b). Some variation is observed as the Moon travels through its orbit with one case starting with a lower percent of trajectories for low Jacobi constants and two of them possessing peaks just before $C = 3.0$. However, all of them have approximately the same upper Jacobi constant cutoff of approximately $C = 3.16$ where the percent of trajectories drops to near zero. The existence of the additional bands of trajectories and the increase in the Jacobi constant where trajectories connecting the Earth and Moon exist in this system raises the question as to where these trajectories come from. These trajectories were plotted in both the Earth–Moon system and the Sun–Earth–Moon system to examine the differences, and a sample of one of these trajectories plotted in both rotating frames is given in Fig. 5-10. As can be seen from the plots, the trajectory ends up in very different places depending on whether the Sun is included in the integration or not. In the Earth–Moon rotating frame with the Sun included, the trajectory travels far away from the system with no close periapses until it approaches the Earth, while the case without the Sun has two relatively close periapses at approximately the lunar distance and ends up far from the Earth. The most telling plots, however, are in the Sun–Earth rotating frame. Here, the characteristic shape of a trajectory using the libration point dynamics of the Sun–Earth system is apparent when the Sun is included. The trajectory travels out toward the L_1 point, lingers there, and then finally falls back

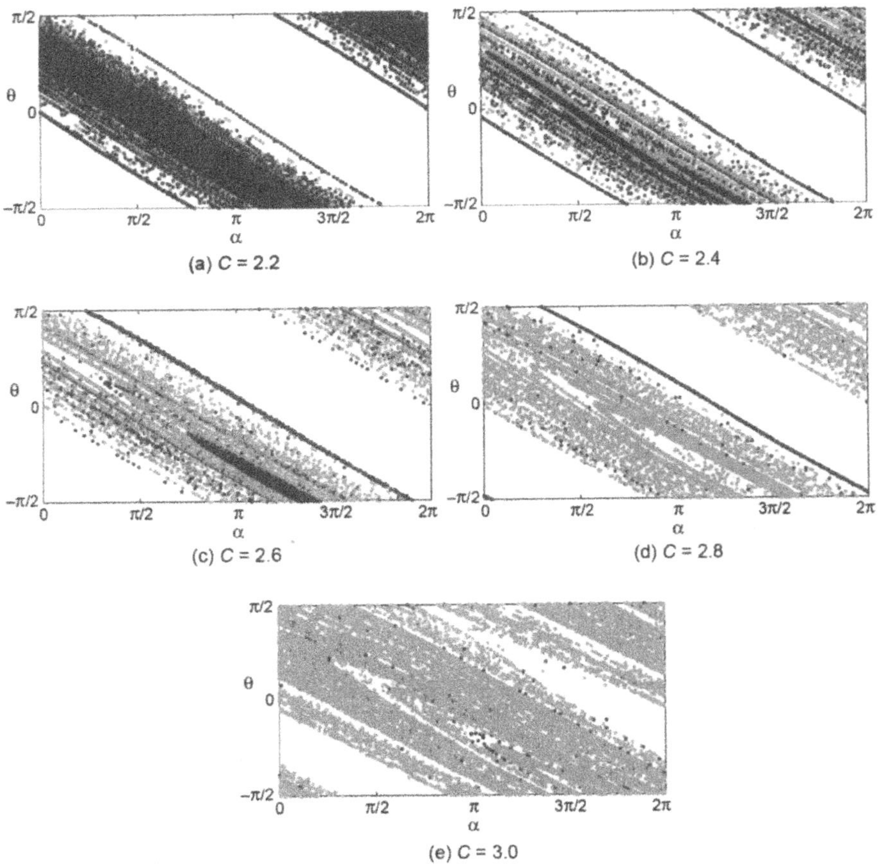

Figure 5-8 Plots showing the origin of each trajectory as a function of the position and orientation of the velocity vectors as the trajectories encounter the Moon. Blue points indicate that the trajectory originated at the Earth and gray that it originated at the Moon. If no point is plotted the integrated trajectory did not encounter either body over the given time span of 200 days (Sun–Earth–Moon Ephemeris system) [192] (Copyright © 2011 by American Astronautical Society Publications Office, San Diego, California (Web Site: http://www.univelt.com), all rights reserved; reprinted with permission of the AAS). *(See insert for color representation of this figure.)*

toward the Earth. Without the Sun, the trajectory stays out near the Moon until it eventually wanders further away from the system, unless there is a lunar flyby.

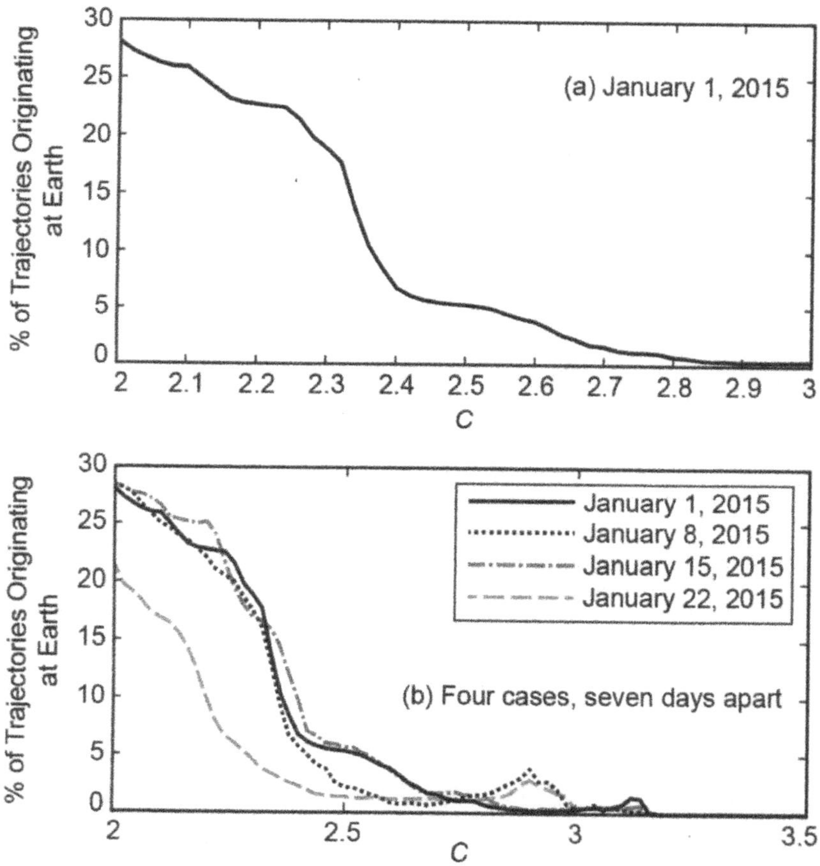

Figure 5-9 Percent of trajectories originating at the Earth for each Jacobi constant (Sun–Earth–Moon Ephemeris system) [192] (Copyright © 2011 by American Astronautical Society Publications Office, San Diego, California (Web Site: http://www.univelt.com), all rights reserved; reprinted with permission of the AAS).

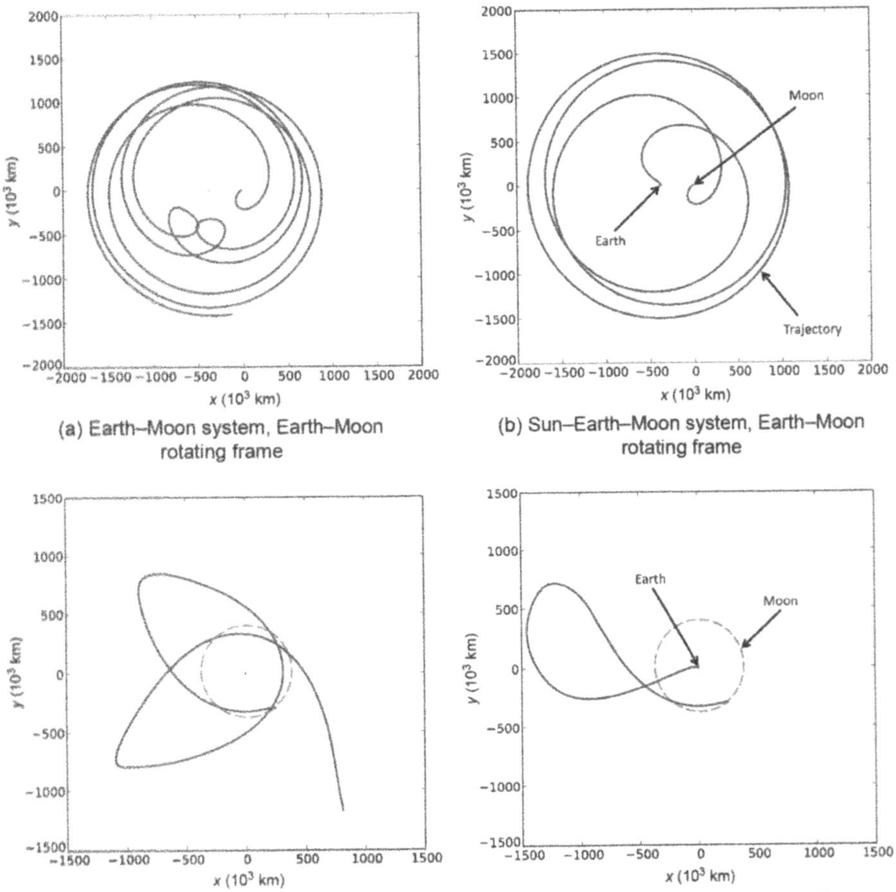

(a) Earth–Moon system, Earth–Moon rotating frame

(b) Sun–Earth–Moon system, Earth–Moon rotating frame

(c) Earth–Moon system, Sun–Earth rotating frame

(d) Sun–Earth–Moon system, Sun–Earth rotating frame

Figure 5-10 Comparison of a single trajectory at $C = 2.8$ ($\alpha \approx 197.5$ deg, $\theta \approx 9.5$ deg) integrated with and without the Sun's gravity in different rotating frames [192] (Copyright © 2011 by American Astronautical Society Publications Office, San Diego, California (Web Site: http://www.univelt.com), all rights reserved; reprinted with permission of the AAS).

5.5 LOW-ENERGY SPATIAL TRANSFERS BETWEEN THE EARTH AND THE LUNAR SURFACE

5.5.1 Trajectories Normal to the Surface

While the planar cases discussed up to this point are quite complicated, it is still possible to plot many of the salient features of the design space given the relatively small dimension of the problem. With the increase in dimension that occurs for the spatial problem, the visualization of the resulting trajectories and their characteristics becomes an even more difficult issue. One of the stated objectives of this analysis is to capture the characteristics of the major trajectory categories while also providing adequate information to evaluate the usefulness of each trajectory. With this objective in mind it is worth noting that if the plots in Figs. 5-4, 5-6, and 5-8 are examined, it can be seen that the majority of the dominant types of trajectories seen in the figures may be captured by making a particular cut at $\theta = 0$. The trajectories obtained with $\theta = 0$ correspond to those trajectories impacting the Moon normal to the lunar surface. As previously mentioned, these types of trajectories are particularly applicable to impactor missions similar to *LCROSS*. Given the results from the planar case, they can also provide a good initial overview of the different categories of Earth–Moon landing trajectories, including those with different flight path angles. The results presented next are restricted to those computed using impacts normal to the surface for an epoch of January 1, 2015. They provide accurate results for impactor-type trajectories, and they also give a good indication of the types of trajectories that may exist for trajectories coming in at other flight path angles.

To allow for easy visualization of the trajectory characteristics, the results are presented for each energy level, which corresponds to a slightly varying velocity magnitude relative to the Moon that depends on the location of the final point on the trajectory at the Moon's surface. The velocity can be used to provide an indication of the change in velocity (ΔV) required for landing, although the specific ΔV will depend on the particular landing trajectory. The regions of the Moon that are accessible for each energy level can be evaluated for particular mission design requirements by using the desired parameters plotted over the surface of the Moon in α and β. α and β are measured in the rotating frame with α positive in the same direction as shown in Fig. 5-3. β is measured like latitude and is positive above the xy plane. Understanding how to connect the trajectory to the Earth becomes more of a challenge in the spatial problem because a large number of possible Earth-relative orientations and methods of injection onto the trans-lunar trajectory are possible. For this reason, a specific set of trajectory characteristics was selected for plotting. The procedure in each case was to begin with the final point on the trajectory with a velocity normal to the lunar surface at the given α and β. The trajectory was then propagated backward in time until it either encountered the Earth or the Moon, or the trajectory duration reached 200 days. For those trajectories not encountering the Earth or the Moon in this time period, a search was then made for the periapse closest to the Earth. Several quantities were then computed using the point at encounter or periapse. They included the periapse radius relative to the Earth, the TOF, the launch energy (C_3), and inclination in the Earth Mean Equator and Equinox of J2000 (EME2000) frame.

Results showing the origin of each trajectory encountering the Moon are given in Fig. 5-11. For these cases three-dimensional effects are included, and it is now possible for the trajectory to miss encounters with the Earth and Moon by traveling above or below them. A significant number of encounters are still observed though, and the features seen for the $\theta = 0$ cases in the planar model may still be observed here where β is 0 deg. Although a significant number of Earth-origin trajectories are observed for low Jacobi constants, as the Jacobi constant increases (energy decreases), the number of Earth–Moon transfers decreases. Once a Jacobi constant of 2.8 is reached, there are no more of these types of trajectories in the Earth–Moon ephemeris model. However, there are still a significant number of Earth–Moon trajectories in the Sun–Earth–Moon ephemeris model. Indeed, a significant number still exist as the Jacobi constant is increased, even above 3.1. As in the planar case, the Sun may be thought of as changing the energy or Jacobi constant of the trajectory while the velocities at the Moon remain the same in each model. This observation emphasizes the need to include the Sun's influence in the trajectory design process, but it raises the question as to what types of Earth–Moon trajectories exist at these energies and how long are their times of flight? It is difficult to answer these questions completely since trajectories are constantly changing with energy, but it is interesting to observe some of the trajectories that exist in the Sun–Earth–Moon system with no corollary in the Earth–Moon system. Two sample trajectories from the line of Earth origin trajectories at $C = 2.6$ that do not exist in the Earth–Moon system are given in Fig. 5-12. The majority of cases found in this line are similar to the trajectory in Fig. 5-12(a), and they exhibit the characteristics of known trajectories designed to utilize the dynamics of the invariant manifolds of libration orbits. They approach the L_1 Lagrange point from the Earth in the Sun–Earth system and then fall away toward the Moon. Although almost all of the trajectories follow this type of orbit, some do have characteristics similar to the trajectory in Fig. 5-12(b). In this case, the Sun's gravity is still influential, but an intermediate flyby is inserted.

It is also interesting to observe the types of trajectories that exist for higher Jacobi constants, or lower velocities, at the lunar surface in the Sun–Earth–Moon system where no analogues in the Earth–Moon system have been found. Several samples are shown along with the trajectory origin plots in Fig. 5-13 to provide an overview of these types of trajectories. Here, an interesting phenomenon occurs. As the Jacobi constant increases to 3.0, the trajectories originating at the Earth are scattered across the map. The majority of the Earth-origin trajectories seem to require multiple flybys of the Earth or the Moon. The sample trajectories shown in Fig. 5-13(a) are intended to be representative of the types of trajectories found across the map. Although a few trajectories, such as those found in the lower left corner of the map, utilize the libration dynamics more directly, the majority seem to require variations on different phasing flybys as shown by the various trajectories. As the Jacobi constant is increased further to a value of 3.1 in Fig. 5-13(b), a line of trajectories appear. These trajectories, as shown in the figure, once again utilize the libration orbit dynamics more directly, sometimes making use of a single flyby along the trajectory. The remaining scattered

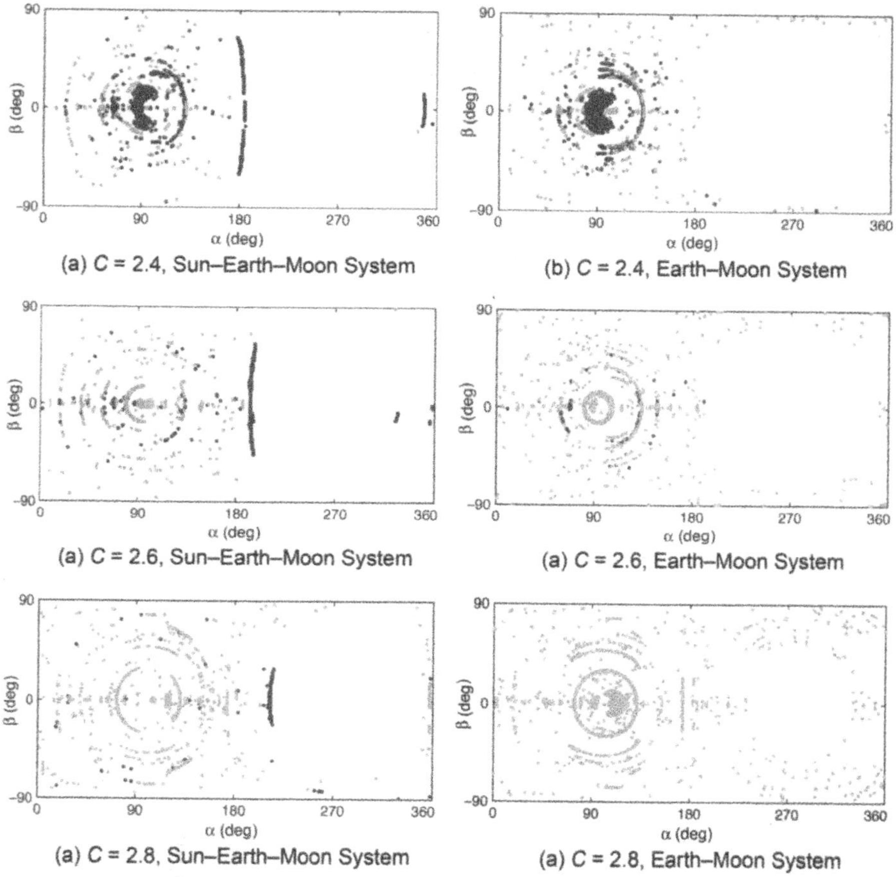

Figure 5-11 Plots showing the origin of the spatial collision trajectories. Black indicates the trajectory originated at the Earth, and gray indicates it originated at the Moon. If it is white, no encounter occurred within 200 days (for epoch of January 1, 2015) [192] (Copyright © 2011 by American Astronautical Society Publications Office, San Diego, California (Web Site: http://www.univelt.com), all rights reserved; reprinted with permission of the AAS, 2006)

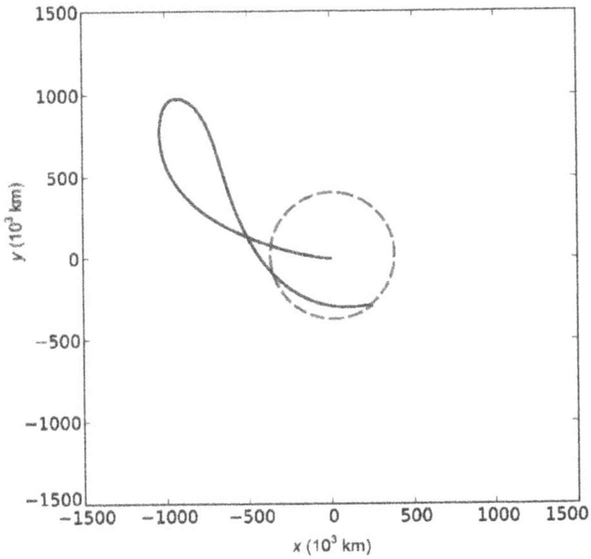

(a) $C = 2.6$, $\alpha \approx 194.5$ deg, $\beta \approx -32.0$ deg

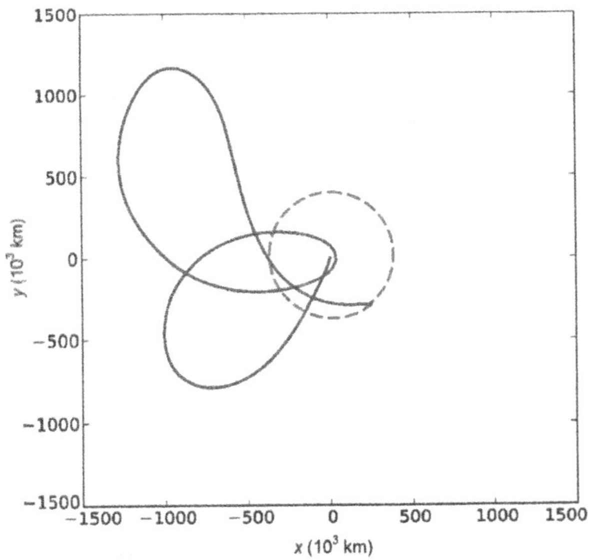

(b) $C = 2.6$, $\alpha \approx 194.5$ deg, $\beta \approx -10.0$ deg

Figure 5-12 Sample trajectories at $C = 2.6$ for the Sun–Earth–Moon system. The trajectories correspond to the line of trajectories not found in the Earth–Moon plots [192] (Copyright © 2011 by American Astronautical Society Publications Office, San Diego, California (Web Site: http://www.univelt.com), all rights reserved; reprinted with permission of the AAS).

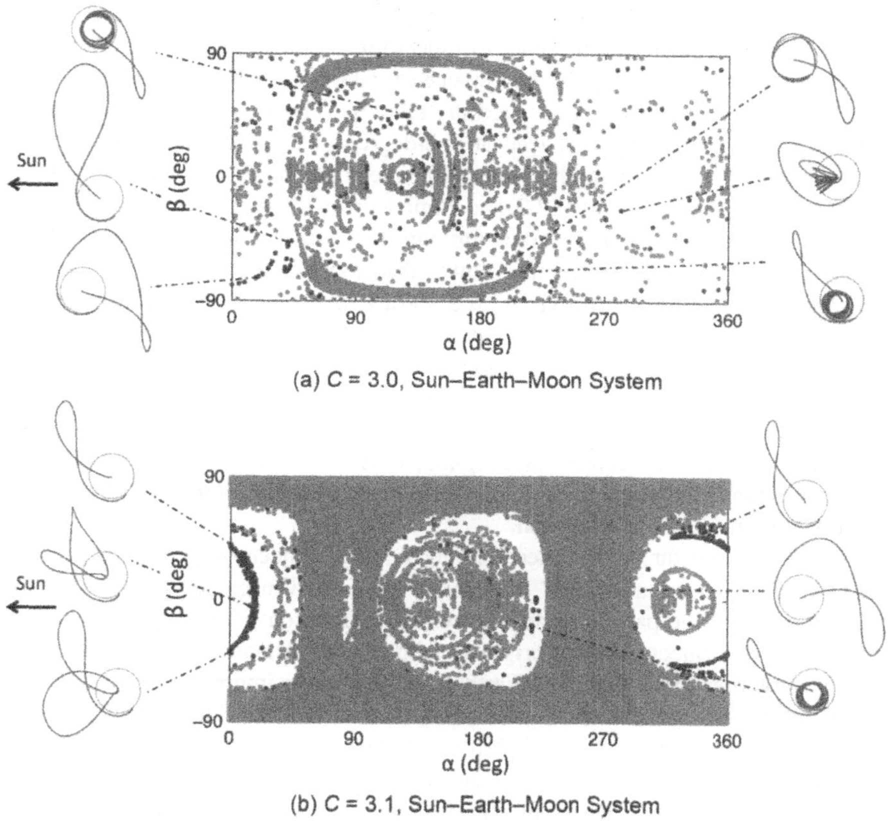

Figure 5-13 Plots showing the origin of the spatial collision trajectories. Black indicates the trajectory originated at the Earth, and gray indicates it originated at the Moon. If it is white, no impact occurred within 200 days. Trajectories are shown for select points in the Earth-centered Sun–Earth rotating frame. The gray circular orbit is the Moon's orbit while the Sun is in the indicated direction. The scale is the same for all trajectories shown, and the trajectories all originate at the Earth [192] (Copyright © 2011 by American Astronautical Society Publications Office, San Diego, California (Web Site: http://www.univelt.com), all rights reserved; reprinted with permission of the AAS).

trajectories found near the center of the plot continue to use multiple gravity flybys to connect the Earth and the Moon.

Another interesting characteristic to include in the analysis is the TOF required for each trajectory originating at the Earth. More specifically, what are the minimum TOF values that may be achieved at each energy? The TOF values provide an indication of whether the trajectories at each Jacobi constant fall more in the category of direct transfers, low-energy transfers, or somewhere in between. The existence of trajectories in the Sun–Earth–Moon system that do not exist in the Earth–Moon system already indicates the presence of trajectories utilizing multi-body effects that would be expected to fall more in the low-energy category. The minimum TOF values for selected Jacobi constants are listed in Table 5-1. These values were computed using a grid with the points spaced at one-degree intervals in each variable. As expected, the TOFs start near the 3-day values seen for the Apollo program's direct transfers for a Jacobi constant of 2.2, and climb to over 100 days for a Jacobi constant of 3.0. It is surprising though that the minimum TOF at a Jacobi constant of 3.1 drops to 78.7 days. Although this point is lower than most of the others at this energy, a number of trajectories still exist in the 90-day time range. The reasons behind this drop in the TOF may be more clearly understood by reexamining the typical trajectories plotted for the Jacobi constants of 3.0 and 3.1 in Fig. 5-13. As mentioned previously, the majority of the trajectories computed for the Jacobi constant of 3.0 required multiple phasing flybys, while the $C = 3.1$ trajectories typically utilize the libration dynamics without these phasing loops. This phenomenon would explain the lower minimum TOF value at $C = 3.1$, since many of the trajectories at this energy actually use a more direct approach.

The analysis so far has focused on categories of trajectories originating at the Earth, with the expectation that trajectories from a given category may often be modified to meet the particular requirements of a mission when they are supplied. Often, however, trajectories that originate within some distance of the Earth may be used by targeting them from low Earth orbit. It is also important to quantify the orbital parameters of the initial conditions of the analyzed trajectories relative to the Earth in order to determine the suitability of the trajectories for particular missions. For example, if a launch from Cape Canaveral is selected, an inclination relative to the Earth's pole of 28.5 deg would be desirable. Particular quantities relevant to mission design are presented next with the objective of presenting an overview of the possible trajectories so that initial estimates may be made for future mission design.

Table 5-1 Minimum TOF values from the computed trajectories originating at the Earth for selected Jacobi constants.

C	TOF (days)
2.2	3.4
2.4	29.8
2.6	58.3
2.7	57.8
2.8	74.0
2.9	94.9
3.0	101.0
3.1	78.7

The analysis here focuses on the Sun–Earth–Moon system so as to encompass the complete range of trajectories.

The closest periapse values obtained over 200 days for selected Jacobi constants in the Sun–Earth–Moon system are plotted in Fig. 5-14. Note that some of the gaps in (a) are Earth intersection trajectories as can be seen be reexamining Fig. 5-13. It can

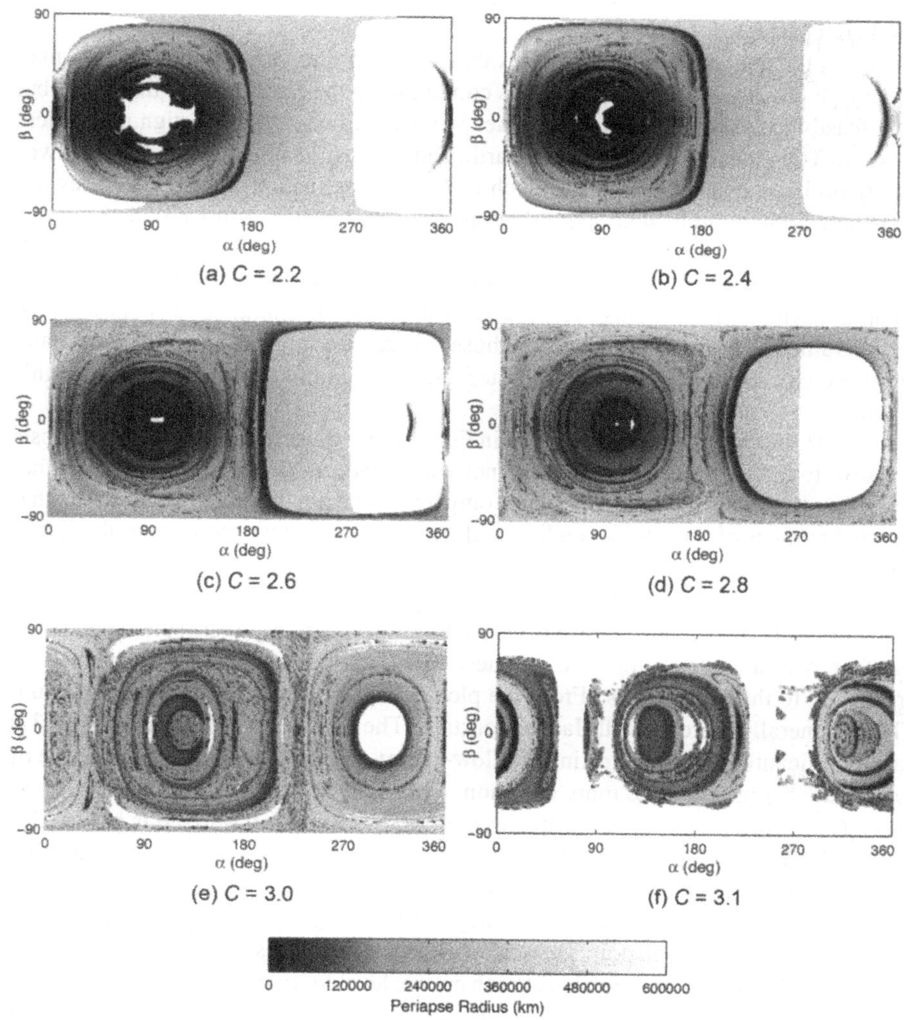

Figure 5-14 Periapse radius values for the computed trajectories plotted over the surface for a range of Jacobi constant values [192] (Copyright © 2011 by American Astronautical Society Publications Office, San Diego, California (Web Site: http://www.univelt.com), all rights reserved; reprinted with permission of the AAS).

immediately be seen from the plots that the majority of the trajectories never come near the Earth. In general, the trajectories originating near $\alpha = 90$ deg produce the most trajectories with periapses closer to Earth. This does shift with Jacobi constant, as was seen in the earlier Europa study [199]. As the Jacobi constant increases and energy decreases, it appears that fewer trajectories come as close to the Earth, but the majority stay near the system. The chaos present in the system can especially be observed for $C = 3.0$, where trajectories very close to each other alternate with low and high periapses.

From the analysis so far, it appears that a large portion of the lunar surface may be physically accessible to trajectories coming from the Earth or near the Earth, but the feasibility of flying these trajectories will depend on mission design parameters such as TOF, launch energy (C_3) at Earth, and inclination. It is uncertain what two-body orbital element parameters (such as C_3 and inclination) mean when they are computed where multi-body perturbations are significant, but this problem may be alleviated by computing these parameters where multi-body effects are minimized. With this objective in mind, only parameters for trajectories with periapses lower than geosynchronous radius are plotted in the following figures. For these plots, the parameters are now included for those trajectories originating at the Earth, and in those cases their values are computed using the initial conditions at the Earth's surface.

The TOFs and C_3 values are plotted in Fig. 5-15 for those trajectories with periapsis relative to the Earth of less than geosynchronous radius. The immediate feature that can be noticed is the sparsity of points compared to the previous plots, confirming that a large number of trajectories ending at the lunar surface never come near the Earth. Indeed, for lower Jacobi constants, the locations between approximately 180 deg and 360 deg have almost no trajectories originating near the Earth. Curiously, around a Jacobi constant of 3.0, the trajectories are more randomly distributed across the surface with a combination of C_3 values. This feature may be partly explained by returning to the TOF values. From the plots, it can be confirmed that the minimum TOFs generally increase with Jacobi constant. The minimum TOF values at $C = 3.0$ are significantly larger, indicating that low-energy trajectories under the influence of chaos are beginning to be more common. Given the variety of trajectory types and the TOFs involved, it is not surprising that more of the lunar surface is potentially covered. Examining the trends in the TOF plots, it may also be observed that longer TOF trajectories appear to exist at each energy level. The lines of long TOF trajectories correspond to low-energy trajectories using the Sun's perturbations and approaching the libration points of the Sun–Earth system. It is also worth noting that a variety of C_3 options are available at each energy level for transfers to the Moon. Even for low Jacobi constants, there still exist some relatively low C_3 options, although the minimum is higher than that found for the higher Jacobi constant cases. It is important to realize that a small change in the landing location can result in a drastic change in the required C_3 even with similar TOFs and the same velocity at the Moon. This fact is important for mission designers, as it may sometimes be possible to move the landing site slightly to improve ΔV, or a similar effect may be obtained by targeting with maneuvers along the trajectory. Trajectory correction maneuvers may

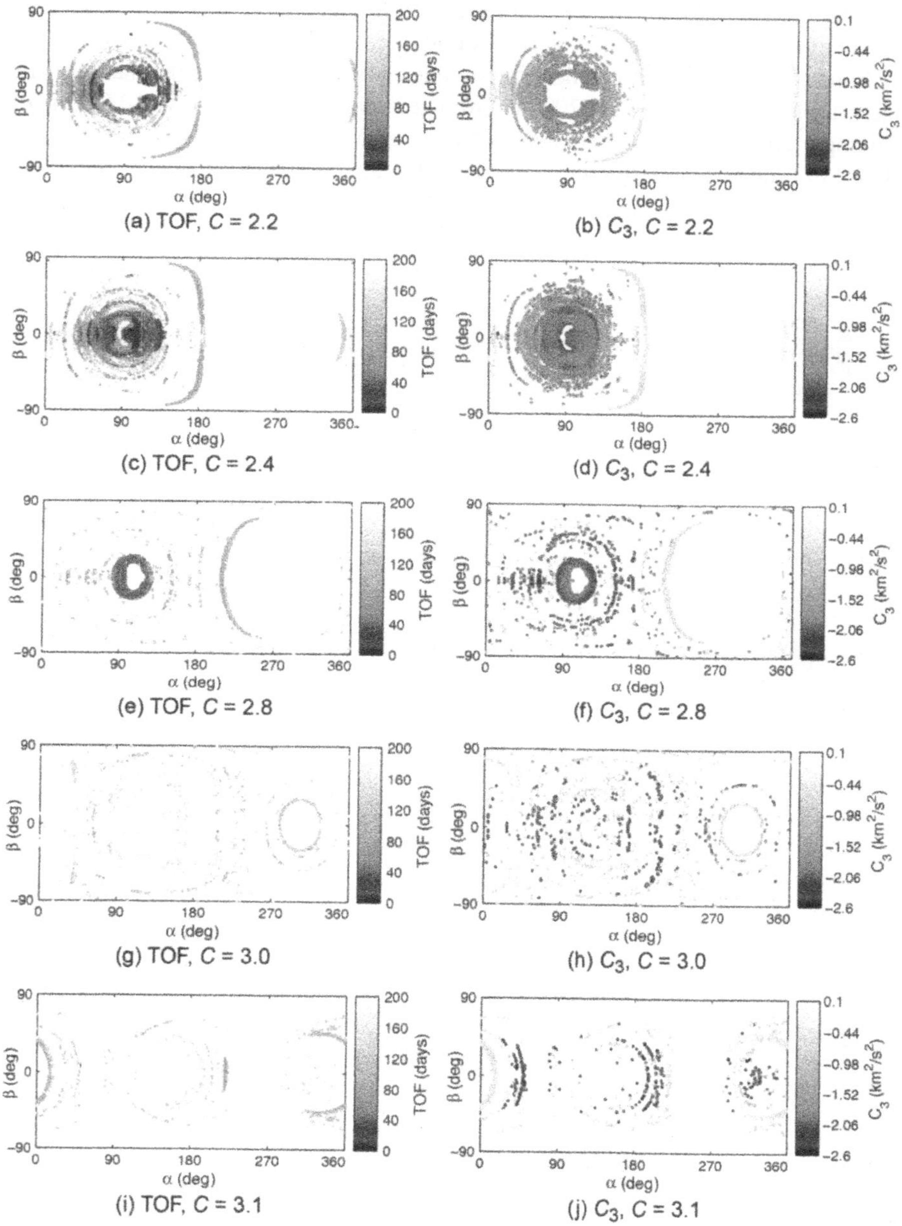

Figure 5-15 TOF and C_3 for each trajectory plotted over the surface for a range of Jacobi constants [192] (Copyright © 2011 by American Astronautical Society Publications Office, San Diego, California (Web Site: http://www.univelt.com), all rights reserved; reprinted with permission of the AAS).

also help aid in reducing the ΔV. In general, it is useful to be aware of the chaotic nature of the design space as seen from these plots.

Finally, it is important for most mission designs to consider the inclination. The inclination results in the EME2000 coordinate frame are given in Fig. 5-16. One of the important features to notice here is that a variety of inclinations are possible. A choice of trajectories exist with the lower inclinations suitable for launch from Cape

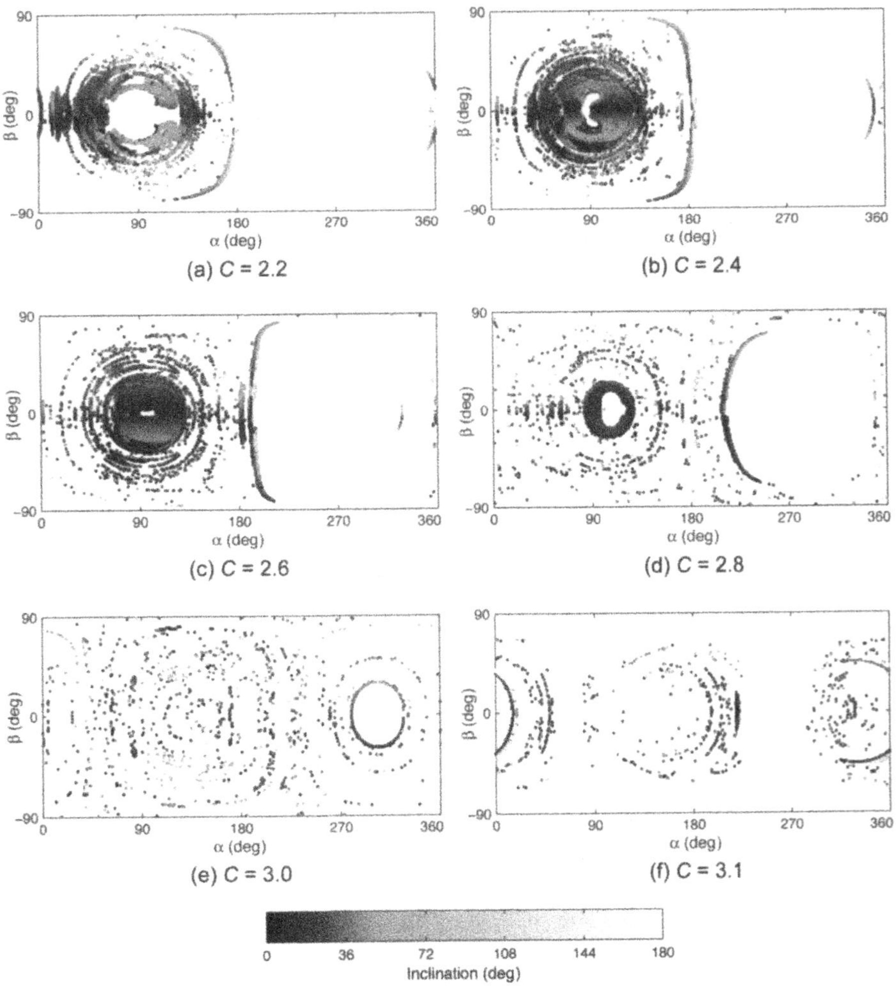

Figure 5-16 Inclination computed relative to the Earth in the EME2000 coordinate frame [192] (Copyright © 2011 by American Astronautical Society Publications Office, San Diego, California (Web Site: http://www.univelt.com), all rights reserved; reprinted with permission of the AAS).

Canaveral. The particular inclinations needed for a mission will depend on the target location on the Moon and the particular constraints of the mission. They are provided here as a sample of the range of the values that are possible.

5.5.2 Trajectories Arriving at Various Angles to the Lunar Surface

The planar CRTBP provides a convenient framework in which to understand and visualize the relationship between invariant manifolds and lunar approach trajectories, but the design of real-world equivalent trajectories often requires a landing at either higher or lower latitudes. Indeed, many of the recently proposed landing sites at the Moon are at northern or southern latitudes [204], and one of the locations that is currently a focus for a lunar lander is more southern latitudes in the Aitken Basin. In this analysis, lunar landing trajectories are analyzed over the three-dimensional surface of the Moon, and the approach geometry of the trajectories in three dimensions is also analyzed.

The landing geometry of trajectories traveling from the Earth to the Moon is of particular importance for mission design. In the previous section and in Anderson and Parker [192, 195] we analyzed trajectories encountering the Moon normal to the surface to determine whether these trajectories originated at the Earth within the previous 200 days. Given this elevation angle constraint, only some locations of the Moon's surface were found to be accessible from the Earth. For this analysis, trajectories were allowed to approach each point on the lunar surface from all directions. These directions were specified relative to the surface at each point. The azimuth angle (Ω) is measured clockwise from north where north is the lunar orbit's North Pole, rather than the Moon's North Pole, to be consistent with the results from the CRTBP. The elevation angle (ϕ) is measured positive above the Moon's surface, with a trajectory encountering the Moon's surface normal to the surface having an elevation angle of 90 deg. (Note that this is different from θ used for the planar case, but it was chosen to be more consistent with typical mission design parameters.) While the previous analysis was ideal for impactors, the trajectories computed here are applicable for a wide range of mission types traveling to the lunar surface. Additional parameters for each trajectory related to the original characteristics relative to the Earth may be computed, but the focus here is on characterizing the approach geometry. For the following analysis the trajectories were computed over the surface of the Moon using 1-deg increments in α and β. The same definition is used for α that is used in the planar problem in Fig. 5-3. As described earlier, β is measured like latitude and is positive above the xy plane. Two different grids were used for the azimuth and elevation angles. In each case, the elevation angle was varied in even increments, and the steps taken in azimuth angle were specified initially for an elevation angle of 0 deg. The number of azimuth points were then decreased with $\cos(\phi)$ so that the number of points decreased with elevation angle. Both a fine grid and a coarser grid were used in this analysis. For the fine grid case, 1-deg increments were taken at 0-deg elevation for Ω, and 1-deg increments were used for elevation. For the coarser grid, 10-deg increments were used for Ω at 0-deg elevation, and 3-deg increments were used for elevation. This coarser grid was found to provide a good approximation that conveyed the overall trends of the fine grid, while allowing for a more reasonable computation time. Even with this coarser grid, computing trajectories over the entire surface in the ephemeris problem for each Jacobi constant required approximately seven days running in parallel on 40 processors. Unless otherwise stated, this coarser grid is the one used throughout the analysis.

As an initial step in the analysis, the set of trajectories was computed in the CRTBP for a Jacobi constant of 2.6. The trajectories were computed for both the fine grid and the coarser grid. Comparing the maximum and minimum elevation angles resulted in trajectories that originate at the Earth as shown in Fig. 5-17. Using the symmetry about the xy plane mentioned earlier, it can be seen that the northern and southern latitudes will be reflected for the elevation plots in the CRTBP. Note that the azimuth angles would need to account for the reflection if they are

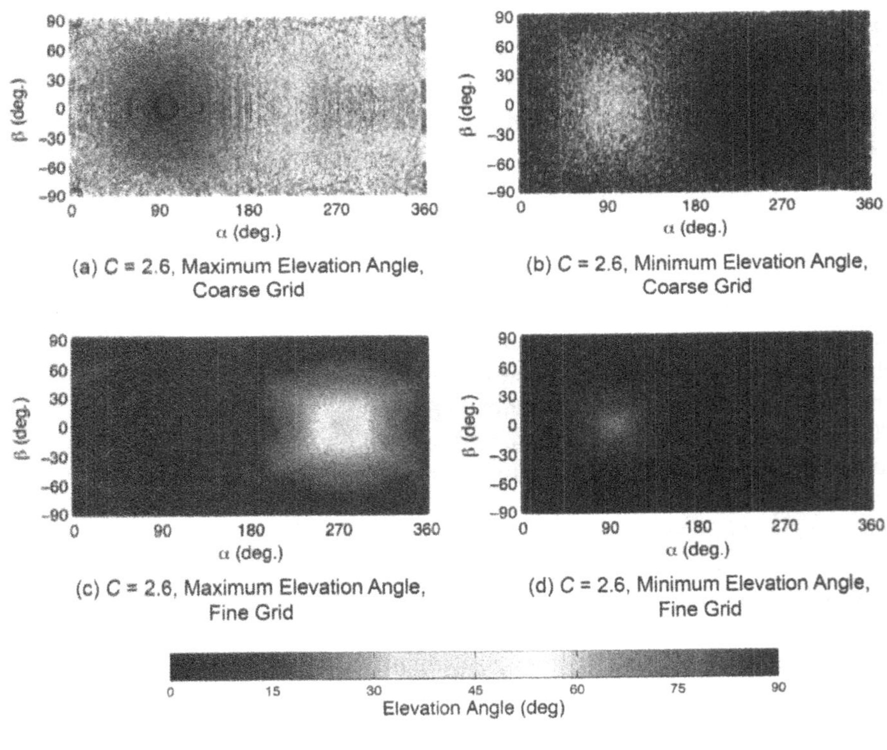

Figure 5-17 Maximum and minimum elevation angles for trajectories originating at the Earth and encountering the Moon at each point on the surface. These cases are computed in the CRTBP for $C = 2.6$. Results from two different grids (in elevation and azimuth angle) are shown [193] (Copyright © 2011 by American Astronautical Society Publications Office, San Diego, California (Web Site: http://www.univelt.com), all rights reserved; reprinted with permission of the AAS). *(See insert for color representation of this figure.)*

plotted, and although similar results would be expected in the ephemeris problem, the variations in the ephemeris require that the northern and southern hemispheres be computed independently. Using this symmetry the values computed for the northern and southern hemispheres were reflected in Fig. 5-17 to save computation time. By comparing the plots, it can be seen that, as might be expected, the finer grid captures more trajectories at higher and lower elevation angles that originate at the Earth, however, the overall trends in the data remain the same for both grids. In each case the range of elevation angles from minimum to maximum is shifted higher near $\alpha = 90$ deg and lower near $\alpha = 270$ deg. Referring back to Fig. 5-3, the 90-deg direction corresponds to the leading edge of the Moon, and the 270-deg direction to the trailing edge. The coarser grid is used in the remainder of this analysis, so it should be remembered that details in the plots may change with a finer grid, but the overall trends can still be observed.

An analysis of trajectories for a Jacobi constant of $C = 2.8$ confirmed our earlier result for trajectories encountering the Moon normal to the surface of the Moon that no Earth-return trajectories were found for this Jacobi constant or higher ones in the CRTBP. However, it is expected that Earth-origin trajectories with velocities consistent with higher Jacobi constants in the CRTBP will exist in the ephemeris problem because these trajectories may use the Sun's perturbations to travel from the Earth to the Moon. Those higher Jacobi constants, especially those approaching the values near C_{L_1} and C_{L_2} are especially relevant for the computation of the invariant manifolds of libration point orbits, which is useful for the comparison later in this study. The elevation angle range results are shown in Fig. 5-18 for Jacobi constants ranging from $C = 2.6$ to 3.1 in the ephemeris problem. Note that, as in Anderson and Parker [192, 195], the Jacobi constant for the ephemeris plots is used as a shorthand for a particular set of velocities computed around the Moon in the CRTBP. These same velocities are attached to the Moon in the ephemeris problem referenced to the instantaneous orbital plane of the Moon's orbit around the Earth. The symmetry used to simplify the computations in the CRTBP is no longer present for the ephemeris problem, and trajectories were directly computed for the entire plot. Once the trajectory is integrated backward from the Moon, the Jacobi constant of the trajectory will vary in both the Earth–Moon and Sun–Earth systems.

Comparing the results from Fig. 5-17 for the Jacobi constant case of 2.6 in the CRTBP and the ephemeris problem results reveals that they are quite similar. The maximum and minimum elevation values still occur at approximately the same locations on the surface for each case. However, several new bands of high-elevation-angle cases occur for the ephemeris case near $\alpha = 180$ deg for the maximum elevation angle case and from approximately $\alpha = 290$ deg to 360 deg. Additional bands also seem to exist for the minimum elevation angle case, especially for high and low latitudes. It is natural to expect from past work that these bands may represent trajectory options that exist as a result of the Sun's influence, and it is interesting that these types of bands remain up through $C = 2.8$ (Figs. 5-18(a) through 5-18(d)). An interesting topic planned for future study is to determine how these characteristics vary with a finer grid. However, the comparison performed here is with the same grid in each case, indicating that these additional trajectories exist.

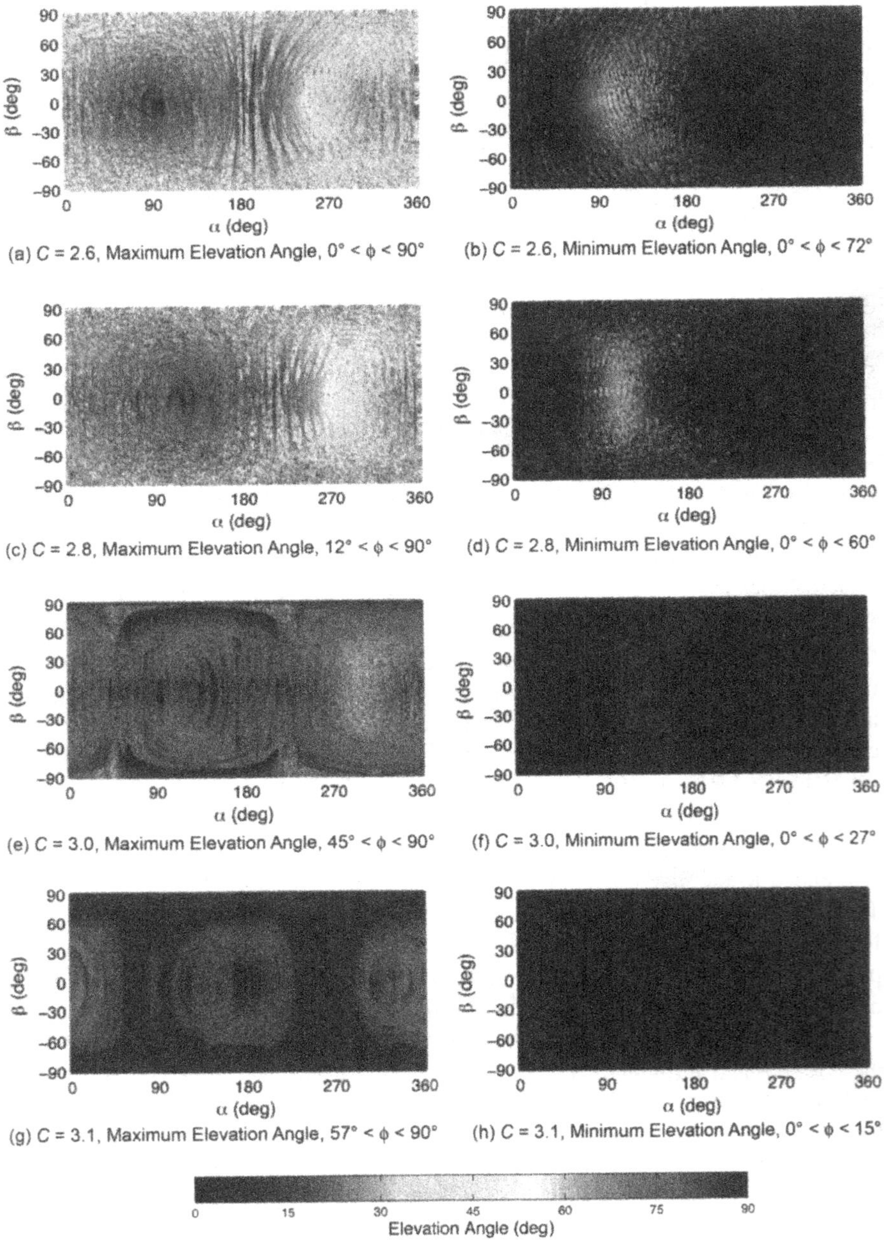

Figure 5-18 The minimum and maximum elevation angles of trajectories originating at the Earth for each point on the lunar surface. These trajectories are computed in the Earth–Moon ephemeris system including the Sun's perturbations [193] (Copyright © 2011 by American Astronautical Society Publications Office, San Diego, California (Web Site: http://www.univelt.com), all rights reserved; reprinted with permission of the AAS). *(See insert for color representation of this figure.)*

As the Jacobi constant increases even more as seen in Figs. 5-18(e) through 5-18(h), the range of elevation angles for lunar landing at each point seems to increase even more. The location of the peaks also seems to shift, and for the maximum elevation angle plots, the peaks move to the right or eastward with increasing Jacobi constant. When a Jacobi constant of 3.1 is reached, the maximum elevation angle for points containing a trajectory originating at the Earth never drops below 57 deg, and the minimum elevation angle for the same points never goes over 15 deg. It is important to mention that although the points look dense across the surface in the plots, this is because of the size of the plot and the points used for plotting. There are individual points on the surface where no Earth-origin trajectory exists for this grid, but there are always nearby points where such a trajectory exists. For real-world mission design, a small ΔV can be used to target slightly different points, and the surface of the Moon is covered in practice for mission design purposes. It has also been found for particular points that if a much finer grid is used, typically some Earth-origin trajectories are found, and these points will be included in future studies. The points with no Earth-origin trajectories for this grid are not included in the elevation angle ranges listed in the plots. These results for higher Jacobi constants agree generally with the normal trajectory cases seen in our previous work [192, 195]. The additional range of geometries available for landing at these energies appears to be a result of the increasingly chaotic nature of the system as the Jacobi constant approaches the values at the L_1 and L_2 libration points. In other words, the trajectories are more able to take advantage of chaos to arrive at different elevation and azimuth angles. This also indicates that these Jacobi constants are of particular interest for comparison with the invariant manifolds of libration orbits. One interesting statistic to examine with a fixed grid is the maximum number of trajectories at a particular point that originate at the Earth. Although this number is generally quite low, there are some points where it peaks. The maximum number of trajectories at a particular point is listed in Table 5-2 for different Jacobi constants. The higher values are found for a Jacobi constant of 2.6 and 3.1. The $C = 2.6$ results include more direct trajectories that still exist in the CRTBP and do not require the Sun's influence, and the $C = 3.1$ results include those trajectories that are heavily influenced by the Sun. The total number of Earth-origin trajectories follows the same trend. These numbers

Table 5-2 Maximum number of Earth-origin trajectories at a single point on the lunar surface for a fixed grid including the corresponding location and the total number of Earth-origin trajectories for various values of Jacobi constant.

Jacobi Constant	Maximum at a Point	Location (α, β)	Total Number
2.6	27	(193 deg, −36 deg)	290,672
2.8	16	(213 deg, −18 deg)	114,684
3.0	14	(225 deg, −11 deg)	162,061
3.1	36	(192 deg, 7 deg)	298,621

are a function of the grid that is being used and can be refined by using a denser grid; however, they do align with the results from the trajectories computed normal to the lunar surface seen in our earlier work.

Because the trajectories are computed in the ephemeris problem for the cases just discussed, the results will naturally vary with the initial epoch of the integration. A sample of the results was computed for four different epochs around the Moon's orbit (with the time intervals each at approximately one-quarter of the Moon's orbit) to determine how they might vary with the initial epoch. Representative results for a Jacobi constant of 2.8 are shown in Fig. 5-19. The salient features of the plots remain generally the same for each epoch in that the maximum values still occur near $\alpha = 90$ deg and the minimum values occur near $\alpha = 270$ deg. The January 7 and 21 cases have more locations with higher elevation angles, especially near $\alpha = 270$ deg, mixed in with lower elevation angle points. These two cases appear better positioned to take advantage of the Sun–Earth libration point dynamics, which could increase

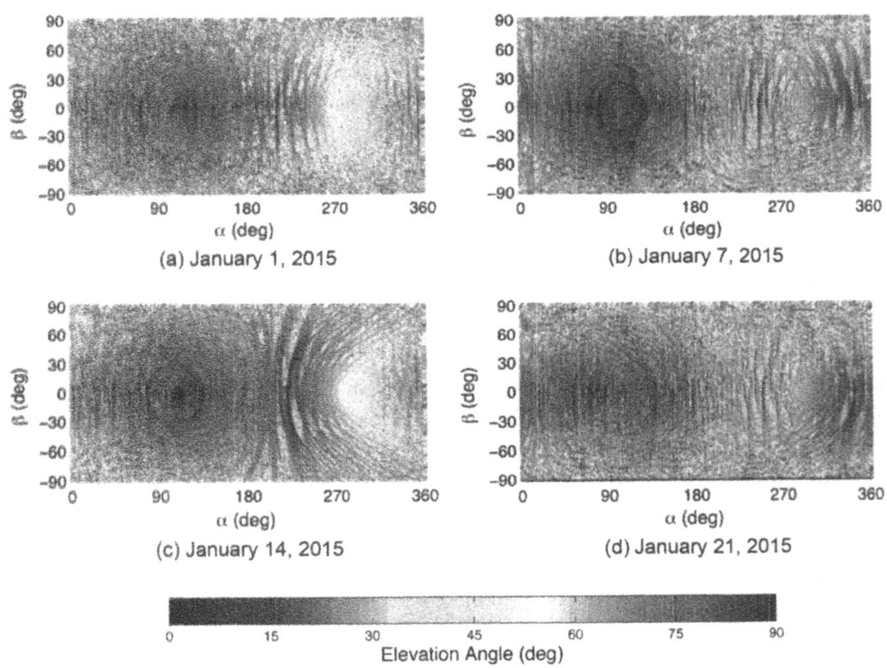

Figure 5-19 Comparison of maximum elevation angle results around the lunar orbit at seven day intervals for $C = 2.8$ [193] (Copyright © 2011 by American Astronautical Society Publications Office, San Diego, California (Web Site: http://www.univelt.com), all rights reserved; reprinted with permission of the AAS). *(See insert for color representation of this figure.)*

the range of elevation angles that may be obtained for approaching the Moon. Overall though, given this comparison, it is expected that the results from this study may be extrapolated to other epochs without drastically changing the outcome.

Another important aspect of the approach for mission design is, of course, the azimuth angle of the trajectory. Plotting this information in a global sense is difficult, but a sample of the types of results obtained for each Jacobi constant may be visualized in Fig. 5-20 for a subset of the points. The azimuth angles are plotted for each point on a grid computed at 30-deg intervals in both α and β. For these plots, the fine grid was used at each point on the surface (which of course produced more trajectory options), and the trajectories were limited to those with $C_3 < 0.0$ km^2/s^2 at the Earth. The orientation of the lines centered on each point indicates the azimuth angle, and the color is used to designate the corresponding elevation angle of each trajectory. Note that the $\pm$ 90-deg cases used azimuths that were rotated differently at each elevation as a result of the transformation used to compute them. So the specific results differ, but they generally show similar trends. It is interesting that there are definite regions where the majority of Earth-origin trajectories appear to have similar elevation angles.

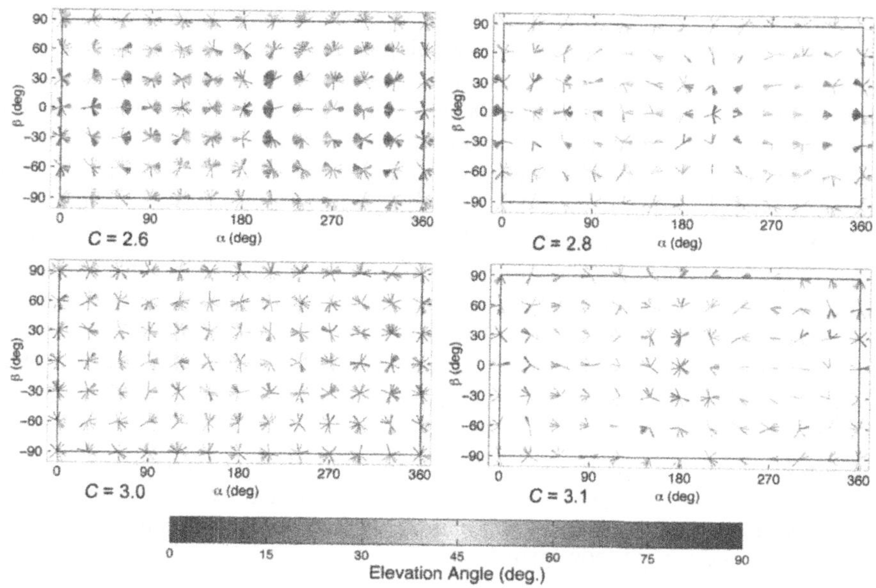

Figure 5-20 Azimuth angles at points on a 30-deg grid on the lunar surface. The plotted lines at each gridpoint are oriented in the proper azimuth direction for each individual trajectory. The color corresponds to the elevation angle of that trajectory. The trajectories shown here are limited to those with $C_3 < 0.0$ km^2/s^2 at the Earth [194] (first published in Ref. [194]; reproduced with kind permission from Springer Science+Business Media B.V.). *(See insert for color representation of this figure.)*

In each case though, there are often just a few high or low trajectories that result in the extremes seen in the elevation angles plots. This fact is worth keeping in mind for mission design since a particular elevation angle may be available in combination with only a few azimuth angles. In general it appears that higher-elevation options are more available as the Jacobi constant increases, although there are typically at least a few low elevation angle options at each point. The combinations of available elevation and azimuth angles are evaluated in more detail for $C = 3.1$ in the following comparison with invariant manifolds, which helps explain the features seen in these plots a little more directly. In general, these plots can provide a broad overview of the available trajectory options.

5.6 TRANSFERS BETWEEN LUNAR LIBRATION ORBITS AND THE LUNAR SURFACE

A general framework and understanding does exist in regard to the relationship between invariant manifolds of unstable orbits and the Moon. (Refer to Section 2.6.10 for more background on invariant manifolds.) Much of the work to design low-energy trajectories from the Earth to the Moon has focused on the use of libration point orbits along with their stable and unstable manifolds [39, 45, 51, 203]. Koon, Lo, Marsden, and Ross examined this problem for the planar case [37], and Parker studied approach cases to lunar libration orbits using invariant manifolds in his dissertation [46]. Baoyin and McInnes analyzed some specific cases of transfers from libration points and planar Lyapunov orbits to the lunar surface [205]. In particular, they searched for the Jacobi constant that would provide complete coverage of the lunar surface by the invariant manifolds of the selected Lyapunov orbit. Von Kirchbach et al. [201] looked at the characteristics of the invariant manifolds of a Lyapunov orbit as they intersected the surface of Europa in the context of the escape problem. Alessi, Gómez, and Masdemont [206] examined the locations of the Moon reachable by the stable manifolds of a range of halo orbits and square Lissajous orbits. They computed the intersections of these invariant manifolds with the surface of the Moon with the expectation that they could be used for astronauts to escape to a libration point orbit if necessary. Anderson [207] examined the approach problem within the context of the invariant manifolds of unstable resonant and Lyapunov orbits as the trajectory ties into the resonances of the Jupiter–Europa endgame problem following invariant manifolds [158, 208–210].

One focus of the transfer to the lunar surface using invariant manifolds is on the final approach from a desired libration orbit to the lunar surface. The problem may be most easily approached using the planar CRTBP and Lyapunov orbits. Two sample Lyapunov orbits found in Anderson and Parker [192, 195] are replotted here in Fig. 5-21. The Jacobi constants for these orbits were chosen so that the invariant manifolds of the Lyapunov orbits just graze the surface of the Moon. The Jacobi constants where the Lyapunov orbits cover the surface of the Moon were computed by Baoyin and McInnes [205] as approximately $C = 3.12185282430647$ for an L_1 Lyapunov orbit and $C = 3.09762627497867$ for an L_2 Lyapunov orbit.

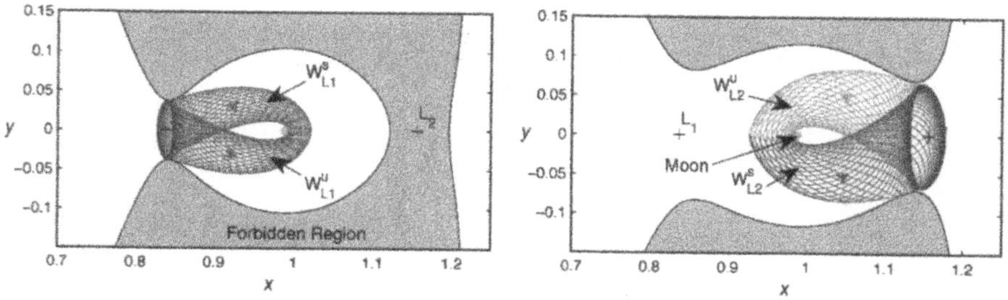

(a) L_1, $C = 3.18258523931748$ (b) L_2, $C = 3.16123125281557$

Figure 5-21 Invariant manifolds of libration orbits computed for Jacobi constants where the manifolds are tangent to the surface of the Moon [193] (Copyright © 2011 by American Astronautical Society Publications Office, San Diego, California (Web Site: http://www.univelt.com), all rights reserved; reprinted with permission of the AAS). *(See insert for color representation of this figure.)*

As a first step in the comparison, the lunar landing geometry of the invariant manifolds of various halo orbits is analyzed. Alessi, Gómez, and Masdemont [206] examined similar trajectories for escaping the surface of the Moon to various halo orbits and summarized the areas on the Moon from which such escape trajectories are possible. We are concerned here with a combination of the landing location along with the landing geometry, therefore, a similar technique to that used in Fig. 5-20 is employed here. In subsequent figures, the intersections of the unstable manifolds of the L_1 halo orbits are indicated by a red point, and the intersections for the L_2 halo orbits are orange points. The azimuth angle and the elevation angle are indicated by the direction and the color of the line segments, respectively.

The results for a halo orbit at $C = 3.1$ are shown in Fig. 5-22. It can be immediately seen that for this energy, the L_1 halo orbit invariant manifolds generally fall on the leading edge of the Moon in its orbit, and the L_2 halo orbit invariant manifolds fall on the trailing edge of the Moon. As expected, the intersections of the northern and southern halo orbits are reflected about $\beta = 0$. The elevation angles are somewhat lower for the L_1 halo orbits than the L_2 halo orbits. All together, the unstable manifolds provide relatively broad coverage of much of the lunar surface, although significant regions are still not intersected by the invariant manifolds. This may be remedied by examining the invariant manifolds at additional energies. The unstable manifold intersections with the Moon can change significantly with the Jacobi constant as can be seen for the intersections plotted with a Jacobi constant of 3.08 in Fig. 5-23. The intersections for the L_1 case have divided into two different regions, and the L_2 intersection case has grown tighter together. It should be reiterated that the unstable manifold intersections can increase if larger time intervals are used for the integration, and these plots focus on short-duration trajectories. The unstable

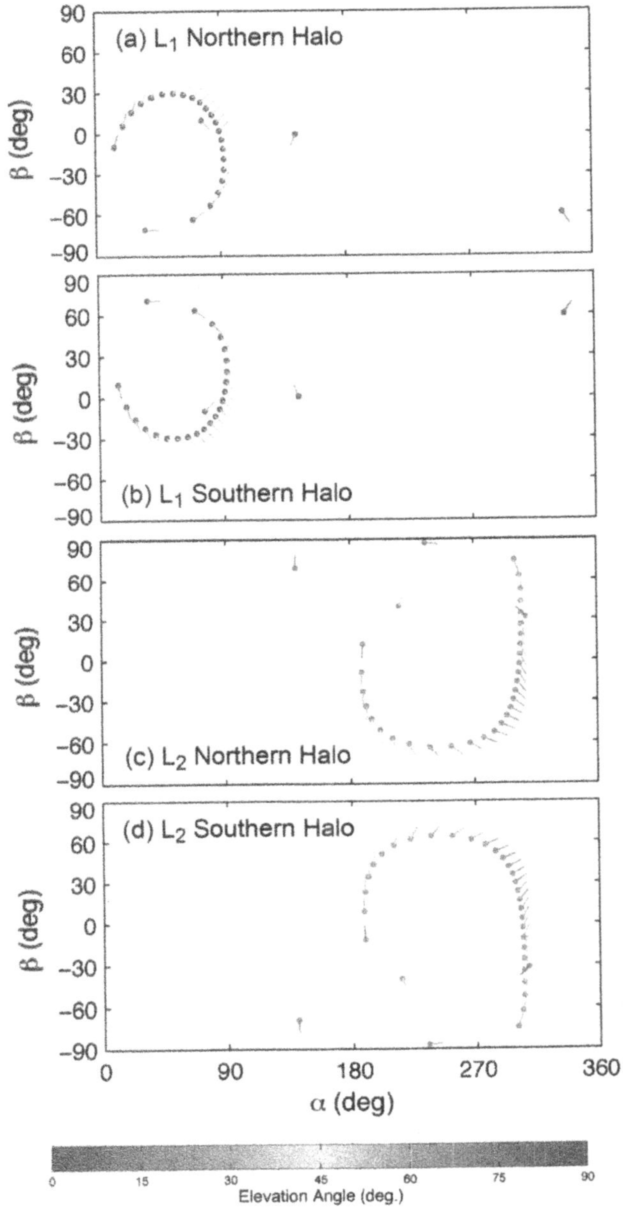

Figure 5-22 Unstable manifold intersections of the specified orbits with the Moon for $C = 3.1$ [193] (Copyright © 2011 by American Astronautical Society Publications Office, San Diego, California (Web Site: http://www.univelt.com), all rights reserved; reprinted with permission of the AAS). *(See insert for color representation of this figure.)*

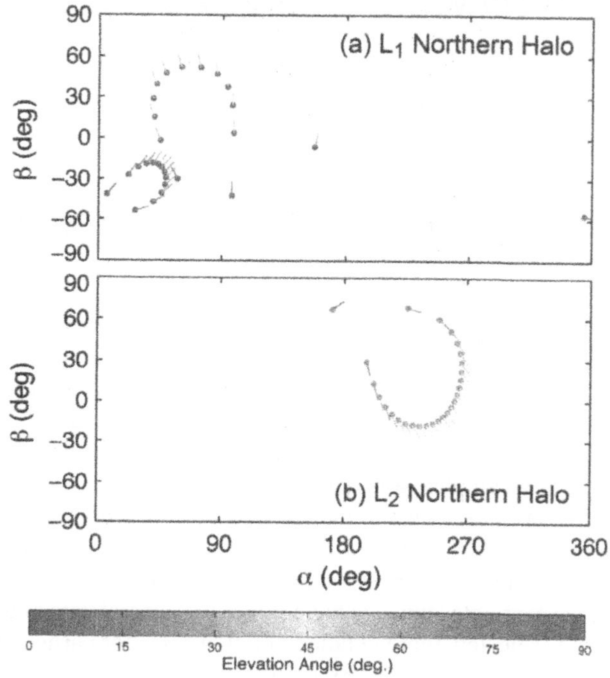

Figure 5-23 Unstable manifold intersections of the specified orbits with the Moon for $C = 3.08$ [193] (Copyright © 2011 by American Astronautical Society Publications Office, San Diego, California (Web Site: http://www.univelt.com), all rights reserved; reprinted with permission of the AAS). *(See insert for color representation of this figure.)*

manifold intersections also change even more as energy continues to change, but these energies appear to provide some of the most direct transfers.

This analysis shows that the unstable manifolds of halo orbits can provide broad coverage for landing at various points on the Moon, although not with the nearly complete coverage found from the previous results. It is also interesting to explore the relationship between the unstable manifolds and these Earth-origin trajectories from the general analysis. A similar examination to the one made for the planar problem would be desired, but the nature of the three-dimensional problem makes this drastically more complex. One possible method for performing this comparison is to examine the origin of the trajectories coming from all azimuth and elevation angles at each point that the unstable manifolds intersect the surface of the Moon. In this case, only one unstable manifold intersection is plotted for each location on the Moon relative to the trajectories coming in from all angles, but it still allows this point to be placed in context of the dynamics indicated by the source of each trajectory.

5.7 TRANSFERS BETWEEN LOW LUNAR ORBITS AND THE LUNAR SURFACE

Information about transfers from low lunar orbits to the lunar surface is provided in Section 4.6 on page 258.

5.8 CONCLUSIONS REGARDING TRANSFERS TO THE LUNAR SURFACE

A wide variety of trajectory options exist for transfer to the lunar surface. These range from more direct trajectories that may be primarily computed using just the influence of the Earth and Moon, to those at lower energies that require the influence of the Sun to compute. The invariant manifolds of libration orbits may be used for transfers to the lunar surface, and in combination with transfers to these libration orbits from the Earth, can be used as a complete transfer from the Earth. The trajectories computed for various Jacobi constants shown in the selected plots in this chapter may be used to obtain an initial idea of the types of trajectories available for different energy regimes. These energy regimes correspond to the constraints dictated by a particular mission, such as the available launch vehicle. Once the general type of trajectory that may be of interest is selected, more detailed initial guesses for particular trajectories may be obtained from the various plots showing trajectories coming in at various angles to the surface or from the invariant manifolds results. A mission designer may then modify and constrain these trajectories, while incorporating the mission design constraints of interest, to compute the final desired trajectory.

CHAPTER 6

OPERATIONS

6.1 OPERATIONS EXECUTIVE SUMMARY

The purpose of this chapter is to address different ways that a low-energy transfer may impact the operations of a spacecraft, compared to conventional lunar transfers. Most conclusions are very straightforward consequences of the fact that low-energy transfers require less change in velocity (ΔV), more time, and have longer link distances during the transfer than direct lunar transfers. For instance, there are fewer demands on the spacecraft's propulsion system and operational schedule, but more demands on the spacecraft's communication capabilities due to the longer distances. The operations team must be able to perform several trajectory correction maneuvers (TCMs) during the trans-lunar cruise, but these maneuvers are typically much more separated in time from launch, lunar arrival, and other maneuvers than they are on conventional lunar transfers.

The majority of discussion in this chapter is devoted to studying the availability and ΔV cost of establishing an extended 21-day launch period for a lunar mission. Conventional lunar missions typically have very constrained launch periods, reflecting the geometry in the Sun–Earth–Moon system. However, low-energy lunar transfers are very flexible and may be adjusted in many ways to accommodate an extended launch period. Several conclusions are drawn from these examinations.

First, the cost of a launch period is dependent on the number of launch days in the period. The examination performed in Section 6.5 estimates that it costs on average 2.5 meters per second (m/s) of ΔV per day added to a launch period; hence, the average 21-day launch period requires about 50 m/s more deterministic ΔV than a 1-day launch period for a given transfer. The cost of the 1-day launch period is dependent on the inclination change that must be performed to inject onto the desirable low-energy transfer from a constrained low Earth orbit (LEO) parking orbit. Section 6.5.7 estimates that it costs approximately 0.97 m/s more transfer ΔV per degree of inclination change that must be performed. The total cost of establishing a 21-day launch period from a 28.5-degree (deg) LEO parking orbit to a given lunar orbit is approximately 71.7 $\pm$ 29.7 m/s (1σ). Thus, to be very conservative when estimating a preliminary ΔV budget for a mission, one may estimate that the ΔV cost to transfer from a 28.5-deg LEO parking orbit to a particular lunar orbit, including a 21-day launch period, will cost approximately 161 m/s, not including statistical costs and/or other deterministic costs. Of course, the 161 m/s accounts for a 3-sigma high value, evaluated from a large set of random mission designs; it is likely that a practical mission may be constructed for significantly less ΔV.

A 21-day launch period does not necessarily have to include 21 consecutive days; in fact, most launch periods constructed in this examination include one or two gaps when the launch operations would have to stand down. The average launch period for the sample set used here requires a total of 27 days; the vast majority of the launch periods may be contained within 40 days.

Finally, it has been found that there is no significant trend between the total launch period ΔV for the sample missions studied here and their reference departure inclination values or their reference transfer durations, except that missions with short durations (< 90 days) require more ΔV to establish an extended launch period, on account of the reduced flexibility of a shorter transfer.

6.2 OPERATIONS INTRODUCTION

This chapter discusses several aspects of a spacecraft mission that must be considered for the low-energy transfers presented in this book to be used in a real mission. Numerous discussions throughout Chapters 3–5 have considered the latitude of the mission's launch site, since that strongly influences the inclination of the parking orbit that may be used in a mission. But other aspects have not been fully discussed, such as which launch vehicle may be used, how to establish a launch period, and what considerations must be made to a spacecraft's design to fly a low-energy transfer.

Sections 6.3 and 6.4 provide information and discussion about which launch sites and launch vehicles are typically used and/or available for lunar missions. Section 6.5 provides a lengthy discussion, analysis, and several algorithms that may be used to generate a 21-day launch period for a given low-energy transfer. The results indicate that simple low-energy transfers may be targeted from nearly any LEO parking orbit with a 21-day launch period for a modest fuel cost on the order of 72 m/s. Section 6.6 discusses issues relevant to navigating a spacecraft while on a low-energy transfer, including the costs of station-keeping and the benefits of having 3–4 months to perform the transfer instead of the conventional 3–6 days. Finally, Section 6.7 presents several considerations that must be made to the spacecraft systems and operations design to accommodate a low-energy lunar transfer.

6.3 LAUNCH SITES

Chapters 3–5 illustrated that low-energy ballistic transfers may be constructed that depart the Earth from parking orbits or direct departures with any orbital inclination. By carefully selecting a particular transfer, one may build a mission that launches from any given launch site and efficiently injects into the ballistic lunar transfer. While this is very important for conventional mission design, Section 6.5 later demonstrates that a mission can actually depart the Earth from virtually any inclination and transfer to a particular lunar arrival for a modest ΔV cost. Still, it is of interest to build a low-energy transfer that is designed to depart the Earth with an inclination that is very similar to the latitude of the mission's launch site so that no sizable orbital plane changes are needed. This is particularly useful for missions with a brief launch period.

Table 6-1 provides a summary of the launch sites that have demonstrated the capability of placing large payloads into orbit. This is not a complete list, but provides a good review of the latitude and longitude of several sites for reference.

6.4 LAUNCH VEHICLES

Many launch vehicles are available to place spacecraft on low-energy lunar transfers. The NASA Launch Services Program (LSP) at Kennedy Space Center coordinates contracts with several launch vehicle providers using NASA Launch Services (NLS) contracts [211]. On September 16, 2010, NASA released the details about the NLS II contacts that were awarded to four launch vehicle providers: Lockheed Martin Space Systems Company of Denver, Colorado; Orbital Sciences Corporation of Dulles, Virginia; Space Exploration Technologies of Hawthorne, California; and United Launch Services, LLC of Littleton, Colorado. This contract includes several families of launch vehicles, including Atlas V, Falcon 9, Pegasus XL, Taurus XL, Athena I, and Athena II. The NLS II contract provides the minimum performance that is contractually obligated by the launch vehicle; a mission may be able to negotiate with the launch vehicle provider to increase the performance of the launch vehicle depending on the mission's requirements [211].

Table 6-2 summarizes the maximum payload capabilities of several launch vehicles injected from Cape Canaveral, Florida, onto low-energy lunar transfers with and without an outbound lunar flyby. The table captures two extreme cases: first, the case where the transfer includes an outbound lunar flyby and requires an injection C_3 of -2.1 kilometers squared over seconds squared (km^2/s^2), which is near the

Table 6-1 The locations of several launch sites that have been used to launch large payloads into orbit, toward the Moon, and/or into Interplanetary space. This is not a complete list, the locations are approximate, and some are representative of several particular launch sites.

Country	Location	Latitude (deg)	Longitude (deg)	Comments
USA	Cape Canaveral Air Force Station, Florida	28.47 N	80.56 W	Interplanetary
USA	Kennedy Space Center, Florida	28.61 N	80.60 W	Lunar
USA	Vandenberg Air Force Base, California	34.77 N	120.60 W	High inclinations
USA	Kodiak Launch Complex, Alaska	57.44 N	152.34 W	Orbital
USA	Mid-Atlantic Regional Spaceport (MARS), Delmarva Peninsula, Virginia	37.83 N	75.48 W	Orbital
USA	Kwajalein Atoll	9.00 N	167.65 E	Orbital
Brazil	Alcântara Launch Center, Maranhão	2.32 S	44.37 W	Orbital
China	Jiuquan Satellite Launch Center	41.12 N	100.46 E	Orbital
China	Xichang Satellite Launch Center	28.25 N	102.03 E	Lunar
French Guiana	Guiana Space Centre, Kourou	5.24 N	52.77 W	Interplanetary
India	Satish Dhawan Space Centre (Sriharikota), Andhra Pradesh	13.74 N	80.24 E	Lunar
Israel	Palmachim Air Force Base	31.88 N	34.68 E	Orbital
Japan	Uchinoura Space Center	31.25 N	131.08 E	Orbital
Japan	Tanegashima Space Center, Tanegashima Island	30.39 N	130.97 E	Orbital
Kazakhstan	Baikonur Cosmodrome, Tyuratam	45.96 N	63.35 E	Interplanetary
Marshall Island	Omelek	9.05 N	167.74 E	Orbital
Russia	Svobodny Cosmodrome, Amur Oblast	51.83 N	128.28 E	Orbital
Russia	Yasny Cosmodrome, Orenburn Oblast	51.21 N	59.85 E	Orbital
Russia	Kapustin Yar Cosmodrome, Astrakhan Oblast	48.58 N	46.25 E	Orbital
Sweden	Esrange, Kiruna	67.89 N	21.10 E	Orbital
Several	Sea Launch / Ocean Odyssey complex	0.0 N	Varies	Orbital

Table 6-2 The payload capabilities of several launch vehicles injected from Cape Canaveral, Florida, onto low-energy lunar transfers with and without an outbound lunar flyby. This information has been captured from the NASA Launch Services (NLS) Program's Launch Vehicle Performance site under the NLS II contract [211].

Launch Vehicle	Maximum Payload Performance (kg)	
	$C_3 = -2.1\ km^2/s^2$	$C_3 = -0.3\ km^2/s^2$
Athena II	395.0	375.0
Falcon 9 Block 1	2125.0	1995.0
Falcon 9 Block 2	2645.0	2515.0
Atlas V 401	3170.0	3050.0
Atlas V 411	4095.0	3955.0
Atlas V 421	4845.0	4680.0
Atlas V 431	5445.0	5265.0
Atlas V 501	2215.0	2110.0
Atlas V 511	3410.0	3285.0
Atlas V 521	4365.0	4215.0
Atlas V 531	5135.0	4965.0
Atlas V 541	5815.0	5625.0
Atlas V 551	6340.0	6140.0

minimum injection energy typically required. The second case presented requires a C_3 of $-0.3\ km^2/s^2$, which is near the maximum injection energy typically required without a lunar flyby. Most missions will fall between these two values: closer to one depending on whether or not the mission aims to fly past the Moon on the outbound segment.

As of September 2011, Orbital Sciences estimates that the Taurus XL may be used to inject as much as 425 kilograms (kg) to a C_3 of $0\ km^2/s^2$, and presumably more to a low-energy lunar transfer. Further, although it is not currently in the NLS II contract, Orbital Sciences estimates that the Taurus II launch vehicle may be able to inject between 920 kg and 1120 kg to a C_3 of $-2.1\ km^2/s^2$ depending on its configuration. The Taurus II's performance drops about 40 kg when injecting payloads to a C_3 of $-0.3\ km^2/s^2$.

In addition, the Pegasus XL launch vehicle may be used to place up to about 470 kg of payload into a 200-km circular parking orbit [212]. A spacecraft could then perform its own trans-lunar injection to transfer to the Moon, much like the proposed *Dust Near Earth* (*DUNE*) mission [146, 213], or similar to the *Interstellar Boundary Explorer* (*IBEX*) mission [214–216].

Other launch vehicles may also be used to inject a spacecraft onto a low-energy lunar transfer, though they do not have a contract with NASA, including the Delta IV family of vehicles. Certainly several international vehicles may be used, assuming the vehicles' guidance algorithms have the capability of targeting such orbital parameters, including the Russian Soyuz and Proton vehicles, Arianespace's Ariane V, China's Long March and CZ vehicles, Japan's H-IIA and H-IIB, and Ukraine's Zenit-3SL, among others. The Indian Space Research Organization's (ISRO's) Polar Satellite Launch Vehicle (PSLV-C11) was used to launch the *Chandrayaan-1* mission to the Moon, though the launch vehicle only injected the spacecraft into a 6-hour orbit about the Earth and the spacecraft performed the remainder of the lunar injection.

6.5 DESIGNING A LAUNCH PERIOD

This section considers how to construct an extended launch period for a low-energy transfer to the Moon. The discussion begins by reviewing several interesting features that exist in the Earth–Moon system and how historical launch periods have been constructed around those features. This provides context for future discussions about designing launch periods for low-energy transfers.

First, the Moon's orbit is nearly circular about the Earth. This means that one may theoretically launch a spacecraft on a conventional, direct transfer with very similar characteristics on any given day. The Moon's elliptical orbit means that the launch energy and transfer duration will vary across the month to some degree, but this is a second-order effect. The largest variation from one day to the next when injecting into a direct direct transfer arises from the obliquity of the Earth relative to the Moon's orbit. The Earth's spin axis is tilted approximately 23.5 deg relative to the ecliptic, and the Moon's orbit has an inclination of about 5.15 deg with respect to the ecliptic. Together, this means that the relative orientation of the Earth's spin axis and the orbit of the Moon may be anywhere from 18.35 deg to 28.65 deg; the orientation of the parking orbit must be adjusted to accommodate this shift. Ultimately this means that the time of day that one must launch shifts from one day to the next, as does the duration of time that the spacecraft coasts in a low Earth parking orbit prior to injecting toward the Moon.

Next, a lunar day is approximately 29.5 Earth days long, which means that the lighting conditions on the Moon roughly repeat every 29.5 days. There are variations on top of this cycle that correspond with where the Moon is in its orbit about the Earth relative to its perigee, and where the Earth is in its orbit about the Sun. The net effect is that if one is interested in viewing a particular lighting condition as one flies by the Moon or impacts the Moon, then one may only be able to launch on a direct transfer one or two days every month. This is very important for missions that aim to land on the surface, including the Apollo missions. The Apollo missions were designed to land on the surface soon after sunrise at the landing site to maximize the amount of sunlit time they had on the surface before needing to ascend. These considerations have a direct effect on the time of arrival at the Moon for any mission, though missions that go into orbit prior to landing/impact can arrive early. The time of arrival is highly correlated with the launch time for direct transfers, since direct transfers have a short transfer duration that cannot be varied much. The time of arrival is loosely correlated with the launch time for low-energy transfers, since low-energy transfers can vary their transfer durations by many days without a large penalty in transfer ΔV.

Another consideration for a mission planner is that many lunar spacecraft are not designed to survive a long eclipse. Lunar eclipses occur roughly every 6 months when the Earth comes directly between the Sun and Moon. The Moon's nonzero orbital inclination relative to the ecliptic means that a lunar eclipse does not occur each and every month, but only occurs when the Moon is near its ascending or descending node when it traverses behind the Earth. Since the Moon's orbit is fixed in inertial space, though subject to perturbations, one of the nodes traverses directly behind the Earth twice per year. If the Moon is near that point in its orbit at that time, then the eclipse will be a full lunar eclipse and any spacecraft on the surface or in a low orbit will traverse through the umbra of the Earth. If the Moon is not near that point in its orbit, then the spacecraft may be able to avoid the shadow, or at least avoid the umbra of the Earth. The *Gravity Recovery and Interior Laboratory* (*GRAIL*) mission was designed with lunar eclipses in mind, since the two *GRAIL* spacecraft were not originally designed to survive an extended passage through shadow. *GRAIL*'s entire science phase was designed to occur between two lunar eclipses in case one of the spacecraft did not survive the following eclipse. This means that *GRAIL*'s launch opportunities do not repeat every month, but only repeat once every six months.

6.5.1 Low-Energy Launch Periods

Low-energy lunar transfers are more flexible than direct lunar transfers since their transfer durations are longer; hence, it is possible to build an extended, 21-day launch period such that every launch opportunity yields a trajectory that a spacecraft can follow that arrives at the Moon at the same time. This is very useful for missions such as *GRAIL* that depend on arriving at the Moon at a particular time of the year or month.

There are often many ways to adjust a trajectory's design so that it may depart the Earth on multiple days, in order to establish a launch period. For this discussion we assume that the trajectory begins with a launch from a particular site into a low, near-circular parking orbit; coasts in the parking orbit for some duration; performs a trans-lunar injection; and then follows a ballistic transfer to the Moon using one or two trajectory correction maneuvers en route to the Moon. Given this trajectory design, several examples of controls include the following:

- Adjust the launch time. By launching at a different time of day, one can change the longitude of the ascending node of the parking orbit that the spacecraft uses prior to its trans-lunar injection.

- Adjust the launch and parking orbit geometry. One may be able to reduce the total transfer ΔV cost and ultimately transfer more payload mass to the Moon by changing the parking orbit's inclination. This reduces the launch vehicle's performance, but it may be worthwhile.

- Adjust the location of the trans-lunar injection maneuver in the parking orbit.

- Adjust the trans-lunar injection maneuver. The maneuver magnitude and/or direction may be adjusted, depending on the control algorithm that operates the maneuver. In the studies presented in this chapter, only the maneuver magnitude is adjusted.

- Add one or more trajectory correction maneuvers in the trans-lunar cruise. These maneuvers may be performed in any direction, though some missions may place constraints on the magnitude or direction of these maneuvers. In the studies presented here, two maneuvers are introduced that may be performed in any direction with any maneuver magnitude, though no two maneuvers may be placed within four days of each other to reduce operations complexity.

- Adjust the lunar arrival conditions as described below.

The available controls upon arriving at the Moon depend on the arrival orbit/geometry and the mission design. Some examples of different missions and their controls include:

Arriving at a lunar libration orbit. Arriving at a lunar libration orbit typically involves a ballistic, asymptotic arrival with a final correction maneuver to ensure that the spacecraft is placed in the target orbit. Controls include:

- Adjust the date/time of arrival. This may vary by mere seconds or by days, depending on the mission's requirements.

- Vary the target libration orbit. It is typically more desirable to maintain a single target libration orbit throughout the launch period, though that depends on the mission.

- Add a libration orbit insertion maneuver, which may vary in magnitude/direction. This is typically much more useful if the target libration orbit is held fixed across a launch period.

Arriving at a low lunar orbit. Arriving at a low lunar orbit typically involves a time-critical lunar-orbit insertion (LOI) maneuver that places the spacecraft into a capture orbit. Controls include:

- Adjust the date/time of the LOI. This may vary by mere seconds or by days, depending on the mission's requirements.

- Adjust the LOI's magnitude and/or direction. Some spacecraft designs require that the maneuver be a fixed-attitude maneuver, a pitch-over maneuver, or a maneuver that rotates about a specified axis at a constant rate. The studies presented here model the maneuver using an impulsive burn and frequently permit the burn to vary in both magnitude and direction.

- Adjust the location of the LOI within the target orbit. This is typically held constant, or varied only a small amount, since the maneuver is much more efficient when performed at the orbit's periapse.

- Adjust the geometry of the capture orbit. The spacecraft's mission design may permit the orbit's argument of periapse to vary, particularly if the goal is to eventually enter a circular orbit. It may also be permissible to vary the inclination or longitude of ascending node of the orbit, though those are typically not varied more than a small amount.

Arriving at the lunar surface. A mission to the lunar surface may be targeting a shallow flight path angle with the goal to land softly, or it may be targeting a steep flight path angle for a targeted impact, similar to the design of the *Lunar Crater Observatory and Sensing Satellite* (*LCROSS*) mission. Some examples of trajectory controls include:

- Adjust the date/time of arrival. This may vary by mere seconds or by days, depending on the mission's requirements.

- Adjust the arrival velocity.

- Adjust the arrival geometry. It may be permissible to vary the flight path angle and/or azimuth of the arrival.

- Adjust the arrival location on the lunar surface.

In addition, it may be possible to incorporate a dramatic shift in a mission's trajectory. For instance, it may be preferable to break a 21-day launch period into two halves, where the early portion of the launch period sends the spacecraft toward the Sun–Earth L_1 vicinity and the second portion implements trajectories that travel near the Sun–Earth L_2 vicinity.

One can see that there are many ways to adjust a trajectory from one launch opportunity to the next in order to establish a launch period. This section presents several scenarios and their corresponding algorithms that may be used to establish a launch period. The algorithms presented here may need to be adjusted for a particular mission, though the results presented here are certainly useful for guiding the early trades for a mission.

6.5.2 An Example Mission Scenario

There are many ways to construct an extended launch period for a low-energy lunar transfer, some of which are outlined above. This section studies one mission design architecture and applies that to a large number of practical cases, in order to generate some useful statistics about that architecture. The design studied here is similar to *GRAIL*'s mission: a spacecraft is launched from a parking orbit that has an inclination of 28.5 deg, for example, one that effectively supports launches from Cape Canaveral, and uses a near-ballistic low-energy transfer to target a low, 100-km, polar orbit about the Moon. The trajectory includes as many as two deterministic trajectory correction maneuvers (TCMs) to assist the construction of a 21-day launch period. The results of the studies presented here for this architecture are (of course) only relevant to very similar missions, but hopefully they shed some light on other low-energy lunar architectures.

Figure 6-1 illustrates one example trajectory taken from the surveys presented in Section 4.4. This trajectory departs the Earth on April 1, 2010, at 05:27 Coordinated Universal Time (UTC) from a 185-km parking orbit with an inclination of approximately 38.3 deg and transfers to the Moon using no maneuvers at all. It arrives at a polar orbit 100 km above the mean radius of the Moon. A launch vehicle may certainly target an outbound inclination of 38.3 deg on that date to inject a spacecraft onto this transfer, but it would suffer a large penalty to its lift capability if it did so from Cape Canaveral, compared to the vehicle's capability to lift payloads to an inclination of 28.5 deg. Further, the launch may slip. This section studies how to adjust the transfer to permit it to depart the Earth from an inclination of 28.5 deg on multiple days. As an example, a new trajectory has been generated using the ballistic transfer shown in Fig. 6-1 as a reference. The new trajectory departs the Earth a full day after the reference, on April 2, 2010, and departs from a 28.5 deg parking orbit. Two maneuvers are required to correct this new outbound trajectory so that it arrives at the same lunar orbit as the reference. Figure 6-2 illustrates the difference between

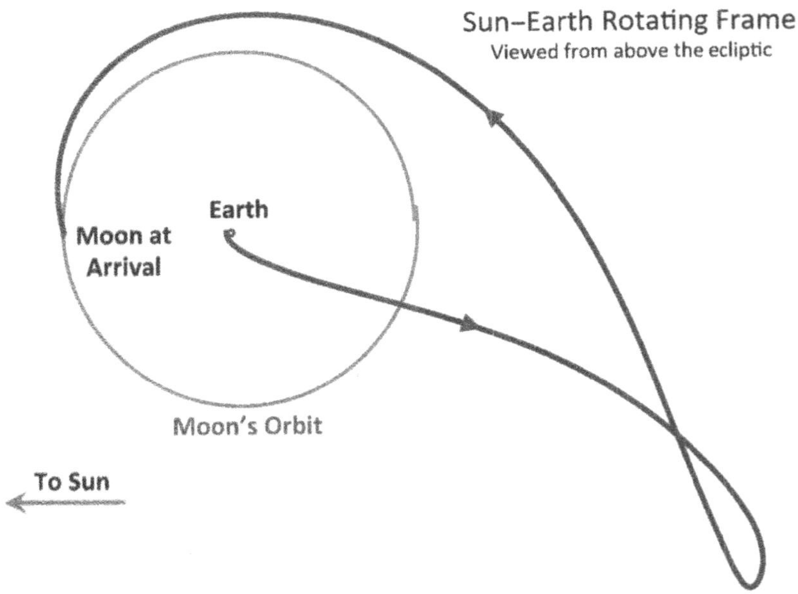

Figure 6-1 An illustration of the example reference low-energy lunar transfer, shown in the Sun–Earth rotating frame from above the ecliptic, where the Sun is fixed to the left [190] (Copyright © 2012 by American Astronautical Society Publications Office, San Diego, California (Web Site: http://www.univelt.com), all rights reserved; reprinted with permission of the AAS).

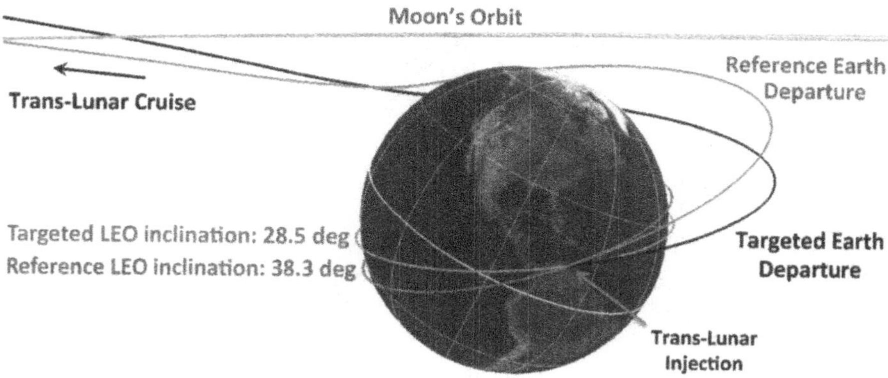

Figure 6-2 The targeted Earth departure compared with the reference Earth departure [191] (*Acta Astronautica* by International Academy of Astronautics, reproduced with permission of Pergamon in the format reuse in a book/textbook via Copyright Clearance Center). *(See insert for color representation of this figure.)*

the Earth departures of the reference trajectory and the newly adjusted trajectory. Figure 6-3 shows the difference between these transfers, as viewed from above in the Sun–Earth rotating reference frame. Finally, Fig. 6-4 shows a top-level view of 21 such trajectories that depart the Earth on 21 different days and all arrive at the Moon at the same time at the same orbit. The details of these trajectories, and whether or not they should vary in any given way, is described later.

The performance of the launch period for this example mission depends on which controls are available. For instance, if one is only permitted to vary the launch time and the trans-lunar injection, while keeping the dates of the trajectory correction maneuvers constant and keeping the geometry of the lunar orbit insertion constant, then the spacecraft must be capable of performing at least 730 m/s of ΔV to reach a 100-km circular polar orbit about the Moon. But if the dates of the TCMs are permitted to vary as well as the magnitude and direction of the lunar orbit insertion, then the spacecraft's fuel budget may be reduced such that it must perform only 706 m/s of ΔV on the most challenging launch day of a 21-day launch period. However, these controls may not be available to the mission. Figure 6-5 illustrates the total Δ V that must be performed for a spacecraft in each of five different launch period configurations. One can see that the launch period ΔV cost may be reduced even by adjusting a single parameter; for instance, Launch Period 3 requires approximately 10.7 m/s less ΔV than Launch Period 2 and the only thing different is that the date of the second TCM is performed 10 days later in each trajectory.

The illustrations shown here are representative of one example lunar mission. This section explores several hundred such missions and characterizes any statistical

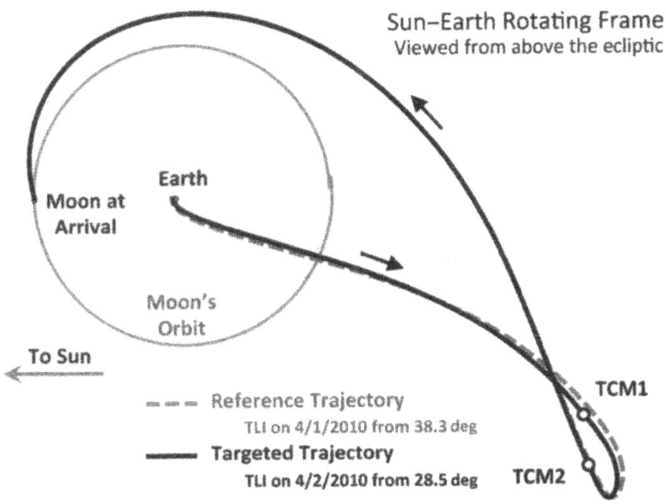

Figure 6-3 The final targeted lunar transfer compared to the reference transfer, viewed in the Sun–Earth rotating frame from above the ecliptic [191] (*Acta Astronautica* by International Academy of Astronautics, reproduced with permission of Pergamon in the format reuse in a book/textbook via Copyright Clearance Center).

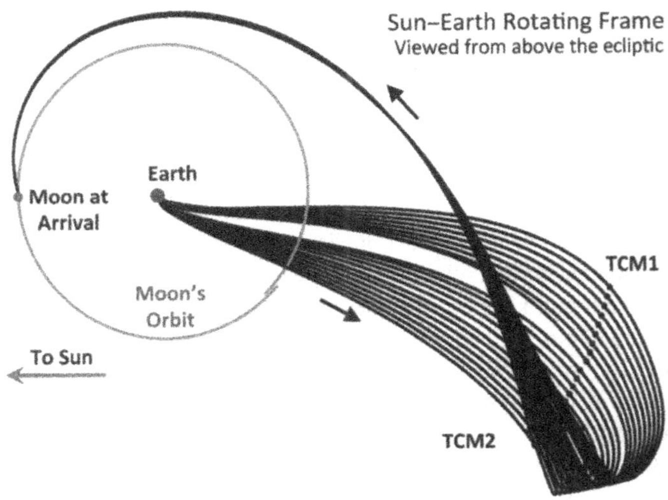

Figure 6-4 An example of 21 trajectories that depart the Earth from 21 different days and all arrive at the Moon at the same time, inserting into the same lunar orbit. The trajectories are viewed in the Sun–Earth rotating frame from above the ecliptic.

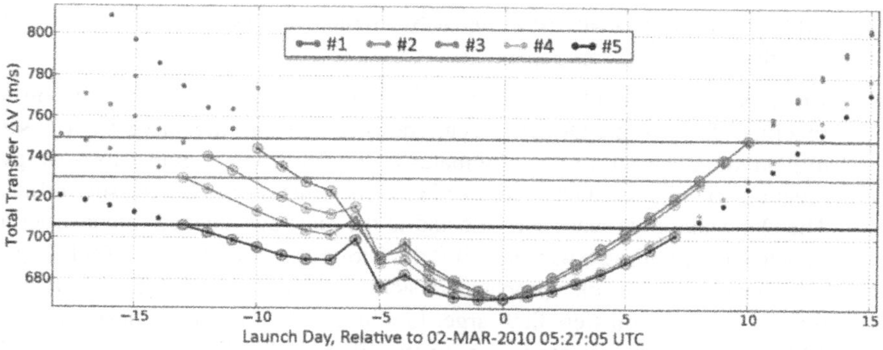

Control	#1	#2	#3	#4	#5	Comments
Launch Time	Vary	Vary	Vary	Vary	Vary	This permits the node of the parking orbit to vary.
Parking Orbit Inc.	Fixed	Fixed	Fixed	Fixed	Fixed	All held the parking orbit inclination fixed at 28.5 deg.
TLI mag, location	Vary	Vary	Vary	Vary	Vary	The magnitude and location of the TLI maneuver.
TCM 1 Epoch	L+21d	L+21d	L+21d	L+21d	Vary	The epoch of TCM1 (most 21 days after launch).
TCM 2 Epoch	L+60d	LOI–40d	LOI–30d	LOI–40d	Vary	The epoch of TCM2, either relative to launch or LOI.
LOI Epoch	Fixed	Fixed	Fixed	Fixed	Fixed	The epoch of the LOI.
LOI ΔV	Fixed	Fixed	Fixed	Vary	Vary	The magnitude and direction of the LOI.
Total ΔV (m/s)	749.2	740.4	729.7	705.9	706.2	Including two TCMs and the full LOI (impulsive).

Figure 6-5 Several example launch periods for the example lunar mission, depending on which controls are fixed, their fixed values, and which controls are permitted to vary. *(See insert for color representation of this figure.)*

findings that provide mission managers rules of thumb for estimating the costs of establishing a launch period for a given low-energy transfer.

6.5.3 Targeting Algorithm

Each lunar mission and its corresponding launch period is constructed here using a straightforward procedure that is described as follows. Once again, this algorithm is formulated for missions to low lunar orbits, though it may be easily modified for other destinations.

Step 1. First, a target lunar orbit is selected and a reference low-energy lunar transfer is constructed. The transfers used here have been taken from the surveys presented in Section 4.4, which provides many more details about these transfers, but to summarize, each transfer targets a low lunar orbit that is constructed by setting its semi-major axis to 1837.4 km, its eccentricity to zero, and its inclination to 90 deg in the International Astronomical Union (IAU) Moon Pole coordinate frame. This defines a circular, polar orbit with an altitude of approximately 100 km. Its longitude of ascending node, Ω, and argument of periapse, ω, are selected from the surveys and can take on a wide variety of combinations.

An impulsive, tangential LOI is applied at the orbit's periapse point on a specified date. The LOI ΔV magnitude is taken from the surveys. It is set to generate a trajectory that originates at the Earth via a simple low-energy transfer: one that contains no close lunar encounters or Earth-phasing orbits. The ΔV value is at least 640 m/s and is the least ΔV needed to construct a transfer that requires fewer than 160 days to reach an altitude of 1000 km or less above the Earth when propagated backward in time. Table 6-3 summarizes several example transfers that target low lunar orbits that each have an Ω of 120 deg; these may be seen in the surveys illustrated in Figs. 4-6, 4-7, and 4-8 and in Table 4-4 in Section 4.4.

Each reference trajectory generated in this study has no maneuvers and does not target any particular Earth orbit when propagated backward in time.

Step 2. Second, the mission's LEO parking orbit and trans-lunar injection time are specified. The LEO parking orbits used in this study are all 185-km circular orbits with inclinations of 28.5 deg, as previously described. The orbit's node, Ω_{LEO}, and the location of the trans-lunar injection (TLI) maneuver about the orbit, ν_{LEO}, are permitted to vary; the TLI is performed impulsively and tangent to the orbit. The values of Ω and ν_{LEO} may initially be set to any arbitrary angle, for example, to 0 deg.

Step 3. If the LOI maneuver is permitted to vary, which it is in the majority of the missions studied here, then the third step is to adjust the low-energy transfer such that its perigee passage occurs at the time of the TLI. This is performed by searching for the smallest change in the LOI $\overrightarrow{\Delta V}$ that results in a new low-energy transfer that originates at the Earth on the date of the TLI, or at

Table 6-3 A summary of the performance parameters of several example simple low-energy lunar transfers. None of these transfers includes any Earth phasing orbits or lunar flybys [190] (Copyright © 2012 by American Astronautical Society Publications Office, San Diego, California (Web Site: http://www.univelt.com), all rights reserved; reprinted with permission of the AAS).

Traj #	Ω (deg)	ω (deg)	ΔV_{LOI} (m/s)	Duration (days)	LEO Inclination (deg) Equatorial	Ecliptic	C_3 (km^2/s^2)
1	120.0	169.2	669.3	83.483	29.441	6.129	−0.723
2	120.0	103.8	692.1	85.287	25.688	34.778	−0.723
3	120.0	70.2	743.9	93.598	57.654	74.955	−0.667
4	120.0	225.3	716.0	93.621	134.322	112.840	−0.657
5	120.0	99.9	697.5	110.060	83.127	61.624	−0.697
6	120.0	186.9	673.2	122.715	23.941	3.088	−0.712

least one that has a perigee on that date even if the perigee altitude is higher than 1000 km. The optimization package sparse nonlinear optimizer (SNOPT) was used for the missions presented here, but other algorithms may certainly be used.

Step 4. The radius of the low-energy transfer with respect to the Earth at a time 20 days after the TLI is noted. The TLI ΔV magnitude, ΔV_{TLI}, is set to a value that takes the Earth-departure trajectory out to that distance at that time. The spacecraft is beyond the orbit of the Moon by that time, assuming no Earth-phasing orbits, and not yet at its apogee.

Step 5. The values of Ω_{LEO} and ν_{LEO} are adjusted to minimize the difference in position between the Earth-departure and the target low-energy transfer at a time 20 days after TLI. After convergence, the algorithm is repeated, this time permitting ΔV_{TLI} to vary as well. It is typically the case that the Earth-departure trajectory will intersect the target low-energy transfer at that time when all three variables are permitted to vary, though it is not necessary. Once again this study implemented the SNOPT package to perform the optimization.

Step 6. Two deterministic maneuvers are added to the trajectory: TCM1 at a time 21 days after TLI, and TCM2 at a time halfway between TCM1 and LOI. It is intentional that the first maneuver be placed near 20 days but not at a value of 20 days in order to improve the performance of the optimization algorithm in the next step [183]. A single-shooting differential corrector (Section 2.6.5.1) may be used to target the values of ΔV_{TCM1} and ΔV_{TCM2} to generate a continuous end-to-end trajectory.

Step 7. Finally, all control parameters are varied using an optimizer to minimize the total transfer ΔV of the trajectory. This study again used the SNOPT package to perform the optimization. The missions generated here permitted eight control variables to vary: the three Earth-departure parameters Ω_{LEO}, ν_{LEO}, and ΔV_{TLI}; the dates of the two trans-lunar maneuvers t_{TCM1} and t_{TCM2}; and the three components of the LOI ΔV, namely, ΔV^x_{LOI}, ΔV^y_{LOI}, and ΔV^z_{LOI}. When the eight parameters are adjusted, an Earth-departure trajectory is generated out to the time of TCM1, a lunar-arrival trajectory is generated backward in time from LOI to the time of TCM2, and a bridge trajectory is generated connecting TCM1 and TCM2 using a single-shooting differential corrector (Section 2.6.5.1). The total transfer ΔV that is minimized includes the sum of the maneuvers that are typically required by the spacecraft, namely, the sum of ΔV_{TCM1}, ΔV_{TCM2}, and ΔV_{LOI}, but not the TLI ΔV. The dates of the TLI and LOI are fixed, and the dates of TCM1 and TCM2 are constrained to be at least four days from any other maneuver to facilitate relaxed spaceflight operations.

When the optimizer has converged, the performance of the trajectory compared with the reference low-energy transfer is recorded. It is often the case that the differential corrector will converge on a local minimum and not the global minimum;

hence, this process is repeated with adjustments in the eight parameters to identify the lowest local minimum possible. This will be discussed more later.

To summarize, this procedure constructs a practical, two-burn, low-energy lunar transfer between a specified Earth departure and a specified lunar arrival. The altitude, eccentricity, and inclination of the Earth parking orbit are specified and fixed, as is the date of the trans-lunar injection maneuver. The target lunar orbit, the LOI position, and the LOI date are all specified and fixed. The TLI maneuver is constrained to be tangential to the parking orbit, though the orientation of the parking orbit may vary; the LOI maneuver is not constrained to be tangential. Finally, the dates of two trans-lunar maneuvers are permitted to vary, which therefore changes their ΔV values.

To illustrate this entire targeting process, Table 6-4 tracks the eight control variables that have been used to generate the adjusted trajectory shown in Figs. 6-2 and 6-3. That is, Table 6-4 shows the steps that were taken to adjust the trajectory from the reference ballistic transfer, which has an Earth departure inclination of 38.3 deg on April 1, 2010, to the desired transfer, which has an Earth departure inclination of 28.5 deg on April 2, 2010. The reference trajectory is summarized in Step 1: the only control variables set are the components of the LOI ΔV. Step 2 does not change any control variables and is hence not shown. Step 3 illustrates the small change in the components of the LOI ΔV vector that are required to shift the timing of the trajectory's perigee from April 1, 2010 to April 2, 2010, coinciding with the TLI maneuver, though the perigee altitude is now 5200 km. The adjustment amounts to a difference of only 3.3 centimeter per second (cm/s) in the LOI ΔV magnitude. Steps 4–6 construct initial guesses for the departure parameters and place two deterministic maneuvers en route to construct a complete end-to-end trajectory. Finally, Step 7 includes the full optimization, where all eight parameters are permitted to vary and the transfer ΔV is minimized.

During Step 4, initial guesses for Ω_{TLI} and ν_{TLI} are needed. In this example they are both set to 0 deg; however, it has been observed that the entire procedure may converge to different local minima using different combinations of initial guesses for these parameters. There are often two local minima that correspond to the typical *short* and *long* coasts for the Earth departure. In addition, the process often converges on different local minima depending on the propagation duration of the initial Earth departure. Research indicates that it is typically computationally efficient to perform Steps 4–6 numerous times with different initial guesses and then send only the best one or two trajectories into Step 7 [190, 191]. This process ensures that the majority of local minima are explored without spending too much time in Step 7, which is by far the most computationally demanding step. It is likely that additional small improvements may be made, but this procedure generates a reliable estimate of the minimum transfer ΔV given a reference lunar transfer.

Taking the preceding into account, this targeting algorithm yields a trajectory that requires only 24.1 m/s of deterministic ΔV to compensate for the change in departure inclination and departure date. This deterministic ΔV will vary throughout a full launch period, but this is a small ΔV penalty compared to the cost of launching into parking orbits at widely varying inclinations.

Table 6-4 The history of the example lunar transfer's control variables as the mission is constructed, where Δt_{TCM1} is the duration of time between TLI and TCM1 and Δt_{TCM2} is the duration of time between TCM1 and TCM2 [190] (Copyright © 2012 by American Astronautical Society Publications Office, San Diego, California (Web Site: http://www.univelt.com), all rights reserved; reprinted with permission of the AAS).

	TLI Parameters			TCM1		TCM2		LOI	Total
Step #	Ω deg	ν deg	ΔV m/s	Δt days	ΔV m/s	Δt days	ΔV m/s	ΔV_x, ΔV_y, and ΔV_z m/s, EME2000	Transfer ΔV, m/s
1	-	-	-	-	-	-	-	−87.728, −271.090, −583.108	-
3	-	-	-	-	-	-	-	−87.732, −271.103, −583.138	-
4	0.00	0.00	3197.44	-	-	-	-	−87.732, −271.103, −583.138	-
5	−25.00	27.18	3196.77	-	-	-	-	−87.732, −271.103, −583.138	-
6	−25.00	27.18	3196.77	21.00	26.10	34.84	6.37	−87.732, −271.103, −583.138	681.500
7	−25.08	27.32	3196.79	20.63	24.09	34.86	0.00	−87.736, −271.118, −583.167	673.155

6.5.4 Building a Launch Period

The process described in 6.5.1 may be repeated for each day in a wide range of dates to identify a practical launch period. The total transfer ΔV typically rises as the TLI date is adjusted further from a reference trajectory's TLI date. For the purpose of these studies, a search is conducted that extends 30 days on either side of the reference trajectory's TLI date and the best, practical, 21-day launch period is identified within those 61 days. The 21 days of opportunities do not have to be consecutive, though they are typically collected in either one or two segments. Since low-energy transfers travel beyond the orbit of the Moon, they may interact with the Moon as they pass by, even if they pass by at a great distance. The Moon may boost or reduce the spacecraft's energy as it passes by, depending on the geometry; typically there is a point in a launch period where the geometry switches and it is often beneficial to avoid launching on one or several days when the geometry is not ideal.

Figure 6-6 illustrates the transfer ΔV cost required to target the reference lunar transfer studied in the previous section as a function of TLI date. Each transfer has been generated using the procedure outlined previously, but with a different TLI date. The trajectories that launch 5–6 days prior to the reference transfer are significantly perturbed by the Moon, though not perturbed enough to break the launch period into two segments. This perturbation is also visible in Fig. 6-4, where a sudden change in the geometry of the transfers appears. One can see that the least expensive 21-day launch period requires a transfer ΔV of approximately 706.2 m/s.

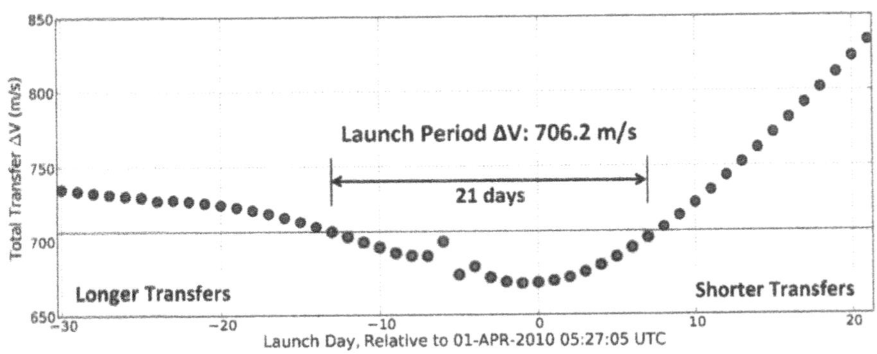

Figure 6-6 An example 21-day launch period, constructed using the reference lunar transfer presented in Fig. 6-1 [190] (Copyright © 2012 by American Astronautical Society Publications Office, San Diego, California (Web Site: http://www.univelt.com), all rights reserved; reprinted with permission of the AAS).

6.5.5 Reference Transfers

A total of 288 reference transfers have been used to generate lunar missions with realistic, 21-day launch periods, each starting from a 28.5-deg LEO parking orbit. These reference trajectories have been randomly sampled from low-ΔV, simple, low-energy transfers presented in the surveys found in Section 4.4. The trajectories target low lunar orbits with any longitude of ascending node and with any argument of periapsis, though the combination of those parameters must yield a satisfactory reference transfer. The transfers arrive at the Moon at any of eight arrival times evenly distributed across a synodic month between July 11, 2010 at 19:41 UTC and August 6, 2010 at 20:37 UTC. The majority of the reference transfers sampled here implement lunar orbit insertion maneuvers with magnitudes between 640 m/s and 750 m/s, though reference transfers have been sampled with LOI ΔV values as high as 1080 m/s. These ΔV values correspond with the full cost of capturing and reducing the orbit to a 100-km circular orbit; although that process typically involves many maneuvers, in this study it will be performed by one maneuver. Finally, reference transfers have been sampled with transfer durations between 65 and 160 days. This collection of reference transfers makes no assumptions about what sort of mission a designer may be interested in, except that each transfer is simple, that is, it includes no Earth-phasing orbits nor lunar flybys, and each transfer targets a polar lunar orbit.

6.5.6 Statistical Costs of Desirable Missions to Low Lunar Orbit

In general, the algorithms described in this section generate successful launch periods with similar characteristics. Figure 6-7 illustrates the total transfer ΔV of several example launch periods that have been generated from these reference transfers. One notices that many of these launch periods include a single main convex ΔV minimum, from which a 21-day launch period is easily identified. Other ΔV curves include two or more local minima. The launch periods are designed to have at most two gaps, where each gap must be less than 14 days in extent. A particular lunar mission may have different requirements dictating the breadth of each segment and/or gap, which will likely change the launch period's cost; the requirements used here are simply representative of a real mission.

It has been found that most 21-day launch periods among the 288 missions studied include the reference launch date, though there are many examples that do not, including two of those shown in Fig. 6-7. In some cases a practical launch period may have extended further than 30 days from the reference launch date and required less total ΔV. A particular lunar mission may certainly relax this constraint, but these extended launch periods are not explored here in order to keep the constraints consistent across every mission studied.

Once again, one also sees frequent lunar perturbations in the 288 launch periods studied, much like the example launch period shown in Section 6.5.4. Since each transfer in a particular launch period departs the Earth in approximately the same direction, the Moon passes near the transfer's outbound leg about once every synodic

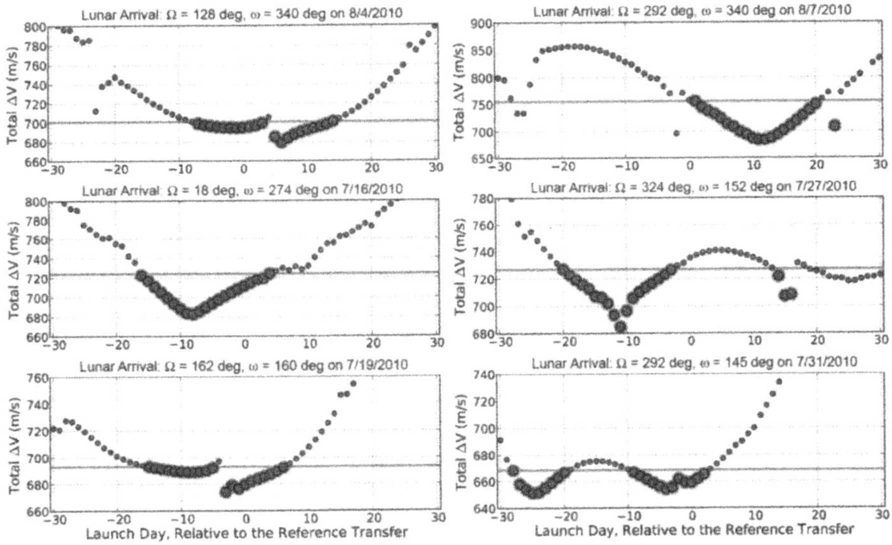

Figure 6-7 Several example curves that illustrate the post-TLI ΔV cost of transferring from a 28.5-deg LEO parking orbit at different TLI dates to a given reference low-energy transfer, including a highlighted 21-day launch period in each case [190] (Copyright © 2012 by American Astronautical Society Publications Office, San Diego, California (Web Site: http://www.univelt.com), all rights reserved; reprinted with permission of the AAS).

month. This causes a brief jump in the launch period. Some transfers do not experience any significant perturbations due to their out-of-plane motion.

Figures 6-8 and 6-9 illustrate two additional views of the six example launch periods shown in Fig. 6-7. Figure 6-8 shows the view of each trajectory in each of the six launch periods as if viewed from above the ecliptic in the Sun–Earth rotating frame, such that the Sun is toward the left in each plot. One notices that some of these launch periods traverse toward the Sun and others traverse away from the Sun. The transfers arrive at the Moon at the exact same point in each mission, but each mission arrives at the Moon at different points in its orbit. The reference transfers are sampled randomly to include a wide variety of different target lunar orbit geometries, arrival times, and transfer durations. Figure 6-9 illustrates the profile of a spacecraft's distance from the Earth over time while traversing each trajectory in each of the six launch periods. One can see that each design involves trajectories with different transfer durations, and trajectories that traverse to different maximum distances. The optimization processes often shift the epochs of the trajectory correction maneuvers, though one can see that the TCM epochs are sometimes shifted more on one trajectory than on its neighbors, depending on the sensitivity of that variable on the trajectory's ΔV costs. Similarly, some TLI parameters are shifted more in one trajectory than its

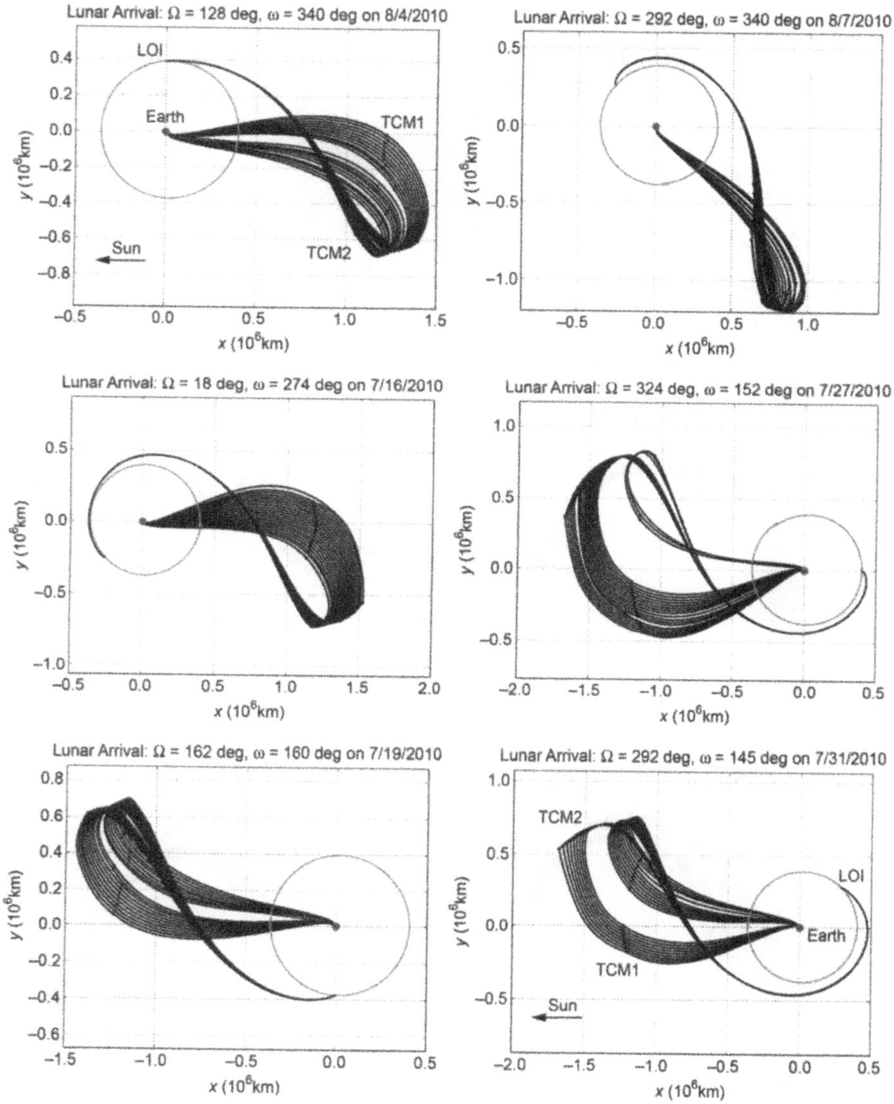

Figure 6-8 Each trajectory in each of the six launch periods illustrated in Fig. 6-7, viewed from above the ecliptic in the Sun–Earth rotating frame. The Moon's orbit is shown for reference.

neighbors. It is likely that a mission designer would use these results to guide further refinements in the optimization of a real mission.

The examples shown in Figs. 6-7–6-9 illustrate six missions; the remainder of this discussion focuses on the random sample of 288 similar missions. Figure 6-10 shows

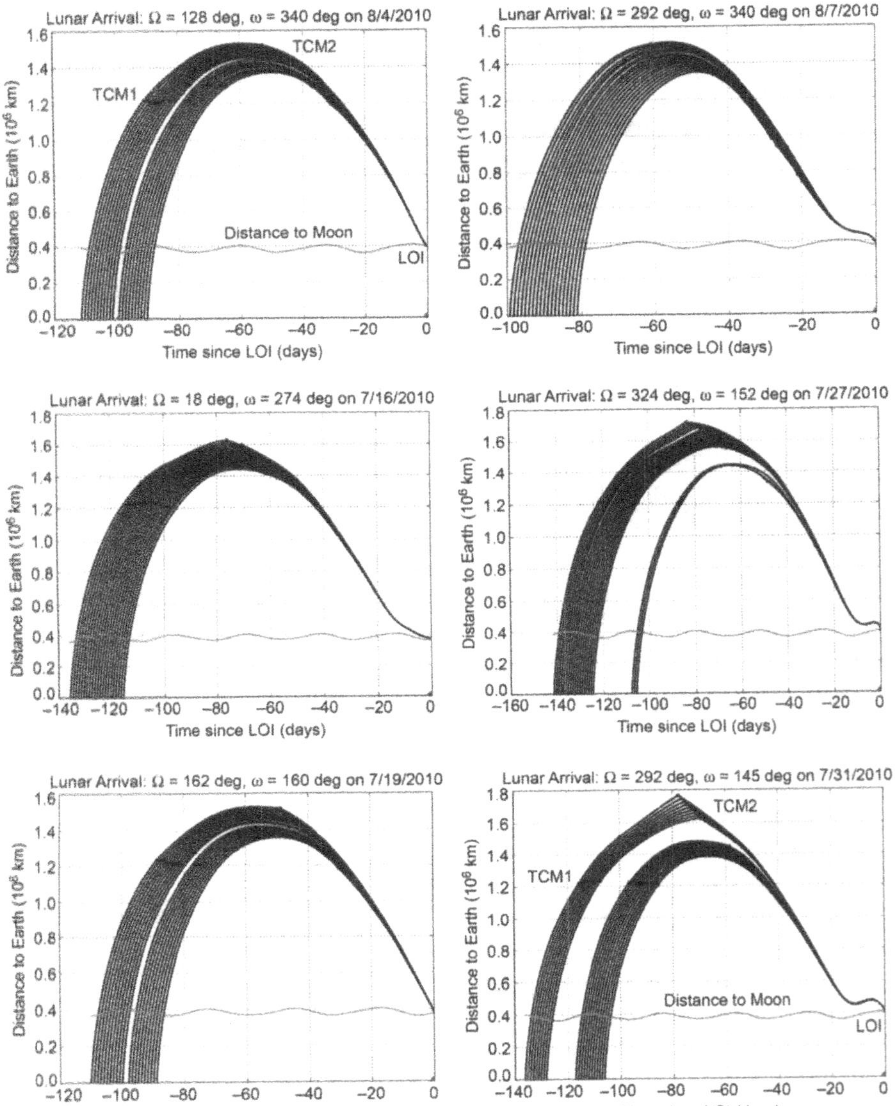

Figure 6-9 The distance from Earth over time for each transfer in the six launch periods shown in Figs. 6-7 and 6-8. The distance to the Moon over time is shown for reference.

the range of the transfer ΔV values that are contained in each 21-day launch period in these 288 missions as a function of their reference transfer ΔV. As an example, the launch period illustrated in Fig. 6-6 was generated using a reference transfer with

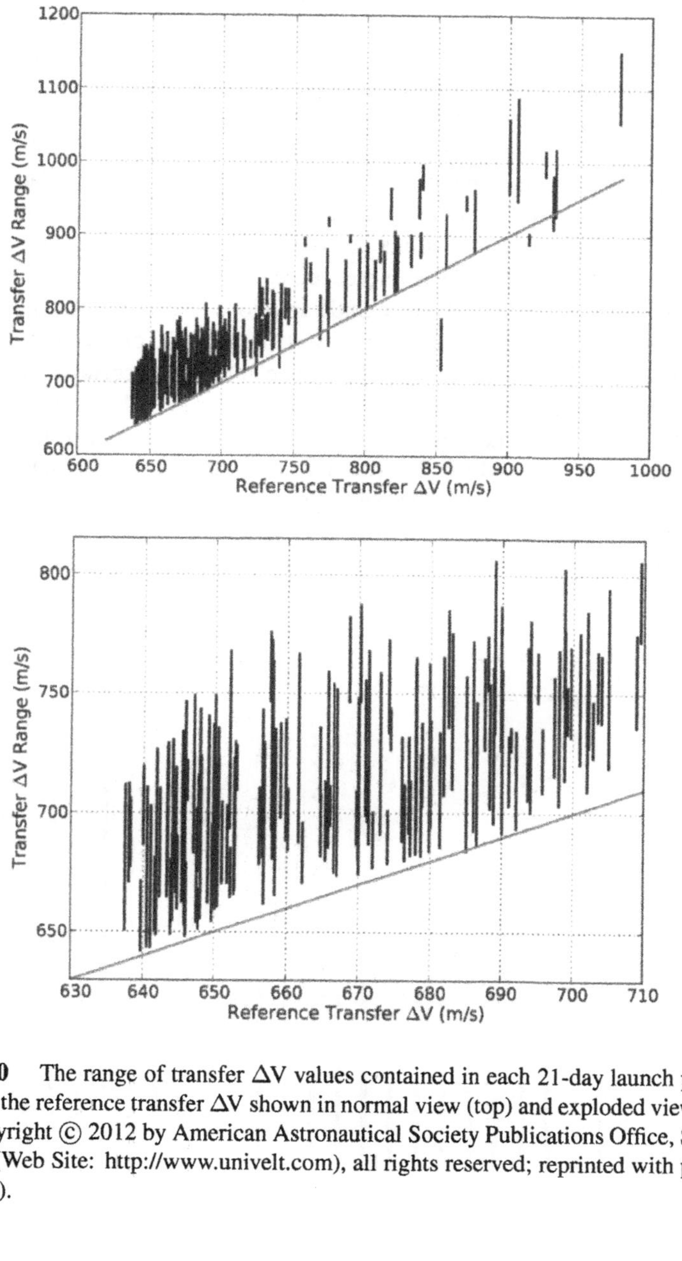

Figure 6-10 The range of transfer ΔV values contained in each 21-day launch period as a function of the reference transfer ΔV shown in normal view (top) and exploded view (bottom) [190] (Copyright © 2012 by American Astronautical Society Publications Office, San Diego, California (Web Site: http://www.univelt.com), all rights reserved; reprinted with permission of the AAS).

a ΔV of 649 m/s (the ordinate of the plots in Fig. 6-10), and the resulting launch period included missions that had transfer ΔV values between 670.6 and 706.2 m/s. One can see that the majority of transfers studied here have reference transfer ΔV values less than 750 m/s, though the transfers sampled include those with reference ΔV values as great as 1080 m/s. The launch period ΔV range often starts above the mission's reference ΔV, since each mission starts from a 28.5-deg LEO parking orbit and the reference transfer typically departs from some other inclination. In a few cases, and one extreme case, the launch period ΔV range starts below the reference ΔV. This is often possible when the reference transfer has a natural Earth departure far from 28.5 deg and a change in the transfer duration reduces the total ΔV. The plots in Fig. 6-10 clearly illustrate that the ΔV cost of establishing a 21-day launch period is highly dependent on the reference transfer's total ΔV. The launch period ΔV cost of these 288 example transfers requires approximately 71.67 $\pm$ 29.71 m/s (1σ) more deterministic ΔV than the transfer's reference ΔV.

The launch periods studied here include missions that depart the Earth on 21 different days, and the launch period ΔV cost is the ΔV of the most expensive transfer in that set. The departure days do not need to be consecutive, as described earlier. In general, increasing the number of launch days included in a launch period increases the ΔV cost of the mission. Figure 6-11 shows a plot of the change in the launch period ΔV cost of the 288 missions studied here as one adds more days to each mission's launch period, relative to the case where each mission has only a single launch day. The line of best fit through these data indicates that on average it requires approximately 2.480 m/s per launch day to add days to a mission's launch

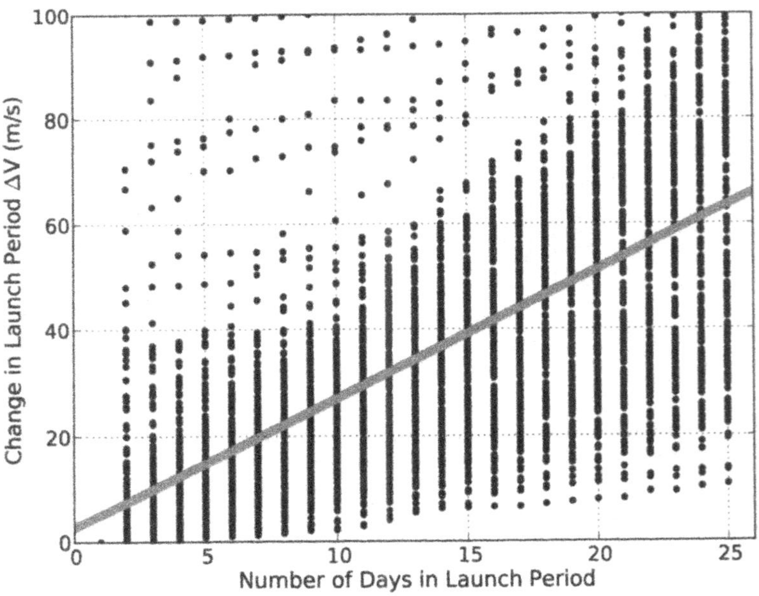

Figure 6-11 The change in the launch period ΔV cost of the 288 missions studied here as a function of the number of days in the launch period. The linear trend has a slope of 2.480 m/s per launch day [190] (Copyright © 2012 by American Astronautical Society Publications Office, San Diego, California (Web Site: http://www.univelt.com), all rights reserved; reprinted with permission of the AAS).

period. There is a significant jump in the launch period ΔV when one moves from a 1-day launch period to a 2-day launch period. This is due to the fact that the Moon's perturbations often produce a single launch day with remarkably low ΔV requirements. The change in a launch period's required ΔV would be more smooth if the effects of lunar perturbations on the Earth-departure leg were ignored.

It has been noted, when studying Fig. 6-7, that a launch period does not necessarily include the reference launch date. However, it is expected that the transfer duration of a reference trajectory may be used to predict a mission's actual transfer duration. Figure 6-12 tracks the range of transfer durations within each 21-day launch period studied here as a function of the mission's reference transfer duration. One can see that the range of transfer durations is indeed correlated with the reference transfer duration. Furthermore, it has been found that the maximum transfer duration of the 288 launch periods is approximately 15.95 ± 8.66 days longer than the mission's reference duration, the minimum transfer duration is approximately 10.91 ± 7.75 days shorter than the reference duration, and the total number of days between the first and final launch date of a given launch period may be estimated at approximately 26.86 ± 6.95 days. Hence, one may predict that a mission's launch period will include 21 of about 27 days, centered on a date several days earlier than the reference launch date, if one constructs a 21-day launch period using the same rules invoked here.

Figure 6-13 tracks the range of ΔV costs associated with each launch period as a function of the duration of the mission's reference transfer. One can see that there is a wide spread of transfer ΔV across the range of durations. As the reference transfer duration drops below 90 days, the launch period ΔV cost climbs, which makes sense because there is less time to perform maneuvers during the shorter transfers. Beyond 90 days, there are launch periods with low ΔV requirements for any transfer duration.

It is expected that the launch period's ΔV cost is dependent upon the reference transfer's natural Earth departure inclination. It is hypothesized that a reference transfer that departs the Earth with an inclination near 28.5 deg will generate a launch period that requires less total ΔV than a reference transfer that departs the Earth with a far different inclination. Figure 6-14 tracks the launch period ΔV cost of the 288 missions constructed here as a function of their reference departure inclination values. The bottom plot in Fig. 6-14 observes the range of transfer ΔV values as a function of the difference between the reference departure inclination value and the target 28.5-deg value. A line has been fit to the maximum ΔV for each launch period using a least-squares approach, which yields the relationship:

$$\text{Launch Period } \Delta V \sim (0.470 \text{ m/s/deg}) \times x + 756.5 \text{ m/s}$$

where x is equal to the absolute value of the difference between the reference departure inclination and 28.5 deg. The sample set of lunar transfers includes low-ΔV and high-ΔV missions, which may swamp any significant relationship between the launch period's ΔV cost and the reference departure inclination. Nevertheless, it is very interesting to observe that the launch period's ΔV cost does not present a strong correlation with the reference departure inclination.

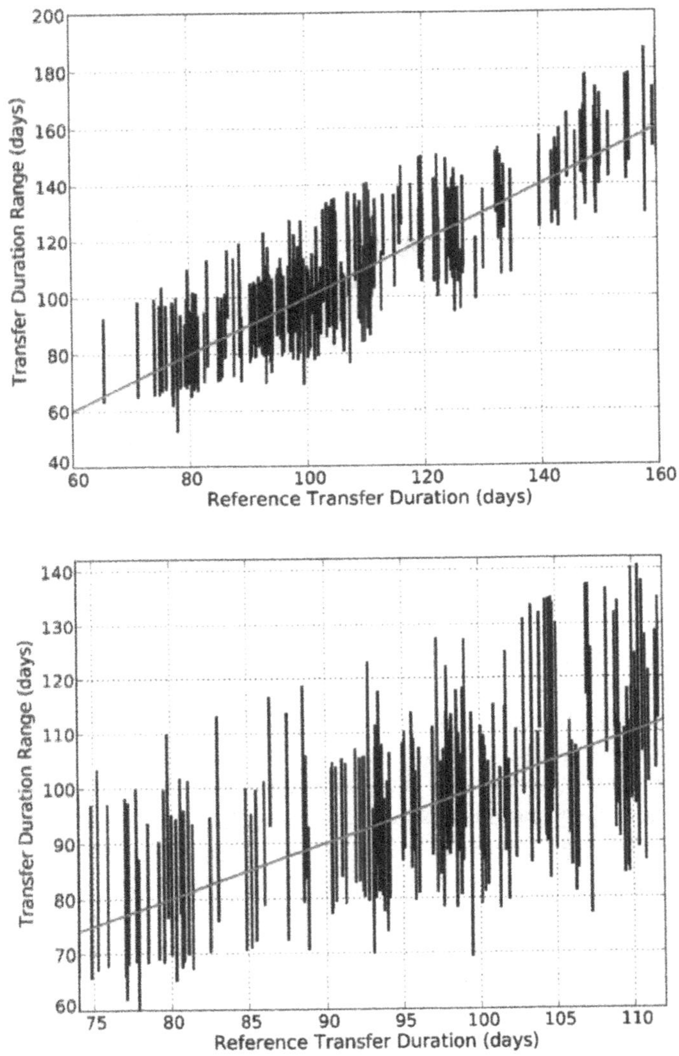

Figure 6-12 The range of transfer durations contained in each 21-day launch period as a function of the reference transfer duration. The plot at the bottom shows an exploded view, focused on transfer durations between 75 and 115 days [190] (Copyright © 2012 by American Astronautical Society Publications Office, San Diego, California (Web Site: http://www.univelt.com), all rights reserved; reprinted with permission of the AAS).

To further test the relationship of a launch period to the reference LEO inclination, each launch period's ΔV has been reduced by its reference ΔV so that each launch period may be more closely compared. Figure 6-15 shows the same two plots as

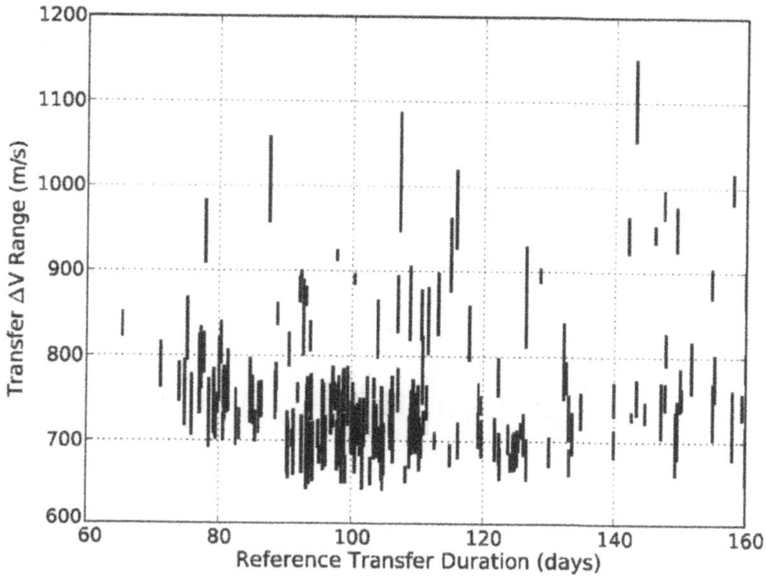

Figure 6-13 The range of transfer ΔV costs contained in each 21-day launch period as a function of the reference transfer's duration [190] (Copyright © 2012 by American Astronautical Society Publications Office, San Diego, California (Web Site: http://www.univelt.com), all rights reserved; reprinted with permission of the AAS).

shown in Fig. 6-14, but with each mission's reference ΔV subtracted from its launch period ΔV range. One can see that the launch period ΔV is not well correlated with the reference departure inclination. The linear fit has a slope of only 0.206 m/s per degree of inclination away from 28.5 deg. It appears that a 21-day launch period absorbs most of the ΔV penalty associated with inclination variations. The natural Earth departure inclination of a transfer certainly varies with transfer duration, and it has already been noticed that the launch period is often not centered about the reference transfer's TLI date. This result is useful, because it indicates that the natural Earth departure inclination is not a good predictor of the launch period ΔV requirement of a reference transfer. The relationship of the low-energy transfer ΔV and the TLI inclination is further explored in the next section.

6.5.7 Varying the LEO Inclination

The results presented previously in this section have only considered missions that begin in a LEO parking orbit at an inclination of 28.5 deg relative to the Equator, corresponding to launch sites such as Cape Canaveral, Florida. Spacecraft missions certainly depart the Earth from other launch sites; launch vehicles from those sites typically deliver the most mass to low Earth orbit if they launch into a parking orbit at an inclination approximately equal to their launch site's latitude. Hence, it

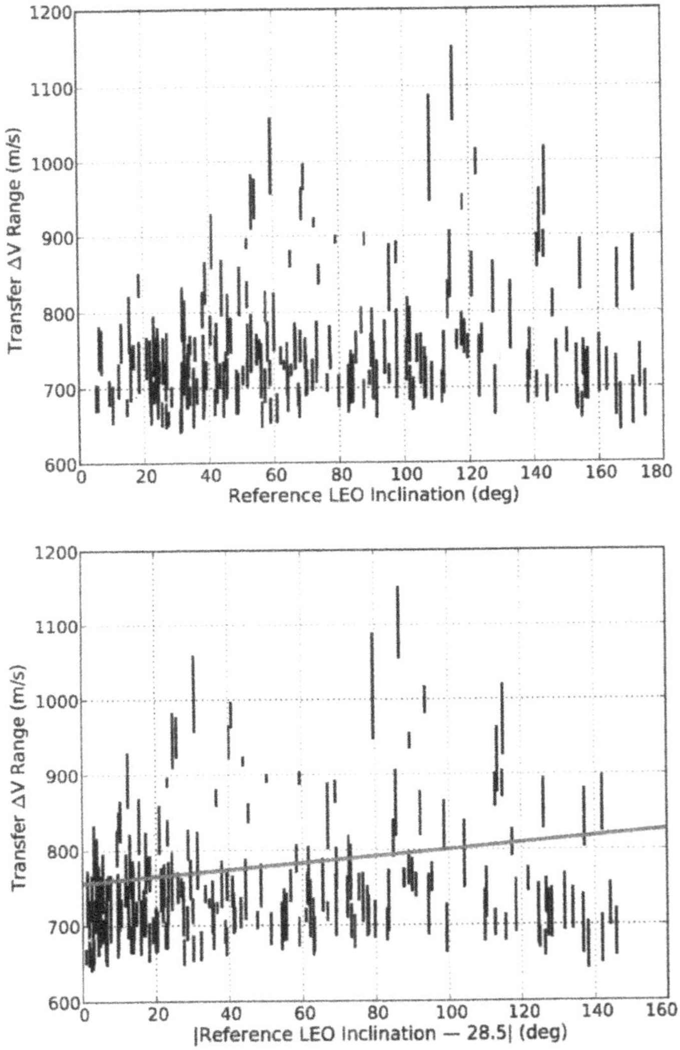

Figure 6-14 The range of transfer ΔV costs contained in each 21-day launch period as a function of the reference transfer's Earth departure inclination (top) and the absolute value of the difference between the reference inclination and 28.5 deg (bottom) [190] (Copyright © 2012 by American Astronautical Society Publications Office, San Diego, California (Web Site: http://www.univelt.com), all rights reserved; reprinted with permission of the AAS).

is of interest to determine the ΔV cost required to depart the Earth from any LEO inclination and transfer to the same lunar orbit using a particular low-energy reference transfer.

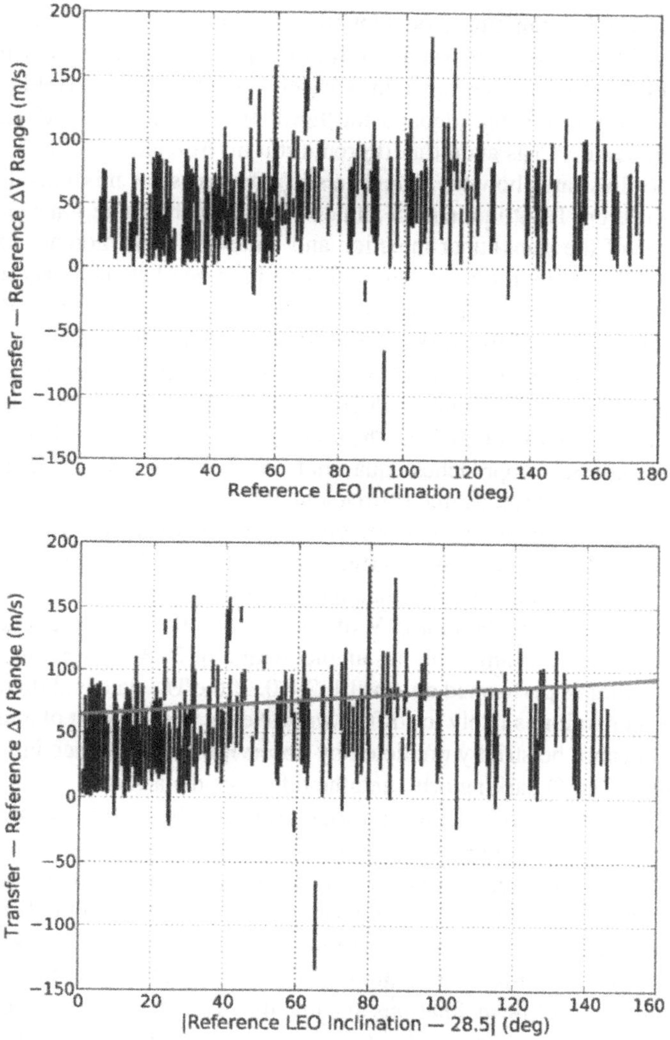

Figure 6-15 The same two plots as shown in Fig. 6-14, but with each mission's reference ΔV subtracted from its 21-day launch period ΔV range [190] (Copyright © 2012 by American Astronautical Society Publications Office, San Diego, California (Web Site: http://www.univelt.com), all rights reserved; reprinted with permission of the AAS).

The algorithms described here have been used to generate missions that depart the Earth from LEO parking orbits at a wide range of inclinations and then target the same reference low-energy transfer discussed earlier (described in Section 6.5.2 and illustrated in Fig. 6-1). The reference trajectory naturally departs the Earth on April 1, 2010, from an orbital inclination of approximately 38.3 deg; hence, a mission that departs the Earth at that time from that orbit requires no deterministic maneuvers en route to the Moon. Upon arrival at the Moon, the reference trajectory requires a 649.0-m/s orbit insertion maneuver to impulsively enter the desired 100-km circular lunar orbit. Any mission that departs the Earth from a different inclination will require deterministic TCMs and/or a different orbit insertion maneuver.

Figure 6-16 illustrates how the deterministic ΔV varies for missions that depart the Earth at different LEO inclination values and target the same lunar orbit. The dates and times of the trans-lunar injection and lunar orbit insertion are fixed. The total transfer ΔV is shown on the top, and the difference between each mission's total ΔV compared to the reference transfer's total ΔV is shown on the bottom. One can see that the ΔV cost of the mission rises as a function of the difference between the mission's departure inclination and the reference transfer's departure inclination. The cost is approximately 0.97 m/s per degree of inclination change for missions with LEO inclinations greater than 20 deg. The transfer cost increases much more rapidly as a mission's departure approaches equatorial. As the departure inclination drops, the system gradually loses a degree of freedom: the LEO parking orbit's ascending node becomes less influential on the geometry of the departure. The ascending node is no longer defined for equatorial departures, and the lunar transfer requires greater than 120 m/s more deterministic ΔV than the reference transfer.

As Fig. 6-16 illustrates, the total ΔV of a mission to the reference lunar orbit is minimized if the LEO parking orbit has an inclination of 38.3 deg, provided that the trans-lunar injection is performed on April 1, 2010. If the TLI date is shifted, then the optimal LEO inclination is likely to shift as well. Hence, the ΔV cost of a full 21-day launch period cannot be strictly predicted by observing the difference in inclination between a desired LEO parking orbit and the reference departure.

Figure 6-17 illustrates three launch periods, corresponding to missions that depart from LEO parking orbits with inclinations of 20, 50, and 80 deg. One can see that the launch period shifts in time, illustrating that the transfer duration may significantly alter the reference trajectory's natural departure inclination. Figure 6-18 illustrates the total transfer ΔV for each launch opportunity of a 21-day launch period departing from a wide range of departure inclinations. One can see that the launch period ΔV is dramatically higher for low inclinations and that the ΔV changes very little from one inclination to another for higher inclination values. It is interesting that the missions with higher inclinations require less ΔV than missions near the reference transfer's departure inclination. The low-ΔV points in the lower left part of the plot correspond to brief opportunities in those launch periods when the Moon passes through an ideal location in its orbit to reduce the transfer ΔV.

6.5.8 Targeting a Realistic Mission to Other Destinations

The algorithms presented in Section 6.5.3 have been applied to the problem of constructing realistic missions to low lunar orbit. The algorithms require little modification for missions to other destinations, such as lunar libration orbits or the lunar surface.

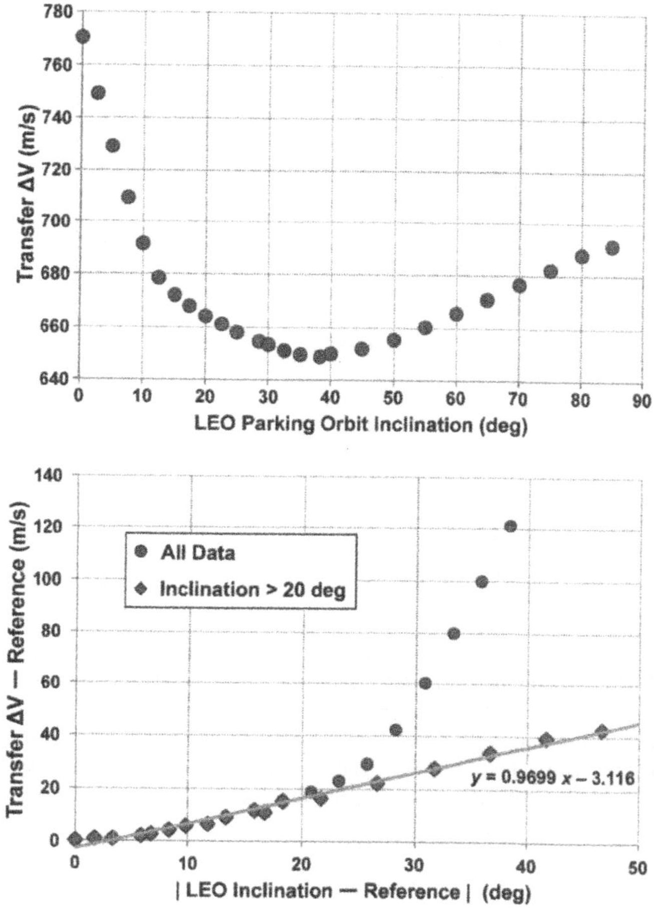

Figure 6-16 How deterministic ΔV varies for different LEO inclination values. Top: The total transfer ΔV for missions that depart the Earth on April 1, 2010, at different inclinations and arrive at the same reference lunar orbit. Bottom: The difference in the total transfer ΔV for these missions compared with the reference low-energy transfer, which departs at an inclination of ~38.3 deg [190] (Copyright © 2012 by American Astronautical Society Publications Office, San Diego, California (Web Site: http://www.univelt.com), all rights reserved; reprinted with permission of the AAS).

Missions to the Lunar Surface. Certainly a mission to the lunar surface may first target an intermediate lunar orbit, such as a low lunar orbit or a lunar libration orbit. Intermediate orbits provide some risk-reduction in the case of a contingency, because one may postpone the landing until the system is fully prepared to land. Alternatively, one may construct a mission that is designed to land immediately upon arrival at the Moon, with the option to divert into a parking orbit of some kind in the event of a

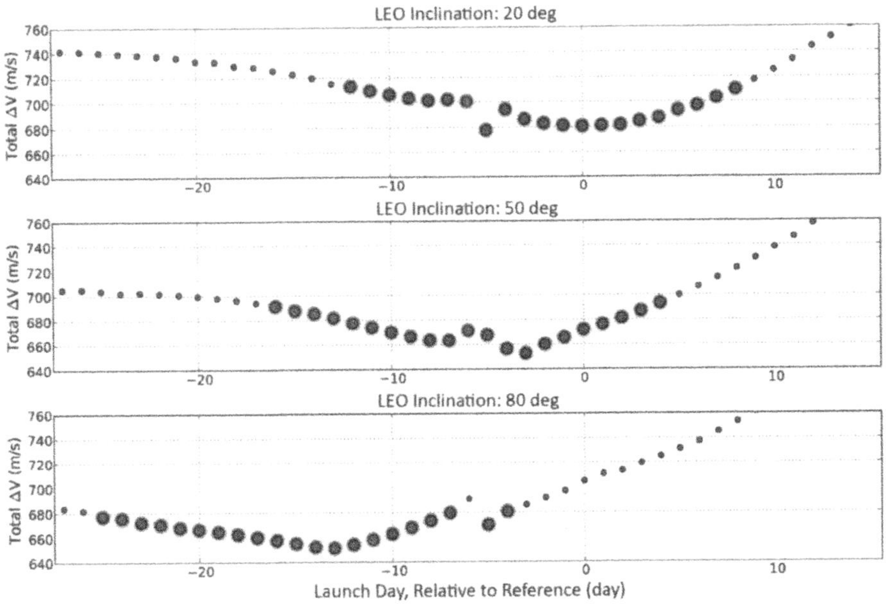

Figure 6-17 Three launch periods for missions to the reference lunar orbit, where each launch period is designed to accommodate a specific LEO inclination; namely, 20 deg (top), 50 deg (middle), and 80 deg (bottom). The Moon perturbs the outbound trajectories for those missions that launch about 5 days before the reference transfer [190] (Copyright © 2012 by American Astronautical Society Publications Office, San Diego, California (Web Site: http://www.univelt.com), all rights reserved; reprinted with permission of the AAS).

contingency. In this scenario, or in the scenario where the mission design has no option but to land immediately, the targeting algorithms described in Section 6.5.3 may be easily modified to accommodate a lander instead of an orbiter.

A lander may be able to adjust its time of arrival or its incoming velocity magnitude, flight path angle, or flight path azimuth. If these parameters must be held fixed, for example, to reduce the complexity, risk, or cost of the design, then one may instead introduce a third trajectory correction maneuver, performed some significant amount of time prior to landing, in order to minimize the total launch period ΔV.

Missions to Lunar Libration Orbits. There are many reasons why a mission to a lunar libration orbit, or other three-body orbit, would benefit by designing a single libration orbit and constructing a launch period that inserted the spacecraft into that same libration orbit. For instance, a mission design team building a lunar lander and/or sample return mission may be interested in focusing their efforts to validate one specific landing sequence, and would have to spend a great deal more effort to support 21 different landing sequences, with varying geometry and timing. It may therefore be less expensive and more reliable to implement a mission that targets

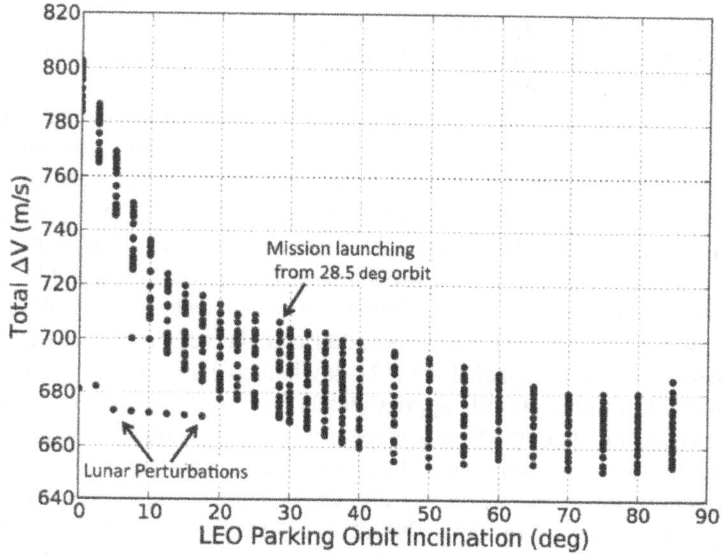

Figure 6-18 The total transfer ΔV for each opportunity of a 21-day launch period for missions to the reference lunar orbit, departing from LEO parking orbits with varying inclination values [190] (Copyright © 2012 by American Astronautical Society Publications Office, San Diego, California (Web Site: http://www.univelt.com), all rights reserved; reprinted with permission of the AAS).

a particular lunar libration orbit, no matter which day it launches on, even if that mission design required slightly higher ΔV budget.

Studies have demonstrated that the algorithms presented here may be used very successfully in conjunction with a libration orbit insertion maneuver [183, 184].

6.5.9 Launch Period Design Summary

The goal of this section is to characterize the ΔV costs associated with building a 21-day launch period for a practical mission to the Moon via a low-energy transfer. We have sampled 288 different low-energy transfers between the Earth and polar orbits about the Moon and have constructed practical 21-day launch periods for each of them, using a 28.5-deg LEO parking orbit and no more than two deterministic maneuvers. The lunar orbits have a wide range of geometries, though they are all polar and have an altitude of approximately 100 km. The reference low-energy transfers include no Earth-phasing orbits nor close lunar flybys, and they require between 65 and 160 days of transfer duration. Each mission has been constructed by using a sequence of steps, varying eight parameters to minimize the transfer ΔV cost. The eight variable parameters include the parking orbit's ascending node, the trans-lunar injection's location in the parking orbit, the trans-lunar injection's ΔV, the times of two deterministic maneuvers en route to the Moon, and three components of the lunar orbit insertion maneuver. All other aspects of the transfer are fixed when building a particular mission.

Several conclusions may be easily drawn from the results presented here. First of all, the cost of a launch period is obviously dependent on the number of launch days in the period. The transfers constructed here demonstrate that it costs on average approximately 2.5 m/s per day added to a launch period; hence, the average 21-day launch period requires about 50 m/s more deterministic ΔV than a 1-day launch period for a given transfer. The cost of a particular launch period may rise nonlinearly as one adds days to the launch period, such that it may be the case that additional days cost exponentially more ΔV or perhaps that additional days do not cost any additional ΔV. The statistical cost of establishing a 21-day launch period to the 288 reference transfers studied in this section is approximately 71.7 ± 29.7 m/s (1σ), where the additional ΔV of more than the 50 m/s is required to accommodate a departure from a 28.5-deg LEO parking orbit. The 21 opportunities in the launch period may be on 21 consecutive days, and frequently are, but typically include one or two gaps. The average launch period for these 288 missions requires a total of 27 days; the vast majority of the launch periods may be contained within 40 days. Finally, we have shown that there is no significant trend between the total launch period ΔV for these 288 missions and their reference departure inclination values or their reference transfer durations, except for short transfers with durations below 90 days.

An additional study has been performed to observe how a mission's ΔV changes as a function of the particular LEO inclination selected. A mission that departs at a particular time requires approximately 0.97 m/s more transfer ΔV per degree of inclination change performed, assuming that the departure inclination is greater than 20 deg. The total transfer ΔV cost increases dramatically as the departure inclination approaches 0 deg. These trends change when considering a full 21-day launch period. The required launch period ΔV is still high for missions that depart from nearly equatorial LEO parking orbits, but the variation in the launch period ΔV is reduced for missions that depart at higher inclinations.

6.6 NAVIGATION

Spacecraft traversing low-energy lunar transfers may be navigated in very similar fashions to those following interplanetary transfers. Indeed, there are many similarities: the trajectories require many weeks, they traverse well beyond the orbit of the Moon, they require trajectory correction maneuvers, etc. There are several differences, including the fact that low-energy lunar transfers remain captured by the Earth, they are not well-modeled by conic sections, and they are unstable. This section discusses how these similarities and differences impact the navigation and operation of the spacecraft during such transfers.

6.6.1 Launch Targets

Launch vehicle operators typically target three parameters when injecting spacecraft onto interplanetary trajectories: one describing the target energy, namely, C_3, and two angular measurements describing the orientation of the departure asymptote, namely, the right ascension and declination of the launch asymptote, RLA and DLA, respectively. Low-energy lunar transfers, conversely, remain captured by the Earth and do not have launch asymptotes. The GRAIL project used two similar target parameters to describe the orientation of the departure ellipse—the right ascension and declination of the instantaneous apogee vector (RAV and DAV, respectively) at the time of the launch vehicle's target interface point (TIP). Combined with the target C_3 parameter, these three targets describe a departure that keeps the expected correction ΔV after the TIP to a minimum.

6.6.2 Station-Keeping

As illustrated in Chapter 2, low-energy lunar transfers are unstable; they depend on a careful balance of the gravitational attraction of the Sun, Earth, and Moon. Any random deviation from the designed trajectories will grow exponentially over time. Therefore, a spacecraft traversing a low-energy lunar transfer in the presence of realistic uncertainties will require TCMs to remain on a desirable course. Fortunately, low-energy lunar transfers are stable enough that maneuvers are typically only needed every 4–8 weeks, though more are needed to support any lunar approach and/or lunar flybys.

The cost of performing statistical corrections on a low-energy lunar transfer may be estimated by considering the stability of trajectories in each region of space that the transfer passes through. First, typical spacecraft missions plan to perform a maneuver soon after the trans-lunar injection in order to clean up any injection errors. For instance, the GRAIL mission planned to have both spacecraft perform a maneuver within a week after injection. Next, the spacecraft spend 1–3 months traversing a region of space far from the Earth, typically near the Sun–Earth L_1 or L_2 points. The stability of this portion of the trajectory may be approximated by measuring the stability of typical Sun–Earth libration orbits—as illustrated in Section 3.4.1. As the spacecraft approach the Moon either for a lunar flyby or for their final lunar approach, the trajectories become more unstable. The stability of the trajectories near the Moon may be approximated by measuring the stability of typical Earth–Moon libration orbits.

There are many ways to measure the stability of a trajectory, but a rather intuitive way is to consider the trajectory's *perturbation doubling time*, that is, the amount of time it takes for a spacecraft to double its distance away from a reference trajectory (see Section 2.6.8.3 on page 80). If at time t_0 a spacecraft is 100 km away from its reference trajectory, then at time $t_0 + \hat{\tau}$ the spacecraft will be approximately 200 km away from its reference, and at time $t_0 + 2\hat{\tau}$ the spacecraft will be approximately 400 km away from its reference, and so on, where $\hat{\tau}$ is the perturbation doubling time.

Typical halo orbits about the Sun–Earth L_1 and L_2 points have $\hat{\tau}$ values of about 17 days. Hence, one may assume that the position error of a spacecraft traversing a low-energy lunar transfer near those orbits will double roughly every $\sim$17 days. Fortunately, this is a rather long time for most spacecraft operations unless the spacecraft has particularly strict flight path requirements. The *GRAIL* mission scheduled two trajectory correction maneuvers per spacecraft while they traversed the region of space near EL_1, though there were deterministic needs for those maneuvers as well. The *Genesis* mission performed maneuvers every couple of months while traversing its EL_1 orbit, requiring only about 10 m/s per year of station-keeping [87]. The *Solar and Heliospheric Observer's* (*SOHO*) spacecraft has demonstrated the ability to remain in orbit about the EL_1 point for even less ΔV. *SOHO*'s first eight station-keeping maneuvers (SKMs) were executed between May 1996 and April 1998, imparting a total ΔV of approximately 4.77 m/s: an average of one maneuver per 99 days with an average maneuver ΔV of only 0.596 m/s [217].

As a spacecraft approaches the Moon, either for the targeted arrival or for a lunar flyby, its trajectory becomes more unstable and the perturbation doubling time shrinks. If the spacecraft arrives at a lunar libration orbit via a transfer such as those presented in Chapter 3 then its stability may be measured by the perturbation doubling time of typical halo orbits about the Earth–Moon libration points. If the spacecraft's destination is a low lunar orbit, the lunar surface, or a flyby, then this measurement is only an approximation and more analysis is needed. Nevertheless, typical halo orbits about the Earth–Moon L_1 and L_2 points have $\hat{\tau}$ values of about 1.4 days. Not surprisingly, a spacecraft's position error doubles about twelve times faster in the Earth–Moon system than it does in the Sun–Earth system. Depending on the mission, it may be necessary to perform maneuvers as often as 1–2 times per week to traverse Earth–Moon libration orbits. Even so, the two *Acceleration, Reconnection, Turbulence, and Electrodynamics of the Moon's Interaction with the Sun* (*ARTEMIS*) spacecraft successfully navigated several months of libration orbits about both the Earth–Moon L_1 and L_2 points, demonstrating that such operations are viable.

There are two fundamentally different strategies that have been implemented when designing the SKMs of historical missions, namely, *tight control* and *loose control*. The term *tight control* describes a strategy where each SKM is designed to bring the spacecraft's trajectory back to a designed reference trajectory. The *International Sun–Earth Explorer-3* (*ISEE-3*) and *Genesis* spacecraft maneuvers are good examples of missions that implemented tight station-keeping control [218, 219]. This strategy is used when the spacecraft's trajectory has particular requirements; for the case of *Genesis*, the trajectory ultimately placed the spacecraft on a course to enter the atmosphere for a landing in Utah. Each SKM is designed such that the resulting trajectory intersects the reference trajectory at the time of the next planned SKM. Conversely, a *loose* station-keeping strategy describes one where a spacecraft may travel anywhere within some wide corridor and the particular route taken is not important. For instance, the *SOHO* spacecraft must remain in orbit about the Sun–Earth L_1 point, but the particular path about the L_1 point is not important. Thus, *SOHO*'s trajectory is re-optimized each time an SKM is designed [217]. *SOHO*'s loose control has resulted in station-keeping ΔV costs just over 2 m/s per year, nearly four times lower than *ISEE-3*'s tight station-keeping control costs.

In summary, one may expect to perform a TCM soon after launch in order to clean up injection errors, followed by TCMs every 4–8 weeks when the spacecraft is traversing the cruise phase far from the Earth or Moon, followed by one or two TCMs per lunar approach. If the spacecraft's itinerary includes lunar libration orbits, or other unstable three-body orbits, then one may expect to perform TCMs every 3–7 days during those phases. The total navigation ΔV cost depends on the spacecraft and its propulsion system's performance, but it is certainly possible to navigate such trajectories for a modest ΔV – on the order of 1–10 m/s per year.

6.6.2.1 Station-Keeping Strategies Numerous station-keeping strategies have been formulated since investigators began applying libration orbits to practical spacecraft mission designs [6, 119]. Most developments have been in support of flight projects and proposals that involved trajectories in the Sun–Earth system [218, 220–230]. More recent investigations have examined station-keeping strategies within the Earth–Moon system, especially with the development and success of the *ARTEMIS* mission [17, 186, 231–234]. Folta et al., surveyed a wide variety of station-keeping strategies with the purpose of applying a desirable strategy to the *ARTEMIS* mission [231]. Ultimately, each of the *ARTEMIS* station-keeping maneuvers was designed using a gradient-based optimizer that ensured the spacecraft would remain on the libration orbit for the next few revolutions. This method kept the total station-keeping fuel cost low without requiring the generation of a reference trajectory. After the maneuvers were designed, a later study found that each of the maneuvers closely aligned with the local stable eigenvector at that point of the spacecraft's orbit [235]. This conclusion has certainly prompted researchers to investigate if the stable eigenvector is a good initial guess for a near-optimal station-keeping strategy.

While *ARTEMIS* employed a *loose* station-keeping strategy very successfully, its strategy was focused on the short term: ensuring that the spacecraft remained on a desirable trajectory for the next few revolutions about the Lagrange point. There is concern that any short-term strategy may fail over the long term, resulting in a trajectory that diverges from the desired orbit. Recent work has applied tools such as the multiple-shooting differential corrector to the goal of achieving a minimum-ΔV long-term station-keeping strategy [234]. This goal is a rich, challenging problem with a wide variety of possible constraints and degrees of freedom available. The solution may differ for each spacecraft mission, with its own operational constraints and desirable mission characteristics.

In the following sections, we study several aspects of the station-keeping problem in order to provide a background for the general problem of station-keeping on a libration orbit. The reader is encouraged to explore different strategies, particularly those surveyed by Folta et al. [231]. We present the results of several analyses, including *tight* and *loose* station-keeping strategies. Typical low-energy lunar transfers are highly constrained, such that there are often not enough degrees of freedom available to the mission designer to employ a loose station-keeping strategy. However, if the goal of the low-energy lunar transfer is to enter a libration orbit and to remain there, then a loose strategy may be beneficial.

6.6.2.2 Station-Keeping Simulations Each of the simulations studied here uses the same set of assumptions, varying only one or two aspects of the station-keeping problem in order to keep the results as comparable as possible. The simulations employ models that are representative of the real solar system, with some simplifications to speed up the computations. The DE421 ephemerides are used to approximate the motion of the Sun, Moon, and planets (Section 2.5.3) and each of the bodies is approximated as a point-mass using the masses presented in Section 2.2. Solar radiation pressure is modeled using a constant solar flux of $1.019794376 \times 10^{17}$ N and a flat plate model where the mass of the spacecraft is 1000 kg and the surface area is 10 m^2. The trajectories are integrated using JPL's Mission-Analysis, Operations, and Navigation Toolkit Environment (MONTE) software (Section 2.7.1).

Each simulation includes a *truth* set of dynamics and an *estimated* set of dynamics, which differ enough to introduce dynamical errors into the navigation problem. The truth set includes the gravitational forces of the Earth, Sun, and Moon and uses a value of 1.00 for the coefficient of radiation of the solar pressure. The estimated set of dynamics also includes the gravitational forces of all of the other planets in the Solar System and uses a value of 1.03 for the coefficient of radiation. These perturbations are somewhat arbitrary and have been selected to approximate the level of accuracy of flight operations.

The reference trajectory for the simulations is a southern halo orbit about the Earth–Moon L_2 point with a z-axis amplitude of approximately 10,000 km (see Section 2.6.6.3 and Section 2.6.9.4). The reference epoch is January 1, 2017. The perturbations depend on the reference epoch, though they will not likely impact the results very much. It is more likely that the choice of orbit will change the results of the simulations.

Each SKM in each of these simulations is generated using a similar process. First, the state of the spacecraft is propagated from one time to the next using the truth dynamics. At the time of a station-keeping maneuver, the estimated state of the spacecraft at that maneuver is computed by taking the truth state at that time and perturbing it with orbit determination errors. The resulting state is used as the initial state for the station-keeping strategy, whatever it may be. Each station-keeping strategy studied here involves propagating estimated trajectories into the future. These trajectories are propagated using the estimated dynamics, which again differ from the truth dynamics. Once a station-keeping ΔV is determined, that ΔV is applied to the *true* spacecraft state. Finally, a maneuver execution error is added to the state as well, and the resulting state is propagated using the truth dynamics. This process is repeated for each station-keeping maneuver in the simulation.

The orbit determination errors are modeled as spherically symmetric distributions, such that each of the three Cartesian position coordinates and each of the three Cartesian velocity components is perturbed using independent Gaussian errors, with zero mean and standard deviations of 100 meters (m) for position and 1 millimeter per second (mm/s) for velocity. Hence, the orbit determination errors may be in any direction, with net 1-σ position uncertainties of approximately 173 m and net 1-σ velocity uncertainties of approximately 1.73 mm/s. These errors are similar in magnitude to those observed by the *ARTEMIS* mission navigators [235]. The maneuver execution error model applies a similar spherically distributed error, such that a Gaussian perturbation of zero mean and 2 mm/s is applied to each of the three components, no matter what size of maneuver it is. Hence, the net 1-σ uncertainty is approximately 3.46 mm/s. The maneuver execution error could be a realization of a burn duration error, an efficiency error, a pressure regulation error, etc. Since it is not clear what is causing the error, the execution error component of the net ΔV is not included in any computation of the average or total station-keeping maneuver ΔV cost presented below.

Finally, each simulation is repeated at least 30 times to generate statistical results.

6.6.2.3 Tight Station-Keeping A very common tight station-keeping strategy is to correct a spacecraft's trajectory in the presence of errors by building each station-keeping maneuver to target the position of the spacecraft's reference trajectory at the time of the following station-keeping maneuver. If all goes well, the station-keeping maneuver will execute perfectly, and the modeled dynamics will perfectly match the true dynamics. In that case, the spacecraft would arrive at the reference trajectory at the time of the next maneuver and perform that maneuver to match its velocity with the reference trajectory. Of course, in reality the spacecraft never arrives precisely on the reference, but must perform another maneuver to correct for additional errors.

Figures 6-19 and 6-20 illustrate this strategy. Figure 6-19 shows a top-down view of the reference halo orbit with a very exaggerated trajectory attempting to follow it. In this case, the SKM are performed at 1-day intervals and the errors are huge, just for visualization purposes. The illustration in Fig. 6-20 shows the difference between the estimated and reference trajectory for a simulation that uses the proper error distributions. The black curve is the truth trajectory, the "x"s indicate the estimated state of the spacecraft at the time of each SKM, and the gray curves illustrate the target trajectories built with the intention to return the spacecraft to the reference.

This tight station-keeping strategy has been applied to a wide range of SKM periods, including periods as short as 0.5 days and as long as 12 days. Figures 6-21 and 6-22 present the resulting range of maneuver ΔV costs. One can see many interesting features in the results. First, Fig. 6-21 presents a clear trend such that the average SKM magnitude grows as the duration of time between maneuvers grows. One exception to this is that if the maneuvers are performed too frequently, the average maneuver magnitude rises as the frequent maneuvers fight their collective execution errors. Second, Fig. 6-22 illustrates that there is a minimum in the total expected station-keeping ΔV cost that occurs at a period of approximately 3 days, requiring slightly less than 2 m/s per year. If maneuvers are performed more frequently, fuel is wasted combating frequent maneuver execution errors. If maneuvers are performed less frequently, then the spacecraft has more time to drift exponentially away from the reference. Third, the relationships between station-keeping ΔV cost and maneuver execution period are very smooth until the maneuvers are executed approximately 7–10 days apart. This duration is slightly longer than half of a revolution period

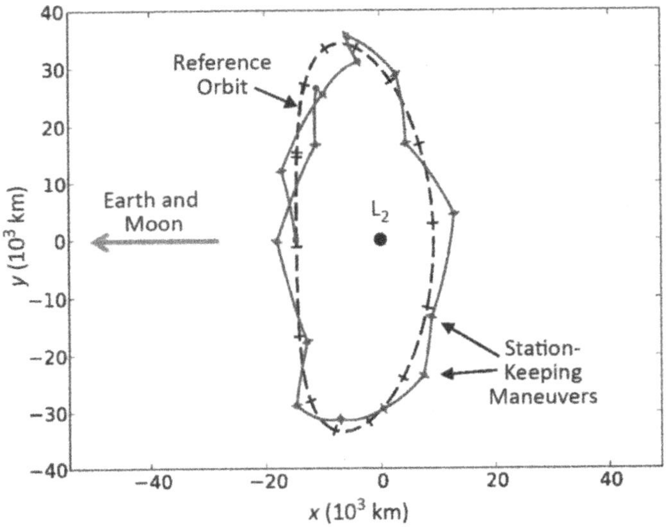

Figure 6-19 A top-down view of a spacecraft following a reference halo orbit using a tight station-keeping strategy in the presence of very large, exaggerated errors. Station-keeping maneuvers are executed once per day.

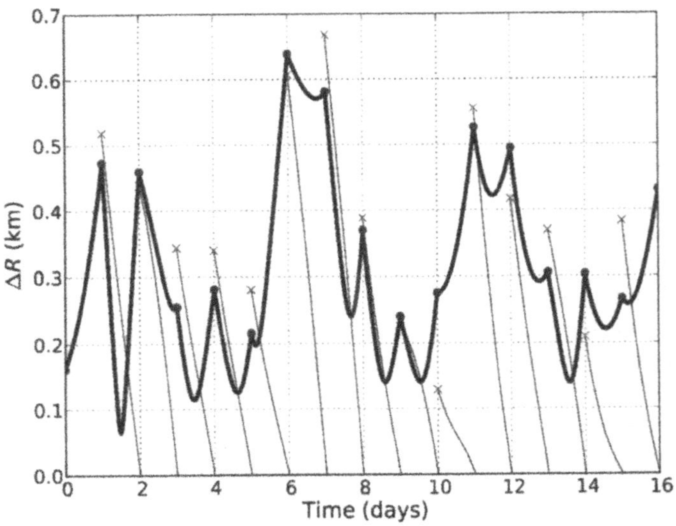

Figure 6-20 The distance between a spacecraft's trajectory and its reference trajectory for an example tight station-keeping scenario, with maneuvers performed once per day.

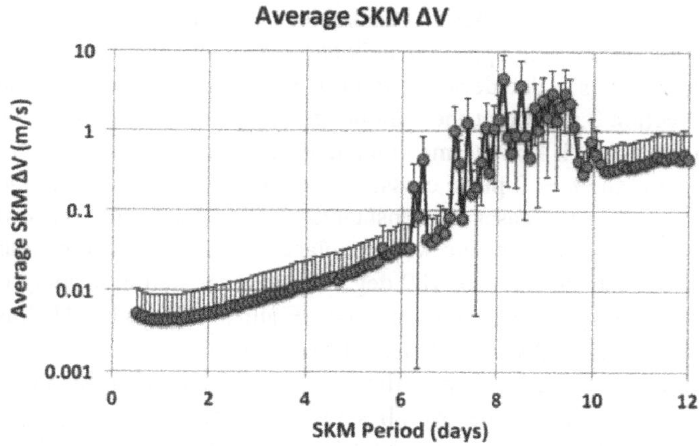

Figure 6-21 The average station-keeping ΔV cost as a function of the duration of time between maneuvers.

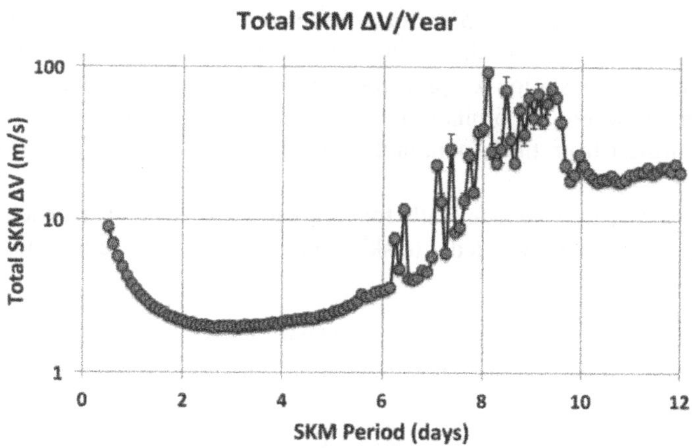

Figure 6-22 The total annual station-keeping ΔV cost as a function of the duration of time between maneuvers.

about the halo orbit. It is hypothesized that the station-keeping sensitivity grows significantly when the target is on the opposite side of the orbit.

Figure 6-22 clearly indicates that if a navigation team intends to reduce the station-keeping cost of a spacecraft on this halo orbit then it is best to perform maneuvers every 2–6 days. From an operational perspective, it is convenient to work on a schedule where a maneuver design cycle is performed every seven days. If the team can support the operational pace, the best strategy may be to design maneuvers every 3.5 days, knowing that if a maneuver is missed then the cost will not grow too high after 7 days. If this is the case, then it is desirable to estimate the total station-keeping cost of a mission performing approximately two maneuvers per orbit.

The next question is to decide where in the orbit to perform those two maneuvers. Recall from Section 2.6.2.3 that the parameter τ may be used to specify a location about a halo orbit, much like the mean anomaly of a conic orbit. We will refer to $\tau = 0$ deg to be at the $y = 0$ plane crossing with positive y-velocity (in the synodic reference frame), and τ increases at a constant rate as the spacecraft traverses the orbit. We have simulated scenarios where we have placed one station-keeping maneuver at a τ value anywhere from 0 deg to 180 deg and the other station-keeping maneuver at a τ value of 180 deg greater than the first. Figures 6-23 and 6-24 illustrate the resulting station-keeping ΔV cost of each of these scenarios.

One can draw several conclusions after observing the relationships presented in Figs. 6-23 and 6-24. First, the overall station-keeping cost is roughly the same order of magnitude anywhere around the orbit, except for the spikes observed near $\tau = 10$ deg and $\tau = 170$ deg. These spikes are rather unexpected features of these curves. The SKMs become very sensitive to variations at those points in the orbit. In contrast, the best places to perform SKMs on this particular halo orbit are at τ values near 30 deg and 150 deg, where the total cost is below 6 m/s per year. Apart from the spikes, the worst locations to perform maneuvers are at τ values of 90 deg and 270 deg, namely, where the orbits extend the furthest from the $y = 0$ plane. It is of interest to note that the station-keeping cost is relatively low at τ values of 0 deg and 180 deg, namely, where the orbits cross the $y = 0$ plane, where they approach the closest and furthest from the Moon, and also where they have their greatest z-axis excursions.

If the mission operations plan calls for frequent small maneuvers, such that it is okay—and perhaps even expected—to skip a maneuver from time to time, then it is

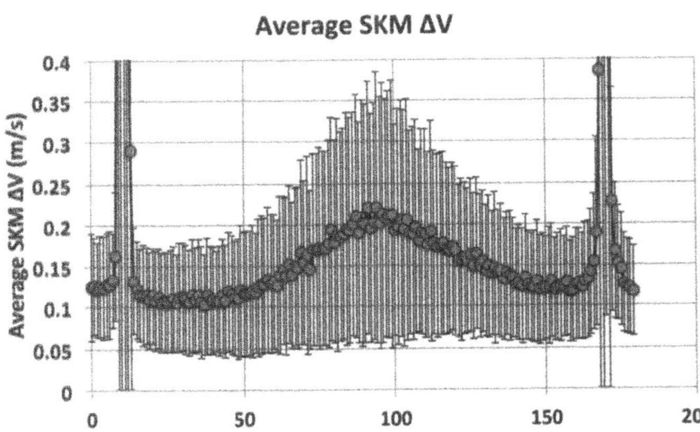

Figure 6-23 The average station-keeping ΔV cost for two maneuvers performed per revolution 180 deg apart in τ, as a function of the τ value of the first maneuver.

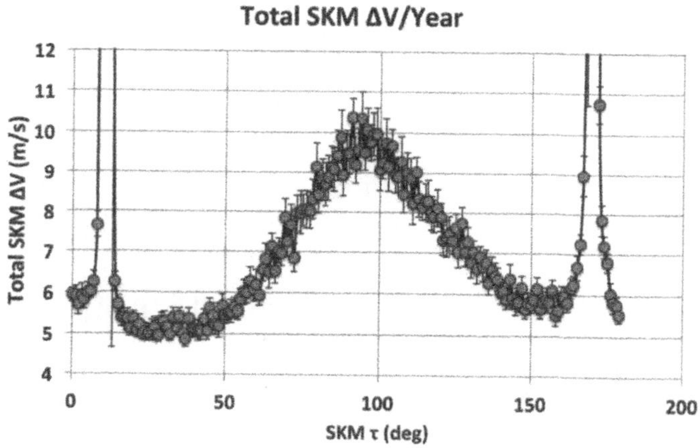

Figure 6-24 The total annual station-keeping ΔV cost for two maneuvers performed per revolution 180 deg apart in τ, as a function of the τ value of the first maneuver.

of interest to measure the station-keeping cost of a slightly different scenario. In this variation, station-keeping maneuvers are planned every 7 days, but each maneuver is targeted to generate a trajectory that would bring the spacecraft to the reference trajectory in only 3.5 days. In a perfect situation, the trajectory would fly past the reference trajectory halfway between each station-keeping maneuver. In reality it will likely fly past the reference trajectory, though at some distance. Figure 6-25 illustrates the distance between the trajectory and reference of one example instance of this scenario. One can see that the position differences pass very close to zero after most of the maneuvers. Further, the maximum excursions from the reference trajectory rise over time. The figure includes a linear fit and a quadratic fit of the maximum excursions over time, and it is clear that both trends are growing.

This strategy may be generalized in order to understand how the cost of station-keeping depends on the station-keeping period and the duration of time between each SKM and the target state. Numerous simulations are studied here, varying the station-keeping period and the target duration in order to study these relationships. Figure 6-26 illustrates a few example scenarios where the SKMs are performed every day, while their targets are 1, 2, 4, and 5 days into the future. This station-keeping period is likely to be far too rapid for any realistic flight operations, but it is easier to see the features of the plots. One can see that the strategy converges for the cases of 1, 2, and 4, but it does not converge if the target is 5 days into the future. In addition, there is a trend that the spacecraft remains further from the reference trajectory if the SKM targets a point further into the future.

Figures 6-27–6-29 illustrate the results of a wide range of scenarios, where the station-keeping period varies from 1 day to 13 days and the target duration varies from 0.5 days to 24 days. Figure 6-27 presents the total annual station-keeping ΔV cost for each combination. Figure 6-28 illustrates the average station-keeping magnitude for

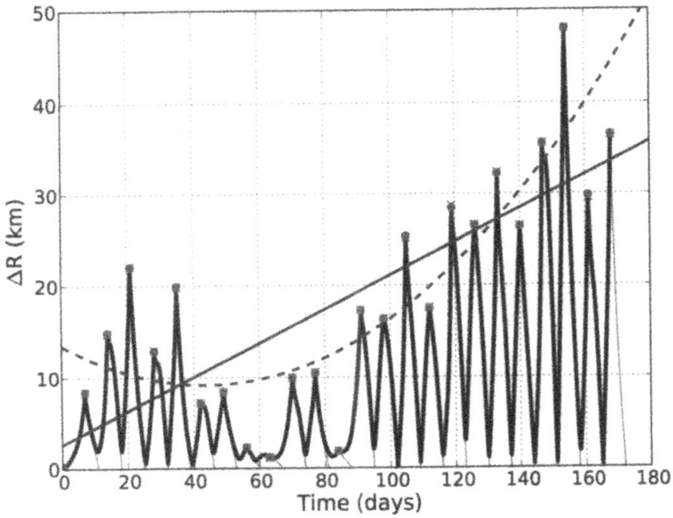

Figure 6-25 The position difference between the simulated trajectory and the reference trajectory for a scenario where SKMs are performed every 7 days, targeting to the reference trajectory at a point 3.5 days later.

each scenario. Figure 6-29 summarizes the average distance between the resulting trajectory and the reference trajectory for each scenario. In each case, the scenarios shaded white exceed the data range and are not viable station-keeping strategies.

One can draw many conclusions studying these charts. First, if one studies the line of solutions that corresponds to the scenarios where the station-keeping period is equal to the target time, one recovers the results shown in Figs. 6-21 and 6-22. These figures also provide further evidence that the station-keeping performance degrades when the station-keeping period and target time are both around 9 days. It is interesting that there are periodic bands of target durations that converge to successful station-keeping strategies for a given station-keeping period, that is, the three near-vertical dark stripes in each figure. When looking back at Fig. 6-26, it is apparent that some target durations yield scenarios where the trajectories must travel farther from the reference trajectory before returning to the reference. If the station-keeping period is too rapid, or set at an undesirable resonant period, then the distance from the reference trajectory at one SKM is greater than the distance at the previous maneuver, and the strategy diverges.

Nevertheless, the performance of the station-keeping strategy does not significantly improve by targeting a point 10 or more days beyond the given SKM. Statistically there is some benefit derived by permitting the target time to be different than the station-keeping period, though it is typically not far from being equal. Figure 6-30 illustrates this by plotting three curves, tracking the station-keeping performance for 1-, 3-, and 7-day station-keeping periods. One can see that the global minimum of

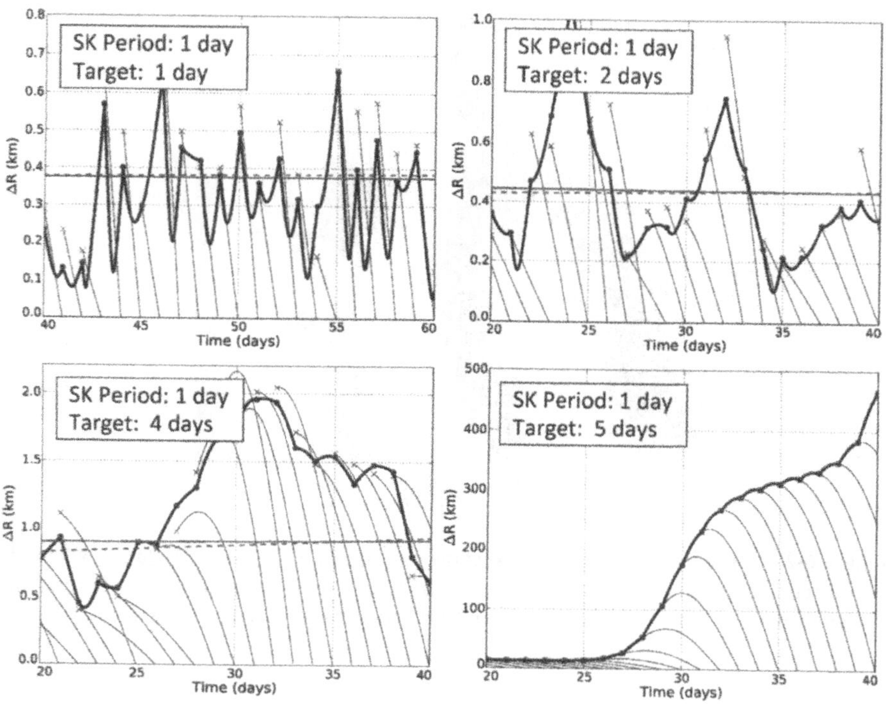

Figure 6-26 The progression in the position difference between the simulated trajectory and the reference trajectory for scenarios where SKMs are performed every day, but their targets are 1, 2, 4, and 5 days into the future.

each curve shown indeed exists toward the right, where the target duration is around 27 days. But the benefits are slight compared to targeting a few days downstream, which is also a more stable and computationally-efficient station-keeping strategy.

6.6.2.4 Loose Station-Keeping A large number of different strategies have been investigated by researchers in order to attempt to reduce the station-keeping ΔV cost. We present one such *loose* strategy, namely, a strategy that keeps the spacecraft in the desired region of space without targeting any sort of reference trajectory. For additional strategies, see for example, Folta et al. [231].

The strategy studied here is designed to work for libration orbits and other trajectories that pierce the $y = 0$ plane with an x-velocity of approximately zero in the synodic reference frame. Halo orbits pierce the $y = 0$ plane orthogonally in the circular restricted three-body problem (CRTBP) and nearly orthogonally in a high-fidelity model of the Solar System. Lissajous orbits are permitted to have some nonzero velocity in the z-axis at those crossings. The loose station-keeping strategy is designed to take advantage of these orbital features.

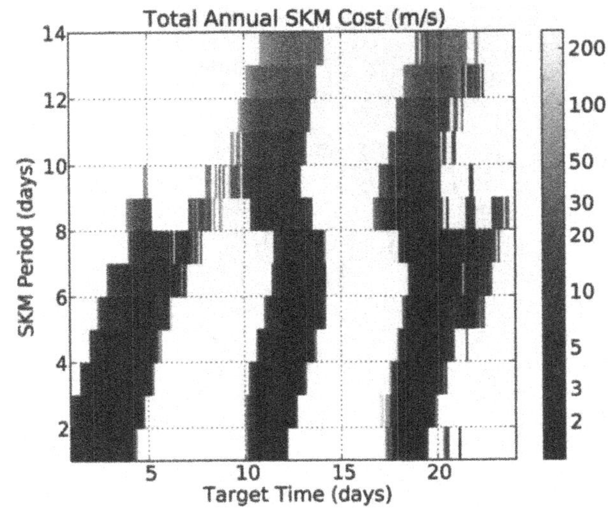

Figure 6-27 The total annual station-keeping cost for a wide range of scenarios, where the x-axis sets the amount of time between each SKM and its target point along the reference trajectory, and the y-axis sets the amount of time between each maneuver and the next.

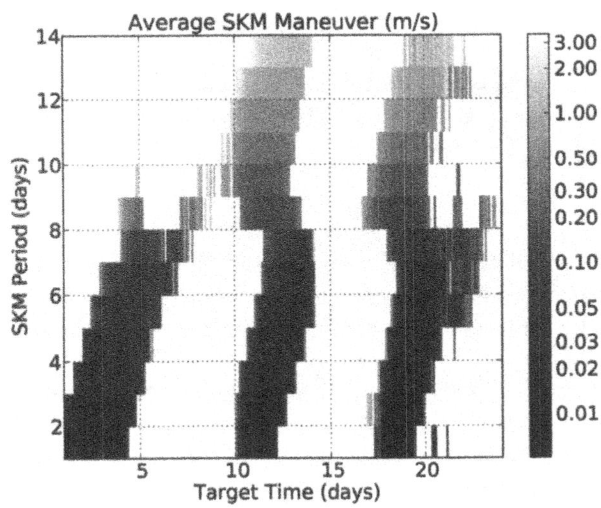

Figure 6-28 The average SKM magnitude for the same trade space given in Fig. 6-27.

The idea is that a given SKM is designed to target a trajectory that pierces the $y = 0$ plane orthogonally at either the next crossing or a subsequent crossing. Doing this ensures that the spacecraft remains in the vicinity of its libration orbit for at

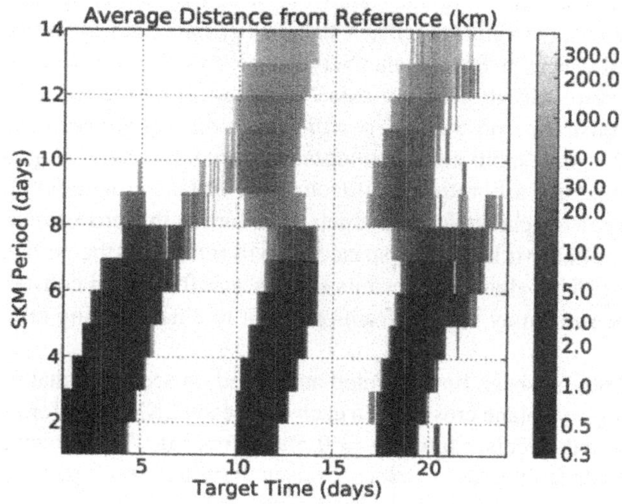

Figure 6-29 The average distance between the trajectory and the reference trajectory for the same trade space given in Fig. 6-27.

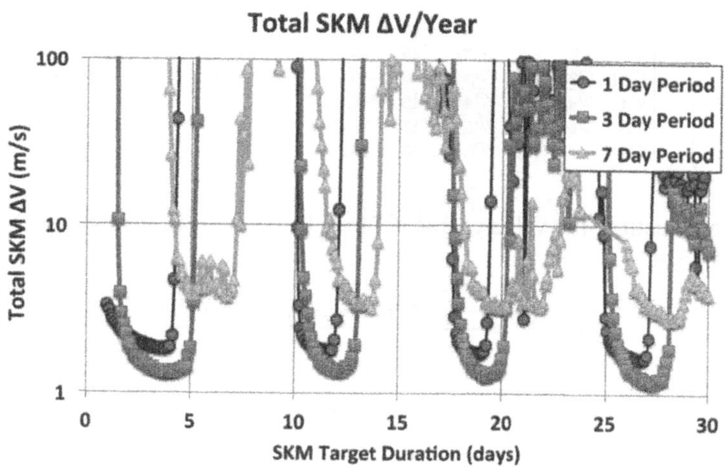

Figure 6-30 The annual station-keeping ΔV cost for three station-keeping periods as functions of the duration between each SKM and the target point along the reference trajectory for that maneuver.

least some time, given the size of the orbit determination and maneuver execution errors. Mission designers typically start the design by targeting the next $y = 0$ plane crossing to have zero velocity in the x-axis; once that design is complete it is used to seed a search for a maneuver that pierces the following $y = 0$ plane crossing with zero velocity in the x-axis. When targeting the second $y = 0$ plane crossing, all constraints on the first $y = 0$ plane crossing are removed. This may be repeated a few times, but modern integrators cannot typically integrate more than two revolutions about a libration orbit (four $y = 0$ plane crossings) into the future accurately enough to achieve further targets. The further this process extends into the future, the more likely it is that the spacecraft will remain on the particular libration orbit of interest. This algorithm permits the spacecraft's Jacobi constant to change; hence, the spacecraft may wander from one orbit to a neighboring orbit in the state space.

This algorithm has been implemented and tested on scenarios that target the first through fourth $y = 0$ plane crossing. In each case, each SKM is performed at a $y = 0$ plane crossing and targets a future $y = 0$ plane crossing. There may be benefit to placing the SKMs at different τ values, or even permitting each maneuver's τ value to vary. But these strategies have not been explored here for brevity.

It has been found that a modified single-shooting differential corrector (Section 2.6.5.1 and Section 2.6.6.2) works very well to generate each SKM rapidly. One formulates the problem by permitting the SKM to be in any direction, targeting a state on the subsequent $y = 0$ plane crossing such that its x-velocity is zero. The following equation is very similar to Eq. (2.40), modified for this application

$$\begin{bmatrix} \delta x_{T/2} \\ 0 \\ \delta z_{T/2} \\ -\dot{x}_{T/2} \\ \delta \dot{y}_{T/2} \\ \delta \dot{z}_{T/2} \end{bmatrix} \approx \begin{bmatrix} \phi_{11} & \phi_{12} & \phi_{13} & \phi_{14} & \phi_{15} & \phi_{16} \\ \phi_{21} & \phi_{22} & \phi_{23} & \phi_{24} & \phi_{25} & \phi_{26} \\ \phi_{31} & \phi_{32} & \phi_{33} & \phi_{34} & \phi_{35} & \phi_{36} \\ \phi_{41} & \phi_{42} & \phi_{43} & \phi_{44} & \phi_{45} & \phi_{46} \\ \phi_{51} & \phi_{52} & \phi_{53} & \phi_{54} & \phi_{55} & \phi_{56} \\ \phi_{61} & \phi_{62} & \phi_{63} & \phi_{64} & \phi_{65} & \phi_{66} \end{bmatrix} (t_{T/2}, t_0) \begin{bmatrix} 0 \\ 0 \\ 0 \\ \delta \dot{x}_0 \\ \delta \dot{y}_0 \\ \delta \dot{z}_0 \\ 0 \end{bmatrix} + \begin{bmatrix} \dot{x} \\ \dot{y} \\ \dot{z} \\ \ddot{x} \\ \ddot{y} \\ \ddot{z} \end{bmatrix} \delta(T/2) \quad (6.1)$$

In this application, the value of $\delta(T/2)$ may be determined from the second line of Eq. (6.1) to be

$$\delta(T/2) = \frac{-\phi_{24}\delta\dot{x}_0 - \phi_{25}\delta\dot{y}_0 - \phi_{26}\delta\dot{z}_0}{\dot{y}} \quad (6.2)$$

Substituting this value into the fourth line of Eq. (6.1) yields

$$-\dot{x}_{T/2} \approx \left(\phi_{44} - \phi_{24}\frac{\ddot{x}}{\dot{y}}\right)\delta\dot{x}_0 + \left(\phi_{45} - \phi_{25}\frac{\ddot{x}}{\dot{y}}\right)\delta\dot{y}_0 + \left(\phi_{46} - \phi_{26}\frac{\ddot{x}}{\dot{y}}\right)\delta\dot{z}_0 \quad (6.3)$$

One now has a choice about how to construct the SKM. Since there are three degrees of freedom and one control, this algorithm works very well for a mission whose maneuvers are constrained. If there are no further constraints, it is typically best to build the maneuver that minimizes the ΔV. We construct the least squares solution as follows

$$M = \left[\phi_{44} - \phi_{24}\frac{\ddot{x}}{\dot{y}}, \quad \phi_{45} - \phi_{25}\frac{\ddot{x}}{\dot{y}}, \quad \phi_{46} - \phi_{26}\frac{\ddot{x}}{\dot{y}} \right] \tag{6.4}$$

$$\begin{bmatrix} \delta\dot{x}_0 \\ \delta\dot{y}_0 \\ \delta\dot{z}_0 \end{bmatrix} = M^T \left(MM^T\right)^{-1} \left(-\dot{x}_{T/2}\right) \tag{6.5}$$

Table 6-5 summarizes the performance of this loose station-keeping strategy for different combinations of maneuver parameters, including the least squares solution $\overrightarrow{\Delta V}_0$, and each case where the maneuver is constrained to be in one Cartesian direction (in the Earth–Moon rotating coordinate frame). Further, Table 6-5 includes information for scenarios that target different target $y = 0$ plane crossings. One can see that the least squares solution performs better than any single-component solution. The z-axis burns did not converge often enough to characterize their performance for the case when the target was the first $y = 0$ plane crossing. The table illustrates very clearly that it is significantly better to target the second or third $y = 0$ plane crossing rather than the first. This makes sense given the amount of oscillation that exists in the system on account of the Moon's noncircular orbit about the Earth–Moon barycenter. These results suggest that targeting the second $y = 0$ plane crossing is the most optimal of these loose station-keeping strategies, applied to these particular constraints, errors, and dynamics.

Table 6-5 A summary of the results of the loose station-keeping strategy explored here.

$y = 0$	ΔV	Avg SKM ΔV (m/s)		Annual SKM ΔV (m/s)		Avg Slope from
Target	Type	Mean	1-σ	Mean	1-σ	Ref (km/day)
1	$\overrightarrow{\Delta V}_0$	0.5317	0.3547	25.6863	0.4614	7.9279
1	ΔV_0^x	0.6116	0.4880	29.3714	0.5992	4.3580
1	ΔV_0^y	1.2691	0.7243	61.2268	1.2253	22.4134
1	ΔV_0^z	Failed to converge				
2	$\overrightarrow{\Delta V}_0$	0.0643	0.0525	3.1067	0.1557	1.9986
2	ΔV_0^x	0.0793	0.0613	3.8287	0.1828	1.7720
2	ΔV_0^y	0.1512	0.1455	7.2116	0.3583	2.7830
2	ΔV_0^z	1.2563	1.0133	60.2222	3.7733	12.8971
3	$\overrightarrow{\Delta V}_0$	0.0667	0.0522	3.2276	0.1837	1.9252
3	ΔV_0^x	0.0846	0.0600	4.0560	0.2046	1.4319
3	ΔV_0^y	0.1536	0.1567	7.3782	0.4482	3.2306
3	ΔV_0^z	1.1862	0.9755	56.6230	2.6863	14.4563

If we compare the loose station-keeping strategy studied here with the tight station-keeping strategy considered earlier, we see that the loose strategy performs better for similar station-keeping periods. However, the tight strategy performs better if a mission can perform maneuvers more frequently, on the order of 3–4 days between maneuvers.

6.6.2.5 Maneuver Execution Errors All of the results presented previously have kept the spacecraft maneuver execution error model the same, namely, set such that each coordinate of a maneuver's execution is perturbed by an error taken from a normal distribution with mean zero and standard deviation of 2 mm/s. This error model is consistent with the errors observed from the *ARTEMIS* mission. Naturally, the station-keeping ΔV budget is dependent on this execution error model. Figure 6-31 presents the annual station-keeping ΔV budget as a function of maneuver execution error for a scenario where SKMs are performed at each $y = 0$ plane crossing, and each maneuver targets the subsequent plane crossing of the reference trajectory. One can see a very linear relationship between the annual ΔV cost and maneuver execution error. The line of best fit of this data is equal to

$$\Delta \text{V} = 1.8705x + 1.9267 \text{ m/s}$$

The curve's linearity is promising in the sense that the station-keeping strategy has kept the trajectory within the vicinity of the reference trajectory enough that linear approximations are valid. One notices also that the curve does flatten out as the maneuver execution error gets very small. It is in this regime that the orbit determination errors begin to dominate the station-keeping performance.

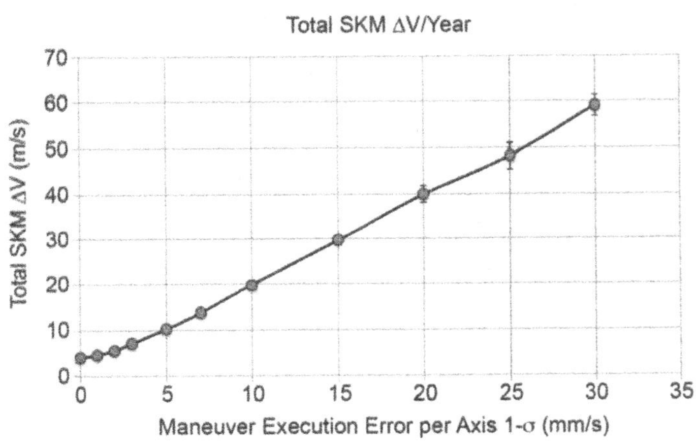

Figure 6-31 The annual station-keeping ΔV cost as a function of maneuver execution errors.

6.7 SPACECRAFT SYSTEMS DESIGN

Several considerations must be made to a spacecraft's design when evaluating low-energy lunar transfers compared to conventional lunar transfers. This discussion is meant to guide further analysis and not to reveal a full list of potential issues that one may have with a low-energy transfer, compared with a conventional transfer.

First, low-energy transfers require much more time than conventional transfers between the Earth and the Moon. This impacts the operations schedule, its risk, and its cost. A low-energy transfer's schedule is typically much more relaxed than a conventional transfer's schedule, which must perform a maneuver within a day or even within hours of injection. A spacecraft operations team has much more time to recover from anomalies and safe-mode events when flying a low-energy transfer. The spacecraft team also has more time to characterize the spacecraft, check out the instruments, outgas, and so forth. The mission may even delay maneuvers as needed. In addition, there is much more time to ensure that a spacecraft is on a proper approach vector when arriving at the Moon via a low-energy transfer than a conventional transfer.

The communications systems for spacecraft traversing low-energy transfers must be capable of reaching out to 1–2 million kilometers, depending on the transfer. This is 3–5 times further than a conventional transfer. This long link distance may require larger ground station antennas, larger spacecraft antennas, and/or more communications power. However, a spacecraft intending to perform its mission objectives at the Moon may not have much data to transmit at its apogee passage, alleviating some of the pressures caused by the long link distance.

A low-energy transfer requires a smaller maneuver when arriving at the Moon, compared with a conventional transfer to the same destination. This fact may benefit a lunar mission in many ways. First, the spacecraft does not require as much fuel and can put more of the ΔV requirements on the launch vehicle rather than the spacecraft. Second, the spacecraft may reduce the amount of gravity losses when performing an insertion into a low lunar orbit using small engines. This was the case for the two *GRAIL* spacecraft, and it could be the case for any lunar landers. Finally, a mission to a lunar libration orbit does not even require a large orbit insertion maneuver, which may open up many design options.

Low-energy transfers commonly traverse through regions of space where the Sun–Earth–spacecraft angle and/or the Sun–spacecraft–Earth angle drops near zero degrees. This characteristic may be detrimental to the communications system on board the spacecraft, though it may only impact the mission for a day or two.

The final consideration presented here is that low-energy transfers typically do not pass through the Van Allen Belts more than once, which may reduce the radiation risks for a lunar spacecraft, compared with a conventional transfer that may implement Earth-phasing orbits.

APPENDIX A

LOCATING THE LAGRANGE POINTS

A.1 INTRODUCTION

The discussion given here, previously authored by Parker [46], is devoted to deriving analytical expressions for the Lagrange points in the circular restricted three-body problem (CRTBP). Szebehely provides more details and a clear description of this derivation [86]. Other authors have provided similar derivations, including Moulton [106] and Broucke et al. [236].

A.2 SETTING UP THE SYSTEM

Let us begin with a system of two masses, m_1 and m_2, such that $m_1 \geq m_2$. Furthermore, each of these masses is orbiting the center of mass of the system in a circle. Then there exist cases where a third body, m_3, of negligible mass can be placed in the system in such a way that the force of gravity from both bodies and the rotational motion in the system balance to produce a configuration that does not change in time with respect to the rotating system. That is, each body rotates about the center of mass at exactly the same rate and is seemingly fixed in the rotating frame of reference. Euler and Lagrange located five of these cases, and those locations have henceforth been known as the five Lagrange points in a three-body system.

To locate the Lagrange points, we begin with the three bodies stationary in the corotating frame of reference. That is

$$\dot{\theta}_1 = \dot{\theta}_2 = \dot{\theta}_3 = \dot{\theta}(t) \tag{A.1}$$

where $\dot{\theta}_i$ is the angular velocity of the body of mass m_i about the center of mass. Furthermore, if the shape of the configuration does not alter over time, the relative distances $r_{12}(t)$, $r_{23}(t)$, and $r_{31}(t)$ are given by

$$\frac{r_{12}(t)}{r_{12}(t_0)} = \frac{r_{23}(t)}{r_{23}(t_0)} = \frac{r_{31}(t)}{r_{31}(t_0)} = f(t) \tag{A.2}$$

So far, there are no constraints on the relative size of the configuration, only on the angular velocity and the shape of the configuration.

Next, we move the origin to the center of mass of the configuration. Then $\vec{R}_i$ describes the vector position of the ith mass, satisfying the constraint

$$\sum_{i=1}^{3} m_i \vec{R}_i = 0 \tag{A.3}$$

Equation (A.3) may be written

$$(m_1 + m_2 + m_3)\vec{R}_1 + m_2(\vec{R}_2 - \vec{R}_1) + m_3(\vec{R}_3 - \vec{R}_1) = 0,$$

or

$$M\vec{R}_1 = -m_2\vec{r}_{12} - m_3\vec{r}_{13} \tag{A.4}$$

where M is equal to the sum of the masses in the system. Squaring this relationship produces

$$M^2R_1^2 = m_2^2r_{12}^2 + m_3^2r_{13}^2 + 2m_2m_3\vec{r}_{12} \bullet \vec{r}_{13} \tag{A.5}$$

where R_i and r_i denote the magnitudes of the vectors $\vec{R}_i$ and $\vec{r}_i$, respectively. Since we know that the relative shape of the configuration does not change, as seen above, we may substitute in the relationships for the relative angles and distances (Eqs. (A.1) and (A.2)) into Eq. (A.5) to find that, in general

$$R_i(t) = R_i(t_0)f(t) \tag{A.6}$$

If F_i is the magnitude of the force per unit mass acting on the mass m_i, then the total force acting on m_i is m_iF_i and the equation of motion of that mass along the direction of the force satisfies

$$m_iF_i = m_i\left(\ddot{R}_i - R_i\dot{\theta}_i^2\right) \tag{A.7}$$

Since all of the particles are rotating at the same rate, we can reduce this relationship to the following

$$m_iF_i = m_i\left[R_i(t_0)\ddot{f}(t) - R_i\dot{\theta}^2\right]$$

or equivalently

$$m_i F_i = R_i m_i \left[\ddot{f}(t)/f(t) - \dot{\theta}^2\right] \tag{A.8}$$

Hence, we have the proportionality relationship

$$F_1 : F_2 : F_3 = R_1 : R_2 : R_3 \tag{A.9}$$

There are two cases that will satisfy the conditions given in Eqs. (A.8) and (A.9). The two cases are

$$\vec{R}_i \times \vec{F}_i = 0 \qquad \text{or} \qquad \vec{R}_i \times \ddot{\vec{R}}_i = 0 \tag{A.10}$$

When we set $i = 1$ and look at the first particle, we have the following force function

$$m_1 \ddot{\vec{R}}_1 = G\left(\frac{m_1 m_2}{r_{12}^3}\vec{r}_{12} + \frac{m_1 m_3}{r_{13}^3}\vec{r}_{13}\right) \tag{A.11}$$

When we take the cross product of $\vec{R}_1$ with each side of Eq. (A.11), we obtain the following expression

$$\vec{R}_1 \times \left(\frac{m_2}{r_{12}^3}\vec{R}_2 + \frac{m_3}{r_{13}^3}\vec{R}_3\right) = 0 \tag{A.12}$$

Using the center of mass relationship given in Eq. (A.3), this can be simplified to

$$m_2 \vec{R}_1 \times \vec{R}_2 \left(\frac{1}{r_{12}^3} - \frac{1}{r_{13}^3}\right) = 0 \tag{A.13}$$

Once, again, there are two similar equations for the other two particles. For Eq. (A.13) to hold, either of the following expressions must be true

$$r_{12} = r_{23} = r_{31} = r \tag{A.14}$$

(the equilateral triangle solution), or

$$\vec{R}_1 \times \vec{R}_2 = \vec{R}_2 \times \vec{R}_3 = \vec{R}_3 \times \vec{R}_1 = 0 \tag{A.15}$$

(the collinear solution).

The triangular and collinear cases are addressed separately in Sections A.3 and A.4.

A.3 TRIANGULAR POINTS

In the equilateral triangle case given in Eq. (A.14), we arrive at the following relationship for the first particle

$$\ddot{\vec{R}}_1 + GM_1 \frac{\vec{R}_1}{R_1^3} = 0 \tag{A.16}$$

where

$$M_1 = \frac{\left(m_2^2 + m_3^2 + m_2 m_3\right)^{3/2}}{\left(m_1 + m_2 + m_3\right)^2} \tag{A.17}$$

This result is the familiar two-body equation of motion. In this case, the first particle moves about the center of mass of the system in any conic orbit as if it had unit mass and a mass of M_1 were placed at the center of mass of the system. Each particle moves in a corresponding trajectory, and the figure remains in an equilateral triangle configuration (although its size may oscillate or grow indefinitely).

A.4 COLLINEAR POINTS

In the collinear case given in Eq. (A.15), we can also first show that each particle's orbit is a conic section. Beginning with the first particle, we can take the collinear axis to be the x axis; the force acting on m_1 is then

$$F_1 = m_2 \frac{(x_2 - x_1)}{x_{12}^3} + m_3 \frac{(x_3 - x_1)}{x_{13}^3} \tag{A.18}$$

But we also know from Eq. (A.6) that

$$x_i(t) = x_i(t_0) f(t)$$

so that

$$F_1 = \frac{1}{f^2}\left[m_2 \frac{(x_2 - x_1)}{x_{12}^3} + m_3 \frac{(x_3 - x_1)}{x_{13}^3}\right]_0 = \frac{\text{constant}}{f^2} \tag{A.19}$$

Since f is proportional to distance, m_1 is acted upon by an inverse-square-law central force. Hence, the particle's orbit is a conic section.

Now we will impose the condition from Eq. (A.9) that

$$F_1 : F_2 : F_3 = x_1 : x_2 : x_3.$$

This condition introduces the proportionality constant A, such that

$$\begin{aligned} F_1 &= A x_1 \\ F_2 &= A x_2 \\ F_3 &= A x_3 \end{aligned} \tag{A.20}$$

or equally

$$\begin{aligned} A x_1 &= m_2 \frac{x_2 - x_1}{x_{12}^3} + m_3 \frac{x_3 - x_1}{x_{13}^3} \\ A x_2 &= m_3 \frac{x_3 - x_2}{x_{23}^3} + m_1 \frac{x_1 - x_2}{x_{21}^3} \\ A x_3 &= m_1 \frac{x_1 - x_3}{x_{31}^3} + m_2 \frac{x_2 - x_3}{x_{32}^3} \end{aligned} \tag{A.21}$$

We are looking for the placement of the particle of mass m_3 with respect to the other two particles such that the relative positions are constant in the rotating frame. The equilibrium positions possible for m_3 are in the arrangements $m_1 - m_3 - m_2$ (case 132), $m_1 - m_2 - m_3$ (case 123), and $m_3 - m_1 - m_2$ (case 312). Each case will be observed separately.

A.4.1 Case 132: Identifying the L$_1$ point

For case 132, we are looking for a positive value of X such that

$$\left.\begin{aligned} X &= \frac{x_2 - x_3}{x_3 - x_1} = \frac{x_{32}}{x_{13}} \\ X + 1 &= \frac{x_2 - x_1}{x_3 - x_1} = \frac{x_{12}}{x_{13}} \end{aligned}\right\} \tag{A.22}$$

We identify X using a series of steps. We first subtract Ax_1 from Ax_3 and Ax_3 from Ax_2 from Eq. (A.21) to arrive at Ax_{13} and Ax_{32}

$$\begin{aligned} Ax_{13} &= -\frac{m_1 + m_3}{x_{13}^2} + m_2\left(\frac{1}{x_{32}^2} - \frac{1}{x_{12}^2}\right) \\ Ax_{32} &= -\frac{m_2 + m_3}{x_{32}^2} + m_1\left(\frac{1}{x_{13}^2} - \frac{1}{x_{12}^2}\right) \end{aligned} \tag{A.23}$$

Using Eq. (A.22), we know that $x_{32} = Xx_{13}$ and $x_{12} = (X+1)x_{13}$. When we substitute these relationships into Eq. (A.23), we find two different relationships for the quantity Ax_{13}^3. When we set them equal and arrange in powers of X, we arrive at Lagrange's quintic equation

$$\begin{aligned} (m_1 + m_3)X^5 + (3m_1 + 2m_3)X^4 + (3m_1 + m_3)X^3 \\ - (3m_2 + m_3)X^2 - (3m_2 + 2m_3)X - (m_2 + m_3) = 0 \end{aligned} \tag{A.24}$$

We can use a quintic solver to solve for X (see Section A.5). Since the coefficients of Eq. (A.24) change sign only once, there can be only one positive real root. We can then use that value for X to determine the relative location of the massless particle, that is, the location of L$_1$, with respect to the other two particles by solving for x_3 in Eq. (A.22)

$$X = \frac{x_2 - x_3}{x_3 - x_1} \qquad \Rightarrow \qquad x_3 = x_1 + \frac{x_2 - x_1}{X + 1} \tag{A.25}$$

A.4.2 Case 123: Identifying the L$_2$ point

For case 123, we are looking for a positive value of X such that

$$\left.\begin{aligned} X &= \frac{x_3 - x_2}{x_2 - x_1} = \frac{x_{23}}{x_{12}} \\ X + 1 &= \frac{x_3 - x_1}{x_2 - x_1} = \frac{x_{13}}{x_{12}} \end{aligned}\right\} \tag{A.26}$$

In order to identify X, we follow a similar derivation as in case 132. We first subtract Ax_2 from Ax_3 and Ax_1 from Ax_2 from Eq. (A.21) to arrive at Ax_{23} and Ax_{12}

$$\begin{aligned} Ax_{23} &= -\frac{m_2 + m_3}{x_{23}^2} + m_1 \left(\frac{1}{x_{12}^2} - \frac{1}{x_{13}^2} \right) \\ Ax_{12} &= -\frac{m_1 + m_2}{x_{12}^2} + m_3 \left(\frac{1}{x_{23}^2} - \frac{1}{x_{13}^2} \right) \end{aligned} \tag{A.27}$$

We then substitute in X and $(X + 1)$ from Eq. (A.26) as before, eliminate Ax_{12}^3 between the resulting equations and arrange in powers of X to produce Lagrange's quintic equation

$$\begin{aligned} &(m_1 + m_2)X^5 + (3m_1 + 2m_2)X^4 + (3m_1 + m_2)X^3 \\ &\qquad - (m_2 + 3m_3)X^2 - (2m_2 + 3m_3)X - (m_2 + m_3) = 0 \end{aligned} \tag{A.28}$$

Once again, we can use a quintic solver to solve for X (see Section A.5), knowing that again there is only one real positive root. We can then use that value for X to determine the relative location of the massless particle, that is, the location of L_2, with respect to the other two particles by solving for x_3 in Eq. (A.26)

$$X = \frac{x_3 - x_2}{x_2 - x_1} \qquad \Rightarrow \qquad x_3 = x_2 + X(x_2 - x_1) \tag{A.29}$$

A.4.3 Case 312: Identifying the L_3 point

For case 312, we are looking for a positive value of X such that

$$\left. \begin{aligned} X &= \frac{x_2 - x_1}{x_1 - x_3} = \frac{x_{12}}{x_{31}} \\ X + 1 &= \frac{x_2 - x_3}{x_1 - x_3} = \frac{x_{32}}{x_{31}} \end{aligned} \right\} \tag{A.30}$$

In order to identify X, we follow a similar derivation as in case 132. We first subtract Ax_1 from Ax_2 and Ax_3 from Ax_1 from Eq. (A.21) to arrive at Ax_{12} and Ax_{31}

$$\begin{aligned} Ax_{31} &= -\frac{m_1 + m_3}{x_{31}^2} + m_2 \left(\frac{1}{x_{12}^2} - \frac{1}{x_{32}^2} \right) \\ Ax_{12} &= -\frac{m_1 + m_2}{x_{12}^2} + m_3 \left(\frac{1}{x_{31}^2} - \frac{1}{x_{32}^2} \right) \end{aligned} \tag{A.31}$$

We then substitute in X and $(X + 1)$ from Eq. (A.30) as before, eliminate Ax_{31}^3 between the resulting equations and arrange in powers of X to produce Lagrange's quintic equation

$$\begin{aligned} &(m_1 + m_3)X^5 + (2m_1 + 3m_3)X^4 + (m_1 + 3m_3)X^3 \\ &\qquad - (m_1 + 3m_2)X^2 - (2m_1 + 3m_2)X - (m_1 + m_2) = 0 \end{aligned} \tag{A.32}$$

Once again, we can use a quintic solver to solve for X (see Section A.5), knowing that again there is only one real positive root. We can then use that value for X to determine the relative location of the massless particle, that is, the location of L_3, with respect to the other two particles by solving for x_3 in Eq. (A.30)

$$X = \frac{x_2 - x_1}{x_1 - x_3} \qquad \Rightarrow \qquad x_3 = x_1 - \frac{x_2 - x_1}{X} \tag{A.33}$$

A.5 ALGORITHMS

The quintics given in Eqs. (A.25), (A.29), and (A.33) provide analytic determinations of the locations of the first, second, and third Lagrange points, respectively, in the circular restricted three-body system. Szebehely outlines a fixed-point iterative scheme that may be implemented to identify the single positive real root of each of the quintic equations [86]. The fourth and fifth Lagrange points make equilateral triangles with the primaries; hence, their locations are easily determined using geometry.

Sections A.5.1–A.5.3 provide pseudo-code that may be used to implement a fixed-point iterative scheme to find the x-coordinate of L_1 – L_3, respectively. The coordinate axis and the definition of μ are defined in Section 2.5.1.

A.5.1 Numerical Determination of L_1

$$\gamma_0 = \left(\frac{\mu(1-\mu)}{3}\right)^{1/3}$$

$$\gamma = \gamma_0 + 1$$

while$|\gamma - \gamma_0| > \text{tol}$

$$\gamma_0 = \gamma$$

$$\gamma = \left(\frac{\mu(\gamma_0 - 1)^2}{3 - 2\mu - \gamma_0(3 - \mu - \gamma_0)}\right)^{1/3}$$

endwhile

$$x_{L_1} = 1 - \mu - \gamma$$

A.5.2 Numerical Determination of L_2

$$\gamma_0 = \left(\frac{\mu(1-\mu)}{3}\right)^{1/3}$$

$$\gamma = \gamma_0 + 1$$

$$\text{while}|\gamma - \gamma_0| > \text{tol}$$

$$\gamma_0 = \gamma$$

$$\gamma = \left(\frac{\mu(\gamma_0 + 1)^2}{3 - 2\mu + \gamma_0(3 - \mu + \gamma_0)}\right)^{1/3}$$

$$\text{endwhile}$$

$$x_{L_2} = 1 - \mu + \gamma$$

A.5.3 Numerical Determination of L_3

$$\gamma_0 = \left(\frac{\mu(1-\mu)}{3}\right)^{1/3}$$

$$\gamma = \gamma_0 + 1$$

$$\text{while}|\gamma - \gamma_0| > \text{tol}$$

$$\gamma_0 = \gamma$$

$$\gamma = \left(\frac{(1-\mu)(\gamma_0 + 1)^2}{1 + 2\mu + \gamma_0(2 + \mu + \gamma_0)}\right)^{1/3}$$

$$\text{endwhile}$$

$$x_{L_3} = -\mu - \gamma$$

REFERENCES

1. *Apollo 11 Flight Plan*, Tech. Rep. AS-506 / CSM-107 / LM-5, Flight Planning Branch, Flight Crew Support Division, NASA, Manned Spacecraft Center (now Johnson Space Center), Houston, Texas, July 1, 1969.

2. J. S. Parker, R. L. Anderson, and A. Peterson, "A Survey of Ballistic Transfers to Low Lunar Orbit," *Proceedings of the AAS/AIAA Spaceflight Mechanics Meetings held February 13–17, 2011, New Orleans, Louisiana* (M. K. Jah, Y. Guo, A. L. Bowes, and P. C. Lai, eds.), vol. 140, Paper AAS 11-277, *Advances in Astronautical Sciences*, AAS/AIAA, Univelt Inc., San Diego, California, pp. 2461–2480, 2011.

3. J. N. Goswami and M. Annadurai, "Chandrayaan-1: India's First Planetary Science Mission to the Moon," *Current Science; Special Section: Chandrayaan-1*, vol. 96, no. 4, pp. 486–491, February 25, 2009.

4. S. B. Broschart, M. J. Chung, S. J. Hatch, J. H. Ma, T. H. Sweetser, S. S. Weinstein-Weiss, and V. Angelopoulos, "Preliminary Trajectory Design for the ARTEMIS Lunar Mission," *Proceedings of the AAS/AIAA Astrodynamics Specialist Conference held August 9–13, 2009, Pittsburgh, Pennsylvania* (A. V. Rao, T. A. Lovell, F. K. Chan, and L. A. Cangahuala, eds.), vol. 134, Paper AAS 09-382, *Advances in Astronautical Sciences*, AAS/AIAA, Univelt Inc., San Diego, California, 2009.

5. A. C. Clarke, *Interplanetary Flight*. Temple Press Books Ltd., London, 1950.

6. R. W. Farquhar, "Station-Keeping in the Vicinity of Collinear Libration Points with an Application to a Lunar Communications Problem," *Space Flight Mechanics*, Vol. 11 of

Science and Technology Series, American Astronautical Society, New York, pp. 519–535, 1966.

7. R. W. Farquhar, "Lunar Communications with Libration-Point Satellites," *Journal of Spacecraft and Rockets*, vol. 4, no. 10, pp. 1383–1384, 1967.
8. K. Hill, *Autonomous Navigation in Libration Point Orbits*. Ph.D. thesis, University of Colorado, Boulder, Colorado, 2007.
9. K. Hill, G. H. Born, and M. W. Lo, "Linked, Autonomous, Interplanetary Satellite Orbit Navigation (LiAISON) in Lunar Halo Orbits," *Proceedings of the AAS/AIAA Astrodynamics Specialist Conference held August 7–11, 2005, South Lake Tahoe, California* (B. G. Williams, L. A. D'Amario, K. C. Howell, and F. R. Hoots, eds.), vol. 123, Paper AAS 05-400, *Advances in Astronautical Sciences*, AAS/AIAA, Univelt Inc., San Diego, California, 2006.
10. K. Hill, M. W. Lo, and G. H. Born, "Linked, Autonomous, Interplanetary Satellite Orbit Navigation (LiAISON)," *Proceedings of the AAS/AIAA Astrodynamics Specialist Conference held August 7–11, 2005, South Lake Tahoe, California* (B. G. Williams, L. A. D'Amario, K. C. Howell, and F. R. Hoots, eds.), vol. 123, Paper AAS 05-399, *Advances in Astronautical Sciences*, AAS/AIAA, Univelt Inc., San Diego, California, 2006.
11. K. Hill, J. S. Parker, G. H. Born, and N. Demandante, "A Lunar L2 Navigation, Communication, and Gravity Mission," *Proceedings of the AIAA/AAS Astrodynamics Specialist Conference held August 2006, Keystone, Colorado*, Paper AIAA 2006-6662, AIAA/AAS, 2006.
12. J. S. Parker, R. L. Anderson, G. H. Born, and K. Fujimoto, *Linked Autonomous Interplanetary Satellite Orbit Navigation (LiAISON) Between Geosynchronous and Lunar Halo Orbits*, Tech. Rep. D-72688 (internal document), Jet Propulsion Laboratory, California Institute of Technology, Pasadena, California, 2012.
13. B. Villac, C. Chow, M. Lo, G. Hintz, and Z. Nazari, "Dynamic Optimization of Multi-Spacecraft Relative Navigation Configurations in the Earth–Moon System," *Proceedings of the AAS George H. Born Symposium held May 13–14, 2010, Boulder, Colorado*, AAS, 2010.
14. V. R. Bond, S. J. Sponaugle, M. F. Fraietta, and S. F. Everett, "Cislunar Libration Point as a Transportation Node for Lunar Exploration," *Proceedings of the 1st AAS/AIAA Spaceflight Mechanics Meeting held February 11–13, 1991, Houston, Texas* (J. K. Soldner, R. K. Misra, L. L. Sackett, and R. Holdaway, eds.), vol. 75, Paper AAS 91-103, *Advances in Astronautical Sciences*, AAS/AIAA, Univelt Inc., San Diego, California, 1991.
15. L. A. D'Amario, *Minimum Impulse Three-Body Trajectories*, Tech. Rep. T-593, Massachusetts Institute of Technology, Boston, Massachusetts, June 1973.
16. R. W. Farquhar, "Future Missions for Libration-point Satellites," *Astronautics & Aeronautics*, pp. 52–56, May 1969.
17. R. W. Farquhar, *The Utilization of Halo Orbits in Advanced Lunar Operations*, Tech. Rep., NASA TN D-6365, Goddard Space Flight Center, Greenbelt, Maryland, July 1971.
18. R. W. Farquhar and D. W. Dunham, "Use of Libration-Point Orbits for Space Observatories," *Observatories in Earth Orbit and Beyond*, Kluwer Academic Publishers, Dordrecht, The Netherlands, pp. 391–395, 1990.
19. C. A. Cross, "Orbits for an Extra-Terrestrial Observatory," *Journal of the British Interplanetary Society*, vol. 13, no. 4, pp. 204–207, 1954.

20. V. A. Egorov, *Certain Problems of Moon Flight Dynamics*, International Physical Index, Inc., New York, 1958.

21. V. G. Fesenkov, "On the Possibility of Capture at Close Passages of Attracting Bodies," *Astronomicheskiy Zhurnal (Astronomical Journal of the Soviet Union)*, vol. 23, no. 1, pp. 45–48, 1946.

22. T. H. Sweetser, "An Estimate of the Global Minimum DV Needed for Earth–Moon Transfer," *Proceedings of the AAS/AIAA Spaceflight Mechanics Meeting held February 1991, Houston, Texas* (J. K. Soldner, A. K. Misra, L. L. Sackett, and R. Holdaway, eds.), vol. 75, Part I, Paper AAS 91-101, *Advances in Astronautical Sciences*, AAS/AIAA, Univelt Inc., San Diego, California, 1991.

23. H. J. Pernicka, D. P. Scarberry, S. M. Marsh, and T. H. Sweetser, "A Search for Low ΔV Earth-to-Moon Trajectories," *The Journal of the Astronautical Sciences*, vol. 43, no. 1, pp. 77–88, January–March 1995.

24. C. Conley, "Low Energy Transit Orbits in the Restricted Three Body Problem," *SIAM Journal of Applied Mathematics*, vol. 16, no. 4, pp. 732–746, 1968.

25. E. A. Belbruno, "Lunar Capture Orbits, a Method of Constructing Earth Moon Trajectories and the Lunar Gas Mission," *Proceedings of the 19th AIAA/DGLR/JSASS International Electric Propulsion Conference held May 1987, Colorado Springs, Colorado*, Paper AIAA 87-1054, 1987.

26. E. A. Belbruno, *Capture Dynamics and Chaotic Motions in Celestial Mechanics*. Princeton University Press, Princeton, New Jersey, 2004.

27. E. A. Belbruno and J. P. Carrico, "Calculation of Weak Stability Boundary Ballistic Lunar Transfer Trajectories," *Proceedings of the AIAA/AAS Astrodynamics Specialist Conference held August 14–17, 2000, Denver, Colorado*, Paper AIAA 2000-4142, AIAA/AAS, 2000.

28. H. Yamakawa, *On Earth–Moon Transfer Trajectory with Gravitational Capture*. Ph.D. thesis, University of Tokyo, Tokyo, Japan, 1992.

29. H. Yamakawa, J. Kawaguchi, N. Ishii, and H. Matsuo, "On Earth–Moon Transfer Trajectory with Gravitational Capture," *Proceedings of the AIAA/AAS Astrodynamics Specialist Conference held August 1993*, Paper AIAA 93-633, AIAA/AAS, 1993.

30. E. A. Belbruno and J. K. Miller, "Sun-Perturbed Earth-to-Moon Transfers with Ballistic Capture," *Journal of Guidance, Control, and Dynamics*, vol. 16, no. 4, pp. 770–774, July–August 1993.

31. V. V. Ivashkin, "On Trajectories of the Earth–Moon Flight of a Particle with Its Temporary Capture by the Moon," *Doklady Physics, Mechanics*, vol. 47, no. 11, pp. 825–827, 2002.

32. V. V. Ivashkin, "On the Earth-to-Moon Trajectories with Temporary Capture of a Particle by the Moon," *Proceedings of the 54th International Astronautical Congress held September 29–October 3, 2003, Bremen, Germany*, Paper IAC-03-A.P.01), pp. 1–9, 2003.

33. V. V. Ivashkin, "On Particle's Trajectories of Moon-to-Earth Space Flights with the Gravitational Escape from the Lunar Attraction," *Doklady Physics, Mechanics*, vol. 49, no. 9, pp. 539–542, 2004.

34. V. V. Ivashkin, "On Trajectories for the Earth-to-Moon Flight with Capture by the Moon," *Proceedings of the International Lunar Conference 2003, Springfield, Virginia* (S. M. Durst, C. T. Bohannan, C. G. Thomason, M. R. Cerney, and L. Yuen, eds.), vol. 108, Paper AAS 03-723, *Science and Technology Series*, American Astronautical Society, pp. 157–166, 2004.

35. E. M. Bollt and J. D. Meiss, "Targeting Chaotic Orbits to the Moon Through Recurrence," *Physics Letters A*, vol. 204, pp. 373–378, August 1995.

36. C. G. Schroer and E. Ott, "Targeting in Hamiltonian Systems that have Mixed Regular/Chaotic Phase Spaces," *Chaos*, vol. 7, no. 4, pp. 512–519, December 1997.

37. W. S. Koon, M. W. Lo, J. E. Marsden, and S. D. Ross, "Shoot the Moon," *Proceedings of the AAS/AIAA Spaceflight Mechanics Meeting held January 23–26, 2000, Clearwater, Florida* (C. A. Kluever, B. Neta, C. D. Hall, and J. M. Hanson, eds.), vol. 105, part 2, Paper AAS 00-166, *Advances in Astronautical Sciences*, AAS/AIAA, Univelt Inc., San Diego, California, pp. 1017–1030, 2000.

38. W. S. Koon, M. W. Lo, J. E. Marsden, and S. D. Ross, "Low Energy Transfers to the Moon," *Celestial Mechanics and Dynamical Astronomy*, vol. 81, no. 1, pp. 63–73, September 2001.

39. J. S. Parker and M. W. Lo, "Shoot the Moon 3D," *Astrodynamics 2005: Proceedings of the AAS/AIAA Astrodynamics Conference held August 7–10, 2005, South Lake Tahoe, California* (B. G. Williams, L. A. D'Amario, K. C. Howell, and F. R. Hoots, eds.), vol. 123, Paper AAS 05-383, *Advances in Astronautical Sciences*, American Astronautical Society, Univelt Inc., San Diego, California, pp. 2067–2086, 2006. website: http://www.univelt.com.

40. J. S. Parker and G. H. Born, "Modeling a Low-Energy Ballistic Lunar Transfer Using Dynamical Systems Theory," *Journal of Spacecraft and Rockets*, vol. 45, no. 6, pp. 1269–1281, San Diego, California, November–December 2008.

41. F. Topputo, M. Vasile, and F. Bernelli-Zazzera, "Interplanetary and Lunar Transfers Using Libration Points," *Proceedings of 18th International Symposium on Space Flight Dynamics held October 11–15, 2004, Munich, Germany*, Paper SP-548, 2004.

42. F. Topputo, M. Vasile, and F. Bernelli-Zazzera, "Low Energy Interplanetary Transfers Exploiting Invariant Manifolds of the Restricted Three-Body Problem," *The Journal of the Astronautical Sciences*, vol. 53, no. 4, pp. 353–372, October–December 2005.

43. K. C. Howell and M. Kakoi, "Transfers Between the Earth–Moon and Sun–Earth Systems using Manifolds and Transit Orbits," *Acta Astronautica*, vol. 59, no. 1–5, pp. 367–380, IAC-05-C1.6.01, 2006.

44. J. S. Parker, "Low-Energy Ballistic Transfers to Lunar Halo Orbits," *Proceedings of the AAS/AIAA Astrodynamics Specialist Conference held 9–13 August 2009, Pittsburgh, Pennsylvania* (A. V. Rao, T. A. Lovell, F. K. Chan, and L. A. Cangahuala, eds.), vol. 135, Paper AAS 09-443, *Advances in Astronautical Sciences, Astrodynamics 2009*, San Diego, CA, AAS/AIAA, Univelt Inc., pp. 2339–2358, 2010.

45. J. S. Parker, "Families of Low-Energy Lunar Halo Transfers," *Proceedings of the 16th AAS/AIAA Spaceflight Mechanics Meetings held January 22–26, 2006, Tampa, Florida* (S. R. Vadali, L. A. Cangahuala, J. P. W. Schumacher, and J. J. Guzman, eds.), vol. 124, Paper AAS 06-132, *Advances in Astronautical Sciences*, AAS/AIAA, Univelt Inc., San Diego, California, 2006.

46. J. S. Parker, *Low-Energy Ballistic Lunar Transfers*. Ph.D. thesis, University of Colorado, Boulder, Colorado, 2007.

47. J. S. Parker, "Monthly Variations of Low-Energy Ballistic Transfers to Lunar Halo Orbits," *Proceedings of the AIAA/AAS Astrodynamics Specialist Conference held August 2–5, 2010, Toronto, Ontario*, Paper AIAA 2010-7963, AIAA/AAS, 2010.

48. G. Mingotti, F. Topputo, and F. Bernelli-Zazzera, "Low-Energy, Low-Thrust Transfers to the Moon," *Celestial Mechanics and Dynamical Astronomy*, vol. 105, pp. 61–74, 2009.

49. G. Mingotti and F. Topputo, "Ways to the Moon: A Survey," *Proceedings of the 21st AAS/AIAA Space Flight Mechanics Meeting held February 13–17, 2011, New Orleans, Louisiana*, Paper AAS 11-283, American Astronautical Society, 2011.

50. A. Ohndorf, B. Dachwald, and E. Gill, "Optimization of Low-Thurst Earth–Moon Transfers using Evolutionary Neurocontrol," *CEC'09 Proceedings of the Eleventh Conference on Congress on Evolutionary, 2009, Piscataway, New Jersey*, IEEE, pp. 358–364, 2009.

51. M. Ozimek and K. Howell, "Low-Thrust Transfers in the Earth–Moon System Including Applications to Libration Point Orbits," *Journal of Guidance, Control, and Dynamics*, vol. 33, no. 2, pp. 533–549, March–April 2010.

52. J. Senent, C. Ocampo, and A. Capella, "Low-Thrust Variable-Specific-Impulse Transfers and Guidance to Unstable Periodic Orbits," *Journal of Guidance, Control, and Dynamics*, vol. 28, no. 2, pp. 280–290, March–April 2005.

53. A. Herman and B. Conway, "Optimal, Low-Thrust, Earth–Moon Orbit Transfers," *Journal of Guidance, Control, and Dynamics*, vol. 21, no. 1, pp. 141–147, January–February 1998.

54. E. M. Alessi, G. Gómez, and J. J. Masdemont, *Low-Energy Transfers in the Earth–Moon System*, Springer, New York, pp. 107–114, 2010.

55. G. Wawrzyniak and K. Howell, "Investigating the Design Space for Solar Sail Trajectories in the Earth–Moon System," *The Open Aerospace Engineering Journal*, vol. 4, pp. 11–29, 2011.

56. L. Lozier, K. Galal, D. Folta, and M. Beckman, "Lunar Prospector Mission Design and Trajectory Support," *Proceedings of the AAS/GSFC International Symposium on Space Flight Dynamics held May 11–15, 1998, Greenbelt, Maryland* (T. H. Stengle, ed.), vol. 100, Part I, Paper AAS 98-323, *Advances in Astronautical Sciences*, American Astronauts Society and Goddard Space Flight Center, 1998.

57. R. Godwin, ed., *Apollo 11—The NASA Mission Reports*, Vol. 1 of *The Apogee Books Space Series*. Collector's Guide Publishing Inc., Burlington, Ontario, Canada, 1999.

58. NASA, "NASA National Space Science Data Center (NSSDC) Lunar and Planetary Home Page," February 2012. http://nssdc.gsfc.nasa.gov/planetary/planets/moonpage.html.

59. A. A. Siddiqi, *Deep Space Chronicle, A Chronology of Deep Space and Planetary Probes, 1958–2000*, NASA SP-2002-4524, Monographs in Aerospace History, Number 24, National Aeronautics and Space Administration, June 2002.

60. R. W. Farquhar, D. Muhonen, and D. Richardson, "Mission Design for a Halo Orbiter of the Earth," *Journal of Spacecraft and Rockets*, vol. 14, no. 3, pp. 170–177, March 1977.

61. D. W. Dunham, "Contingency Plans for the ISEE-3 Libration Point Mission," *Proceedings of the AIAA/AAS Astrodynamics Specialist Conference*, Paper AIAA 79-129, AIAA/AAS, 1979.

62. "NASA's High Energy Astrophysics Science Archive Research Center," web page, Goddard Space Flight CenterGreenbelt, Maryland, April 2012. http://heasarc.gsfc.nasa.gov/.

63. "NASA's Polar, Wind and Geotail Missions," web page, Goddard Space Flight Center, Greenbelt, Maryland, August 2011. http://pwg.gsfc.nasa.gov/.

64. C. Simó, G. Gómez, J. Llibre, and R. Martinez, "Station Keeping of a Quasiperiodic Halo Orbit Using Invariant Manifolds," *Second International Symposium on Flight Dynamics:*

Proceedings of the ESA Symposium held October 20–23, 1986, Darmstadt, Germany, Paper ESA SP-255, European Space Agency, pp. 65–70, 1986.

65. D. W. Dunham, S. J. Jen, C. E. Roberts, A. W. Seacord, P. J. Sharer, D. C. Folta, and D. P. Muhonen, "Transfer Trajectory Design for the SOHO Libration-Point Mission," *Proceedings of the 43rd Congress of the International Astronautical Federation held August 28–September 5, 1992, Washington, District of Columbia*, Paper IAF 92-0066, 1992.

66. R. M. Bonnet and F. Felici, "Overview of the SOHO Mission," *Advances in Space Research*, vol. 20, pp. 2207–2218, 1997.

67. G. Gómez, A. Jorba, J. Llibre, R. Martinez, J. Masdemont, and C. Simó, *Dynamics and Mission Design near Libration Points*, Vol. I–IV. World Scientific Publishing Co., Singapore, 2001.

68. NASA, "Glenn Research Center Home," Cleveland, Ohio, March 2007. http://www.nasa.gov/centers/glenn/.

69. P. Murdin, "ACE (Advanced Composition Explorer)," *Encyclopedia of Astronomy and Astrophysics*, no. 4518, Goddard Space Flight Center, Greenbelt, Maryland, 2000.

70. C. L. Bennett, "The Microwave Anisotropy Probe (MAP) Mission," *Bulletin of the American Astronomical Society, 189th AAS Meeting held January 1997, Toronto, Ontario*, vol. 28, Paper AAS 88.05, American Astronomical Society, p. 1391, 1996.

71. K. C. Howell, B. T. Barden, and M. W. Lo, "Application of Dynamical Systems Theory to Trajectory Design for a Libration Point Mission," *Journal of the Astronautical Sciences*, vol. 45, no. 2, pp. 161–178, April–June 1997.

72. M. W. Lo, B. G. Williams, W. E. Bollman, D. S. Han, Y. S. Hahn, J. L. Bell, E. A. Hirst, R. A. Corwin, P. E. Hong, K. C. Howell, B. Barden, and R. Wilson, "Genesis Mission Design," *The Journal of the Astronautical Sciences*, vol. 49, no. 1, pp. 169–184, January–March 2001.

73. G. L. Pilbratt, T. Prusti, A. M. Heras, S. Leeks, A. P. Marston, and R. Vavrek, "Herschel Space Observatory," *Proceedings of the American Astronomical Society Meeting 204*, Paper 01-81, 2004.

74. "Herschel Mission Home Page," web page, Jet Propulsion Laboratory, California Institute of Technology, Pasadena, California, August 2011. http://herschel.jpl.nasa.gov/.

75. "ESA Science and Technology," web page, European Space Agency, August 2011. http://sci.esa.int/.

76. "Wilkinson Microwave Anisotropy Probe," web page, Goddard Space Flight Center, Greenbelt, Maryland, March 2007. http://map.gsfc.nasa.gov/.

77. "Genesis: Search for Origins," web page, Jet Propulsion Laboratory, California Institute of Technology, Pasadena, California, March 2007. http://genesismission.jpl.nasa.gov/.

78. J. P. Gardner, "The James Webb Space Telescope," *Large Telescopes and Virtual Observatory: Visions for the Future, Proceedings of the 25th meeting of the IAU held July 2003, Sydney, Australia*, International Astronomical Union, 2003.

79. C. Beichman, G. Gómez, M. W. Lo, J. Masdemont, and L. Romans, "Searching for Life with the Terrestrial Planet Finder: Lagrange Point Options for a Formation Flying Interferometer," *Advances in Space Research*, vol. 34, no. 3, pp. 637–644, 2004.

80. E. A. Belbruno and J. Miller, *A Ballistic Lunar Capture Trajectory for the Japanese Spacecraft Hiten*, Tech. Rep. IOM 312/90.4-1731-EAB, (internal document), Jet Propulsion Laboratory, California Institute of Technology, Pasadena, California, 1990.

81. K. Uesugi, "Japanese first double Lunar swingby mission 'HITEN'," *Acta Astronautica*, vol. 25, no. 7, pp. 347–355, 1991.

82. B. H. Foing and G. R. Racca, "The ESA SMART-1 Mission to the Moon with Solar Electric Propulsion," *Advances in Space Research*, vol. 23, no. 11, pp. 1865–1870, 1999.

83. R. B. Roncoli and K. K. Fujii, "Mission Design Overview for the Gravity Recovery and Interior Laboratory (GRAIL) Mission," *Proceedings of the AIAA/AAS Astrodynamics Specialist Conference held August 2–5, 2010, Toronto, Ontario*, Paper AIAA 2010-8383, AIAA/AAS, 2010.

84. M. J. Chung, S. J. Hatch, J. A. Kangas, S. M. Long, R. B. Roncoli, and T. H. Sweetser, "Trans-Lunar Cruise Trajectory Design of GRAIL (Gravity Recovery and Interior Laboratory) Mission," *Proceedings of the AIAA/AAS Astrodynamics Specialist Conference held August 2–5, 2010, Toronto, Ontario*, Paper AIAA 2010-8384, AIAA/AAS, 2010.

85. S. J. Hatch, R. B. Roncoli, and T. H. Sweetser, "GRAIL Trajectory Design: Lunar Orbit Insertion through Science," *Proceedings of the AIAA/AAS Astrodynamics Specialist Conference held August 2–5, 2010, Toronto, Ontario*, Paper AIAA 2010-8385, AIAA/AAS, 2010.

86. V. Szebehely, *Theory of Orbits: The Restricted Problem of Three Bodies*. Academic Press, New York, 1967.

87. K. E. Williams, G. D. Lewis, R. S. Wilson, C. E. Helfrich, and C. L. Potts, "Genesis Earth Return: Refined Strategies and Flight Experience," *Proceedings of the AAS/AIAA Astrodynamics Specialist Conference held January 23–27, 2005, Copper Mountain, Colorado* (D. A. Vallado, M. J. Gabor, and P. N. Desai, eds.), vol. 120, Paper AAS 05-116, *Advances in Astronautical Sciences*, AAS/AIAA, Univelt Inc., San Diego, California, 2005.

88. A. S. Konopliv, S. W. Asmar, E. Carranza, W. L. Sjogren, and D. N. Yuan, "Recent Gravity Models as a Result of the Lunar Prospector Mission," *International Journal of Solar System Studies*, vol. 150, no. 1, pp. 1–18, 2001.

89. P. K. Seidelmann, V. K. Abalakin, M. Bursa, M. E. Davies, C. DeBergh, J. H. Lieske, J. Oberst, J. L. Simon, E. M. Standish, P. Stooke, and P. C. Thomas, "Report of the IAU/IAG Working Group on Cartographic Coordinates and Rotational Elements of the Planets and Satellites: 2000," *Celestial Mechanics and Dynamical Astronomy*, vol. 82, pp. 83–110, 2002.

90. E. M. Standish, *JPL Planetary and Lunar Ephemerides, DE405/LE405*, Tech. Rep. IOM 312.F-98-048, Jet Propulsion Laboratory, California Institute of Technology, Pasadena, California, August 1998.

91. R. B. Roncoli, *Lunar Constants and Models Document*. JPL D-32296, Jet Propulsion Laboratory, California Institute of Technology, Pasadena, California, September 23, 2005.

92. P. K. Seidelmann, ed., *Explanatory Supplement to the Astronomical Almanac*. University Science Books, Sausalito, California, 1992.

93. E. M. Standish, "Time Scales in the JPL and CfA Ephemerides," *Astronomy and Astrophysics*, vol. 336, pp. 381–384, 1998.

94. C. Ma, E. F. Arias, T. M. Eubanks, A. L. Fey, A. M. Gontier, C. S. Jacobs, O. J. Sovers, B. A. Archinal, and P. Charlot, "The International Celestial Reference Frame as Realized

by Very Long Baseline Interferometry," *The Astronomical Journal*, vol. 116, pp. 516–546, July 1998.

95. M. Feissel and F. Mignard, "The Adoption of ICRS on 1 January 1998: Meaning and Consequences," *Astronomy and Astrophysics*, vol. 331, letter to the Editor, pp. L33–L36, 1998.

96. B. A. Archinal, M. F. A'Hearn, E. Bowell, A. Conrad, G. J. Consolmagno, R. Courtin, T. Fukushima, D. Hestroffer, J. L. Hilton, G. A. Krasinsky, G. Neumann, J. Oberst, P. K. Seidelmann, P. Stooke, D. J. Tholen, P. C. Thomas, and I. P. Williams, "Report of the IAU Working Group on Cartographic Coordinates and Rotational Elements: 2009," *Celestial Mechanics and Dynamical Astronomy*, vol. 109, pp. 101–135, 2011.

97. D. A. Vallado, *Fundamentals of Astrodynamics and Applications, Second Edition*. Microcosm Press, El Segundo, California, and Kluwer Academic Publishers, Dordrecht-Boston-London, 2001.

98. G. Gómez, W. S. Koon, M. W. Lo, J. E. Marsden, J. Masdemont, and S. D. Ross, "Invariant Manifolds, the Spatial Three-Body Problem and Space Mission Design," *Proceedings of the AIAA/AAS Astrodynamics Specialist Meeting held August 2001, Quebec City, Canada*, Paper AIAA 01-301, AIAA/AAS, 2001.

99. A. Miele, "Theorem of Image Trajectories in the Earth–Moon Space," *Astronautica Acta*, vol. 6, no. 51, pp. 225–232, 1960.

100. W. M. Folkner, J. G. Williams, and D. H. Boggs, *The Planetary Network Progress Report 42-178*, Tech. Rep., Jet Propulsion Laboratory, California Institute of Technology, Pasadena, California, August 15, 2009. http://ipnpr.jpl.nasa.gov/progress_report/.

101. J. S. Parker, K. E. Davis, and G. H. Born, "Chaining Periodic Three-Body Orbits in the Earth–Moon System," *ACTA Astronautica*, vol. 67, pp. 623–638, 2010.

102. G. H. Darwin, "Periodic Orbits," *Acta Mathematica*, vol. 21, pp. 99–242, 1897.

103. G. H. Darwin, "Periodic Orbits," *Scientific Papers*, vol. 4, Cambridge University Press, Cambridge, Massachusetts, 1911.

104. G. W. Hill, "Review of Darwin's Periodic Orbits," *Astronomical Journal*, vol. 18, no. 423, p. 120, 1898.

105. H. C. Plummer, "On Oscillating Satellites–1," *Monthly Notices of the Royal Astronomical Society*, vol. 63, no. 8, pp. 436–443, 1903.

106. F. R. Moulton, "Periodic Orbits," *Carnegie Institute of Washington Publications*, no. 161, Washington, 1920.

107. E. Strömgren, "Connaissance actuelle des orbites dans le problème des trios corps," *Copenhagen Observatory Publications*, no. 100, 1935. Also *Bulletin Astronomique* vol. 9, pp. 87–130, 1935.

108. R. A. Broucke, *Periodic Orbits in the Restricted Three-Body Problem with Earth–Moon Masses*, Tech. Rep. 32-1168, Jet Propulsion Laboratory, California Institute of Technology, Pasadena, California, 1968.

109. M. Hénon, "Exploration Numérique du Problème des Trois Corps, (I), Masses Egales, Orbites Périodiques," *Annales d'Astrophysique*, vol. 28, no. 3, pp. 499–511, 1965.

110. M. Hénon, "Exploration Numérique du Problème des Trois Corps, (II), Masses Egales, Orbites Périodiques," *Annales d'Astrophysique*, vol. 28, no. 6, pp. 992–1007, 1965.

111. M. Hénon, "Exploration Numérique du Problème des Trois Corps, (III), Masses Egales, Orbites Non Périodiques," *Bulletin Astronomique*, vol. 1, no. 1, pp. 57–80, 1966.

112. M. Hénon, "Exploration Numérique du Problème des Trois Corps, (IV), Masses Egales, Orbites Non Périodiques," *Bulletin Astronomique*, vol. 1, no. 2, pp. 49–66, 1966.

113. M. Hénon, "Numerical Exploration of the Restricted Problem. V., Hill's Case: Periodic Orbits and Their Stability," *Astronomy & Astrophysics*, vol. 1, pp. 223–238, 1969.

114. R. F. Arenstorf, "Existence of Periodic Solutions Passing Near Both Masses of the Restricted Three-Body Problem," *AIAA Journal*, vol. 1, p. 238, 1963.

115. C. L. Goudas, "Three Dimensional Periodic Orbits and their Stability," *Icarus*, vol. 2, pp. 1–18, 1963.

116. T. A. Bray and C. L. Goudas, "Doubly Symmetric Orbits about the Collinear Lagrange Points," *The Astronomical Journal*, vol. 72, no. 2, pp. 202–213, March 1967.

117. T. A. Bray and C. L. Goudas, "Three Dimensional Periodic Oscillations about L1, L2, L3," *Advances in Astronomy and Astrophysics*, vol. 5, pp. 71–130, 1967.

118. R. Kolenkiewicz and L. Carpenter, "Stable Periodic Orbits about the Sun Perturbed Earth–Moon Triangular Points," *AIAA Journal*, vol. 6, no. 7, pp. 1301–1304, 1968.

119. R. W. Farquhar, *The Control and Use of Libration-Point Satellites*. Ph.D. thesis, Department of Aeronautics and Astronautics, Stanford University, Stanford, California, 1968.

120. R. W. Farquhar and A. A. Kamel, "Quasi-Periodic Orbits about the Translunar Libration Point," *Celestial Mechanics*, vol. 7, no. 4, pp. 458–473, June 1973.

121. J. V. Breakwell and J. V. Brown, "The Halo Family of 3-Dimensional Periodic Orbits in the Earth–Moon Restricted 3-Body Problem," *Celestial Mechanics*, vol. 20, pp. 389–404, November 1979.

122. K. C. Howell, "Three-Dimensional, Periodic, 'Halo' Orbits," *Celestial Mechanics*, vol. 32, no. 1, pp. 53–71, 1984.

123. D. L. Richardson, "Analytical Construction of Periodic Orbits about the Collinear Points," *Celestial Mechanics*, vol. 22, pp. 241–253, 1980.

124. K. C. Howell and H. J. Pernicka, "Numerical Determination of Lissajous Trajectories in the Restricted Three-Body Problem," *Celestial Mechanics*, vol. 41, pp. 107–124, 1988.

125. G. Gómez, J. Masdemont, and C. Simó, "Quasihalo Orbits Associated with Libration Points," *The Journal of the Astronautical Sciences*, vol. 46, no. 2, pp. 135–176, 1998.

126. "Seeking a Human Spaceflight Program Worthy of a Great Nation," *Review of U.S. Human Spaceflight Plans Committee*, National Aeronautics and Space Administration, September 2009.

127. K. Hill, M. W. Lo, and G. H. Born, "Liaison Navigation in the Sun–Earth–Moon Four-Body Problem," *Proceedings of the AAS/AIAA 16th Spaceflight Mechanics Meetings held January 22–26, 2006, Tampa, Florida* (S. R. Vadali, L. A. Cangahuala, J. P. W. Schumacher, and J. J. Guzman, eds.), vol. 124, Paper AAS 06-221, *Advances in Astronautical Sciences*, AAS/AIAA, Univelt Inc., San Diego, California, 2006.

128. K. Hill and G. H. Born, "Autonomous Interplanetary Orbit Determination Using Satellite-to-Satellite Tracking," *AIAA Journal of Guidance, Control, and Dynamics*, vol. 30, no. 3, pp. 679–686, May–June 2007.

129. T. S. Parker and L. O. Chua, *Practical Numerical Algorithms for Chaotic Systems.* Springer-Verlag, New York, 1989.

130. S. H. Strogatz, *Nonlinear Dynamics and Chaos.* Perseus Books Publishing, L.L.C., Cambridge, Massachusetts, 1994.

131. A. H. Nayfeh and B. Balachandran, *Applied Nonlinear Dynamics.* John Wiley and Sons, Inc., New York, 1995.

132. H. J. Pernicka, *The Numerical Determination of Lissajous Orbits in the Circular Restricted Three-Body Problem,* master's thesis, Purdue University, West Lafayette, Indiana, December 1986.

133. H. J. Pernicka, *The Numerical Determination of Nominal Libration Point Trajectories and Development of a Station-Keeping Strategy.* Ph.D. thesis, Purdue University, West Lafayette, Indiana, May 1990.

134. R. Wilson, *Derivation of Differential Correctors Used in GENESIS Mission Design,* Tech. Rep. JPL IOM 312.I-03-002 (internal document), Jet Propulsion Laboratory, California Institute of Technology, Pasadena, California, 2003.

135. R. Wilson, *Trajectory Design in the Sun–Earth–Moon Four Body Problem.* Ph.D. thesis, Purdue University, West Lafayette, Indiana, December 1998.

136. R. S. Wilson and K. C. Howell, "Trajectory Design in the Sun–Earth–Moon System Using Lunar Gravity Assists," *Journal of Spacecraft and Rockets,* vol. 35, no. 2, pp. 191–198, 1998.

137. K. C. Howell, B. T. Barden, R. S. Wilson, and M. W. Lo, "Trajectory Design Using a Dynamical Systems Approach with Application to Genesis," *Proceedings of the AAS/AIAA Astrodynamics Specialist Conference held August 4–7, 1997, Sun Valley, Idaho* (F. R. Hoots, B. Kaufman, P. J. Cefola, and D. B. Spencer, eds.), vol. 97, Paper AAS 97-709, *Advances in Astronautical Sciences,* AAS/AIAA, Univelt Inc., San Diego, California, 1998.

138. K. C. Howell and R. S. Wilson, "Trajectory Design in the Sun–Earth–Moon System Using Multiple Lunar Gravity Assists," *Proceedings of the AIAA/AAS Astrodynamics Specialist Conference held August 1996,* Paper AIAA 96-3642, San Diego, California, AIAA/AAS, 1996.

139. D. L. Richardson and N. D. Cary, "A Uniformly Valid Solution for Motion of the Interior Libration Point for the Perturbed Elliptic-Restricted Problem," *Proceedings of the AAS/AIAA Astrodynamics Specialist Conference held July 28–30, 1975, Nassau, Bahamas* (W. F. Powers, H. E. Rauch, B. D. Tapley, and C. E. Velez, eds.), vol. 33, Paper AAS 75-021, *Advances in Astronautical Sciences,* AAS/AIAA, Univelt Inc., San Diego, California, 1976.

140. M. W. Lo and J. S. Parker, "Unstable Resonant Orbits Near Earth and Their Applications in Planetary Missions," *AIAA/AAS Astrodynamics Specialist Conference,* vol. 14, Providence, RI, AIAA/AAS, August 16–19, 2004.

141. R. C. Paffenroth, E. J. Doedel, and D. J. Dichmann, "Continuation of Periodic Orbits Around Lagrange Points and AUTO2000," *Proceedings of the AIAA/AAS Astrodynamics Specialist Conference held July 2001, Quebec City,* Paper AIAA 01-303, AIAA/AAS, 2001.

142. R. A. Broucke, "Stability of Periodic Orbits in the Elliptic, Restricted Three-Body Problem," *AIAA Journal,* vol. 7, no. 6, pp. 1003–1009, June 1969.

143. R. H. Battin, *An Introduction to the Mathematics and Methods of Astrodynamics*. American Institute of Aeronautics and Astronautics, Inc., New York, 1987.

144. D. J. Scheeres, D. Han, and Y. Hou, "Orbit Determination Uncertainty Distributions and Mappings in an Unstable Halo Orbit," *The Interplanetary Network Progress Report 42-146*, Jet Propulsion Laboratory, California Institute of Technology, Pasadena, California, April–June 2001. http://ipnpr.jpl.nasa.gov/progress_report/.

145. R. Russell, "Global Search for Planar and Three-Dimensional Periodic Orbits near Europa," *Proceedings of the AAS/AIAA Astrodynamics Specialist Conference held August 7–11, 2005, South Lake Tahoe, California* (B. G. Williams, L. A. D'Amario, K. C. Howell, and F. R. Hoots, eds.), vol. 123, Paper AAS 05-290, *Advances in Astronautical Sciences*, AAS/AIAA, Univelt Inc., San Diego, California, 2006.

146. J. S. Parker and G. H. Born, "Mission Design Options for the DUNE Spacecraft," *Proceedings of the AAS Byron Tapley Special Symposium held February 1, 2008, Austin, Texas*, American Astronautical Society, 2008.

147. G. Gómez, W. S. Koon, J. E. Marsden, J. Masdemont, and S. D. Ross, "Connecting Orbits and Invariant Manifolds in the Spatial Restricted Three-Body Problem," *Nonlinearity*, vol. 17, no. 5, pp. 1571–1606, September 2004.

148. S. D. Ross, *Cylindrical Manifolds and Tube Dynamics in the Restricted Three-Body Problem*. Ph.D. thesis, California Institute of Technology, Pasadena, California, April 2004.

149. R. L. Anderson, *Low Thrust Trajectory Design for Resonant Flybys and Captures Using Invariant Manifolds*. Ph.D. thesis, University of Colorado, Boulder, Colorado, 2005.

150. E. Kolemen, N. J. Kasdin, and P. Gurfil, "Quasi-Periodic Orbits of the Restricted Three-Body Problem Made Easy," *Proceedings of the New Trends in Astrodynamics and Applications III held August 16–18, 2006, Princeton, New Jersey*, vol. 886, *American Institute of Physics Conference Series*, pp. 68–77, February 2007.

151. W. S. Koon, M. W. Lo, J. E. Marsden, and S. D. Ross, "Heteroclinic Connections Between Periodic Orbits and Resonance Transitions in Celestial Mechanics," *Chaos*, vol. 10, no. 2, pp. 427–469, 2000.

152. R. P. McGehee, *Some Homoclinic Orbits for the Restricted Three Body Problem*. Ph.D. thesis, University of Wisconsin, Madison, Wisconsin, 1969.

153. E. A. Belbruno and B. G. Marsden, "Resonance Hopping in Comets," *The Astronomical Journal*, vol. 113, no. 4, pp. 1433–1444, April 1997.

154. D. Wilczak and P. Zgliczyński, "Heteroclinic Connections Between Periodic Orbits in Planar Restricted Circular Three Body Problem. Part II," *Communications in Mathematical Physics*, vol. 259, no. 3, pp. 561–576, November 2005.

155. E. Canalias, A. Delshams, J. J. Masdemont, and P. Roldán, "The Scattering Map in the Planar Restricted Three Body Problem," *Celestial Mechanics and Dynamical Astronomy*, vol. 95, pp. 155–171, 2006.

156. E. Canalias and J. J. Masdemont, "Homoclinic and Heteroclinic Transfer Trajectories between Lyapunov Orbits in the Sun–Earth and Earth–Moon Systems," *Discrete and Continuous Dynamical Systems*, vol. 14, pp. 261–279, 2006.

157. R. L. Anderson and M. W. Lo, "The Role of Invariant Manifolds in Low Thrust Trajectory Design (Part II)," *Proceedings of the AIAA/AAS Astrodynamics Specialist Conference held*

August 16–19, 2004, Providence, Rhode Island, Paper AIAA 2004-5305, AIAA/AAS, 2004.

158. R. L. Anderson and M. W. Lo, "Role of Invariant Manifolds in Low-Thrust Trajectory Design," *Journal of Guidance, Control, and Dynamics*, vol. 32, no. 6, pp. 1921–1930, November–December 2009.

159. K. Davis, R. Anderson, D. Scheeres, and G. Born, "The Use of Invariant Manifolds for Transfers Between Unstable Periodic Orbits of Different Energies," *Celestial Mechanics and Dynamical Astronomy*, vol. 107, no. 4, pp. 471–485, 2010.

160. K. Davis, R. Anderson, D. Scheeres, and G. Born, "Optimal Transfers Between Unstable Periodic Orbits Using Invariant Manifolds," *Celestial Mechanics and Dynamical Astronomy*, vol. 109, no. 3, pp. 241–264, 2011.

161. M. W. Lo and S. D. Ross, *SURFing the Solar System: Invariant Manifolds and the Dynamics of the Solar System*, JPL IOM 312/97 (internal document), Jet Propulsion Laboratory, California Institute of Technology, Pasadena, California, 1997.

162. M. W. Lo and J. S. Parker, "Chaining Simple Periodic Three-Body Orbits," *Proceedings of the AAS/AIAA Astrodynamics Specialist Conference held August 7–11, 2005, South Lake Tahoe, California* (B. G. Williams, L. A. D'Amario, K. C. Howell, and F. R. Hoots, eds.), vol. 123, Paper AAS 05-380, *Advances in Astronautical Sciences*, AAS/AIAA, Univelt Inc., San Diego, California, 2006.

163. C. Robinson, *Dynamical Systems: Stability, Symbolic Dynamics, and Chaos*, Second Edition. CRC Press, LLC, Boca Raton, Florida, 1999.

164. F. T. Krogh, *Notes on a New Package for Ordinary Differential Equations*, Computing Memorandum 361 (internal document), Jet Propulsion Laboratory, Section 914, California Institute of Technology, Pasadena, California, May 1974.

165. F. T. Krogh, *Variable Order Adams Method for Ordinary Differential Equations*, Tech. Rep., California Institute of Technology, 2010 Math à la Carte, Inc., http://mathalacarte.com/cb/mom.fcg/ya64, Tujunga, California, October 1975.

166. F. T. Krogh, *Design of a New Variable Order Adams Code*, Computing Memorandum 558, D-79149 (internal document), Jet Propulsion Laboratory, California Institute of Technology, Pasadena, California, June 1998.

167. E. Fehlberg, *Classical Fifth-, Sixth-, Seventh-, and Eighth-Order Runge-Kutta Formulas with Stepsize Control*, NASA Technical Report TR R-287, Marshall Space Flight Center, Huntsville, Alabama, October 1968.

168. P. E. Gill, W. Murray, and M. A. Saunders, *User's Guide for SNOPT 7.1: A Fortran Package for Large-Scale Nonlinear Programming*, Numerical Analysis Report NA 04-1, Department of Mathematics, University of California, San Diego, La Jolla, California, 2004.

169. P. E. Gill, W. Murray, and M. A. Saunders, "SNOPT: An SQP Algorithm for Large-Scale Constrained Optimization," *SIAM Review*, vol. 47, no. 1, pp. 99–131, 2005.

170. S. Flanagan, T. Drain, T. Ely, and T. Martin-Mur, "Navigation and Mission Analysis Software for the Next Generation of JPL Missions," *Proceedings of the 16th International Symposium on Space Flight Dynamics held December 3–7, 2001, Pasadena, California*, Jet Propulsion Laboratory, California Institute of Technology, 2001. http://trs-new.jpl.nasa.gov/dspace/bitstream/2014/36810/1/01-1301.pdf.

171. C. H. Acton, "Ancillary Data Services of NASA's Navigation and Ancillary Information Facility," *Planetary and Space Science*, vol. 44, no. 1, pp. 65–70, 1996.

172. M. Li and J. Zheng, "Indirect Transfer to the Earth–Moon L1 Libration Point," *Celestial Mechanics and Dynamical Astronomy*, vol. 108, no. 2, pp. 203–213, 2010.

173. T. N. Edelbaum, *Libration Point Rendezvous*, N70-26184, Report No. 70–12, Analytical Mechanics Associates, Inc., Cambridge, Massachusetts, February 1970.

174. J. S. Parker and G. H. Born, "Direct Lunar Halo Orbit Transfers," *Spaceflight Mechanics 2007: Proceedings of the 17th AAS/AIAA Spaceflight Mechanics Conference held January 28–February 1, 2007, Sedona, Arizona* (M. R. Akella, J. W. Gearhart, R. H. Bishop, and A. J. Treder, eds.), vol. 127, Paper AAS 07-229, *Advances in Astronautical Sciences*, American Astronautical Society (AAS), American Astronautical Society Publications Office, San Diego, California, pp. 1923–1945, 2007. Website: http://www.univelt.com.

175. J. S. Parker and G. H. Born, "Direct Lunar Halo Orbit Transfers," *Journal of the Astronautical Sciences*, vol. 56, no. 4, pp. 441–476, October–December 2008.

176. R. R. Rausch, *Earth to Halo Orbit Transfer Trajectories*, master's thesis, Purdue University, West Lafayette, Indiana, August 2005.

177. D. P. Gordon, *Transfers to Earth–Moon L_2 Halo Orbits Using Lunar Proximity and Invariant Manifolds*, master's thesis, Purdue University, West Lafayette, Indiana, August 2008.

178. E. M. Alessi, G. Gómez, and J. J. Masdemont, "Two-Manoeuvres Transfers Between LEOs and Lissajous Orbits in the Earth–Moon System," *Advances in Space Research*, vol. 45, no. 10, pp. 1276–1291, 2010.

179. G. Winter, J. Périaux, M. Galán, and P. Cuesta, eds., *Genetic Algorithms in Engineering and Computer Science*. John Wiley and Sons Ltd., New York, 1995.

180. M. Avriel, *Nonlinear Programming Analysis and Methods*. Dover Publications, Inc., Mineola, New York, 2003.

181. C. A. C. Coello, D. A. V. Veldhuizen, and G. B. Lamont, *Evolutionary Algorithms for Solving Multi-Objective Problems*. Kluwer Academic/Plenum Publishers, New York, 2002.

182. V. M. Alexeyev, "New Examples of Capture in the Three-Body Problem," *Soviet Astronomy*, vol. 6, no. 4, pp. 565–572, 1963.

183. J. S. Parker, "Targeting Low-Energy Ballistic Lunar Transfers," *Proceedings of the AAS George H. Born Special Symposium held May 13–14, 2010, Boulder, Colorado*, American Astronautical Society, 2010.

184. J. S. Parker, "Targeting Low-Energy Ballistic Lunar Transfers," *Journal of the Astronautical Sciences*, vol. 58, no. 3, part I, pp. 311–334, July–September 2011.

185. G. Gómez and J. Masdemont, "Some Zero Cost Transfers Between Libration Point Orbits," *Proceedings of the AAS/AIAA Spaceflight Mechanics Meeting held January 23–26, 2000, Clearwater, Florida* (C. A. Kluever, B. Neta, C. D. Hall, and J. M. Hanson, eds.), vol. 105, Paper AAS 00-177, *Advances in Astronautical Sciences*, AAS/AIAA, Univelt Inc., San Diego, California, 2000.

186. D. Folta and F. Vaughn, "A Survey of Earth–Moon Libration Orbits: Stationkeeping Strategies and Intra-Orbit Transfers," *Proceedings of the AIAA/AAS Astrodynamics Specialist Conference and Exhibit held August 16–19, 2004, Providence, Rhode Island*, Paper AIAA 2004-4741, AIAA/AAS, 2004.

187. M. Kato, S. Sasaki, and Y. Takizawa, "The Kaguya Mission Overview," *The Kaguya Mission to the Moon* (A. Matsuoka and C. T. Russell, eds.), vol. 154, *Space Science Reviews*, Springer, New York, New York, pp. 3–19, 2011.

188. M. Beckman, "Mission Design for the Lunar Reconnaissance Orbiter," *Proceedings of the 29th Annual AAS Rocky Mountain Guidance and Control Conference held February 4–8, 2006, Breckenridge, Colorado* (S. D. Jolly and R. D. Culp, eds.), vol. 125, Paper AAS 07-057, *Advances in Astronautical Sciences*, AAS, Univelt Inc., San Diego, California, 2007.

189. F. Martikan and F. Santora, eds., *Lunar Flight Handbook*, Vol. 2 of *Space Flight Handbooks*. NASA, Office of Scientific and Technical Information, Washington, DC, 1963.

190. J. S. Parker and R. L. Anderson, "Targeting Low-Energy Transfers to Low Lunar Orbit," *Proceedings of the AAS/AIAA Astrodynamics Specialist Conference held July 31–August 4, 2011, Girdwood, Alaska* (H. Schaub, B. C. Gunter, R. P. Russell, and W. T. Cerven, eds.), vol. 142, Paper AAS 11-459, *Advances in Astronautical Sciences*, AAS/AIAA, Univelt Inc., San Diego, California, pp. 847–866, 2012.

191. J. S. Parker and R. L. Anderson, "Targeting Low-Energy Transfers to Low Lunar Orbit," *ACTA Astronautica*, vol. 84, pp. 1–14, March–April 2013.

192. R. L. Anderson and J. S. Parker, "A Survey of Ballistic Transfers to the Lunar Surface," *Proceedings of the AAS/AIAA Spaceflight Mechanics Meetings held February 13–17, 2011, New Orleans, Louisiana* (M. K. Jah, Y. Guo, A. L. Bowes, and P. C. Lai, eds.), vol. 140, Paper AAS 11-278, *Advances in Astronautical Sciences*, AAS/AIAA, Univelt Inc., San Diego, California, pp. 2481–2500, 2011.

193. R. L. Anderson and J. S. Parker, "Comparison of Low-Energy Lunar Transfer Trajectories to Invariant Manifolds," *Proceedings of the AAS/AIAA Astrodynamics Specialist Conference held July 31–August 4, 2011, Girdwood, Alaska* (H. Schaub, B. C. Gunter, R. P. Russell, and W. T. Cerven, eds.), vol. 142, Paper AAS 11-423, *Advances in Astronautical Sciences*, AAS/AIAA, Univelt Inc., San Diego, California, pp. 333–352, 2012.

194. R. L. Anderson and J. S. Parker, "Comparison of Low-Energy Lunar Transfer Trajectories to Invariant Manifolds," *Celestial Mechanics and Dynamical Astronomy*, vol. 115, no. 3, pp. 311–331, February 2013. DOI 10.10075 10569-012-9466-3, published online February 16, 2013.

195. R. L. Anderson and J. S. Parker, "Survey of Ballistic Transfers to the Lunar Surface," *Journal of Guidance, Control, and Dynamics*, vol. 35, no. 4, pp. 1256–1267, July–August 2012.

196. Mission Evaluation Team, *Apollo 11 Mission Report*, NASA SP-238, National Aeronautics and Space Administration, Washington, District of Columbia, 1971.

197. R. W. Easton, "Regularization of Vector Fields by Surgery," *Journal of Differential Equations*, vol. 10, pp. 92–99, 1971.

198. R. McGehee, "Triple Collision in the Collinear Three-Body Problem," *Inventiones Mathematicae*, vol. 27, pp. 191–227, 1974.

199. R. L. Anderson and M. W. Lo, "Virtual Exploration by Computing Global Families of Trajectories with Supercomputers," *Proceedings of the Advances in the Astronautical Sciences, Spaceflight Mechanics* (D. A. Vallado, M. J. Gabor, and P. N. Desai, eds.), vol. 120, Part II, Paper AAS 05-220, Copper Mountain, Colorado, American Astronautical Society, Univelt Inc., San Diego, California, pp. 1855–1874, 2005.

200. B. F. Villac and D. J. Scheeres, "Escaping Trajectories in the Hill Three-Body Problem and Applications," *Journal of Guidance, Control, and Dynamics*, vol. 26, no. 2, pp. 224–232, March–April 2003.

201. C. Von Kirchbach, H. Zheng, J. Aristoff, J. Kavanagh, B. F. Villac, and M. W. Lo, "Trajectories Leaving a Sphere in the Restricted Three Body Problem," *Advances in the Astronautical Sciences, Spaceflight Mechanics, Proceedings of the AAS/AIAA 15th Space Flight Mechanics Meeting, January 23–27, 2005, Copper Mountain, Colorado* (D. A. Vallado, M. J. Gabor, and P. N. Desai, eds.), vol. 120, Part II, Paper AAS 05-221, American Astronautical Society, Univelt Inc., San Diego, California, pp. 1875–1902, 2005.

202. D. S. Cooley, K. F. Galal, K. Berry, L. Janes, G. Marr, J. Carrico, and C. Ocampo, "Mission Design for the Lunar CRater Observation and Sensing Satellite (LCROSS)," *Proceedings of the AIAA/AAS Astrodynamics Specialist Conference held August 2–5, 2010, Toronto, Ontario*, Paper AIAA-2010-8386, 2010.

203. M. W. Lo and M.-K. J. Chung, "Lunar Sample Return via the Interplanetary Superhighway," *Proceedings of the AIAA/AAS Astrodynamics Specialist Meeting held August 5–8, 2002, Monterey, California*, Paper AIAA 2002-4718, 2002.

204. *NASA's Exploration Systems Architecture Study: Final Report*, NASA TM-2005-214062, National Aeronautics and Space Administration, November 2005.

205. H. Baoyin and C. R. McInnes, "Trajectories to and from the Lagrange Points and the Primary Body Surfaces," *Journal of Guidance, Control, and Dynamics*, vol. 29, no. 4, pp. 998–1003, July–August 2006.

206. E. M. Alessi, G. Gómez, and J. J. Masdemont, "Leaving the Moon by Means of Invariant Manifolds of Libration Point Orbits," *Communications in Nonlinear Science and Numerical Simulation*, vol. 14, no. 12, pp. 4153–4167, December 2009.

207. R. L. Anderson, "Approaching Moons from Resonance via Invariant Manifolds," *Proceedings of the 22nd AAS/AIAA Space Flight Mechanics Meeting held January 29–February 2, 2012, Charleston, South Carolina*, Paper AAS 12-136, 2012.

208. R. L. Anderson and M. W. Lo, "Dynamical Systems Analysis of Planetary Flybys and Approach: Planar Europa Orbiter," *Journal of Guidance, Control, and Dynamics*, vol. 33, no. 6, pp. 1899–1912, November–December 2010.

209. R. L. Anderson and M. W. Lo, "A Dynamical Systems Analysis of Planetary Flybys and Approach: Ballistic Case," *The Journal of the Astronautical Sciences*, vol. 58, no. 2, pp. 167–194, April–June 2011.

210. R. L. Anderson and M. W. Lo, "Flyby Design using Heteroclinic and Homoclinic Connections of Unstable Resonant Orbits," *Proceedings of the 21st AAS/AIAA Space Flight Mechanics Meeting held February 13–17, 2011, New Orleans, Louisiana*, Paper AAS 11-125, 2011.

211. "NASA Launch Services Program: Launch Vehicle Performance Website," web page, Flight Dynamics Branch of the NASA Launch Services Program at Kennedy Space Center, September 2010. http://elvperf.ksc.nasa.gov/elvMap/.

212. *Pegasus User's Guide*, Release 7.0, Orbital Sciences Corporation, Huntington Beach, California, April 2010. http://www.orbital.com/NewsInfo/Publications/pegasus_ug.pdf (accessed February 5, 2013).

213. J. S. Parker and G. H. Born, "Trajectory Options for DUNE, a Near-Earth Dust Telescope," *Journal of the Astronautical Sciences*, vol. 56, no. 3, pp. 287–309, July–September 2008.

214. D. McComas, F. Allegrini, P. Bochsler, M. Bzowski, M. Collier, H. Fahr, H. Fichtner, P. Frisch, H. Funsten, S. Fuselier, G. Gloeckler, M. Gruntman, V. Izmodenov, P. Knappenberger, M. Lee, S. Livi, D. Mitchell, E. Möbius, T. Moore, D. Reisenfeld, E. Roelof, N. Schwadron, M. Wieser, M. Witte, P. Wurz, and G. Zank, "The Interstellar Boundary Explorer (IBEX)," *Proceedings of the Physics of the Outer Heliosphere: Third International IGPP Conference held February 8–13, 2004, Riverside, California* (V. Florinski, N. V. Pogorelov, and G. P. Zank, eds.), Paper AIP CP719, pp. 162–181, 2004.

215. D. J. McComas, F. Allegrini, L. Bartolone, P. Bochsler, M. Bzowski, M. Collier, H. Fahr, H. Fichtner, P. Frisch, H. Funsten, S. Fuselier, G. Gloeckler, M. Gruntman, V. Izmodenov, P. Knappenberger, M. Lee, S. Livi, D. Mitchell, E. Möbius, T. Moore, S. Pope, D. Reisenfeld, E. Roelof, H. Runge, J. Scherrer, N. Schwadron, R. Tyler, M. Wieser, M. Witte, P. Wurz, and G. Zank, "The Interstellar Boundary Explorer (IBEX) Mission," *Proceedings of the Solar Wind 11/SOHO 16 Conference: Connecting Sun and Heliosphere, September 2005, Whistler, Canada*, Paper ESA SP-592, p. 689, 2005.

216. D. J. McComas, F. Allegrini, L. Bartolone, P. Bochsler, M. Bzowski, M. Collier, H. Fahr, H. Fichtner, P. Frisch, H. Funsten, S. Fuselier, G. Gloeckler, M. Gruntman, V. Izmodenov, P. Knappenberger, M. Lee, S. Livi, D. Mitchell, E. Möbius, T. Moore, S. Pope, D. Reisenfeld, E. Roelof, H. Runge, J. Scherrer, N. Schwadron, R. Tyler, M. Wieser, M. Witte, P. Wurz, and G. Zank, "The Interstellar Boundary Explorer (IBEX): Update at the End of Phase B," *Proceedings of the Physics of the Inner Heliosheath Conference* (J. Heerikhuisen, V. Florinski, G. P. Zank, and N. V. Pogorelov, eds.), Paper CP 858, IGPP, American Institute of Physics, 2006.

217. C. E. Roberts, "The SOHO Mission L1 Halo Orbit Recovery From the Attitude Control Anomalies of 1998," *Proceedings of the Libration Point Orbits and Application Conference held June 10–14, 2002, Parador d'Aiguablava, Girona, Spain*, 2002.

218. D. P. Muhonen, "Accelerometer-Enhanced Orbit Control Near the Sun–Earth L_1 Libration Point," *Proceedings of the AIAA 21st Aerospace Sciences Meeting held January 10–13, 1983, Reno, Nevada*, Paper AIAA-83-0018, AIAA, 1983.

219. N. G. Smith, K. E. Williams, R. C. Wiens, and C. E. Rasbach, "Genesis—The Middle Years," *Proceedings of the Aerospace Conference held March 8–15, 2003, Big Sky, Montana*, vol. 1, Paper 213, IEEE, 2003.

220. K. C. Howell and T. M. Keeter, "Station-Keeping Strategies for Libration Point Orbits: Target Point and Floquet Mode Approaches," *Proceedings of the AAS/AIAA Spaceflight Mechanics Conference 1995* (R. Proulx, J. Liu, P. Seidelmann, and S. Alfano, eds.), vol. 89, *Advances in Astronautical Sciences*, AAS/AIAA, Univelt Inc., San Diego, California, 1995.

221. K. E. Williams, B. T. Barden, K. C. Howell, M. W. Lo, and R. S. Wilson, "Genesis Halo Orbit Stationkeeping Design," *Proceedings of the International Symposium: Spaceflight Dynamics, June 2000, Biarritz, France*, 2000.

222. D. W. Dunham and C. E. Roberts, "Stationkeeping Techniques for Libration-Point Satellites," *Journal of the Astronautical Sciences*, vol. 49, no. 1, pp. 127–144, January–March, 2001.

223. D. Rohrbaugh and C. Schiff, "Stationkeeping Approach for the Microwave Anisotropy Probe (MAP)," *Proceedings of the AIAA/AAS Astrodynamics Specialist Conference held August 2002, Monterey, California*, Paper AIAA 2002-4429, AIAA/AAS, 2002.

224. M. Limon, *Wilkinson Microwave Anisotropy Probe (WMAP): Explanatory Supplement*, Tech. Rep. 1.0, Goddard Space Flight Center, Greenbelt, Maryland, February 11, 2003.

225. L. Janes and M. Beckman, "Optimizing Stationkeeping Maneuvers for James Webb Space Telescope," *Proceedings of the Goddard Flight Mechanics Symposium, Greenbelt, Maryland*, 2005.

226. C. Renault and D. Scheeres, "Statistical Analysis of Control Maneuvers in Unstable Orbital Environments," *Journal of Guidance, Control, and Dynamics*, vol. 26, no. 5, pp. 758–769, September–October 2003.

227. B. Marchand and K. Howell, "Formation Flight Near L1 and L2 in the Sun–Earth–Moon Ephemeris System Including Solar Radiation Pressure," *Proceedings of the AAS/AIAA Astrodynamics Specialist Conference held August 3–7, 2003, Big Sky, Montana* (J. d. Lafontaine, A. J. Treder, M. T. Soyka, and J. A. Sims, eds.), vol. 116, Paper AAS 03-596, *Advances in Astronautical Sciences*, AAS/AIAA, Univelt Inc., San Diego, California, 2004.

228. K. C. Howell and S. C. Gordon, "Orbit Determination Error Analysis and a Station-Keeping Strategy for Sun–Earth L1 Libration Point Orbits," *Journal of the Astronautical Sciences*, vol. 42, no. 2, pp. 207–228, April–June 1994.

229. G. Gómez, J. Llibre, R. Martinez, and C. Simó, *Dynamics and Mission Design near Libration Points, Vol I: Fundamentals: The Case of Collinear Libration Points*. World Scientific Monograph Series, World Scientific Publishing Ltd., Singapore, 2001.

230. K. C. Howell and H. J. Pernicka, "Stationkeeping Method for Libration Point Trajectories," *Journal of Guidance, Control, and Dynamics*, vol. 16, no. 1, pp. 151–159, January–February 1993.

231. D. Folta, T. Pavlak, K. Howell, M. A. Woodard, and D. W. Woodfork, "Stationkeeping of Lissajous Trajectories in the Earth–Moon System with Applications to ARTEMIS," *Proceedings of the 20th AAS/AIAA Space Flight Mechanics Meeting held February 14–17, 2010, San Diego, California* (D. Mortari, T. F. Starchville, Jr., A. J. Trask, and J. K. Miller, eds.), vol. 136, Paper AAS 10-113, *Advances in Astronautical Sciences*, AAS/AIAA, Univelt Inc., San Diego, California, 2010.

232. D. Hoffman, *Stationkeeping at the Colinear Equilibrium Points of the Earth–Moon System*, Tech. Rep. NASA JSC-26189, Johnson Space Center, September 1993.

233. G. Gómez, K. Howell, J. Masdemont, and C. Simó, "Station-Keeping Strategies for Translunar Libration Point Orbits," *Proceedings of the AAS/AIAA Spaceflight Mechanics Meeting held February 9–11, 1998, Monterey, California* (J. W. Middour, L. L. Sackett, L. D'Amario, and D. V. Byrnes, eds.), vol. 99, Paper AAS 98-168, *Advances in Astronautical Sciences*, AAS/AIAA, Univelt Inc., San Diego, California, 1998.

234. T. Pavlak and K. C. Howell, "Strategy for Long-Term Libration Point Orbit Stationkeeping in the Earth–Moon System," *Proceedings of the AAS/AIAA Astrodynamics Specialist Conference held July 31–August 4, 2011, Girdwood, Alaska* (H. Schaub, B. C. Gunter, R. P. Russell, and W. T. Cerven, eds.), vol. 142, Paper AAS 11-516, *Advances in Astronautical Sciences*, AAS/AIAA, Univelt Inc., San Diego, California, 2012.

235. D. Folta, M. Woodard, and D. Cosgrove, "Stationkeeping of the First Earth–Moon Libration Orbiters: The ARTEMIS Mission," *Proceedings of the AAS/AIAA Astrodynamics Specialist Conference held July 31–August 4, 2011, Girdwood, Alaska* (H. Schaub, B. C. Gunter, R. P. Russell, and W. T. Cerven, eds.), vol. 142, Paper AAS 11-515, *Advances in Astronautical Sciences*, AAS/AIAA, Univelt Inc., San Diego, California, 2012.

236. R. A. Broucke, E. Davoust, J. D. Anderson, H. Lass, and L. Blitzer, "Periodic solutions about the collinear Lagrangian solution in the general problem of three bodies," *Celestial Mechanics*, vol. 24, pp. 63–82, May 1981.

TERMS

ΔV	change in velocity, delta velocity
Ω	ascending node
3BSOI	three-body sphere of influence
ACE	*Advanced Composition Explorer*
ACS	attitude control system
AAS	American Astronautical Society
AAS	American Astronomical Society
AIAA	American Institute of Aeronautics and Astronautics
AMMOS	Advanced Multi-Mission Operations System
ARTEMIS	*Acceleration, Reconnection, Turbulence and Electrodynamics of the Moon's Interaction with the Sun*
AU	astronomical unit, ~149,600,000 kilometers
BIPM	Bureau International des Poids et Mesures
BLT	ballistic lunar transfer

C_3	launch injection energy parameter
CH-1	*Chandrayaan-1*
cm/s	centimeter per second
CRTBP	circular restricted three-body problem
DAV	declination of apogee vector
DE	developmental ephemerides, e.g., DE421
deg	degree
DLA	declination of launch asymptote
DPO	distant prograde orbit
DRO	distant retrograde orbit
DSN	Deep Space Network
DUNE	*Dust Near Earth*
EDL	entry, descent, and landing
EL_1 / EL_2	Sun–Earth Lagrange point 1 / 2
EM	Earth–Moon
EME2000	Earth Mean Equator and Equinox of J2000
EMO2000	Earth Mean Orbit of J2000
EPO	Earth-phasing orbits
ET	Ephemeris Time, also called Dynamical Time
FPA	flight path angle
FPAz	flight path azimuth angle
GEO	geosynchronous Earth orbit
GM	gravitational constant × mass
GPS	Global Positioning System
GRAIL	*Gravity Recovery and Interior Laboratory*
HGS-1	name given to *AsiaSat 3* after *AsiaSat 3* failed to get a correct orbit and was transferred to Hughes Global Services, Inc.

IAG	International Association of Geodesy
IAU	International Astronomical Union
IBEX	*Interstellar Boundary Explorer*
ICE	*International Cometary Explorer*
ICRF	International Celestial Reference Frame
ISEE-3	*International Sun–Earth Explorer-3*
ISRO	Indian Space Research Organization
ISTP	International Solar Terrestrial Physics
J2000	currently used standard equinox for January 1, 2000
JPL	Jet Propulsion Laboratory
JSC	Johnson Space Center
km	kilometer
km/s	kilometers per second
km^2/s^2	kilometers squared per second squared
km^3/s^2	cubic kilometer per second squared
KSC	Kennedy Space Center
L_1	Lagrange point 1, between the two primary bodies
L_2	Lagrange point 2, on the far side of the smaller primary
L_3	Lagrange point 3, on the far side of the larger primary
L_4	Lagrange point 4, leading the smaller primary in its orbit about the barycenter
L_5	Lagrange point 5, trailing the smaller primary in its orbit about the barycenter
LCROSS	*Lunar Crater Observation and Sensing Satellite*
L.E.	low energy
LEO	low-Earth orbit
LL_1 / LL_2	Earth–Moon lunar Lagrange point 1 / 2
LLO	low-lunar orbit
LOI	lunar-orbit insertion
LPABF	lunar principal-axis body-fixed

LRO	*Lunar Reconnaissance Orbiter*
LPO	libration-point orbit
LSP	Launch Services Program
LST	Local Solar Time
LTool	Libration Point Mission Design Tool
LTST	Local True Solar Time
MARS	Mid-Atlantic Regional Spaceport
MGSS	Multimission Ground System and Services Office
MI	manifold insertion
mm/s	millimeters per second
mo	month
MONTE	Mission-analysis, Operations, and Navigation Toolkit Environment
m/s	meters per second
MUSES	*Mu Space Engineering Spacecraft (Hiten) A*
NASA	National Aeronautics and Space Administration
NLS	NASA Launch Services
OLST	Orbit Local Solar Time
PSLV	Polar Satellite Launch Vehicle
RAV	right ascension of apogee vector
RFK78	Runge-Kutta-Fehlberg seventh-order (integrator)
RLA	right ascension of launch asymptote
SE	Sun–Earth
SELENE	*Selenological and Engineering Explorer*
SI	Système International
SKM	station-keeping maneuver
SMART-1	*Small Missions for Advanced Research in Technology 1*

SNOPT	sparse nonlinear optimizer
SOHO	*Solar and Heliospheric Observatory*
SOI	sphere of influence
SQP	sequential quadratic programming
SRM	state relationship matrix
TAI	Temps Atomique International / International Atomic Time
TCM	trajectory correction maneuver
TDB	Barycentric Dynamic Time
THEMIS	*Time History of Events and Macroscale Interactions during Substorms*
TIP	targeting interface point
TLC	trans-lunar cruise
TLI	trans-lunar injection
TOF	time of flight
TT	Terrestrial Time
USA	United States of America
USSR	Union of Soviet Socialist Republics (Soviet Union)
UT	Universal Time
UTC	Coordinated Universal Time
UTTR	Utah Test and Training Range
VLBI	very long baseline interferometry
WMAP	*Wilkinson Microwave Anistropy Probe*
WSB	weak stability boundary
yr	year

Constants

AU	astronomical unit	1.49597871×10^8	km
c	speed of light	299,792.458	km/s
C	Jacobi constant (see Eq. 2.6)		
D_m	mean distance between the Earth and Moon	384,400	km
R_e	mean equatorial radius of the Earth	6378.1363	km
R_m	mean equatorial radius of the Moon	1737.4	km
G	universal gravitational constant	6.67300×10^{-20}	$km^3/s^2/kg$
GM_e	gravitational parameter of the Earth	398,600.432897	km^3/s^2
GM_m	gravitational parameter of the Moon	4902.800582	km^3/s^2
GM_{em}	gravitational parameter of the Earth–Moon Barycenter	403,503.233479	km^3/s^2
GM_s	gravitational parameter of the Sun	$1.32712440 \times 10^{11}$	km^3/s^2
μ_{em}	three-body constant of the Earth–Moon system	0.0121505856	
μ_{se}	three-body constant of the Sun–Earth/Moon system	$3.04042339 \times 10^{-6}$	

Index

Printed and bound by CPI Group (UK) Ltd, Croydon, CR0 4YY
17/09/2026
14964289-0001